# Diversity and Ecology of Fungi from Underexplored and Extreme Environments

# Diversity and Ecology of Fungi from Underexplored and Extreme Environments

Guest Editors

**Daniela Isola**
**Francesc Xavier Prenafeta Boldú**

MDPI

Basel • Beijing • Wuhan • Barcelona • Belgrade • Novi Sad • Cluj • Manchester

*Guest Editors*
Daniela Isola
Department of Agriculture and Forest Sciences
Tuscia University
Viterbo
Italy

Francesc Xavier Prenafeta Boldú
Sustainability in Biosystems
Institute of Agrifood Research and Technology (IRTA)
Caldes de Montbui
Spain

*Editorial Office*
MDPI AG
Grosspeteranlage 5
4052 Basel, Switzerland

This is a reprint of the Special Issue, published open access by the journal *Journal of Fungi* (ISSN 2309-608X), freely accessible at: https://www.mdpi.com/journal/jof/special_issues/3XPJJP15C4.

For citation purposes, cite each article independently as indicated on the article page online and as indicated below:

| Lastname, A.A.; Lastname, B.B. Article Title. *Journal Name* **Year**, *Volume Number*, Page Range. |
|---|

**ISBN 978-3-7258-4629-0 (Hbk)**
**ISBN 978-3-7258-4630-6 (PDF)**
**https://doi.org/10.3390/books978-3-7258-4630-6**

Cover image courtesy of Daniela Isola

# Contents

# About the Editors

**Daniela Isola**

Daniela Isola is a Contract Researcher at the University of Tuscia. She is a biologist with a background in toxicology, pharmacology, and both classical and molecular microbiology, gained during her degree and subsequent specializations. During her specialization in bioinformatics and later her PhD in Biological Evolution and Biochemistry, she began studying black meristematic fungi—extremotolerant organisms resistant to chemical and physical stressors. Her research has primarily focused on black fungi from both natural and polluted environments, as well as on the biodeterioration of cultural heritage. Her work has contributed to advancing knowledge on fungal diversity, the characterization of their metabolic traits and ecological niches, and the development of eco-sustainable methods for their control. Recently, she has begun exploring the biodeterioration of new and bio-based materials.

**Francesc Xavier Prenafeta Boldú**

Francesc Xavier Prenafeta Boldú is a Senior Research Scientist at the Institute of Agrifood Research and Technology (IRTA) in Catalonia, Spain. He holds a PhD in microbial biotechnology from Wageningen University (The Netherlands) and conducted postdoctoral research at the Westerdijk Fungal Biodiversity Institute, focusing on the phylogeny and physiology of black fungi. His current research centers on the application of filamentous fungi and other microorganisms in sustainable bioprocesses, including bioremediation, microbiome engineering, and circular bioeconomy strategies. He has coordinated and contributed to numerous national and European projects on the environmental valorization of organic waste and the role of microbial communities in soil health. His recent work explores the taxonomy and functional diversity of extremophilic and endophytic fungi for use in agriculture and biotechnology.

# Preface

This Reprint explores the diversity and functions of fungi from extreme environments, aiming to highlight current research and stimulate future work. Motivated by their scientific and biotechnological relevance, it is addressed to researchers in fungal biology, environmental microbiology, and related fields.

**Daniela Isola and Francesc Xavier Prenafeta Boldú**
*Guest Editors*

Journal of *Fungi*

MDPI

*Editorial*

# Diversity and Ecology of Fungi from Underexplored and Extreme Environments

**Daniela Isola [1,*] and Francesc X. Prenafeta-Boldú [2,*]**

1 Department of Agriculture and Forest Sciences (DAFNE), University of Tuscia, Via S. Camillo de Lellis, 01100 Viterbo, Italy
2 Program of Sustainability in Biosystems, Institute of Agrifood Research and Technology (IRTA), 08140 Caldes de Montbui, Spain
* Correspondence: isola@unitus.it (D.I.); francesc.prenafeta@irta.cat (F.X.P.-B.)

## 1. Introduction

Fungi represent one of the most diverse and ecologically important groups of organisms on Earth, yet much of their biodiversity remains unknown and unexplored [1]. Fungi do not only inhabit the obvious niches that we associate with soil and decaying matter but also thrive in some of the most challenging environments on the planet [2]. Their prolonged evolutionary history has equipped them with strategies and adaptations that allow them to colonize ecological frontiers where few other organisms survive. These environments may result from anthropogenic pressures such as pollution or habitat degradation or may be naturally extreme due to arid, saline, cold, or nutrient-deprived conditions. As a result, extremotolerant and extremophilic fungi represent an emerging research field in mycology [3], both for their relevance to applied sciences and for the novelty of their taxonomic and functional diversity.

This Special Issue presents a collection of nine research articles and one review contributing to the growing field of extremotolerant and extremophilic fungi. The purpose of this Editorial is not only to briefly summarize the compiled contributions but also to provide a broader interpretive context, one that references past research and anticipates directions for future research.

## 2. Overview of Published Articles

A first group of contributions focuses on fungi isolated from anthropogenically altered environments and engineered biosystems. The resilience of fungal communities in the face of induced abiotic stressors is exemplified in the work by Wang et al., who investigate dark septate endophytes and their symbiotic interactions with *Astragalus membranaceus*—a perennial herb used in traditional Chinese medicine—under drought and cadmium stress. Their findings underscore the potential of fungal endophytes in enhancing plant resistance in degraded or contaminated soils. Further in the realm of bioremediation and melanized fungi, Medina-Armijo and coworkers asses the metallotolerance and biosorption capacity of a collection of black fungi exposed to arsenate and chromate, providing promising leads for fungal applications in detoxifying environments polluted with heavy metals and metalloids.

Interestingly, the conditions under which some of those melanized fungal species were repeatedly isolated—particularly in association with plant roots [4]—suggest a possible overlap between the ecological roles of black fungi/black yeasts and dark septate endophytes (DSEs). While the resilience and stress tolerance of melanized fungi make them valuable in biotechnology and plant symbiosis, they also raise concerns regarding their potential to threaten subterranean cultural heritage [4,5]. Moreover, DSEs and human-pathogenic black fungi share key evolutionary and physiological traits, raising intriguing questions about the continuum between plant mutualism/commensalism and opportunism in melanized fungal lineages [6]. These findings point toward an emerging convergence between very distinct research areas. The ecological and industrial significance of industrial fungi is addressed in the work by Lappe-Oliveras et al., who characterize the genetic and phenotypic variability of *Kluyveromyces marxianus* from traditional Mexican agave-based beverages, highlighting the evolutionary and functional traits of yeasts adapted to specific ethanol-producing fermentation processes. Certain fungi can also thrive in completely anaerobic fermentation systems, such as methanogenic bioreactors. Obi and colleagues unveil a surprisingly diverse fungal community that coexists with bacterial consortia, pointing to previously unrecognized roles of fungi in anaerobic ecosystems. These findings are particularly relevant given the current growth and industrial scaling of the biomethane sector, where understanding the full complexity of microbial consortia could enhance process efficiency and stability.

A second group of studies explore fungal communities thriving in natural ecological niches often considered inhospitable because of salinity and aridity. Guerra-Mateo et al. survey the Ascomycete diversity in Western Mediterranean marine sediments, employing direct plating and flocculation techniques to isolate new taxa. The study by Barnés-Guirado and co-authors adds to this narrative by introducing a new genus—*Dactyliodendromyces*—from a hypersaline lagoon, contributing to the taxonomic refinement of the Microascaceae family. This Special Issue also sheds light on saxicolous fungi, with Sastoque et al. describing several novel species from rock-inhabiting fungal communities in Tarragona Province, Spain. Similarly, Gross and colleagues investigate xerophilic Aspergillaceae in the mound nests of *Formica obscuripes*, highlighting how fungal diversity in these microhabitats is shaped by aridity and suggesting potential ecological functions within insect–fungus interactions. At the opposite end in terms of natural extremes, two geographically remote studies extend the scope of this Special Issue to polar ecosystems. Marchetta and colleagues use metabarcoding to profile microfungal communities in Arctic and Antarctic lakes, revealing a remarkable diversity and the presence of cryptic or uncultured lineages, offering insights into the ecological functioning of fungi in cold aquatic environments. This Special Issue concludes with a comprehensive review by Rajakaruna et al. that explores the pharmaceutical and ecological potential of seagrass-associated endophytic and epiphytic fungi, providing a critical synthesis of current knowledge and future research directions in marine mycology.

## 3. Outlook and Prospects

It is important to acknowledge that despite substantial progress, our fundamental understanding of fungal species remains incomplete. New species continue to be isolated and described at a steady pace, yet a considerable knowledge gap on fungal biodiversity persists. Not all fungi can be readily cultured, and standard isolation methods often favor fast-growing or competitive species, while ecologically specialized groups may require prolonged incubation periods or tailored techniques [7]. This highlights the continued relevance of traditional microbiology, even in an era dominated by high-throughput approaches. Culture collections thus remain indispensable—not only as repositories for biotechnologically promising strains but also as foundational resources for calibrating and interpreting large-scale molecular datasets [8]. These collections serve as critical bridges between classical taxonomy and modern omics-driven investigations, enabling a more holistic and accurate understanding of fungal diversity.

Looking ahead, the research on extremophilic and extremotolerant fungi is poised to develop in several promising directions. Advances in environmental sequencing, single-cell genomics, and metabolomics will continue to uncover cryptic diversity and functional potential in fungal communities from poorly explored habitats, from deep ecosystems to the upper limits of the biosphere, and the colonization of artificial or contaminated substrates. Understanding the molecular and physiological mechanisms that underpin fungal survival under different types of stressors, besides those already covered here—e.g., high pressure, ionizing radiation, extreme pH values—will not only illuminate evolutionary strategies but also inform biomimetic and biotechnological innovations.

At the same time, interdisciplinary approaches that integrate ecology, phylogenetics, and materials science are expected to grow in relevance, particularly in the context of climate change, bioremediation, and sustainable resource development. Greater emphasis on symbiotic interactions—whether with plants, insects, bacteria, or other fungi—may also reveal how extremophilic fungi contribute to ecosystem resilience and co-adaptation [9]. Importantly, as the interest in applications using extremotolerant and extremophilic fungi expands across fields such as synthetic biology, medical mycology, pharmacology, the preservation of cultural heritage, and symbiosis-driven agriculture, it is crucial for research to remain grounded in ecological knowledge. Taxonomic accuracy will also be essential in translating extremophile biology into efficient and safe solutions. We hope that this Special Issue will stimulate further research into the untapped potential of fungi from underexplored and extreme environments and foster interdisciplinary dialog across the fields of mycology, ecology, environmental sciences, and biotechnology.

**Conflicts of Interest:** The authors declare no conflicts of interest.

**List of Contributions:**

1. Wang, D.; Xie, Y.; Zhang, W.; Yao, L.; He, C.; He, X. Study on the biological characteristics of dark septate endophytes under drought and cadmium stress and their effects on regulating the stress resistance of *Astragalus membranaceus*. *JoF* **2024**, *10*, 491. https://doi.org/10.3390/jof10070491.
2. Medina-Armijo, C.; Isola, D.; Illa, J.; Puerta, A.; Viñas, M.; Prenafeta-Boldú, F.X. The metallotolerance and biosorption of As (V) and Cr (VI) by black fungi. *JoF* **2024**, *10*, 47. https://doi.org/10.3390/jof10010047.
3. Lappe-Oliveras, P.; Avitia, M.; Sánchez-Robledo, S.D.; Castillo-Plata, A.K.; Pedraza, L.; Baquerizo, G.; Le Borgne, S. Genotypic and phenotypic diversity of *Kluyveromyces marxianus* isolates obtained from the elaboration process of two traditional Mexican alcoholic beverages derived from agave: Pulque and henequen (*Agave fourcroydes*) mezcal. *JoF* **2023**, *9*, 795 https://doi.org/10.3390/jof9080795.
4. Obi, L.U.; Roopnarain, A.; Tekere, M.; Zhou, J.; Li, H.; Wang, Y.; Zhang, Y.; Adeleke, R.A. Dynamics and Insights into the Unique Ecological Guild of Fungi in Bacteria-Bioaugmented Anaerobic Digesters. *JoF* **2025**, *11*, 56. https://doi.org/10.3390/jof11010056.
5. Guerra-Mateo, D.; Cano-Lira, J.F.; Fernández-Bravo, A.; Gené, J. Sunken riches: Ascomycete diversity in the Western Mediterranean Coast through direct plating and flocculation, and description of four new taxa. *JoF* **2024**, *10*, 281. https://doi.org/10.3390/jof10040281.
6. Barnés-Guirado, M.; Stchigel, A.M.; Cano-Lira, J.F. A new genus of the Microascaceae (Ascomycota) family from a hypersaline lagoon in Spain and the delimitation of the genus *Wardomyces*. *JoF* **2024**, *10*, 236. https://doi.org/10.3390/jof10040236.
7. Sastoque, A.P.; Cano-Lira, J.F.; Stchigel, A.M. Diversity of rock-inhabiting fungi in Tarragona Province, Spain. *JoF* **2024**, *10*, 170. https://doi.org/10.3390/jof10030170.
8. Gross, R.M.; Geer, C.L.; Perreaux, J.D.; Maharaj, A.; Du, S.; Scott, J.A.; Untereiner, W.A. Xerophilic Aspergillaceae Dominate the Communities of Culturable Fungi in the Mound Nests of the Western Thatching Ant (*Formica obscuripes*). *JoF* **2024**, *10*, 735. https://doi.org/10.3390/jof10110735.

9. Marchetta, A.; Papale, M.; Rappazzo, A.C.; Rizzo, C.; Camacho, A.; Rochera, C.; Azzaro, M.; Urzì, C.; Giudice, A.L.; De Leo, F. A deep insight into the diversity of microfungal communities in Arctic and Antarctic lakes. *JoF* **2023**, *9*, 1095. https://doi.org/10.3390/jof9111095.
10. Rajakaruna, O.; Wijayawardene, N.N.; Udagedara, S.; Jayasinghe, P.K.; Gunasekara, S.S.; Boonyuen, N.; Bamunuarachchige, T.C.; Ariyawansa, K.G. Exploring Fungal Diversity in Seagrass Ecosystems for Pharmaceutical and Ecological Insights. *JoF* **2024**, *10*, 627. https://doi.org/10.3390/jof10090627.

## References


1. Cunningham, J.A.; Padamsee, M.; Wilson, S.; Costello, M.J. Fungi species description rates confirm high global diversity and suggest half remain unnamed. *Front. Biogeogr.* **2024**, *16.2*, e62358. [CrossRef]
2. Gostinčar, C.; Zalar, P.; Gunde-Cimerman, N. No need for speed: Slow development of fungi in extreme environments. *Fun Biol. Rev.* **2022**, *39*, 1–14. [CrossRef]
3. Tiquia-Arashiro, S.M.; Grube, M. *Fungi in Extreme Environments: Ecological Role and Biotechnological Significance*; Springer Nature Switzerland AG: Berlin/Heidelberg, Germany, 2019. [CrossRef]
4. Isola, D.; Bartoli, F.; Morretta, S.; Caneva, G. The Roman houses of the Caelian Hill (Rome, Italy): Multitemporal evaluation of biodeterioration patterns. *Microorganisms* **2023**, *11*, 1770. [CrossRef] [PubMed]
5. Alonso, L.; Pommier, T.; Simon, L.; Maucourt, F.; Doré, J.; Dubost, A.; Van, V.T.; Minard, G.; Moro, C.V.; Douady, C.J.; et al. Microbiome analysis in Lascaux Cave in relation to black stain alterations of rock surfaces and collembola. *Environ. Microbiol. Rep.* **2023**, *15*, 80–91. [CrossRef] [PubMed]
6. Quan, Y.; Deng, S.; Prenafeta-Boldú, F.X.; Mayer, V.E.; Muggia, L.; Cometto, A.; Vicente, V.A.; da Silva, N.M.; Grisolia, M.E.; Song, Y.; et al. The origin of human pathogenicity and biological interactions in Chaetothyriales. *Fungal Divers.* **2023**, *125*, 99–120. [CrossRef]
7. Dziurzynski, M.; Gorecki, A.; Pawlowska, J.; Istel, L.; Decewicz, P.; Golec, P.; Styczynski, M.; Poszytek, K.; Rokowska, A.; Gorniak, D.; et al. Revealing the diversity of bacteria and fungi in the active layer of permafrost at Spitsbergen Island (Arctic)–combining classical microbiology and metabarcoding for ecological and bioprospecting exploration. *Sci. Total Environ.* **2003**, *856*, 159072. [CrossRef] [PubMed]
8. Isola, D.; Lee, H.J.; Chung, Y.J.; Zucconi, L.; Pelosi, C. Once upon a Time, There Was a Piece of Wood: Present Knowledge and Future Perspectives in Fungal Deterioration of Wooden Cultural Heritage in Terrestrial Ecosystems and Diagnostic Tools. *JoF* **2024**, *10*, 366. [CrossRef] [PubMed]
9. Martínez-García, L.B.; De Deyn, G.B.; Pugnaire, F.I.; Kothamasi, D.; van der Heijden, M.G. Symbiotic soil fungi enhance ecosystem resilience to climate change. *Glob. Change Biol.* **2017**, *23*, 5228–5236. [CrossRef] [PubMed]

Journal of
*Fungi*

MDPI

*Article*

# A New Genus of the *Microascaceae* (Ascomycota) Family from a Hypersaline Lagoon in Spain and the Delimitation of the Genus *Wardomyces*

**María Barnés-Guirado, Alberto Miguel Stchigel * and José Francisco Cano-Lira**

Mycology Unit, Medical School, Universitat Rovira i Virgili, C/Sant Llorenç 21, 43201 Reus, Spain; maria.barnes@urv.cat (M.B.-G.); jose.cano@urv.cat (J.F.C.-L.)
* Correspondence: albertomiguel.stchigel@urv.cat; Tel.: +34-977759341

**Abstract:** The Saladas de Sástago-Bujaraloz is an endorheic and arheic complex of lagoons located in the Ebro Basin and protected by the Ramsar Convention on Wetlands. Due to the semi-arid climate of the region and the high salinity of their waters, these lagoons constitute an extreme environment. We surveyed the biodiversity of salt-tolerant and halophilic fungi residents of the Laguna de Pito, a lagoon belonging to this complex. Therefore, we collected several samples of water, sediments, and soil of the periphery. Throughout the study, we isolated 21 fungal species, including a strain morphologically related to the family *Microascaceae*. However, this strain did not morphologically match any of genera within this family. After an in-depth morphological characterization and phylogenetic analysis using a concatenated sequence dataset of four phylogenetically informative molecular markers (the internal transcribed spacer region (ITS) of the nuclear ribosomal DNA (nrDNA); the D1-D2 domains of the 28S gene of the nuclear ribosomal RNA (LSU); and a fragment of the translation elongation factor 1-alpha (*EF-1α*) and the β-tubulin (*tub2*) genes), we established the new genus *Dactyliodendromyces*, with *Dactyliodendromyces holomorphus* as its species. Additionally, as a result of our taxonomic study, we reclassified the paraphyletic genus *Wardomyces* into three different genera: *Wardomyces sensu stricto*, *Parawardomyces* gen. nov., and *Pseudowardomyces* gen. nov., with *Parawardomyces ovalis* (formerly *Wardomyces ovalis*) and *Pseudowardomyces humicola* (formerly *Wardomyces humicola*) as the type species of their respective genera. Furthermore, we propose new combinations, including *Parawardomyces giganteus* (formerly *Wardomyces giganteus*) and *Pseudowardomyces pulvinatus* (formerly *Wardomyces pulvinatus*).

**Keywords:** endorheic; extremophiles; fungi; halophilic; halotolerant; phylogeny; Sordariomycetes; taxonomy

## 1. Introduction

Extreme environments are habitats where physical and/or chemical conditions are exceptionally hostile to the survival and proliferation of most life forms [1]. Typically, these harsh conditions are determined by remarkably high or low ("extreme") temperatures or pH, a high salt concentration, osmolarity, hydrostatic pressure, UV radiation, low water activity, or a high concentration of toxic compounds, such as organic solvents and heavy metals [1,2]. Organisms that can survive and proliferate in such environmental conditions are known as "extremophiles", and their ability to thrive relies on their distinctive biochemical machinery and physiology [3] and on their cell ultrastructure and composition [4].

Endorheic basins and their associated lagoons and lakes are landlocked hydrogeological structures that do not drain into large waterbodies (rivers, oceans), only experiencing water loss through evapotranspiration and percolation to the underground [5]. Commonly found in arid and semi-arid continental regions, these lagoons and lakes tend to be saline due to the gradual salt accumulation resulting from thousands of years of evaporative processes [5]. Though some of the world's biggest lakes are endorheic (i.e., Great Salt Lake in North America, the Caspian and Aral Seas in Central Asia, and Lake Titicaca in South America), most of these sorts of lagoons and lakes are temporary, alternating dry and wet phases, as the evaporative process exceeds water inputs, mostly represented by precipitations [5,6]. Species inhabiting these lagoons are adapted to the arid/semi-arid climate, periodical droughts, and high salt concentrations, thus making them extremophiles [6,7].

The Saladas de Sástago-Bujaraloz is an endorheic and arheic (with no circulation of superficial water) complex of lagoons located at the center of the Ebro Basin, ~70 km southeast of Zaragoza city (Aragon Community) in the northeast of Spain [8,9]. The lagoon complex comprises more than one hundred basins positioned over a platform that is around 400 m.a.s.l. and more than 100 m above the Ebro River. The local climate is semi-arid, characterized by low rainfall rates and high evaporation rates that contribute to the biphasic wet/dry states and salt accumulation [9,10]. Due to its unique characteristics, 26 of these lagoons, representing the best-conserved and most representative, are protected by the Ramsar Convention on Wetlands [11], thus being part of the 10% of Spanish Ramsar sites that are inland saline wetlands [9]. Although some studies focused on the flora and the fauna, and the bacterial and archaeal microbiota [9,12–14] were conducted at the Saladas de Sástago-Bujaraloz, no fungal studies have been conducted yet.

The most commonly isolated fungi from hypersaline lagoons and lakes belong to the families *Aspergillaceae*, *Cladosporiaceae*, *Hypocreaceae*, *Pleosporaceae*, *Saccharomycetaceae*, and *Teratosphaeriaceae*, with members of the *Microascaceae* family being less frequently recovered [15–18]. The *Microascaceae* family was established by Luttrell (1951) to accommodate the genus *Microascus* [19]. Morphologically, members of the *Microascaceae* have asexual states predominantly characterized by the production of annellidic conidiogenous cells, forming unicellular or (more rarely) bicellular conidia, and sexual states producing closed or perithecial ascomata within soon evanescent asci and triangular, reniform, or lunate ascospores with or without germ pores [20,21]. This family includes fungi isolated from soil, decaying plant material, and air, and several species are pathogens for animals, including mammals and humans [20]. The *Microascaceae* currently comprises 23 genera, including *Acaulium*, *Brachyconidiellopsis*, *Cephalotrichum*, *Enterocarpus*, *Fairmania*, *Gamsia*, *Kernia*, *Lomentospora*, *Lophotrichus*, *Microascus*, *Parascedosporium*, *Petriella*, *Pseudallescheria*, *Pseudoscopulariopsis*, *Pithoascus*, *Polycytella*, *Rhexographium*, *Rhinocladium*, *Scedosporium*, *Scopulariopsis*, *Wardomyces*, *Wardomycopsis*, and *Yunnania*, and about 300 species [20–24]. The taxonomic position of the genus *Canariomyces* is controversial, as the phylogenetic analysis conducted by Wang et al. [25] correctly placed the genus in the *Chaetomiaceae* family, but Wang et al. [22] retain the genus in the *Microascaceae* family based on another molecular study. Among them, some species belonging to the genera *Lomentospora*, *Scedosporium*, and *Scopulariopsis* are frequently involved as pathogens in opportunistic infections in humans [26–28]. Since the reorganization of the family structure by Sandoval-Denis et al. [20], no further taxonomic adjustments have been made, except for Su et al. [29], who revised the genera *Acaulium* and *Kernia*. In addition, several new species belonging to this family have been described in more recent works [22,30–33].

During a survey on the fungal diversity of soils, lake sediments, and hypersaline waters carried out at the Laguna de Pito (one of the lagoons of Saladas de Sástago-Bujaraloz), we isolated several fungal taxa, including a strain showing morphological features of the *Microascaceae* family but not matching any previously described genera.

The main aim of this study was to show the fungal diversity inhabiting the Laguna de Pito, as well as to characterize phenotypically and to determine the phylogenetic placement of such fungal strains and other morphologically related taxa in the *Microascaceae*.

## 2. Materials and Methods

### 2.1. Sampling and Fungal Isolation

We collected several samples of water, sediments, and soil from the surrounding areas of Laguna de Pito in January 2022. This lagoon covers approximately 50 ha., dries intermittently, and is surrounded by fields designated for the cultivation of cereals; its conservation status was reported as good [34]. The salinity of the water samples, measured by an Aokuy refractometer (Shenzhenshi Jinshenghe Shangmao Youxiangongsi, Guangdong, China), was 50‰ *w/v*, and the pH measured with SRSE water test strips (Tepcom GmbH & Co., KG, Bendorf, Germany) was 7.8. The samples were transferred to 100 mL sterile plastic containers and were transported whilst being refrigerated (at 4–7 °C) to the laboratory. To maximize the diversity of isolated fungi, the following culture media were employed: 18% of glycerol agar (G18; 2.5 g peptone, 5 g dextrose, 0.5 g $KH_2PO_4$, 0.25 g $MgSO_4$, 90 mL glycerol, 7.5 g agar–agar, 410 mL distilled water; [35]), potassium acetate agar (5 g potassium acetate, 1.25 g yeast extract, 0.5 g dextrose, 15 g agar–agar, 500 mL distilled water; [36]), potato dextrose agar (PDA; Laboratorios Conda S.A., Madrid, Spain; [37]) supplemented with 10% NaCl, and 2% malt extract agar (MEA; Difco Inc., Detroit, MI, USA; [38]) plus 30% glycerol. Moreover, sediment samples were activated with acetic acid following the modified protocol of Furuya and Naito [39,40]. All culture media were supplemented with 250 mg/L of L-chloramphenicol to prevent the development of bacteria. Sediment samples were vigorously shaken in the same containers they were collected in and were settled for 1 min. Once settled, water was removed by decantation and the sediment was poured onto several layers of sterile filter paper placed over plastic trays until dry [41]. Approximately, one gram of dried sediments and soil samples was sprinkled onto all the types of culture media in 90 mm Petri dishes. Different volumes of water for each of the samples (5, 15, and 30 mL) were filtered through a filter membrane of 0.45 μm diameter (Millipore SA, Molsheim, France) using a vacuum pump. Later, the filter membranes were placed onto the different culture media in 90 mm Petri dishes. Every sample was cultured by duplicate, being incubated in darkness at 15 °C and 37 °C, respectively. Plates were examined daily for up to two months by using a stereomicroscope. Each colony developed was transferred to 55 mm Petri dishes containing oatmeal agar (OA; 15 g filtered oat flakes, 7.5 g agar, 500 mL tap water; [38]) by using sterile disposable tuberculin-type needles, and these colonies were incubated at room temperature until axenic cultures of each isolate were obtained. Fungal strains suspected to be novel species or pertaining to uncommon taxa were deposited in the culture collection of the Faculty of Medicine of Reus (FMR; Reus, Tarragona Province, Spain), and the ex-type strains and the herborized specimens (as holotypes) were deposited at the Westerdijk Fungal Biodiversity Institute (CBS; Utrecht, The Netherlands).

### 2.2. Phenotypic Study

The macroscopic characterization of the colonies was performed on OA, MEA, PDA, and potato carrot agar (PCA; 10 g potato, 10 g carrot, 6.5 g agar, 500 mL distilled water) after incubation for 7–14 d at 25 °C in darkness [37,38]. The color description of the colonies was made according to Kornerup and Wanscher [42]. Cardinal growth temperatures were determined on PDA, ranging from 5 to 40 °C at 5 °C intervals, with an additional measurement at 37 °C.

The microscopic characterization of vegetative and reproductive structures was carried out by using fungal material from the colonies grown on OA under the same conditions as specified for macroscopic characterization. Measurements of at least 30 of the structures were taken from slide mountings using Shear's medium (3 g potassium acetate, 60 mL glycerol, 90 mL ethanol 95%, 150 mL distilled water; [43]) and using an Olympus BH-2 bright field microscope (Olympus Corporation, Tokyo, Japan). Micrographs were taken employing a Zeiss Axio-Imager M1 light microscope (Zeiss, Oberkochen, Germany) with a DeltaPix Infinity × digital camera using Nomarski differential interference contrast.

### 2.3. DNA Extraction, Amplification, and Sequencing

Total genomic DNA was extracted from colonies grown on PDA for 7 to 10 days at 25 ± 1 °C in darkness following the modified protocol of Müller et al. [44] and quantified using a Nanodrop 2000 (Thermo Scientific, Madrid, Spain). For each fungal strain, we amplified the molecular marker that allowed for the most accurate preliminary identification according to the bibliography. The internal transcribed spacers (ITS) region and the D1-D2 domains of the 28S nrRNA (LSU) were amplified using the primer pairs ITS5/ITS4 [45] and LR0R/LR5 [46], respectively. Fragments of the translation elongation factor 1α (*EF-1α*) and the β-tubulin (*tub2*) genes were amplified using the primer pairs 983F/2218R and EF-728F/EF-986R [47,48] and BT2a/BT2b [49]. For our strain of interest, we amplified the following markers: ITS, LSU, *tub2*, and *EF-1α* (using the 983F/2218R set of primers). Single-band PCR products were stored at −20 °C and sequenced at Macrogen Europe (Macrogen Inc., Madrid, Spain) with the same amplification primers. Lastly, the software SeqMan v. 7.0.0 (DNAStarLasergene, Madison, WI, USA) was employed to edit and assemble the consensus sequences.

### 2.4. Phylogenetic Analysis

The sequences obtained were compared with all the sequences available at the National Center for Biotechnology Information (NCBI) database using the Basic Local Alignment Search Tool (BLAST; https://blast.ncbi.nlm.nih.gov/Blast.cgi, accessed on 11 October 2023) to obtain a preliminary molecular identification of each isolate. A maximum level of identity (MLI) of ≥98% was considered to allow for species-level identification [50]. Single and combined phylogenetic analyses of all the specific molecular markers mentioned above were initially conducted by performing a sequences alignment with the software MEGA (Molecular Evolutionary Genetics Analysis) v. 7.0. [51] using the ClustalW algorithm [52] and refining with MUSCLE [53] or/and manually, if needed. Subsequently, the phylogenetic reconstruction was made by maximum likelihood (ML) and the Bayesian Inference (BI) methods were made by two different software, RAxML-HPC2 on XSEDE v. 8.2.12 [54] software on the online CIPRES Science gateway portal [55] and MrBayes v.3.2.6 [56], respectively. The best substitution model for all the gene matrices was settled by the software from CIPRES Science gateway portal (ML) and by jModelTest v.2.1.3 following the Akaike criterion (BI) [55,57]. Regarding the ML analysis, phylogenetic support for internal branches was established by 1000 ML bootstrapped pseudoreplicates, being considered significant bootstrap support (bs) values ≥70 [58]. Regarding the BI analysis, 5 million Markov Chain Monte Carlo (MCMC) generations were used, with four runs (three heated chains and one cold chain), and samples were stored every 1000 generations. To calculate the 50% majority rule consensus tree and posterior probability values (pp), the first 25% of samples were discharged, and pp values of ≥0.95 were considered significant [59]. The resulting phylogenetic trees were plotted using FigTree v.1.3.1 (http://tree.bio.ed.ac.uk/software/figtree/, accessed on 11 October 2023). The DNA sequences and the sequence alignments generated in this study were deposited in GenBank (Table 1) and in TreeBASE (https://treebase.org, accessed on 11 October 2023), respectively. The novel taxa have been registered in MycoBank (https://www.mycobank.org/, accessed 23 March 2023 for *Dactyliodendromyces holomorphus* gen. et sp. nov.).

**Table 1.** Fungi and nucleotide sequences of the molecular markers used to build the phylogenetic trees.

| Taxon | Strain Number | Source | Origin | Sequence Accession Number | | | |
|---|---|---|---|---|---|---|---|
| | | | | LSU | ITS | *EF-1α* | *tub2* |
| *Acaulium acremonium* | CBS 290.38 | Skin of a horse | København, Denmark | LN851001 | LM652456 | HG380362 | LN851108 |
| *Acaulium acremonium* | MUCL 8274 [ET] | Wheat field soil | Schleswig-Holstein, Germany | LN851002 | LM652457 | LN851056 | LN851109 |
| *Acaulium albonigrescens* | IHEM 18560 [ET] | Litter treated with urea | Nemuro-shi, Japan | LN851004 | LM652389 | LN851058 | LN851111 |
| *Acaulium caviariformis* | CBS 536.87 [ET] | Decaying meat | Flemalle, Belgium | LN851005 | LM652392 | LN851059 | LN851112 |
| *Cephalotrichum asperulus* | CBS 127.22 | Seed | Wageningen, The Netherlands | LN851006 | LN850959 | LN851060 | LN851113 |
| *Cephalotrichum asperulus* | CBS 582.71 [IT] | Soil | Buenos Aires, Argentina | LN851007 | LN850960 | LN851061 | LN851114 |
| *Cephalotrichum brevistipitatum* | CBS 157.57 [T] | Tuber | Wageningen, The Netherlands | LN851031 | LN850984 | LN851084 | LN851138 |
| *Cephalotrichum columnare* | CBS 159.66 [T] | Dung of hare | Johannesburg, South Africa | LN851010 | LN850963 | LN851064 | LN851117 |
| *Cephalotrichum cylindricum* | CBS 448.51 | Timber | Bekker, South Africa | LN851011 | LN850964 | LN851065 | LN851118 |
| *Cephalotrichum cylindricum* | UAMH 1348 [ET] | Seed of sorghum | KS, USA | LN851012 | LN850965 | LN851066 | LN851119 |
| *Cephalotrichum dendrocephalum* | CBS 528.85 [IT] | Cultivated soil | Basrah, Iraq | LN851013 | LN850966 | LN851067 | LN851120 |
| *Cephalotrichum gorgonifer* | CBS 635.78 [ET] | Hair | The Netherlands | LN851024 | LN850977 | LN851077 | LN851131 |
| *Cephalotrichum gorgonifer* | UAMH 3585 | Mushroom compost | Spruce Grove, AB, Canada | LN851025 | LN850978 | LN851078 | LN851132 |
| *Cephalotrichum hinnuleum* | CBS 289.66 [T] | Dung of deer | Tasmania, Australia | LN851032 | LN850985 | LN851085 | LN851139 |
| *Cephalotrichum microsporum* | CBS 523.63 [ET] | Wheat field soil | Schleswig-Holstein, Germany | LN851014 | LN850967 | LN851068 | LN851121 |
| *Cephalotrichum microsporum* | UAMH 9365 | Indoor air | Peace River, AB, Canada | LN851015 | LN850968 | LN851069 | LN851122 |
| *Cephalotrichum nanum* | CBS 191.61 [ET] | Dung of deer | Richmond Park, SRY, ENG, UK | LN851016 | LN850969 | LN851070 | LN851123 |
| *Cephalotrichum nanum* | UAMH 9126 | Dung of bison | Elk Island National Park, AB, Canada | LN851017 | LN850970 | LN851071 | LN851124 |
| *Cephalotrichum purpureofuscum* | UAMH 9209 | Indoor air | Pemberton, BC, Canada | LN851018 | LN850971 | LN851072 | LN851125 |
| *Cephalotrichum stemonitis* | CBS 103.19 [NT] | Seed | Wageningen, The Netherlands | LN850952 | LN850951 | LN850953 | LN850954 |
| *Cephalotrichum stemonitis* | CBS 180.35 | Unknown | Unknown | LN851019 | LN850972 | LN851073 | LN851126 |
| *Cephalotrichum verrucisporum* | CBS 187.78 | Dune soil | Katijk, The Netherlands | LN851033 | LN850986 | LN851086 | LN851140 |

**Table 1.** *Cont.*

| Taxon | Strain Number | Source | Origin | Sequence Accession Number | | | |
|---|---|---|---|---|---|---|---|
| | | | | LSU | ITS | *EF-1α* | *tub2* |
| ***Dactyliodendromyces holomorphus*** | CBS 149968 | Lagoon sediment | Zaragoza, Spain | OR141719 | OR141718 | OR142400 | OR142401 |
| *Fairmania singularis* | CBS 414.64 | Laboratory contaminant | Tokyo, Japan | LN851035 | LM652442 | LN851088 | LN851142 |
| *Fairmania singularis* | CBS 505.66 $^{ET}$ | Barrel bottom | Kittery Point, ME, USA | LN851036 | LN850988 | LN851089 | LN851143 |
| *Gamsia aggregata* | CBS 251.69 $^{IT}$ | Dung of carnivore | Wycamp Lake, MI, USA | LN851037 | LM652378 | LN851090 | LN851144 |
| *Gamsia columbina* | CBS 546.69 $^{T}$ | Milled *Oryza sativa* | Osaka, Japan | LN851041 | LM652379 | LN851094 | LN851148 |
| *Gamsia columbina* | CBS 233.66 $^{ET}$ | Sandy soil | Giessen, Germany | LN851039 | LN850990 | LN851092 | LN851146 |
| ***Parawardomyces ovalis*** | CBS 234.66 $^{T}$ | Wheat field soil | Schleswig-Holstein, Germany | LN851050 | LN850996 | LN851101 | LN851155 |
| ***Parawardomyces giganteus*** | CBS 746.69 $^{T}$ | Insect frass in a dead log | Coldwater, ON, Canada | LN851045 | LM652411 | LN851096 | LN851150 |
| ***Pseudowardomyces pulvinatus*** | CBS 112.65 $^{T}$ | Salt marsh | CHS, ENG, UK | LN851051 | LN850997 | LN851102 | LN851156 |
| ***Pseudowardomyces humicola*** | CBS 369.62 $^{IT}$ | Soil in tropical greenhouse | Guelph, ON, Canada | LN851046 | LN850993 | LN851097 | LN851151 |
| *Scopulariopsis brevicaulis* | MUCL 40726 $^{T}$ | Indoor air | Scandia, AB, Canada | LN851042 | LM652465 | HG380363 | LM652672 |
| *Microascus longirostris* | CBS 196.61 $^{NT}$ | Wasp's nest | Kittery Point, ME, USA | LN851043 | LM652421 | LM652566 | LM652634 |
| *Wardomyces anomalus* | CBS 299.61 $^{ET}$ | Air cell of egg | Ottawa, ON, Canada | LN851044 | LN850992 | LN851095 | LN851149 |
| *Wardomyces inflatus* | CBS 216.61 $^{IT}$ | Wood, *Acer* sp. | Sainte-Cécile-de-Masham, QC, Canada | LN851047 | LM652496 | LN851098 | LN851152 |
| *Wardomyces inflatus* | CBS 367.62 $^{NT}$ | Greenhouse soil | Heverlee, Belgium | LN851048 | LN850994 | LN851099 | LN851153 |
| *Wardomycopsis humicola* | CBS 487.66 $^{IT}$ | Soil | Guelph, ON, Canada | LM652554 | LM652497 | LN851103 | LN851157 |
| *Wardomycopsis humicola* | FMR 3993 | Sediment of Ter River | Girona, Spain | LN851052 | LN850998 | LN851104 | LN851158 |
| *Wardomycopsis humicola* | FMR 13592 | Soil | Reus, Spain | LN851053 | LN850999 | LN851105 | LN851159 |
| *Wardomycopsis inopinata* | FMR 10305 | Soil | Bagan, Myanmar | LN851054 | LM652498 | LN851106 | LN851160 |
| *Wardomycopsis inopinata* | FMR 10306 | Soil | Bagan, Myanmar | LN850956 | LN850955 | LN850957 | LN850958 |
| *Wardomycopsis litoralis* | CBS 119740 $^{T}$ | Beach soil | Castellon, Spain | LN851055 | LN851000 | LN851107 | LN851161 |

CBS, CBS-KNAW Westerdijk Fungal Biodiversity Institute (Utrecht, the Netherlands). FMR, Facultat de Medicina Reus (URV—Reus—Spain). IHEM, BCCM/IHEM Belgian Fungi Collection: Human and Animal Health. MUCL, BCCM/MUCL Belgian Agro-food and Environmental Fungal Collection. UAMH, University of Alberta Mold Herbarium and Culture Collection (Edmonton, Canada). New taxa are in **bold**. $^{ET}$, ex-epitype strain. $^{IT}$, ex-isotype strain. $^{NT}$, ex-neotype strain. $^{T}$, ex-type strain.

## 3. Results

The fungi isolated from various substrates collected in Laguna de Pito are listed in Table 2, which also includes their extremophilic character (if previously reported and/or determined during the development of our study).

Notably, among all the fungi identified, it is noteworthy that the strain FMR 20493 displayed a percentage of identity of 99.4% with *Cephalotrichum dendrocephalum* CBS 528.85 (GenBank MH873591.1; identities = 836/841; no gaps) in a BLAST search using the LSU. However, the closest hit using the ITS was *Wardomyces pulvinatus* CBS 803.69 (GenBank MH859434.1; identities = 434/447 (97.09%); two gaps), using *EF-1α* was *Wardomyces pulvinatus* CBS 112.65 and *Wardomyces humicola* CBS 369.62 (GenBank LN851102.1 and LN851097.1; for both cases: identities = 856/876 (97.7%); no gaps), and for the *tub2*, it was *Wardomyces anomalus* CBS 299.61 (GenBank LN851149.1; identities = 422/470 (89.8%); three gaps), all of them with a percentage of identity below 98%.

In the analysis involving species within the *Microascaceae*, the individual dataset for ITS, LSU, *EF-1α*, and *tub2* showed no conflicts related to the tree topologies for the 70% reciprocal bootstrap trees; thus, a multi-gene analysis was performed. The final concatenated dataset included 43 ingroup strains belonging to the genera *Acaulium*, *Cephalotrichum*, *Gamsia*, *Fairmania*, *Wardomyces* and *Wardomycopsis*, and *Microascus longirostris* CBS 196.61 and *Scopulariopsis brevicaulis* MUCL 40726 as the outgroup. The alignment encompassed a total of 3019 characters, including gaps (661 for ITS, 843 for LSU, 965 for *EF-1α*, and 550 for *tub2*), 732 of them parsimony informative (203 for ITS, 73 for LSU, 190 for *EF-1α*, and 266 for *tub2*) and 960 of them being variable sites (293 for ITS, 91 for LSU, 264 for *EF-1α*, and 312 for *tub2*). The tree obtained through the BI analysis was both congruent and similar in topology to the one obtained by ML analysis. Regarding the BI analysis, GTR + G, GTR + G + I, GTR + G + I, and HKY + G + I were selected as the models that fitted the best for ITS, LSU, *EF-1α*, and *tub2*, respectively. The support values showed slight differences between the two analysis methods, making the ML bootstrap support values lower than the BI posterior probabilities.

The phylogenetic analysis (Figure 1) revealed six fully supported clades representing the genera *Cephalotrichum* (clade I), *Gamsia* (clade VI), *Acaulium* (clade VII), *Wardomycopsis* (clade VIII), and *Fairmania* (clade IX). However, the species of the genus *Wardomyces* were placed in three fully supported independent terminal clades: clade II, comprising *Wardomyces anomalus* (the type species of the genus) and *Wardomyces inflatus*; clade III, including *Wardomyces giganteus* (basionym *Microascus giganteus*) and *Wardomyces ovalis*; and clade IV, composed of *Wardomyces humicola* and *Wardomyces pulvinatus*. Furthermore, our strain CBS 149968 was placed as an independent terminal clade itself (clade V). Therefore, clades III, IV, and V represent three novel genera, also supported by phenotypic features.

*Taxonomy*

*Microascaceae* Luttrell et Malloch, Mycologia 62:734 (1970). Mycobank MB 81001.

*Ascomata* globose, pyriform or irregular in shape, dark brown to black, hairy, rarely bare, arising from coiled ascocarp initials, with or without an ostiole; *asci* arising singly or in chains on the ascogenous hyphae, without croziers, ovoid to globose, soon evanescent; *ascospores* reddish brown to coppery-colored, one-celled, with a germ pore at one or both ends, dextrinoid when young, smooth- and thin-walled.

*Dactyliodendromyces* Barnés-Guirado, Cano & Stchigel, gen. nov. MycoBank MB 848097.

*Etymology*. From Greek *δακτύλιος*- (daktýlios), anything ring-shaped, -*δένδρον*- (déndron), tree, and -*μύκητας* (mýkitas), fungus, because the fungus produces tree-like conidiophores bearing anellidic conidiogenous cells.

**Table 2.** Fungal taxa recovered from Laguna de Pito and their extremophilic properties.

| Taxon | Strain Nr [1] | Identity Percentage (%) | GenBank Accession Nr [2] | Markers Used | Source | Extremophilic Features Reported | References |
|---|---|---|---|---|---|---|---|
| *Acrostalagmus luteoalbus* | 19813 * | 99.69 | KP050692 | ITS | water | Alkali-tolerant | [60] |
| *Actinomucor elegans* | 19823 * | 99.79 | AY243954 | ITS | water | Thermotolerance (strain-dependent) | [61] |
| *Alternaria alternata* | 20034 * | 100 | KP124364 | ITS | sediment | Halotolerant; alkali-tolerant | [62] |
| *Alternaria chlamydospora* | 20037 * | 100 | MG020753 | ITS | sediment | Acidophilic; alkali-tolerant; psychrotolerant; xerotolerant | [63] |
| *Aspergillus amstelodami* | 20038 * | 99.22 | MT820427 | *tub2* | water | Xerophilic; thermotolerant | [64,65] |
| *Aspergillus calidoustus* | 19423 *, 19820 | 99.7 | LT798990 | *tub2* | water | Thermotolerant | [64] |
| *Aspergillus intermedius* | 19821 * | 100 | LT671082 | *tub2* | water | Osmophilic; thermotolerant; xerophilic | [66] |
| *Aspergillus montevidensis* | 20492 * | 100 | KF499570 | *tub2* | sediment | Halotolerant | [67] |
| *Aspergillus versicolor* | 19427 *, 20659 | 100 | ON807694 | *tub2* | sediment | Halotolerant | [68] |
| *Cephalotrichiella penicillata* | 20498 * | 100 | NR_153893 | ITS | sediment | Not reported | [69] |
| *Chaetomium grande* | 20036 * | 99.30 | KT214731 | *tub2* | sediment | Thermotolerant | [70] |
| *Cladosporium europaeum* | 19425 *, 20499 | 100 | HM148294 | *EF-1α* | water | Halotolerant | Our study |
| *Dactyliodendromyces holomorphus* | 20493 * | On text | | ITS, LSU, *EF-1α*, *tub2* | sediment | Not reported | Our study |
| *Epicoccum italicum* | 20044 * | 100 | MN983956 | *tub2* | sediment | Psychrotolerant; halotolerant | [71] |
| *Fusarium culmorum* | 20248 * | 99.85 | KT008433 | *EF-1α* | sediment | Not reported | |
| *Malbranchea zuffiana* | 20033 * | 98.90 | MH869293 | ITS | sediment | Halotolerant | Our study |
| *Ovatospora amygdalispora* | 20322 * | 99.41 | MZ343030 | *tub2* | sediment | Not reported | |
| *Parachaetomium truncatulum* | 20041 *, 20495 | 99.77 | HM365298 | *tub2* | sediment | Thermotolerant | [70] |
| *Penicillium egyptiacum* | 20328 *, 20324, 20323, 20331, 20337 | 100 | JX996851 | *tub2* | sediment | Psychrotolerant; non-thermotolerant; xerotolerant | [72] |
| *Sordaria fimicola* | 19587 * | 99.61 | MH860820 | ITS | sediment | Striking stimulation of ascospore germination by acetate | [73] |
| *Stachybotrys chartarum* | 19808 * | 100 | KU846678 | ITS | sediment | Halotolerant | [74] |

* Strain sequenced. [1] FMR, Facultat de Medicina Reus (URV—Reus—Spain). [2] Nucleotide sequence for which the best match has been recorded.

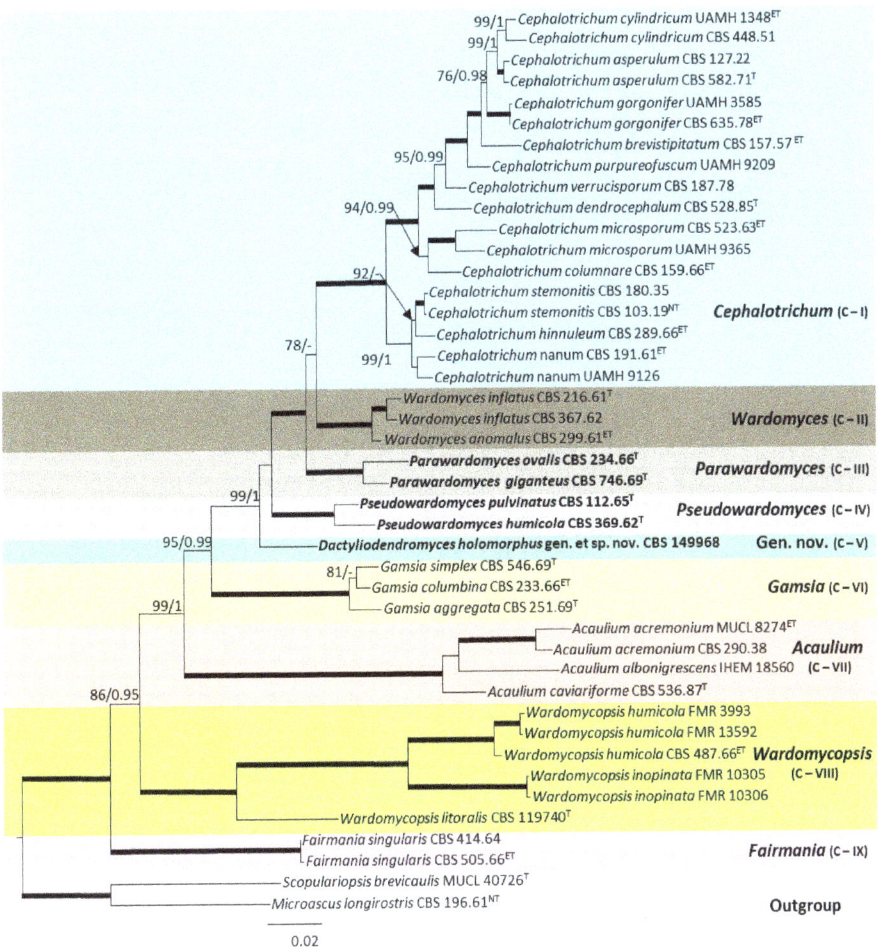


**Figure 1.** Maximum likelihood phylogenetic tree obtained by combining ITS, LSU, *EF-1α*, and *tub2* sequences from 43 representative taxa of the *Microascaceae*. RAxML bootstrap support (BS) values and Bayesian posterior probabilities (PP) greater than 70% and 0.95, respectively, are shown above the branches. Fully supported branches (100% BS/1 PP) are indicated as broad lines. Novel genera are indicated in bold. The tree was rooted to *Microascus longirostris* CBS 196.61 and *Scopulariopsis brevicaulis* MUCL 40726. $^{T}$ = Ex-type; $^{ET}$ = Ex-epitype; $^{NT}$ = Ex-neotype.

*Description: Hyphae* hyaline to subhyaline, septate, smooth-walled to asperulate, thin-walled, branched, sometimes aggregated and frequently anastomosing. *Asexual morph—Conidiophores* macronematous, penicillate, branching up to three times, subhyaline to pale brown or pale olivaceous. *Conidiogenous cells* annellidic, mono- or polyblastic, terminal, discrete, flask-shaped, ventricose, with a short terminal neck. *Conidia* solitary or disposed in short basipetal chains on the conidiogenous cell, one-celled, pale brown to brown, smooth- and thick-walled, ovoid to lenticular, flattened at the base, without germ slits or pores, secession schizolytic. *Sexual morph—Ascomata* erumpent, dark greyish brown when mature, ostiolate, setose, globose to subglobose, neck short, cylindrical; setae dark brown, septate; peridium superficially areolate when young, becoming carbonaceous with the age, of *textura angularis*, formed by an outer wall of dark brown polygonal cells, and an inner wall of hyaline to pale brown polygonal cells. *Asci* 8-spored, broad ellipsoidal to ovoid, soon evanescent. *Ascospores* one-celled, apricot to pale orange, hearth- or kidney-shaped, small, with a terminal germ pore.

*Type species*: *Dactyliodendromyces holomorphus* Barnés-Guirado, Cano & Stchigel, sp. nov. MycoBank MB 848098.

*Dactyliodendromyces holomorphus* Barnés-Guirado, Cano & Stchigel, sp. nov. MycoBank MB 848098 (Figure 2).

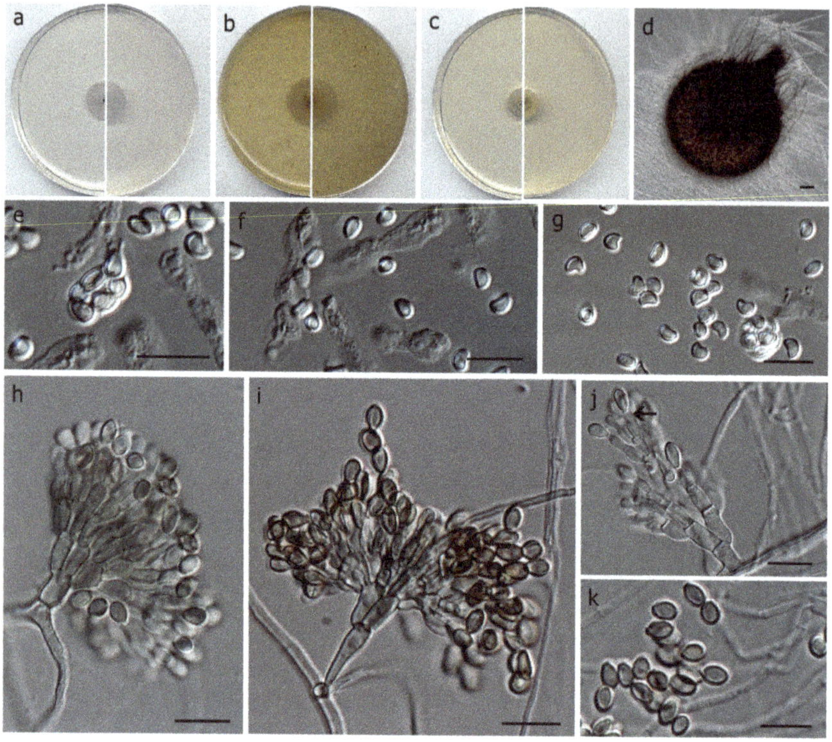

**Figure 2.** *Dactyliodendromyces holomorphus* CBS 149968. Colonies on PCA (**a**), OA (**b**), and PDA (c) after two weeks at 25 + 1 °C (surface, left; reverse, right); (**d**) young ascoma; (**e**) asci; (**f**) catenated asci; (**g**) ascospores; (**h**,**i**) Conidiophores; (**j**) Conidiogenous cells, arrow points out annelid's rings; (**k**) conidia. Scale bars: (**e**–**k**) = 10 µm; (**d**) = 25 µm.

*Etymology*. From Greek *όλος*—(ólos), the whole, and -*μορφή* (morfí), form, because the fungus produces both sexual and asexual morphs.

*Description: Hyphae* hyaline to subhyaline, septate, smooth-walled to asperulate, thin-walled, branched, sometimes aggregated and frequently anastomosing, 1.0–3.0 µm wide. *Asexual morph*—*Conidiophores* macronematous, penicillate, 1–3-branched, smooth- and thick-walled, subhyaline to pale brown or pale olivaceous, 20–55 µm long, 2.0–3.0 µm wide at the base; *branches* smooth- and thick-walled, 2–3.5 µm wide and 4–3.5 µm long, primary branches bearing 2 to 4 secondary branches, secondary branches bearing 1 to 2 tertiary branches, and the terminal branches bearing 1 to 5 conidiogenous cells. *Conidiogenous cells* annellidic, mono- or polyblastic, terminal, discrete, smooth- and thick-walled, hyaline to pale brown or pale olivaceous, flask-shaped, ventricose with a short terminal neck, 3.5–6 × 1–3 µm, bearing a terminal conidium or conidia disposed in short chains. *Conidia* one-celled, pale brown to brown, smooth- and thick-walled, ovoid to lenticular, 5–8 × 2–3 µm, rounded at the apex and flattened at the base, without germ slits or pores, secession schizolytic. *Sexual morph*—*Ascomata* erumpent, usually formed at the periphery of the colony, hyaline to pale brown when young, becoming dark greyish brown to very dark brown when mature, areolate when young, ostiolate, tomentose, setose, piriform, body without the neck globose to subglobose, 211–327 × 220–336 µm; *neck* short, up to 55 µm, cylindrical, 36–41 × 26.5–35 µm; *setae* smoky olivaceous brown to dark brown, septate, smooth- and thick-walled, needle-shaped but with a sinuous wall, 16.5–450 × 1–5 µm, mostly tapering and paler towards the top; *peridium* at first areolate, of *textura angularis*, becoming carbonaceous with the age, formed by an outer wall of dark brown polygonal cells, and an inner wall of hyaline to pale brown polygonal cells; not easily breakable under external pressure. *Asci* 8-spored, broadly ellipsoidal to ovoid, 10 × 6–8 µm, soon evanescent, catenated when young. *Ascospores* one-celled, apricot to pale orange, smooth- and thin-walled, heart-shaped to kidney-shaped but flattened at one side, 3–4 × 2.5–3 × 2 µm, with an inconspicuous germ pore at one of the extremes.

*Culture characteristics* (after 14 d at 25 °C)—Colonies on PDA reaching 11 mm diam., convex, smooth texture, cerebriform, white (1A1), undulate, sporulation absent; reverse white, (1A1), and olive brown (4E6) at centre, white (1A1) towards periphery, soluble pigment absent. Colonies on PCA reaching 18 mm diam., slightly raised at centre, flattened at the edges, granulose, smooth, grey (30F1) at centre, white (1A1) at the edges, filamentous margins, sporulation (conidiophores) moderate; reverse white (1A1), soluble pigment absent. Colonies on OA reaching 24 mm diam., flattened, granulose, smooth, grey (30F1) at centre, white (1A1) towards periphery, filamentous margins, sporulation moderate to abundant (conidiophores); reverse olive (3F5) at centre, white (1A1) towards periphery, soluble pigment absent. Non-halophilic nor highly halotolerant (does not grow above 10% *w*/*v* NaCl). *Cardinal temperatures of growth*: minimum 5 °C, optimum 20 °C, maximum 30 °C.

*Specimen*: CBS 149968. Spain, Aragon community, Zaragoza province, Laguna de Pito (41°24′44.2″ N 0°09′02.2″ W), isolated from lagoon sediment, 17 January 2022, collected by María Barnés Guirado, Alan Omar Granados Casas, Alberto Miguel Stchigel Glikman and José F. Cano-Lira, isolated by María Barnés Guirado, holotype CBS H-25252.

*Diagnosis*: The sexual morph of the genus *Dactyliodendromyces* resembles those species of the *Microascaceae* producing heart-shaped ascospores: *Acaulium albonigrescens*, *Fairmania singularis*, and *Wardomyces giganteus*, as well as several species of *Microascus* and *Scopulariopsis*. Nevertheless, *Dactyliodendromyces* produces ostiolate ascomata with a superficially areolate peridium and true setae, features not seen in the other taxa. On the other hand, the asexual morph of genus *Dactyliodendromyces* differs from *Wardomyces*, and the newly proposed genera *Parawardomyces* and *Pseudowardomyces*, in having annellidic conidiogenous cells, which are holoblastic in all of them. The genus *Gamsia* differs from *Dactyliodendromyces* by the formation of hyaline, mostly undifferentiated and unbranched conidiophores bearing polyblastic and anellidic conidiogenous cells, which are dematiaceous, well-developed, penicillated, and bear exclusively annellidic conidiogenous cells in *Dactyliodendromyces*. Consequently, the asexual morph of *Dactyliodendromyces* is morphologically more similar to *Acaulium*, *Cephalosporium*, *Fairmania*, and *Wardomycopsis*. However, *Dactyliodendromyces* is easily discriminated from *Fairmania* and *Wardomycopsis* because these genera produce conidia with longitudinal striations or germ slits (not seen in *Dactyliodendromyces*), and from *Cephalosporium*, because it produces conidiophores grouped in synnemata (which are absent in *Dactyliodendromyces*), and also lacks a sexual morph (present in *Dactyliodendromyces*). *Acaulium*, in comparison to the *Dactyliodendromyces*, produces more simple hyaline conidiophores, which are dematiaceous and penicillate in the new genus.

*Pseudowardomyces* Barnés-Guirado, Stchigel & Cano, gen. nov. Mycobank MB 851965.

*Etymology*: From Greek ψευδο- (psevdo), false, because of its morphological resemblance to the genus *Wardomyces*.

*Description: Hyphae* hyaline, septate, smooth- and thin-walled, branched, sometimes aggregated and frequently anastomosing. *Conidiophores* hyaline, macronematous, mostly branched, bi- to terverticillate, with a stipe of short to medium length. *Conidiogenous cells* holoblastic, terminal or subterminal, globose to barrel-shaped, producing one to three conidia per cell. *Conidia* two-celled, smooth- and thick-walled, navicular, slightly constricted at the septum, upper cell ovoid with a truncate base, subacute at the apex, dark brown, with a longitudinal pale-colored germ slit, basal cell smaller and hyaline, irregularly barrel-shaped to campaniform, secession rhexolitic. *Sexual morph* not observed.

*Type species*: *Pseudowardomyces humicola* (Hennebert & G.L. Barron) Barnés-Guirado, Stchigel & Cano, comb. nov. MycoBank MB851997. *Wardomyces humicola* Hennebert & G.L. Barron, Can. J. Bot. 40: 1209 (1962). [Basionym].

*Other species*: *Pseudowardomyces pulvinatus* (Marchal) Barnés-Guirado, Stchigel & Cano, comb. nov. MycoBank MB851998. *Echinobotryum pulvinatum* Marchal, Bull. Soc. R. Bot. Belg. 34(no. 1): 139 (1895). [Basionym].

*Diagnosis*: The genus *Pseudowardomyces* is morphologically similar to the genus *Wardomyces* but differs in the production of more complex conidiophores. On the other hand, *Pseudowardomyces* differs morphologically from *Parawardomyces* by the production of two-celled conidia (unicellular in *Parawardomyces*).

*Parawardomyces* Barnés-Guirado, Stchigel & Cano gen. nov. MycoBank MB851964.

*Etymology*: From Greek *παρα-* (para) due to its morphological resemblance to the genus *Wardomyces*.

*Description*: *Hyphae* hyaline, septate, smooth- and thin-walled, branched. Asexual morph *Conidiophores* micronematous to semi-macronematous, monoverticillate, short-stipitate, hyaline. *Conidiogenous cells* holoblastic, terminal, short, cylindrical to barrel-shaped, producing usually one to three, sometimes more conidia per cell. *Conidia* one-celled, hyaline to pale brown, solitary, ellipsoid to cylindrical with a rounded apex and a truncate base, smooth-walled, with a longitudinal pale-colored germ slit, secession schizolytic. Scopulariopsis-like synanamorph present. *Conidiophores* micronematous to macronematous, biverticillate, stipitate, penicillate. *Conidiogenous cells* annellidic, flask-shaped, solitary, in whorls of three to five on the vegetative hyphae, or on verticillate metulae. *Conidia* one-celled, smooth- and thick-walled, pyriform to ovoid, basally truncate, in basipetal chains. *Sexual morph—Ascomata* dark brown, ostiolate, setose, subglobose to globose; neck long, cylindrical, setose; *peridial wall* of *textura angularis*. *Asci* subglobose to globose, thin-walled, non-stipitate, eight-spored, soon evanescent. *Ascospores* one-celled, subhyaline to pale orange, smooth-walled, kidney-shaped, flattened laterally, with a germ pore at each end, germinating by means of germ tubes through one or both pores.

*Type species*: *Parawardomyces ovalis* (W. Gams) Barnés-Guirado, Stchigel & Cano, comb. nov. MycoBank MB 851999. *Wardomyces ovalis* W. Gams, Trans. Br. mycol. Soc. 51(5): 798 (1968). [Basionym].

*Other species*: *Parawardomyces giganteus* (Malloch) Barnés-Guirado, Stchigel & Cano, comb. nov. MycoBank MB 852000. *Microascus giganteus* Malloch, Mycologia 62(4): 731 (1970). [Basionym].

*Diagnosis*: *Parawardomyces* differs from *Pseudowardomyces* and *Wardomyces* by the production of unicellular conidia (bicellular in the latter genera), and by a scopulariopsis-like synanamorph (absent in the other two genera). In addition, one of their species (*Parawardomyces gigantea*) produces a microascus-like sexual morph, which is absent in both *Pseudowardomyces* and *Wardomyces*.

Due to the reassignment of some species that belonged to the genus *Wardomyces* into other genera, we have amended it as follows.

*Wardomyces* F.T. Brooks & Hansf. Trans. Br. mycol. Soc. 8(3): 137 (1923). MycoBank MB10433.

*Description: Hyphae* hyaline, branched, sometimes aggregated and septate. *Conidiophores* semi-macronematous, mononematous, mostly biverticillate, sometimes terverticillate, short-stipitate, straight, hyaline, smooth, and branched. *Conidiogenous cells* polyblastic, determinate, ampulliform, doliiform, or irregularly shaped. *Conidia* solitary, ovoid, sometimes pointed at the apical end, ellipsoidal to slightly cylindrical, truncated at the base, brown or blackish brown, smooth, with a longitudinal germ slit, aseptate, secession schizolytic. *Sexual morph* not observed.

*Type species: Wardomyces anomalus* F.T. Brooks & Hansf. [as 'anomala'], Trans. Br. mycol. Soc. 8(3): 137 (1923). MycoBank MB256937.

Other species: *Wardomyces inflatus* (Marchal) Hennebert. MycoBank MB341001. *Trichosporum inflatum* Marchal, Champ. copr. Belg. 7: 142 (1896). [Basionym].

*Diagnosis*: *Wardomyces* differs from *Parawardomyces* and *Pseudowardomyces* by presenting semi-macronematous, short-stipitate, mostly biverticillate conidiophores. Moreover, it differs from *Parawardomyces* because does not present the scopulariopsis-like synanamorph and lacks a sexual morph, and from *Pseudowardomyces* in the conidial shape, the absence of a septum and its schizolytic secession.

## 4. Discussion

Although some fungi found during the course of this study, such as *Alternaria alternata*, *Aspergillus amstelodami*, and *Aspergillus versicolor*, have been previously reported in hypersaline lagoons and lakes [75], there are no reports of these identified species in endorheic lagoons in Spain. Notably, the extremophilic nature of several species belonging to globally distributed genera recovered in this study, such as *Aspergillus*, *Cladosporium*, and *Penicillium*, has been documented in earlier studies [76–79]. Additionally, most of the identified species have been previously reported as extremophilic or extremotolerant. For instance, *Aspergillus amstelodami*, *Aspergillus intermedius*, and *Penicillium egyptiacum* exhibit osmophilic, xerophilic, or xerotolerant behavior [64,72], while *Parachaetomium truncatulum*, *Aspergillus calidoustus*, and *Chaetomium grande* display thermotolerance [64,70].

Particularly interesting are the findings of *Chaetomium grande* and *Parachaetomium truncatulum*, which represent two new reports for Spain and Europe, also being the first time that this species has been isolated from lake sediments [70,80–82]. Furthermore, *Acrostalagmus luteoalbus* can thrive in soils with high pH values (alkali-tolerant), and *Aspergillus versicolor* and *Stachybotrys chartarum* exhibit halotolerance [60,68,74]. Some species have not been previously reported as extremophilic or extremotolerant, yet our study reveals their ability to grow in up to 10% $w/v$ NaCl, such as *Cladosporium europaeum* and *Malbranchea zuffiana*. Surprisingly, we recovered *Cephalotrichiella penicillata*, *Fusarium culmorum*, and *Ovatospora amygdalispora*, taxa that do not exhibit any extremophilic/extremotolerant characteristics and, consequently, should be considered as non-specialized.

Based on both phylogenetic analysis and phenotypic features, we introduced the new monotypic genus *Dactyliodendromyces*, isolated from a sediment sample from *Laguna de Pito*. This fungus does not exhibit a strong halotolerant behavior and belongs to the *Microascaceae* family. While members of *Microascaceae* have a global distribution, only a few have been initially discovered in Spain, such as *Wardomycopsis litoralis* and *Pseudoscopulariopsis schumacheri* [20,83]. Although most of the members of *Microascaceae* are not usually isolated from extreme environments [29,84–86], some species of the genera *Microascus* and *Scopulariopsis* have been isolated from halophyte plants and salt marshes, respectively [87,88]. Moreover, only two species of this family were first isolated from salty habitats: *Wardomyces pulvinatus* and *Wardomycopsis litoralis* [83,89]. *Dactyliodendromyces holomorphus* differs from its closely related genera *Gamsia*, *Parawardomyces*, and *Pseudowardomyces* by the production of the holomorph in vitro. The only exception is *Parawardomyces giganteus*, which also produces a sexual morph, yet they do not morphologically resemble each other [19]. The sexual morph of *D. holomorphus* consists of short-naked ostiolate ascomata with true setae and a tomentose, carbonaceous peridium, with these features being uncommon among the *Microascaeae* [24,85,90,91].

Despite previous attempts to separate the genus *Wardomyces* into different genera, nowadays, it is still considered a paraphyletic genus [92]. However, based on our phylogenetic analysis using the ITS-LSU-*EF-1α*-*tub2* markers, *Wardomyces* could be segregated into three different genera. Furthermore, based on the morphological features, *Wardomyces* produces complex conidiophores, bi- to terverticillate, and one-celled conidia, whereas *Parawardomyces* is characterized by the production of monoverticillate conidiophores, and a scopulariopsis-like and a mammaria-like synanamorph, whereas *Pseudowardomyces* is distinguished by the production of bi-celled conidia.

To date, there is limited information on fungi isolated from endorheic lagoons in Europe, including Spain. Therefore, this study makes a significant contribution to the understanding of mycobiota in such environments by documenting the discovery of a new genus of the order Microascales and several rare taxa, particularly from the *Chaetomiaceae* (order Sordariales) family. It is noteworthy that no extremely halophilic fungi were identified.

**Author Contributions:** Conceptualization, A.M.S. and J.F.C.-L.; methodology, M.B.-G., A.M.S. and J.F.C.-L.; software, M.B.-G. and J.F.C.-L.; validation, A.M.S. and J.F.C.-L.; formal analysis, M.B.-G., A.M.S. and J.F.C.-L.; investigation, M.B.-G.; resources, J.F.C.-L.; data curation, M.B.-G.; writing—original draft preparation, M.B.-G.; writing—review and editing, M.B.-G., A.M.S. and J.F.C.-L.; visualization, M.B.-G., A.M.S. and J.F.C.-L.; supervision, A.M.S. and J.F.C.-L.; project administration, J.F.C.-L.; funding acquisition, J.F.C.-L. All authors have read and agreed to the published version of the manuscript.

**Funding:** This work was supported by the Spanish Ministerio de Economía y Competitividad, grant CGL2017-88094-P.

**Institutional Review Board Statement:** Not applicable.

**Informed Consent Statement:** Not applicable.

**Data Availability Statement:** Data are contained within the article.

**Acknowledgments:** M.B.-G. is grateful to University Rovira i Virgili and Diputación de Tarragona for a Martí-Franqués doctoral grant.

**Conflicts of Interest:** The authors declare no conflicts of interest.

## References


1. Rampelotto, P.H. Extremophiles and Extreme Environments. *Life* **2013**, *3*, 482–485. [CrossRef] [PubMed]
2. D'Elia, L.; del Mondo, A.; Santoro, M.; de Natale, A.; Pinto, G.; Pollio, A. Microorganisms from Harsh and Extreme Environments: A Collection of Living Strains at ACUF (Naples, Italy). *Ecol. Quest.* **2018**, *29*, 63–74. [CrossRef]
3. Santra, H.K.; Banerjee, D. Extremophilic Microbes and their Extremozymes for Industry and Allied Sectors. In *Microbiomes of Extreme Environments*, 1st ed.; Yadav, A.N., Rastegari, A.A., Yadav, N., Eds.; CRC Press, Taylor and Francis Group: Boca Raton, FL, USA, 2021; Volume 1, pp. 1–109.
4. Blachowicz, A.; Chiang, A.J.; Elsaesser, A.; Kalkum, M.; Ehrenfreund, P.; Stajich, J.E.; Torok, T.; Wang, C.C.C.; Venkateswaran, K. Proteomic and Metabolomic Characteristics of Extremophilic Fungi under Simulated Mars Conditions. *Front. Microbiol.* **2019**, *10*, 1013. [CrossRef] [PubMed]
5. Yapiyev, V.; Sagintayev, Z.; Inglezakis, V.J.; Samarkhanov, K.; Verhoef, A. Essentials of Endorheic Basins and Lakes: A Review in the Context of Current and Future Water Resource Management and Mitigation Activities in Central Asia. *Water* **2017**, *9*, 798. [CrossRef]
6. van den Broeck, M.; Waterkeyn, A.; Rhazi, L.; Grillas, P.; Brendonck, L. Assessing the Ecological Integrity of Endorheic Wetlands, with Focus on Mediterranean Temporary Ponds. *Ecol. Indic.* **2015**, *54*, 1–11. [CrossRef]
7. Nicolet, P.; Biggs, J.; Fox, G.; Hodson, M.J.; Reynolds, C.; Whitfield, M.; Williams, P. The Wetland Plant and Macroinvertebrate Assemblages of Temporary Ponds in England and Wales. *Biol. Conserv.* **2004**, *120*, 261–278. [CrossRef]
8. Blanché, C.; Molero, J. Las cubetas arreicas al sur de Bujaraloz (Valle del Ebro). *Contribución a su estudio fitocenológico. Lazaroa* **1986**, *9*, 277–299.
9. Castañeda, C.; Herrero, J.; Conesa, J.A. Distribution, morphology and habitats of saline wetlands: A case study from Monegros, Spain. *Geol. Acta* **2013**, *11*, 371–388. [CrossRef]
10. Castañeda, C.; García-Vera, M.Á. Water Balance in the Playa-Lakes of an Arid Environment, Monegros, NE Spain. *Hydrogeol. J.* **2008**, *16*, 87–102. [CrossRef]
11. Ramsar Convention Secretariat. Designating Ramsar Sites: Strategic Framework and guidelines for the future development of the List of Wetlands of International Importance. In *Ramsar Handbooks for the Wise Use of Wetlands*, 4th ed.; Ramsar Convention Secretariat: Gland, Switzerland, 2010; pp. 17–116.
12. Ribera, I.; Blasco-Zumeta, J. Biogeographical links between steppe insects in the Monegros region (Aragón, NE Spain), the eastern Mediterranean, and central Asia. *J. Biogeogr.* **1998**, *25*, 969–986. [CrossRef]
13. Vives i Durán, J.; Vives i Noguera, E. Coleópteros halófilos de Los Monegros. *Bol. Asoc. Esp. Entom.* **1978**, *2*, 205–214.
14. Casamayor, E.O.; Triadó-Margarit, X.; Castañeda, C. Microbial Biodiversity in Saline Shallow Lakes of the Monegros Desert, Spain. *FEMS Microbiol. Ecol.* **2013**, *85*, 503–518. [CrossRef] [PubMed]

15. Azpiazu-Muniozguren, M.; Perez, A.; Rementeria, A.; Martinez-Malaxetxebarria, I.; Alonso, R.; Laorden, L.; Gamboa, J.; Bikandi, J.; Garaizar, J.; Martinez-Ballesteros, I. Fungal Diversity and Composition of the Continental Solar Saltern in Añana Salt Valley (Spain). *J. Fungi* **2021**, *7*, 1074. [CrossRef] [PubMed]
16. Perl, T.; Kis-Papo, T.; Nevo, E. Fungal Biodiversity in the Hypersaline Dead Sea: Extinction and Evolution. *Biol. J. Linn. Soc.* **2017**, *121*, 122–132. [CrossRef]
17. Gostinčar, C.; Gunde-Cimerman, N. Understanding Fungi in Glacial and Hypersaline Environments. *Annu. Rev. Microbiol.* **2023**, *77*, 89–109. [CrossRef] [PubMed]
18. Georgieva, M.L.; Bilanenko, E.N.; Ponizovskaya, V.B.; Kokaeva, L.Y.; Georgiev, A.A.; Efimenko, T.A.; Markelova, N.N.; Kuvarina, A.E.; Sadykova, V.S. Haloalkalitolerant Fungi from Sediments of the Big Tambukan Saline Lake (Northern Caucasus): Diversity and Antimicrobial Potential. *Microorganisms* **2023**, *11*, 2587. [CrossRef] [PubMed]
19. Malloch, D. New Concepts in the *Microascaceae* Illustrated by Two New Species. *Mycologia* **1970**, *62*, 727–740. [CrossRef]
20. Sandoval-Denis, M.; Guarro, J.; Cano-Lira, J.F.; Sutton, D.A.; Wiederhold, N.P.; de Hoog, G.S.; Abbott, S.P.; Decock, C.; Sigler, L.; Gené, J. Phylogeny and Taxonomic Revision of *Microascaceae* with Emphasis on Synnematous Fungi. *Stud. Mycol.* **2016**, *83*, 193–233. [CrossRef]
21. Lackner, M.; de Hoog, G.S.; Yang, L.; Ferreira Moreno, L.; Ahmed, S.A.; Andreas, F.; Kaltseis, J.; Nagl, M.; Lass-Flörl, C.; Risslegger, B.; et al. Proposed Nomenclature for *Pseudallescheria, Scedosporium* and Related Genera. *Fungal Divers.* **2014**, *67*, 1–10. [CrossRef]
22. Wang, M.M.; Yang, S.Y.; Li, Q.; Zheng, Y.; Ma, H.H.; Tu, Y.H.; Wei, L.; Cai, L. *Microascaceae* from the marine environment, with descriptions of six new species. *J. Fungi* **2024**, *10*, 45. [CrossRef]
23. Wijayawardene, N.N.; Hyde, K.D.; Dai, D.Q.; Sánchez-García, M.; Goto, B.; Saxena, R.; Erdoğdu, M.; Selcuk, F.; Rajeshkumar, K.; Aptroot, A.; et al. Outline of Fungi and fungus-like taxa—2021. *Mycosphere* **2022**, *13*, 53–453. [CrossRef]
24. Sandoval-Denis, M.; Gené, J.; Sutton, D.A.; Cano-Lira, J.F.; de Hoog, G.S.; Decock, C.A.; Wiederhold, N.P.; Guarro, J. Redefining *Microascus, Scopulariopsis* and Allied Genera. *Persoonia* **2016**, *36*, 1–36. [CrossRef] [PubMed]
25. Wang, X.W.; Han, P.J.; Bai, F.Y.; Luo, A.; Bensch, K.; Meijer, M.; Kraak, B.; Han, D.Y.; Sun, B.D.; Crous, P.W.; et al. Taxonomy, Phylogeny and Identification of *Chaetomiaceae* with Emphasis on Thermophilic Species. *Stud. Mycol.* **2022**, *101*, 121–243. [CrossRef] [PubMed]
26. Ramirez-Garcia, A.; Pellon, A.; Rementeria, A.; Buldain, I.; Barreto-Bergter, E.; Rollin-Pinheiro, R.; De Meirelles, J.V.; Xisto, M.I.D.S.; Ranque, S.; Havlicek, V.; et al. *Scedosporium* and *Lomentospora*: An Updated Overview of Underrated Opportunists. *Med. Mycol.* **2018**, *56*, 102–125. [CrossRef] [PubMed]
27. Liu, W.; Feng, R.Z.; Jiang, H.L.; Schildgen, O. *Scedosporium* spp. Lung Infection in Immunocompetent Patients: A Systematic Review and MOOSE-Compliant Meta-Analysis. *Medicine* **2019**, *98*, 41–49. [CrossRef] [PubMed]
28. Chen, S.C.A.; Halliday, C.L.; Hoenigl, M.; Cornely, O.A.; Meyer, W. *Scedosporium* and *Lomentospora* Infections: Contemporary Microbiological Tools for the Diagnosis of Invasive Disease. *J. Fungi* **2021**, *7*, 23. [CrossRef]
29. Su, L.; Zhu, H.; Niu, Y.; Guo, Y.; Du, X.; Guo, J.; Zhang, L.; Qin, C. Phylogeny and Taxonomic Revision of *Kernia* and *Acaulium*. *Sci. Rep.* **2020**, *10*, 10302. [CrossRef]
30. Zhang, Z.F.; Zhou, S.Y.; Eurwilaichitr, L.; Ingsriswang, S.; Raza, M.; Chen, Q.; Zhao, P.; Liu, F.; Cai, L. Culturable Mycobiota from Karst Caves in China II, with Descriptions of 33 New Species. *Fungal Divers.* **2021**, *106*, 29–136. [CrossRef]
31. Crous, P.W.; Boers, J.; Holdom, D.; Osieck, E.R.; Steinrucken, T.V.; Tan, Y.P.; Vitelli, J.S.; Shivas, R.G.; Barrett, M.; Boxshall, A.G.; et al. Fungal Planet Description Sheets: 1383–1435. *Persoonia* **2022**, *48*, 261–371. [CrossRef]
32. Crous, P.W.; Cowan, D.A.; Maggs-Kölling, G.; Yilmaz, N.; Larsson, E.; Angelini, C.; Brandrud, T.E.; Dearnaley, J.D.W.; Dima, B.; Dovana, F.; et al. Fungal Planet Description Sheets: 1112–1181. *Persoonia* **2020**, *45*, 251–409. [CrossRef]
33. Preedanon, S.; Suetrong, S.; Srihom, C.; Somrithipol, S.; Kobmoo, N.; Saengkaewsuk, S.; Srikitikulchai, P.; Klaysuban, A.; Nuankaew, S.; Chuaseeharonnachai, C.; et al. Eight Novel Cave Fungi in Thailand's Satun Geopark. *Fungal. Syst. Evol.* **2023**, *12*, 1–30. [CrossRef] [PubMed]
34. Castañeda, C. Las Saladas Del Sur de Monegros: Facies, Régimen Hídrico y Estado Actual. Ph.D. Thesis, Universidad de Zaragoza, Zaragoza, Spain, 2004.
35. Hocking, A.D.; Pitt, J.I. Dichloran-Glycerol Medium for Enumeration of Xerophilic Fungi from Low-Moisture Foods. *Appl. Environ. Microbiol.* **1980**, *39*, 488–492. [CrossRef] [PubMed]
36. Börner, G.V.; Cha, R.S. Induction and Analysis of Synchronous Meiotic Yeast Cultures. *Cold Spring Harb. Protoc.* **2015**, *10*, 908–913. [CrossRef] [PubMed]
37. Hawksworth, D.L.; Kirk, P.M.; Sutton, B.C.; Pegler, D.N. *Ainsworth & Bisby's Dictionary of the Fungi*, 8th ed.; CAB International: Oxon, UK, 1995; p. 616.
38. Samson, R.A.; Houbraken, J.; Thrane, U.; Frisvad, J.C.; Andersen, B. *Food and Indoor Fungi*, 2nd ed.; CBS-KNAW Fungal Biodiversity Centre: Utrecht, The Netherlands, 2010; pp. 1–475.
39. Furuya, K.; Naito, A. An effective method for isolation of *Boothiella tetraspora* from soil. *Trans. Mycol. Soc.* **1979**, *20*, 309–311.
40. Furuya, K.; Naito, A. Stimulation of ascospore germination by phenolic compounds in members of the Sordariaceae. *Trans. Mycol. Soc.* **1980**, *21*, 77–85.
41. Ulfig, K.; Guarro, J.; Cano, J.; Genie, J.; Vidal, R.; Figueras, M.J. General Assessment of the Occurrence of Keratinolytic Fungi in River and Marine Beach Sediments of Catalonian Waters (Spain). *Water Air Soil Poll.* **1997**, *94*, 275–287. [CrossRef]
42. Kornerup, A.; Wanscher, J.H. *Methuen Handbook of Colour*, 3rd ed.; Methuen: London, UK, 1978.

43. Chupp, C. Further notes on double cover-glass mounts. *Mycologia* **1940**, *32*, 269–270. [CrossRef]
44. Müller, F.M.; Werner, K.E.; Kasai, M.; Francesconi, A.; Chanock, S.J.; Walsh, T.J. Rapid extraction of genomic DNA from medically important yeasts and filamentous fungi by high-speed cell disruption. *J. Clin. Microbiol.* **1998**, *36*, 1625–1629. [CrossRef]
45. White, T.J.; Bruns, T.; Lee, S.J.W.T.; Taylor, J. Amplification and direct sequencing of fungal ribosomal RNA genes for phylogenetics. In *PCR Protocols: A Guide to Methods and Applications*, 1st ed.; Innis, M.A., Gelfand, D.H., Sninsky, J.J., White, T.J., Eds.; Academic Press: San Diego, CA, USA, 1990; pp. 315–322.
46. Vilgalys, R.; Hester, M. Rapid genetic identification and mapping of enzymatically amplified ribosomal DNA from several *Cryptococcus* species. *J. Bacteriol.* **1990**, *172*, 4238–4246. [CrossRef]
47. Rehner, S.A.; Buckley, E. A *Beauveria* phylogeny inferred from nuclear ITS and EF1-α sequences: Evidence for cryptic diversification and links to *Cordyceps* teleomorphs. *Mycologia* **2005**, *97*, 84–98. [CrossRef]
48. Iturrieta-González, I.; García, D.; Gené, J. Novel Species of *Cladosporium* from Environmental Sources in Spain. *MycoKeys* **2021**, *77*, 1–25. [CrossRef] [PubMed]
49. Glass, N.L.; Donaldson, G.C. Development of Primer Sets Designed for Use with the PCR To Amplify Conserved Genes from Filamentous Ascomycetes. *Appl. Environ. Microbiol.* **1995**, *61*, 1323–1330. [CrossRef] [PubMed]
50. Torres-Garcia, D.; García, D.; Cano-Lira, J.F.; Gené, J. Two Novel Genera, *Neostemphylium* and *Scleromyces* (*Pleosporaceae*) from Freshwater Sediments and Their Global Biogeography. *J. Fungi* **2022**, *8*, 868. [CrossRef] [PubMed]
51. Tamura, K.; Stecher, G.; Peterson, D.; Filipski, A.; Kumar, S. MEGA7: Molecular evolutionary genetics analysis version 7.0. *Mol. Biol. Evol.* **2013**, *30*, 2725–2729. [CrossRef] [PubMed]
52. Thompson, J.D.; Higgins, D.G.; Gibson, T.J. CLUSTAL W: Improving the sensitivity of progressive multiple sequence alignment through sequence weighting, position-specific gap penalties and weight matrix choice. *Nucleic Acids Res.* **1994**, *22*, 4673–4680. [CrossRef] [PubMed]
53. Edgar, R.C. MUSCLE: Multiple sequence alignment with high accuracy and high throughput. *Nucleic Acids Res.* **2004**, *32*, 1792–1797. [CrossRef]
54. Stamatakis, A. RAxML version 8: A tool for phylogenetic analysis and post-analysis of large phylogenies. *Bioinformatics* **2014**, *30*, 1312–1313. [CrossRef]
55. Miller, M.A.; Pfeifferm, W.; Schwartz, T. The CIPRES science gateway: Enabling high-impact science for phylogenetics researchers with limited resources. In Proceedings of the 1st Conference of the Extreme Science and Engineering Discovery Environment: Bridging from the Extreme to the Campus and Beyond, Chicago, IL, USA, 16–20 July 2012; Association for Computing Machinery: New York, NY, USA, 2012; pp. 1–8.
56. Ronquist, F.; Teslenko, M.; Van Der Mark, P.; Ayres, D.L.; Darling, A.; Höhna, S.; Larget, B.; Liu, L.; Suchard, M.A.; Huelsenbeck, J.P. Mrbayes 3.2: Efficient Bayesian Phylogenetic Inference and Model Choice across a Large Model Space. *Syst. Biol.* **2012**, *61*, 539–542. [CrossRef]
57. Darriba, D.; Taboada, G.L.; Doallo, R.; Posada, D. JModelTest 2: More Models, New Heuristics and Parallel Computing. *Nat. Methods* **2012**, *9*, 772. [CrossRef]
58. Hillis, D.M.; Bull, J.J. An Empirical Test of Bootstrapping as a Method for Assessing Confidence in Phylogenetic Analysis. *Syst. Biol.* **1993**, *42*, 182–192. [CrossRef]
59. Hespanhol, L.; Vallio, C.S.; Costa, L.M.; Saragiotto, B.T. Understanding and Interpreting Confidence and Credible Intervals around Effect Estimates. *Braz. J. Phys. Ther.* **2019**, *23*, 290–301. [CrossRef] [PubMed]
60. Grum-Grzhimaylo, A.A.; Georgieva, M.L.; Bondarenko, S.A.; Debets, A.J.M.; Bilanenko, E.N. On the Diversity of Fungi from Soda Soils. *Fungal Divers.* **2016**, *76*, 27–74. [CrossRef]
61. Nguyen, T.T.T.; Jung, H.Y.; Lee, Y.S.; Voigt, K.; Lee, H.B. Phylogenetic Status of Two Undescribed Zygomycete Species from Korea: *Actinomucor elegans* and *Mucor minutus*. *Mycobiology* **2017**, *45*, 344–352. [CrossRef] [PubMed]
62. Moubasher, A.H.; Ismail, M.A.; Hussein, N.A.; Gouda, H.A. Enzyme Producing Capabilities of Some Extremophilic Fungal Strains Isolated from Different Habitats of Wadi El-Natrun, Egypt. Part 2: Cellulase, Xylanase and Pectinase. *Eur. J. Biol. Res.* **2016**, *6*, 103–111.
63. Ismail, M.A.; Hussein; Abdel-Sater, M.A.; Sayed, R.M. Evaluation of physiological and biochemical characteristics of *Alternaria* species isolated from soil in Assiut Governorate, Egypt, in addition to dichotomous key to the encountered species. *Assiut Univ. J. Bot. Microbiol.* **2020**, *49*, 34–59.
64. Kozakiewicz, Z.; Smith, D. Physiology of *Aspergillus*. In *Biotechnology Handbooks*, 7th ed.; Smith, J.E., Ed.; Springer: Boston, MA, USA, 1994; pp. 23–40.
65. Butinar, L.; Zalar, P.; Frisvad, J.C.; Gunde-Cimerman, N. The Genus *Eurotium*—Members of Indigenous Fungal Community in Hypersaline Waters of Salterns. *FEMS Microbiol. Ecol.* **2005**, *51*, 155–166. [CrossRef] [PubMed]
66. Chen, A.J.; Hubka, V.; Frisvad, J.C.; Visagie, C.M.; Houbraken, J.; Meijer, M.; Varga, J.; Demirel, R.; Jurjević, Ž.; Kubátová, A.; et al. Polyphasic Taxonomy of *Aspergillus* Section *Aspergillus* (Formerly *Eurotium*), and Its Occurrence in Indoor Environments and Food. *Stud. Mycol.* **2017**, *88*, 37–135. [CrossRef] [PubMed]
67. Liu, K.H.; Ding, X.W.; Narsing Rao, M.P.; Zhang, B.; Zhang, Y.G.; Liu, F.H.; Liu, B.B.; Xiao, M.; Li, W.J. Morphological and Transcriptomic Analysis Reveals the Osmo-adaptive Response of Endophytic Fungus *Aspergillus montevidensis* ZYD4 to High Salt Stress. *Front. Microbiol.* **2017**, *8*, 1789. [CrossRef]

68. Sklenář, F.; Glässnerová, K.; Jurjević, Ž.; Houbraken, J.; Samson, R.A.; Visagie, C.M.; Yilmaz, N.; Gené, J.; Cano, J.; Chen, A.J.; et al. Taxonomy of *Aspergillus* Series *Versicolores*: Species Reduction and Lessons Learned about Intraspecific Variability. *Stud. Mycol.* **2022**, *102*, 53–93. [CrossRef] [PubMed]
69. Crous, P.W.; Shivas, R.G.; Quaedvlieg, W.; van der Bank, M.; Zhang, Y.; Summerell, B.A.; Guarro, J.; Wingfield, M.J.; Wood, A.R.; Alfenas, A.C.; et al. Fungal Planet Description Sheets: 214–280. *Persoonia* **2014**, *32*, 184–306. [CrossRef]
70. Asgari, B.; Rasoul, Z. The Genus *Chaetomium* in Iran, a Phylogenetic Study Including Six New Species. *Mycologia* **2011**, *103*, 863–882. [CrossRef] [PubMed]
71. Garmendia, G.; Alvarez, A.; Villarreal, R.; Martínez-Silveira, A.; Wisniewski, M.; Vero, S. Fungal Diversity in the Coastal Waters of King George Island (Maritime Antarctica). *World J. Microbiol. Biotechnol.* **2021**, *37*, 142–154. [CrossRef] [PubMed]
72. Pitt, J.I. An Appraisal of Identification Methods for *Penicillium* Species: Novel Taxonomic Criteria Based on Temperature and Water Relations. *Mycologia* **1973**, *65*, 1135–1157. [CrossRef] [PubMed]
73. Bretzloff, C.W. The Growth and Fruiting of *Sordaria fimicola*. *Am. J. Bot.* **1954**, *41*, 58–67. [CrossRef]
74. Paiva, D.S.; Fernandes, L.; Pereira, E.; Trovão, J.; Mesquita, N.; Tiago, I.; Portugal, A. Exploring Differences in Culturable Fungal Diversity Using Standard Freezing Incubation—A Case Study in the Limestones of Lemos Pantheon (Portugal). *J. Fungi* **2023**, *9*, 501. [CrossRef] [PubMed]
75. Kis-Papo, T.; Grishkan, I.; Oren, A.; Wasser, S.P.; Nevo, E. Spatiotemporal diversity of filamentous fungi in the hypersaline Dead Sea. *Mycol. Res.* **2001**, *6*, 749–756. [CrossRef]
76. Gunde-Cimerman, N.; Zalar, P. Extremely Halotolerant and Halophilic Fungi Inhabit Brine in Solar Salterns Around the Globe. *Food Technol. Biotechnol.* **2014**, *52*, 170–179.
77. Sayed, A.M.; Hassan, M.H.A.; Alhadrami, H.A.; Hassan, H.M.; Goodfellow, M.; Rateb, M.E. Extreme Environments: Microbiology Leading to Specialized Metabolites. *J. Appl. Microbiol.* **2020**, *128*, 630–657. [CrossRef]
78. Martinelli, L.; Zalar, P.; Gunde-Cimerman, N.; Azua-Bustos, A.; Sterflinger, K.; Piñar, G. *Aspergillus atacamensis* and *A. salisburgensis*: Two New Halophilic Species from Hypersaline/Arid Habitats with a phialosimplex-like Morphology. *Extremophiles* **2017**, *21*, 755–773. [CrossRef]
79. Gostinčar, C.; Zalar, P.; Gunde-Cimerman, N. No Need for Speed: Slow Development of Fungi in Extreme Environments. *Fungal Biol. Rev.* **2022**, *39*, 1–14. [CrossRef]
80. Abdel-Azeem, A.M.; Blanchette, R.A.; Held, B.W. New Record of *Chaetomium grande* Asgari & Zare (*Chaetomiaceae*) for the Egyptian and African Mycobiota. *Phytotaxa* **2018**, *343*, 283–288. [CrossRef]
81. Mehrabi, M.; Asgari, B. Description of *Allocanariomyces* and *Parachaetomium*, Two New Genera, and *Achaetomium aegilopis* sp. nov. in the *Chaetomiaceae*. *Mycol. Prog.* **2020**, *19*, 1415–1427. [CrossRef]
82. Zhang, Y.; Wu, W.; Cai, L. Polyphasic Characterisation of *Chaetomium* Species from Soil and Compost Revealed High Number of Undescribed Species. *Fungal Biol.* **2017**, *121*, 21–43. [CrossRef] [PubMed]
83. Silvera-Simón, C.; Gené, J.; Cano, J.; Guarro, J. *Wardomycopsis litoralis*, a New Soil-Borne Hyphomycete from Spain. *Mycotaxon* **2008**, *105*, 195–202.
84. Sun, B.; Zhou, Y.; Chen, A.J. Two New *Microascus* Species with Spinous Conidia Isolated from Pig Farm Soils in China. *Mycoscience* **2020**, *61*, 190–196. [CrossRef]
85. Tazik, Z.; Rahnama, K.; Iranshahi, M.; White, J.F.; Soltanloo, H. A New Species of *Pithoascus* and First Report of This Genus as Endophyte Associated with *Ferula ovina*. *Mycoscience* **2020**, *61*, 145–150. [CrossRef]
86. Abrantes, R.A.; Refojo, N.; Hevia, A.I.; Fernández, J.; Isla, G.; Córdoba, S.; Dávalos, M.F.; Lubovich, S.; Maldonado, I.; Davel, G.O.; et al. *Scedosporium* spp. From Clinical Setting in Argentina, with the Proposal of the New Pathogenic Species *Scedosporium americanum*. *J. Fungi* **2021**, *7*, 160. [CrossRef]
87. Calabon, M.S.; Jones, E.B.G.; Promputtha, I.; Hyde, K.D. Fungal Biodiversity in Salt Marsh Ecosystems. *J. Fungi* **2021**, *7*, 648. [CrossRef]
88. Jalili, B.; Bagheri, H.; Azadi, S.; Soltani, J. Identification and Salt Tolerance Evaluation of Endophyte Fungi Isolates from Halophyte Plants. *Int. J. Environ. Sci. Technol.* **2020**, *17*, 3459–3466. [CrossRef]
89. Dickinson, C.H. The Genus *Wardomyces*. *Trans. Br. Mycol. Soc.* **1964**, *47*, 321. [CrossRef]
90. Lee, J.H.; Ten, L.N.; Lee, S.Y.; Jung, H.Y. Novel Fungal Species Belonging to the Genus *Acaulium* Isolated from *Riptortus clavatus* (Heteroptera: Alydidae) in Korea. *Korean J. Mycol.* **2021**, *49*, 477–486. [CrossRef]
91. Woudenberg, J.H.C.; Meijer, M.; Houbraken, J.; Samson, R.A. *Scopulariopsis* and scopulariopsis-like Species from Indoor Environments. *Stud. Mycol.* **2017**, *88*, 1–35. [CrossRef]
92. Morelet, M. Micromycètes du var et d'ailleurs (2me Note). *Ann. Soc. Sci. Nat. Archèol. Toulon du Var* **1969**, *21*, 104–106.

Journal of
*Fungi*

MDPI

*Article*

# Sunken Riches: Ascomycete Diversity in the Western Mediterranean Coast through Direct Plating and Flocculation, and Description of Four New Taxa

**Daniel Guerra-Mateo , José F. Cano-Lira , Ana Fernández-Bravo and Josepa Gené ***

Unitat de Micologia i Microbiologia Ambiental, Facultat de Medicina i Ciències de la Salut and Institut Universitari de Recerca en Sostenibilitat, Canvi Climàtic i Transició Energètica (IU-RESCAT), Universitat Rovira i Virgili, 43201 Reus, Spain; daniel.guerra@urv.cat (D.G.-M.); jose.cano@urv.cat (J.F.C.-L.); ana.fernandez@urv.cat (A.F.-B.)

* Correspondence: josepa.gene@urv.cat; Tel.: +34-977759359

**Abstract:** The Mediterranean Sea stands out as a hotspot of biodiversity, whose fungal composition remains underexplored. Marine sediments represent the most diverse substrate; however, the challenge of recovering fungi in culture hinders the precise identification of this diversity. Concentration techniques like skimmed milk flocculation (SMF) could represent a suitable solution. Here, we compare the effectiveness in recovering filamentous ascomycetes of direct plating and SMF in combination with three culture media and two incubation temperatures, and we describe the fungal diversity detected in marine sediments. Sediments were collected at different depths on two beaches (Miracle and Arrabassada) on the Spanish western Mediterranean coast between 2021 and 2022. We recovered 362 strains, and after a morphological selection, 188 were identified primarily with the LSU and ITS barcodes, representing 54 genera and 94 species. *Aspergillus*, *Penicillium*, and *Scedosporium* were the most common genera, with different percentages of abundance between both beaches. Arrabassada Beach was more heterogeneous, with 42 genera representing 60 species (Miracle Beach, 28 genera and 54 species). Although most species were recovered with direct plating (70 species), 20 species were exclusively obtained using SMF as a sample pre-treatment, improving our ability to detect fungi in culture. In addition, we propose three new species in the genera *Exophiala*, *Nigrocephalum*, and *Queenslandipenidiella*, and a fourth representing the novel genus *Schizochlamydosporiella*. We concluded that SMF is a useful technique that, in combination with direct plating, including different culture media and incubation temperatures, improves the chance of recovering marine fungal communities in culture-dependent studies.

**Keywords:** *Ascomycota*; culture media; isolation; marine fungi; multi-locus phylogeny; taxonomy

## 1. Introduction

Fungi constitute a significant portion of the global genetic diversity, estimated to encompass around 2.5 million species [1]. Currently, only around 155,000 species have been formally described, suggesting that over 90% of species remain undiscovered by science [2]. The marine environment is no exception to the dearth of knowledge on fungal diversity. This environment was considered poor for fungi throughout history based on the traditional approach to detecting fungal species, i.e., culturing on prepared media or incubated samples coupled with identification based exclusively on microscopy. However, after great sampling effort, the establishment of fungal barcodes for identification, and the development of culture-independent approaches like metabarcoding, the perception has shifted, and the concealed diversity of marine fungi is gradually being unraveled [3]. In fact, although the number of marine fungi is low in comparison to terrestrial ecosystems, the count of species identified in marine habitats continues to rise, with over 1900 species currently documented worldwide [4,5]. The number of marine fungal species and their respective descriptions can be accessed in repositories such as marine fungi (https://www.marinefungi.org/ (accessed on 20 January 2024); Ref. [6]).

In this context, the Mediterranean Sea stands out as a biodiversity hotspot [7]. Biodiversity hotspots are endangered habitats that have become international priorities for conservation efforts due to their great biodiversity and their role as repositories of undiscovered taxa [8]. In the Mediterranean Sea, various substrates have been studied in search of fungi, including water and sediment [9], driftwood and seagrasses [10,11], and even invertebrate animals [12]. Sediments have been the most studied substrate, revealing that the fungal community is dominated by the Phylum *Ascomycota* [9,11,13]. The unique geographic location and relatively higher temperature of the Mediterranean Sea compared to the oceans suggest that it may host a distinctive fungal community [14].

The prevalence of *Ascomycota* in Mediterranean sediments has been confirmed by metabarcoding, although only a small fraction of the detected ascomycetes could be confidently assigned to known species [15]. The effectiveness of fungal identification using this powerful technique ultimately relies on the quality and availability of barcodes in public databases [16]. Unfortunately, merely half of the known marine fungi are represented through DNA barcodes of either terrestrial or marine origin in GenBank [14]. Fungal DNA barcodes are derived from pure cultures, emphasizing the need for ongoing efforts in marine mycology to improve the tools for fungal detection in culture [17].

While yeasts are a well-documented group of marine fungi, owing to their ease of cultivation, knowledge regarding filamentous fungi is limited [14]. The challenges in culturing filamentous marine fungi stem from either the fungus's incompatibility with synthetic media due to its lifestyle (i.e., endophytes, parasitic) or their low abundance in the environment. Given that the marine habitat is composed primarily of seawater, which can disperse fungal propagules, a potential solution to enhance the chance of culturing marine fungi could be to concentrate the environmental sample.

Skimmed milk flocculation (SMF) is a concentration technique designed to collect the microbial content within an environmental sample in flocs of skimmed milk. This technique was initially developed to improve the detection of viruses in seawater [18–20] and has subsequently been demonstrated to be successful in the detection of bacteria and protozoa [21]. In addition, SMF is recognized as a repeatable, cost-effective technique with broad applicability, such as monitoring water quality [21–23]. However, despite its proven efficacy in other microbial groups, the application of this technique for the detection of fungi remains unexplored.

In this work, we aimed to detect in culture as much diversity as possible of filamentous ascomycetes from marine sediments of the western Mediterranean Sea basin (Spain, Catalonia, Tarragona province). To accomplish this, we employed a dual approach involving direct plating of the sediment and a pre-treatment using SMF on the sample. We describe the culturable community of filamentous ascomycetes and assess the impact of each approach on detecting fungal diversity. This evaluation is complemented by the use of different culture media and isolation temperatures. In addition, we delineated the taxonomy of several interesting specimens under a consolidated species concept, resulting in the description of a new genus and four new species that belong to different classes of *Ascomycota*.

## 2. Materials and Methods

### 2.1. Study Area and Sampling

The Tarragona province is located on the western side of the Mediterranean Sea in the southern part of Catalonia, Spain. This area is known as the "golden coast" for its natural sandy beaches, which are a key attraction for tourism [24]. Notably, the port of Tarragona serves as a docking point for tourist cruise ships, ranking as the fifth most important harbor in Spain [25]. Home to a diverse marine community, this area features a range of fish species, marine invertebrates such as Mollusca, and even prairies of flowering plants like Cymodocea nodosa [26–29]. In this work, we focused on Miracle and Arrabassada Beaches, located in front of the city of Tarragona, adjacent to the port (Figure 1).

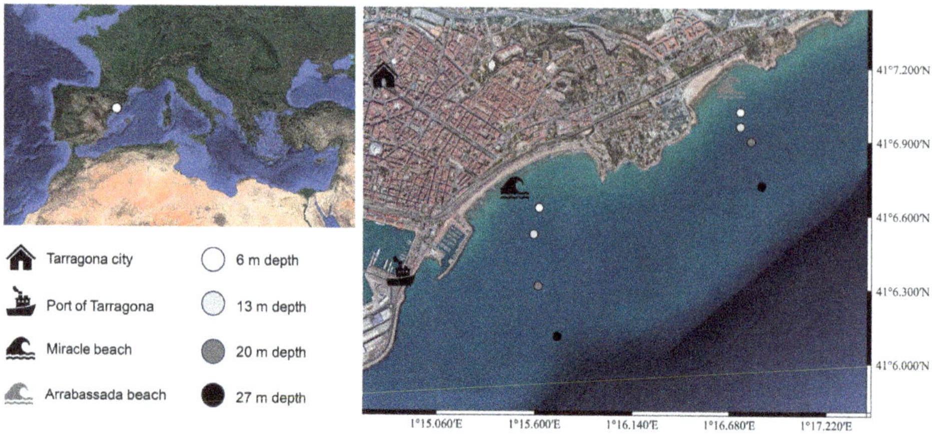


**Figure 1.** Collection sites at the golden coast of Tarragona, western Mediterranean Sea (Miracle and Arrabassada Beaches, Catalonia, Spain). Map was created using QGIS (v3.34.5; Ref. [30]).

Sediment samples were collected on these two beaches through 2021 and 2022. Each beach underwent two sampling events, with Miracle Beach sampled in June and October 2021 and Arrabassada Beach in February and June 2022. Four collection sites were determined based on the sediment grain size and water column depth, with the first point at 6 m of depth (sand), the second at 13 m (sand), the third at 20 m (transition between sand and silt), and the fourth point at 27 m of depth (silt). Four sub-samples were collected at each point, 15 cm below the seabed surface, using 50 mL plastic tubes. Throughout the collection process, tubes were stored in a refrigerated container and subsequently transported to the laboratory for immediate processing.

## 2.2. Direct Plating, Flocculation Pre-Treatment, Culturing, and Isolation of Fungi

The sediments recovered from each depth (collection sites) were treated separately. However, the sub-samples from each collection site were analyzed together. For this, each set of sub-samples was mixed in a container, vigorously shaken, and then the sediment was poured onto plastic trays with several layers of sterile filter paper to remove excess water. Subsequently, each mixed sediment sample underwent two distinct approaches for fungal isolation: direct plating into sterile Petri dishes (direct plating) and plating after a skimmed milk flocculation (SMF) pre-treatment.

The SMF pre-treatment aimed to improve the probability of detecting microbial content within the collected sediments, following the methodology described by Calgua et al. [18–20] and Rusiñol et al. [21]. In summary, 10 g of sediment from each sample was added to 1 L of sterile distilled water, treating each sample in separate containers. A skimmed milk solution was prepared by dissolving 1 g of skimmed milk powder (Difco-Becton, Dickinson and Company, Le Pont de Claix, France) in 100 mL of artificial seawater (33.33 g of sea salts), adjusting the pH to 3.5. Then, 10 mL of this solution was added to each sample to obtain a final concentration of 0.01% of skimmed milk. The samples were stirred for 16 h at room temperature to allow the flocs to settle by gravity. After removing the supernatants, the sediment was collected, transferred to 500 mL centrifuge containers, and centrifuged at 8000× $g$ for 30 min at 4 °C. Pellets were suspended in 5 mL of 0.2 M phosphate buffer, pH 7.5 (1:2, $v/v$ of 0.2 M $Na_2HPO_4$ and 0.2 M $NaH_2PO_4$) and stored at −20 °C.

Moreover, we used three culture media to recover the broadest culturable ascomycete diversity within marine sediments: dichloran rose bengal chloramphenicol agar (DRBC; 5 g peptone, 10 g glucose, 1 g $KH_2PO_4$, 0.5 g $MgSO_4$, 25 mg rose bengal, 200 mg chloramphenicol, 2 mg dichloran, 15 g agar, 1 L distilled water), 3% malt extract agar supplemented with sea water (SWMEA3%; 30 g malt extract, 5 g mycological peptone, 15 g agar,

1 L seawater) as a favorable medium for the isolation of marine fungi [31], and potato dextrose agar (PDA; Condalab, Madrid, Spain) supplemented with 2 g/L of cycloheximide (PDA+C) to detect strains resistant to this protein synthesis inhibitor, a frequent trait among various groups of *Ascomycota* [32]. To inhibit bacterial growth, 5 mL of chloramphenicol (15 g/L ethanol) was added to both SWMEA3% and PDA+C culture media.

For direct plating, the procedure involved culturing sediments on SWMEA3% and PDA+C as follows: 1 g of sediment from each sampling point was distributed across two Petri dishes per medium and mixed with the melted medium at 45 °C. In the case of DRBC, only 0.5 g of sediment from each sampling point was distributed across two Petri dishes to address fast-growing fungi. A similar methodology was applied to the sediment pre-treated with SMF, with the exception that 1 mL of floccule from each sampling point was distributed across two Petri dishes in the case of both SWMEA3% and PDA+C, and 0.5 mL of floccule was mixed with DRBC. This step was conducted in duplicate for both approaches. Subsequently, a set of the primary plates from the different culture media was incubated at 22–24 °C, while the other set was incubated at 15 °C to detect fungi capable of growing at lower temperatures. The plates were incubated in darkness and examined under the stereomicroscope for 5–8 weeks.

Pure cultures were obtained on PDA from fragments of the colonies or conidia in the primary plates using a sterile dissection needle. These cultures were used to provide a preliminary morphological identification. Strains that did exhibit sporulation were further subcultured to potato carrot agar (PCA; 20 g potato, 20 g carrot, 15 g agar, 1 L distilled water) or oatmeal agar (OA; 30 g oatmeal, 15 g agar, 1 L distilled water) to promote sporulation.

### 2.3. Morphological Analyses

All strains were carefully examined and compared morphologically with each other for presumed generic identification or species complexes when possible. Due to the large number of strains recovered from the sediments, we only selected a representative subset of strains for molecular identification. In addition, among the strains morphologically identical or similar, a maximum of three strains per collection point were selected for sequencing based on the conserved morphological traits frequently displayed between phylogenetically close ascomycete species [33]. Strains that remained sterile in vitro were also included in the molecular analyses, at the very least, to elucidate their potential affiliation with a fungal group.

For the putative novel species, macroscopic characterization of the colonies was conducted using different culture media, i.e., PDA, MEA (20 g malt extract, 15 g agar, 1 L distilled water), and OA, unless otherwise specified in the species description. Color notations in descriptions are followed by Kornerup and Wanscher [34]. Microscopic characterization was made on OA after 14 days at 25 °C in darkness, unless otherwise specified in the description. Reproductive structures were mounted with lactic acid and observed under an Olympus BH-2 bright-field microscope (Olympus Corporation, Tokyo, Japan). The descriptions were based on a minimum of 30 measurements of the relevant structures to provide size ranges. A Zeiss Axio-Imager M1 light microscope (Zeiss, Oberkochen, Germany) with a DeltaPix Infinity digital camera was used to develop photomicrographs. Photoplates were assembled using Photoshop CS6 v13.0.

The ability of each novel species to grow at different temperatures was also assessed on PDA from 5 to 40 °C at intervals of 5 °C, including measurements at 37 °C.

The strains recovered throughout the study were preserved in the culture collection of the Faculty of Medicine in Reus (FMR, Reus, Catalonia, Spain). In addition, the strains of novel or rare fungi were deposited at the Westerdijk Fungal Biodiversity Institute in Utrecht (CBS, Utrecht, The Netherlands), together with holotypes (i.e., dry colonies on the most appropriate media for their sporulation) and cultures of the selected type strains. The nomenclatural novelties were registered in MycoBank (https://www.mycobank.org/; accessed on 6 February 2024).

### 2.4. DNA Extraction, PCR Amplification, Sequencing, and Strain Identification

The strains selected based on morphology were further delineated through DNA barcodes. Genomic DNA was extracted using the modified protocol of Müller et al. [35] and quantified with a Nanodrop 2000 (Thermo Scientific, Madrid, Spain). Barcoding primers were selected based on the preliminary morphological identification and the most recent phylogenetic studies of the respective groups. The primer pairs and PCR conditions used for the different fungal groups recovered in this study are specified in Table 1. PCR products were purified and sequenced at Macrogen Corp. Europe (Madrid, Spain) using the same primers employed for amplification. Consensus sequences were assembled using SeqMan v. 7.0.0 (DNAStar Lasergene, Madison, WI, USA).

**Table 1.** Sets of primer combinations were used for DNA amplification and sequencing for each fungal family.

| Locus | Primer Pairs | Taxa | Annealing Temp. (°C) | Orientation | Primer Reference |
|---|---|---|---|---|---|
| ITS–LSU (LSU D1-D3 region) | ITS5 | Every strain | 53 | Forward | [36] |
| | LR5 | | | Reverse | [37] |
| β-tubulin (*tub*2) | Bt2a | *Aspergillaceae, Bionectriaceae, Microascaceae, Trichocomaceae* | 56 | Forward | [38] |
| | Bt2b | | | Reverse | |
| | Btub2f | *Trichocomaceae* | 56 | Forward | [39] |
| | Btub4R | | | Reverse | |
| | T1 | *Chaetomiaceae, Schizotheciaceae* | 56 | Forward | [40] |
| | Btub4R | | | Reverse | [39] |
| Traslation elongation factor-1α (*tef*1-α) | EF-983F | *Bionectriaceae, Plectosphaerellaceae* | 57 | Forward | [41] |
| | EF-2218R | | | Reverse | |
| | EF-728F | *Cladosporiaceae, Nectriaceae* | 57 | Forward | [42] |
| | EF-986R | | | Reverse | |
| RNA polymerase II second largest subunit (*rpb*2) | RPB2-5F | *Bionectriaceae, Chaetomiaceae, Cucurbitariaceae, Plectosphaerellaceae, Schizotheciaceae, Thyridariaceae* | 60 | Forward | [43] |
| | RPB2-7R | | | Reverse | |
| SSU | NS1 | *Incertae sedis* | 53 | Forward | [36] |
| | NS4 | | | Reverse | |

The resulting sequences were compared with those available at the National Center for Biotechnology Information (NCBI) using the Basic Local Alignment Search Tool (BLAST; https://blast.ncbi.nlm.nih.gov/Blast.cgi, accessed on 6 October 2023). A maximum similarity level of >98% with ≥90% sequence coverage was used to determine related GenBank sequences. The accessions of these sequences were carefully sorted to determine those associated with type strains of accepted species. Additional barcodes were sequenced to resolve phylogenetically close taxa. Strains were assigned a species name based on morphology and the percentage of similarity to type strains. Similarity values <98% were considered indicative of potential unknown fungi, and these strains were subjected to further analyses. Each strain was delineated using the barcodes listed in Table S1 (Supplementary Materials).

### 2.5. Phylogenetic Analyses

The strains that could represent potential unknown fungi were further analyzed under a consolidated species concept, combining morphological and phylogenetic data [44]. Alignments were constructed, including the sequences with the closest percentage of identity from the BLAST search and sequences from the most recent phylogenetic studies of the respective groups, namely *Exophiala* [45], *Nigrocephalum* [46], *Queenslandipenidiella* [44], and the family *Schizotheciaceae* [47,48]. Sequences were aligned using the ClustalW algorithm [49] in MEGA (Molecular Evolutionary Genetics Analysis) software v. 6.0 [50] and refined with

MUSCLE [51] or adjusted manually as needed. The analyses included more than one genetic region based on the previous phylogenies of each group [44–48]. Each region was aligned individually before being combined into a single dataset. The phylogenetic concordance among phylogenies was visually assessed to identify incongruent results among clades with high statistical support. Following the confirmation of concordance, the separate alignments were concatenated into a single data matrix in MEGA [50].

Maximum likelihood (ML) analyses were conducted using the CIPRES Science Gateway portal v. 3.3 (https://www.phylo.org/, accessed on 5 December 2023; Ref. [52]) and RAxML-HPC2 on ACCESS v. 8.2.12 [53] with the default GTR substitution matrix and 1000 rapid bootstrap replications. Bootstrap support (bs) ≥70 was considered significant [54]. Bayesian analyses were performed using MrBayes v. 3.2.6 [55]. The best substitution model for each locus was estimated using jModelTest v. 2.1.3 following the Akaike criterion [56,57]. Markov chain Monte Carlo sampling (MCMC) was performed for 10 million generations using four simultaneous chains (one cold chain and three heated chains) starting from a random tree topology. Trees were sampled every 1000th generation or until the run was stopped automatically when the average standard deviation of split frequencies fell below 0.01. The first 25% of the trees were discarded as the burn-in phase of each analysis, and the remaining trees were used to calculate posterior probabilities (pp). A pp value of ≥0.95 was considered significant [58]. The resulting trees were plotted using FigTree v. 1.3.1 (http://tree.bio.ed.ac.uk/software/figtree/, accessed on 5 December 2023). The DNA sequences generated in this study were deposited in GenBank (Table S1 in Supplementary Materials), and the alignments of the novel species were submitted to Zenodo (https://doi.org/10.5281/zenodo.10658628, accessed on 20 May 2023). Information about representative strains and outgroups used in the phylogenetic analyses is provided in Table S2 (Supplementary Materials).

### 2.6. Representation of Diversity Data

The relative abundance of the detected diversity was visualized at the family and genus levels to analyze trends in diversity across beaches and collection sites. Strains representing known genera but not assigned to any existing family, as well as unidentified strains due to lack of sporulation or inability to extract DNA for molecular identification, were labeled as *incertae sedis*. We used node networks to represent diversity at the species level using the software Cytoscape v3.10.2 [59]. This way, we could represent the diversity detected at each beach, assess the performance of direct plating and the SMF pre-treatment to detect ascomycetes, and we could also represent the connection between species, culture media, and temperature.

## 3. Results

Among the marine sediments processed through direct plating and the SMF pre-treatment, we recovered 362 strains of filamentous ascomycetes, with 190 strains from Miracle Beach and 172 strains from Arrabassada Beach. Following morphological examination, 188 strains (52%) were selected for sequencing (Table S1 in Supplementary Materials), encompassing all the morphological diversity detected, with the majority identified at the genus level.

### 3.1. Diversity at the Family and Genus Level

The recovered strains represented 29 families of ascomycetes, with *Aspergillaceae* and *Microascaceae* being the most prevalent on both beaches (Figure 2A). The families detected accounted for 54 genera, 42 identified at Arrabassada Beach and 28 at Miracle Beach, with 16 genera detected at both beaches (Figure 2B).

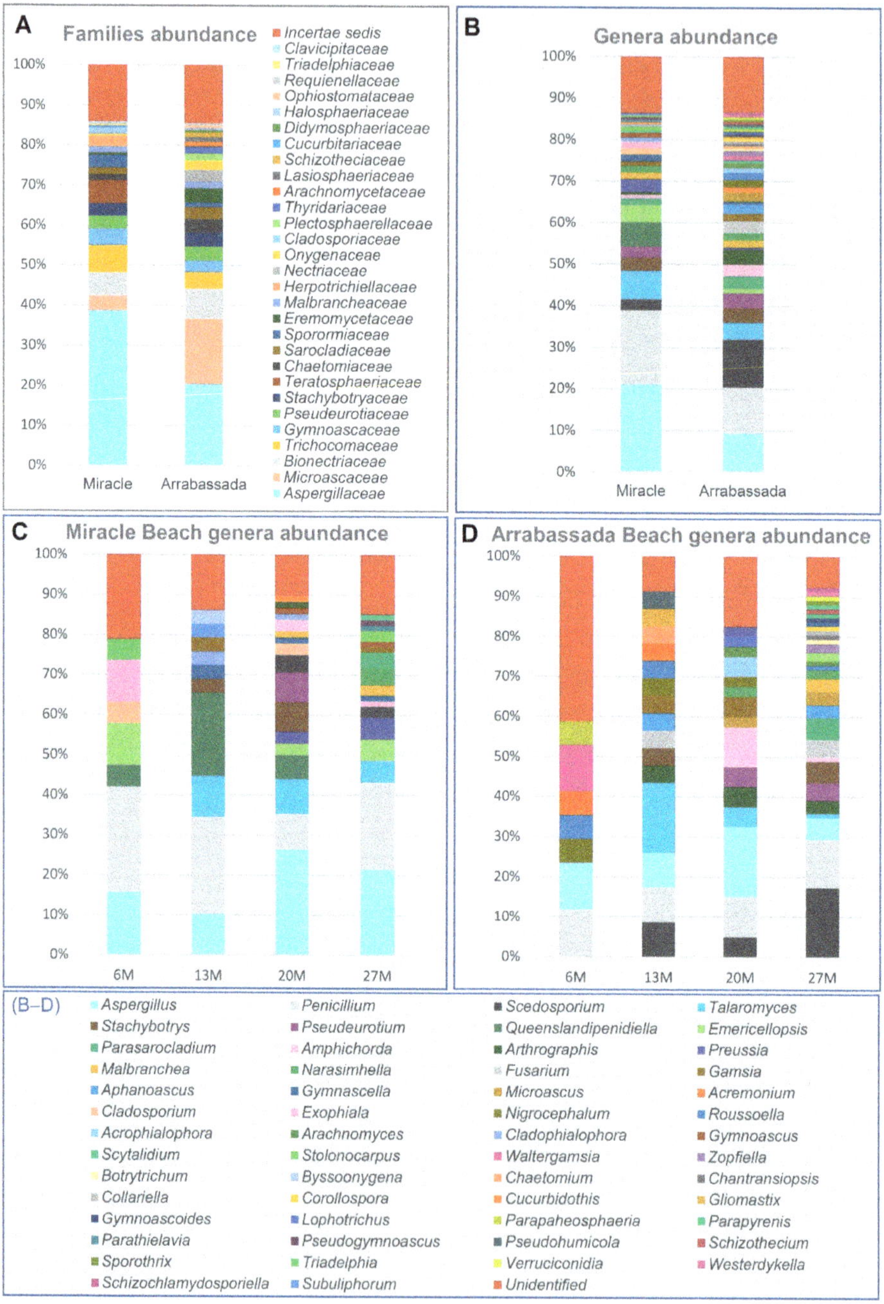


**Figure 2.** Relative abundance of filamentous ascomycetous taxa was detected in marine sediments from Miracle and Arrabassada Beaches. (**A**) Diversity is detected at each beach at the family level. (**B**) Diversity is detected at each beach at the genus level. (**C**) Genera detected at Miracle Beach, sorted by the depth of the collection point. (**D**) Genera detected at Arrabassada Beach, sorted by the depth of the collection point.

Considering both beaches, the genera most frequently detected were *Aspergillus* (15%), *Penicillium* (14%), and *Scedosporium* (7%). Miracle Beach is characterized by the prevalence of *Aspergillus* (21%) and *Penicillium* (18%), followed by genera like *Talaromyces* (7%), *Queenslandipenidiella* (6%) and *Emericellopsis* (4%), which were detected in lower frequency. In contrast, Arrabassada Beach exhibited more heterogeneity, with *Scedosporium* (12%) being the most frequently detected genus, followed by *Penicillium* (11%), *Aspergillus* (9%), *Talaromyces* (4%), and *Amphichorda* (3%). Regarding the genera exclusively recovered from each of the beaches, *Byssoonygena, Cladophialophora, Cladosporium, Cucurbidothis, Exophiala, Gymnoascus, Parathielavia, Pseudogymnoascus, Queenslandipenidiella, Stolonocarpus,* and *Triadelphia* were detected in Miracle Beach, whereas *Acrophialophora, Aphanoascus, Arachnomyces, Botryotrichum, Chaetomium, Chantransiopsis, Collariella, Corollospora, Fusarium, Gliomastix, Gymnoascoideus, Lophotrichus, Microascus, Nigrocephalum, Paraphaeosphaeria, Parapyrenis, Pseudohumicola, Roussoella, Schizothecium, Sporothrix, Verruciconidia, Waltergamsia, Westerdykella,* and *Zopfiella* were exclusively recovered from Arrabassada Beach.

The complete list of detected genera at each depth for both beaches can be accessed in Figure 2C,D. Taxa belonging to *Aspergillus* and *Penicillium* were detected at all collection sites on both beaches. At Miracle Beach (Figure 2C), most of the diversity was observed at depths of 20 and 27 m. Fifteen genera were exclusively recovered from these two collection sites, including *Amphichorda* and *Pseudeurotium*, which were exclusively isolated at 20 m of depth, and *Parasarocladium* and *Pseudogymnoascus* exclusively at 27 m of depth. At Arrabassada Beach (Figure 2D), *Amphichorda* and *Pseudeurotium* were recovered at both 20 and 27 m, while *Parasarocladium* was exclusively isolated again at 27 m of depth. Twelve genera were exclusively recovered from 27 m depth in this beach, including *Schizothecium* and *Westerdykella*, which were absent in Miracle Beach.

It is noteworthy that, on both beaches, the diversity detected increased with the depth of collection, particularly below 13 m. In fact, genera recovered from depths of 20 and 27 m represented approximately 60% of the total diversity.

### 3.2. *Diversity at the Species Level*

#### 3.2.1. Species Detected in Miracle and Arrabassada Beaches

The strains selected for molecular analyses represented 94 species, 60 from Arrabassada Beach and 54 from Miracle Beach, with 20 species detected at both beaches. The complete list of species can be accessed in Figure 3.

Among these species, 21 could not be assigned to any extant taxa and require further analyses to ensure their correct identification. However, in the present study, we have resolved the taxonomy of four of them as putative new species. The strains FMR 19606 and FMR 19607 were morphologically identified as members of the genus *Exophiala* and likely belonging to the *jeanselmei*-clade due to their subcylindrical, pale olivaceous, annellidic conidiogenous cells.

The strains FMR 19852, FMR 20069, and FMR 20174 were recognized as belonging to *Nigrocephalum* based on their dark-pigmented colonies, subcylindrical and phialidic conidiogenous cells, with conspicuous collarettes, producing conidia in dark-pigmented slimy heads.

The set of nine marine strains (FMR 19473–FMR 19481) was identified as similar to the genus *Queenslandipenidiella* based on their brown and macronematous conidiophores with polyblastic conidiogenous cells that give rise to at least two sets of ramoconidia.

On the contrary, the strain FMR 20114 could not be recognized as any known genus since it only produced intercalary chlamydospore-like cells.

Since a BLAST search using different barcoding regions of these strains could not assign them to any known species of the mentioned genera, we proceeded to their phenotypic and genetic characterization. The phylogenetic delineation and morphological description of these new taxa are provided below, and they are proposed as *Exophiala* (*E.*) *littoralis*, *Nigrocephalum* (*N.*) *paracollariferum*, *Queenslandipenidiella* (*Q.*) *verrucosa*, and *Schizochlamydosporiella* (*S.*) *marina* in the taxonomy section.

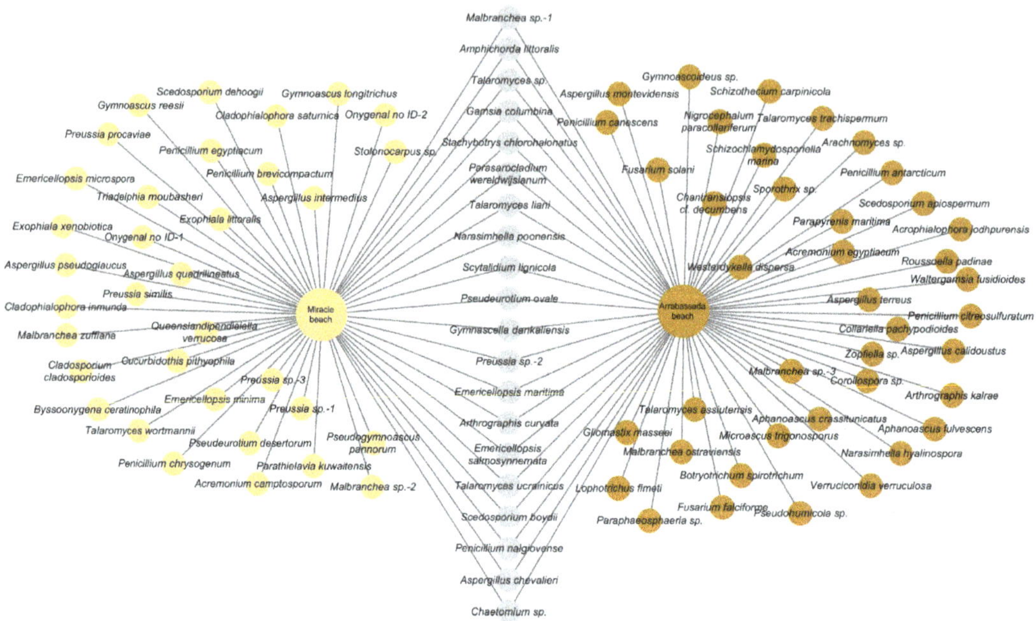

Figure 3. Network representing the species of filamentous ascomycetes detected in marine sediments from Miracle (cream circles) and Arrabassada beaches (brown circles). Grey circles represent species recovered from both areas.

### 3.2.2. Direct Plating and SMF Pre-Treatment: Effect of Culture Media and Temperature

To assess the performance of direct plating vs. SMF pre-treatment on diversity, the species detected at both beaches were analyzed in combination. As a result, 70 species were detected through direct plating and 42 species through SMF, with 17 species being detected through both approaches. The species exclusively detected by each approach are shown in Figure 4. Of note is that the flocculation pre-treatment enabled the isolation of 20 species that were not obtained through direct plating, increasing the number of species detected by 25%.

In addition to direct plating and SMF, we used three different culture media incubated at two temperatures to further increase the chance of detecting filamentous ascomycetes. The complete list of species detected through direct plating and SMF considering the medium of isolation can be accessed in Figure 4A. A total of 61 species were detected with SWMEA3%, 40 with DRBC, and 19 with PDA+C. Species like *Penicillium antarcticum*, *Q. verrucosa*, and *Stachybotrys chlorohalonatus* were isolated in the three media, with up to 19 species detected in more than one medium. Around 90% of the diversity was recovered with SWMEA3% and DRBC.

The medium that enabled the recovery of most of the diversity was SWMEA3%, with 41 species obtained exclusively from it. DRBC enabled the detection of different fungal species that overlapped with SWMEA3% by only 30%. Species from the genus *Talaromyces* were exclusively detected through DRBC. The medium PDA+C provided the least number of species. However, this was the only medium that enabled the recovery of onygenalean fungi like *Aphanoascus crassitunicatus*, *Gymnoascoideus* sp., *Gymnoascus longitrichus*, *Malbranchea zuffiana*, and *Narasimhella poonensis*. Regarding the effect of temperature (Figure 4B), a total of 69 species were detected at 15 °C and 59 species at 25 °C. Of these, 37 were exclusively detected at 15 °C, and 27 were exclusively detected at 25 °C.

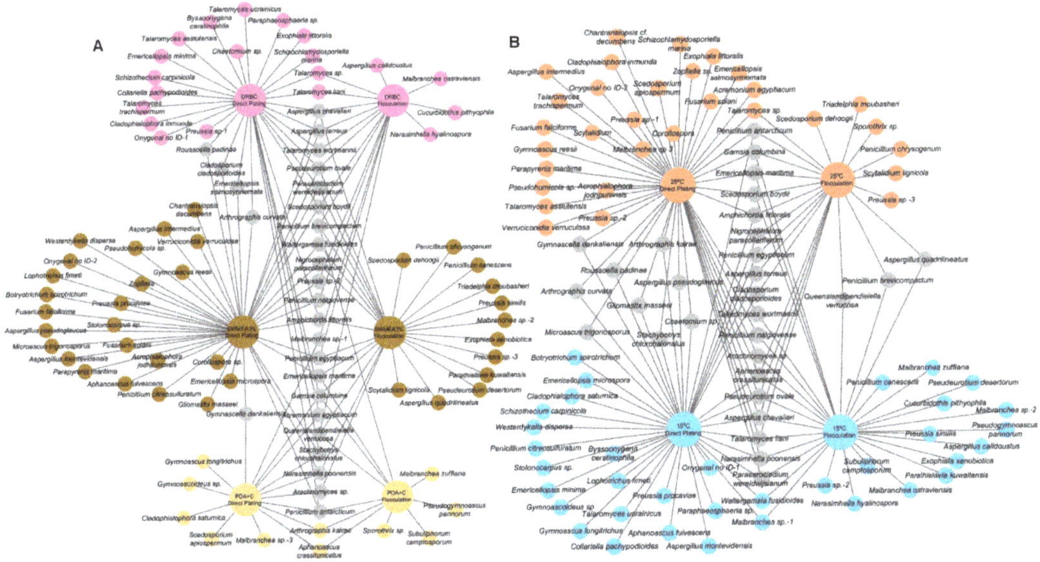


**Figure 4.** Network representing the species of filamentous ascomycetes detected in marine sediments from Miracle and Arrabassada Beaches through direct plating and flocculation pre-treatment sorted by (**A**) culture media and (**B**) temperature. Grey circles represent species recovered from more than one medium or at both temperatures. The species exclusively recovered from either DRBC, SWMEA3%, PDA+A, 25 °C or 15 °C are highlighted in pink, brown, cream, orange and blue circles, respectively.

### 3.3. Phylogeny

The unidentified strains of *Exophiala* (FMR 19606 and FMR 19607), *Nigrocephalum* (FMR 19852, FMR 20069, and FMR 20174), *Queenslandipenidiella* (FMR 19473–FMR 19481), and the strain FMR 20114 that could not be assigned to any known genera of *Ascomycota* were subjected to phylogenetic analyses with different gene markers depending on the fungal group to which they were related. The number of conserved, variable, and parsimony informative sites for each alignment, together with the models used in each analysis, are shown in Table 2.

A BLAST search using the ITS and LSU regions confirmed the relationship of the *Exophiala* marine strains with the *jeanselmei*-clade, showing a 99% sequence similarity with LSU sequences regarding those of the reference strain of *E. oligosperma* (CBS 725.88). However, for the ITS region, the maximum percentage of similarity was around 90% with the species *E. lamphunensis* (CMU 404) and *E. saxicola* (CMU 415). For species delineation, phylogenetic analyses were conducted using ITS and LSU sequences of the species of the *jeanselmei*-clade. Sequences of each gene marker were aligned individually (Figures S1 and S2 in Supplementary Materials), and after confirming the absence of incongruences, these sequences were concatenated into a single matrix. The final alignment comprised 27 sequences representing ex-type and reference strains of the species of the *jeanselmei*-clade and the outgroups selected (*E. dehoogii* CBS 149779 and *E. palmae* UPCB 86822). The resulting phylogenetic tree (Figure 5) resolved the unidentified strains of *Exophiala* in an independent and well-supported lineage separated from the other species of the *jeanselmei*-clade, which represents a novel species for the genus *Exophiala*.

**Table 2.** Overview of the alignment parameters and phylogenetic models used for each genetic region in *Exophiala*, *Nigrocephalum*, *Queenslandipenidiella*, and *Schizochlamydosporiella*.

| Dataset | Parameters | *Exophiala* | *Nigrocephalum* | *Queenslandipenidiella* | *Schizochlamydosporiella* |
|---|---|---|---|---|---|
| ITS | Lenth (bp) | 625 | 511 | 529 | 545 |
| | Conserved sites | 382 | 374 | 284 | 376 |
| | Variable sites | 226 | 130 | 240 | 159 |
| | Parsimony informative | 147 | 93 | 204 | 75 |
| | BI model | SYM+I+G | GTR+G | GTR+I+G | SYM+G |
| LSU | Lenth (bp) | 873 | 791 | 751 | 827 |
| | Conserved sites | 801 | 714 | 568 | 730 |
| | Variable sites | 60 | 77 | 180 | 94 |
| | Parsimony informative | 38 | 42 | 139 | 45 |
| | BI model | K80+I+G | K80+I | SYM+I+G | K80+I |
| *rpb*2 | Lenth (bp) | - | 743 | - | 840 |
| | Conserved sites | - | 484 | - | 460 |
| | Variable sites | - | 259 | - | 380 |
| | Parsimony informative | - | 219 | - | 223 |
| | BI model | - | GTR+I+G | - | GTR+I+G |
| *tef*1-a | Lenth (bp) | - | 787 | - | - |
| | Conserved sites | - | 624 | - | - |
| | Variable sites | - | 163 | - | - |
| | Parsimony informative | - | 132 | - | - |
| | BI model | - | GTR+G | - | - |
| Concatenated | Lenth (bp) | 1498 | 2832 | 1280 | 2212 |
| | Conserved sites | 1183 | 2196 | 852 | 1566 |
| | Variable sites | 286 | 629 | 420 | 633 |
| | Parsimony informative | 185 | 486 | 343 | 343 |

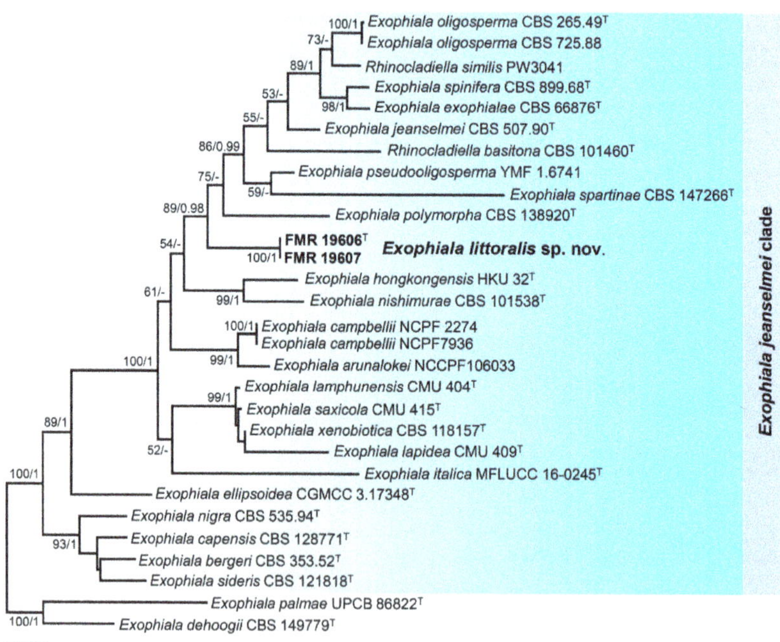


**Figure 5.** Phylogenetic tree inferred from a concatenated alignment of ITS and LSU sequences of 29 strains representing the *Exophiala jeanselmei*-clade. Numbers at the branches indicate support values (RAxML-BS/BI-PP) above 70%/0.95. The tree is rooted in *Exophiala dehoogii* CBS 149779 and *Exophiala palmae* UPCB 86822. The new species is highlighted in bold. $^{T}$ indicates ex-type strains. The scale bar represents the expected number of changes per site.

The marine strains of *Nigrocephalum* were also analyzed through a BLAST search using the ITS and LSU regions. The identification at the genus level was confirmed through the LSU region with a 99% similarity with the ex-type strain of *N. collariferum* (CBS 124586), the type species of the genus *Nigrocephalum*. The ITS sequences showed a 97% similarity with the same strain. For species delineation, we performed phylogenetic analyses with sequences of the ITS, LSU, *tef*1-α, and *rpb*2 loci. Each locus was aligned individually, and the resulting tree topologies were similar and without incongruences (Figures S3–S6 in Supplementary Materials). The alignments were, therefore, concatenated into a single matrix. The final alignment included 20 strains that comprised the marine strains, the type species of *Nigrocephalum*, and ex-type or reference strains of species of other genera of *Plectosphaerellaceae* related to *Nigrocephalum*, as well as the selected outgroups for the analyses (*Phialoparvum bifurcatum* CBS 299.70B and *Plectosphaerella cucumerina* CBS 137.33). In the phylogenetic tree (Figure 6), the marine strains of *Nigrocephalum* formed an independent and fully supported sister clade with that representative of *N. collariferum*, but with enough phylogenetic distance (99% LSU, 97% ITS, 97% for *tef*1-α, and 90% for *rpb*2) to be considered a new species for the genus *Nigrocephalum*.

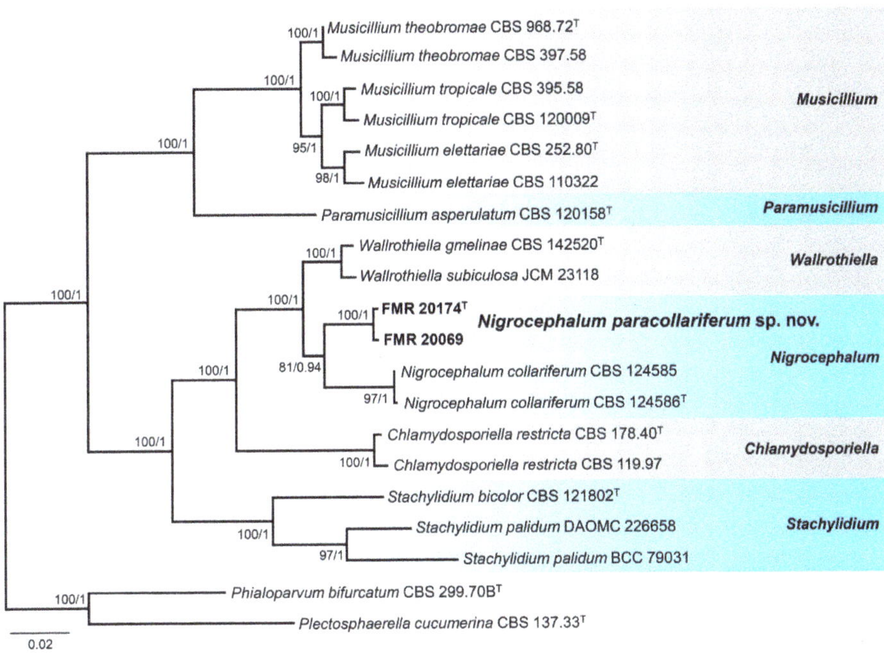


**Figure 6.** Phylogenetic tree was inferred from a concatenated alignment of ITS, LSU, *tef*1-α, and *rpb*2 sequences of 20 strains representing *Nigrocephalum* and related genera within *Plectosphaerellaceae*. Numbers at the branches indicate support values (RAxML-BS/BI-PP) above 70%/0.95. The tree is rooted in *Phialoparvum bifurcatum* CBS 299.70B and *Plectosphaerella cucumerina* CBS 137.33. The new species is highlighted in bold. $^{T}$ indicates ex-type strains. The scale bar represents the expected number of changes per site.

The BLAST search of the strains morphologically identified as *Queenslandipenidiella* with LSU and ITS sequences confirmed their placement into the family *Teratosphaeriaceae* and showed a similarity of around 96% with *Q. kurandae* (CBS 121715) using the LSU region and around 88% with the same strain using the ITS barcode. While the strains FMR 19473–FMR 19477 and 19481 were genetically identical, the strains FMR 19478 and FMR 19479 showed slight genetic variability. Therefore, only the strains FMR 19477, FMR 19478, and FMR 19479 were included in the phylogenetic analyses with the ITS and LSU regions to prevent branch imbalance in the resulting phylogenetic tree. Since separate alignments of each region and the resulting tree did not show incongruences (Figures S7 and S8 in Supplementary Materials), they were concatenated into a single matrix. The final alignment comprised 29 strains representing genera in the *Teratosphaeriaceae* and the outgroups *Ramularia* (*R.*) *eucalypti* (CBS 120726) and *R. endophylla* (CBS 113265). The resulting phylogenetic tree resolved the marine strains as an independent lineage, which was considered representative of a new species of the genus *Queenslandipenidiella* (Figure 7).

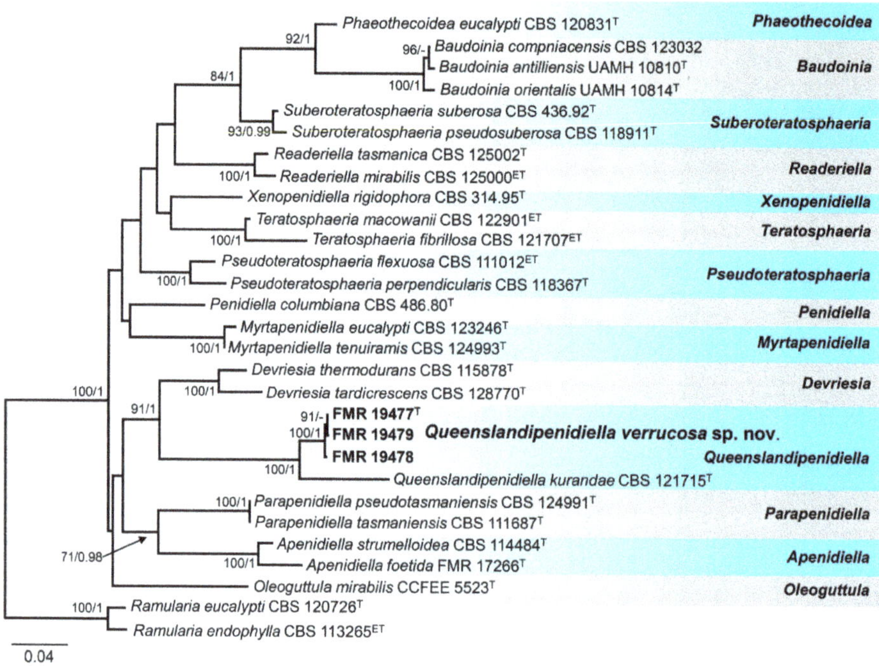


**Figure 7.** Phylogenetic tree was inferred from a concatenated alignment of ITS and LSU sequences of 27 strains representing *Queenslandipenidiella* and related genera within *Teratosphaeriaceae*. Numbers at the branches indicate support values (RAxML-BS/BI-PP) above 70%/0.95. The tree is rooted in *Ramularia eucalypti* CBS 120726 and *Ramularia endophylla* CBS 113265. The new species is highlighted in bold. $^{T}$ and $^{ET}$ indicate ex-type and ex-epitype strains, respectively. The scale bar represents the expected number of changes per site.

Finally, a BLAST search with ITS and LSU sequences revealed that the marine strain FMR 20114 was related to members of the family *Schizotheciaceae*. The highest percentage of similarity was obtained with sequences of the ex-type strains *Apiosordaria* (*A.*) *microcarpa* (CBS 692.82: 95% ITS and 98% LSU) and *Apodus deciduus* (CBS 506.70: 90% ITS and 98% LSU). For a more precise identification, in addition to the phylogenetic analyses with ITS and LSU, we also analyzed the sequences of the *rpb2* gene. Once the lack of incongruences was confirmed (Figures S9–S11 in Supplementary Materials), individual alignments were concatenated into a single matrix. The final alignment comprised 29 strains, which, in addition to the unidentified strains, included ex-type and reference strains of the species belonging to the *Schizotheciaceae* family and the selected outgroups (*Amesia atrobrunnea* CBS 379.66 and *Triangularia bambusae* CBS 352.33). The resulting phylogenetic tree resolved the strain FMR 20114 as an independent lineage within the *Schizotheciaceae* (Figure S12 in Supplementary Materials). However, due to the great phylogenetic distance between different genera and the lack of ITS or *rpb2* sequences for numerous species in the family (Table S2 in Supplementary Materials), we performed an additional phylogenetic analysis only with the strains of the species phylogenetically close to FMR 20114 and with the three gene markers. The resulting tree (Figure 8) resolved the strain FMR 20114 in a singleton distant branch, phylogenetically distant from its counterpart *A. microcarpa* (98% LSU, 95% ITS, 81% *tub2*, 77% *rpb2*), suggesting an undescribed genus for the family that is proposed as *Schizochlamydosporiella* in the taxonomy section.

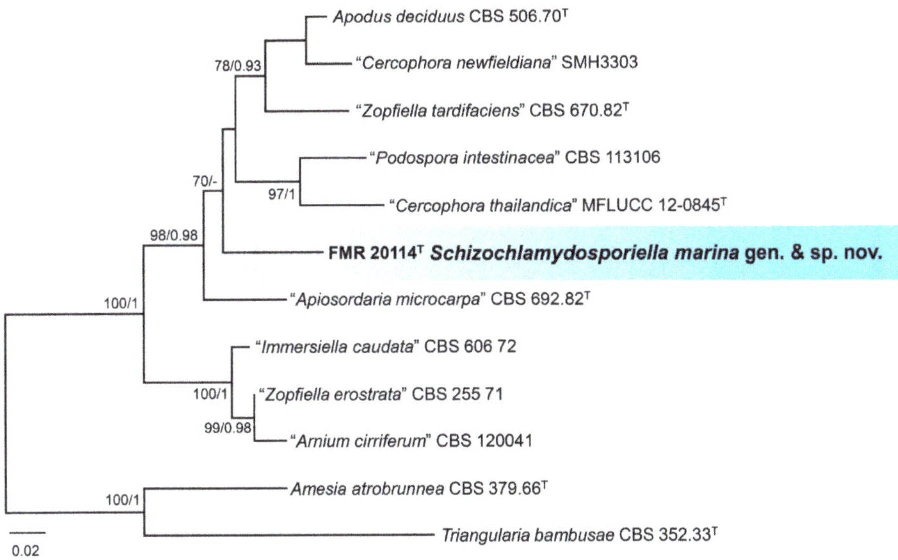


**Figure 8.** Phylogenetic tree was inferred from a concatenated alignment of ITS, LSU, and *rpb2* sequences of 12 strains representing a clade in *Schizotheciaceae*. Numbers at the branches indicate support values (RAxML-BS/BI-PP) above 70%/0.95. The tree is rooted in *Amesia atrobrunnea* CBS 379.66 and *Triangularia bambusae* CBS 352.33. The new species is highlighted in bold. $^{T}$ indicates ex-type strains. Quote marks indicate strains with an unresolved taxonomy. The scale bar represents the expected number of changes per site.

### 3.4. Taxonomy

***Exophiala littoralis*** Guerra-Mateo, Cano & Gené, sp. nov. Figure 9.

MycoBank 852016

*Etymology*. The name refers to the area where the species was collected, the western Mediterranean coast.

*Type*. Spain, Catalonia, Mediterranean coast, Tarragona, Platja del Miracle, 41°6′19″ N, 1°15′37″ E, from sediments at 6 m depth, June 2021, *G. Quiroga-Jofre* and *D. Guerra-Mateo* (holotype CBS H-25351, ex-type FMR 19606, CBS 151312).

*Classification*. *Eurotiomycetes*, *Chaetothyriales*, *Herpotrichiellaceae*.

*Saprobic* on marine sediments at 6 m of depth. *Asexual morph* on OA at 25 °C. *Mycelium* is composed of septate, branched, smooth- and thick-walled, pale olivaceous to brown, 2–3 µm wide hyphae; sparse torulose hyphae are present on PDA. *Conidiophores* micronematous are often reduced to conidiogenous cells growing directly from vegetative hyphae. *Conidiogenous cells* are annellidic, mostly lateral, arising in acute angles, terminal and intercalary also present, cylindrical, tapering terminally in an irregular annellated zone, pale olivaceous to brown, (6–)10–20(–26) × 1.5–3 µm. *Conidia* are one-celled, smooth-walled, pale olivaceous, subcylindrical to clavate, 3–6 × 1.5–2 µm, often with a truncate and prominent basal scar. Budding cells were not observed. *Sexual morph* is unknown.

*Culture characteristics* (after 14 days at 25 °C). Colonies on the PDA attaining 27–28 mm diam., umbonate, felty, dark green (30F6) at center to grayish green (30E5) towards the periphery, margin entire; reverse dark green (30F4–30F6). On MEA, reaching 26–27 mm diam., umbonate, floccose, pale gray (1B1) at the center to brownish orange (5C5) towards the periphery, margin entire; reverse grayish brown (5E3) to brownish orange (5C5) at the periphery. On OA, attaining 22 mm diam., slightly umbonate, floccose turning velvety towards the periphery, dark gray (1E1), margin entire; reverse dark gray (1E1). The diffusible pigment was not observed on any medium.

*Cardinal temperatures for growth.* Minimum 5 °C (2 mm), optimum 30 °C (30 mm), maximum 37 °C (2 mm).

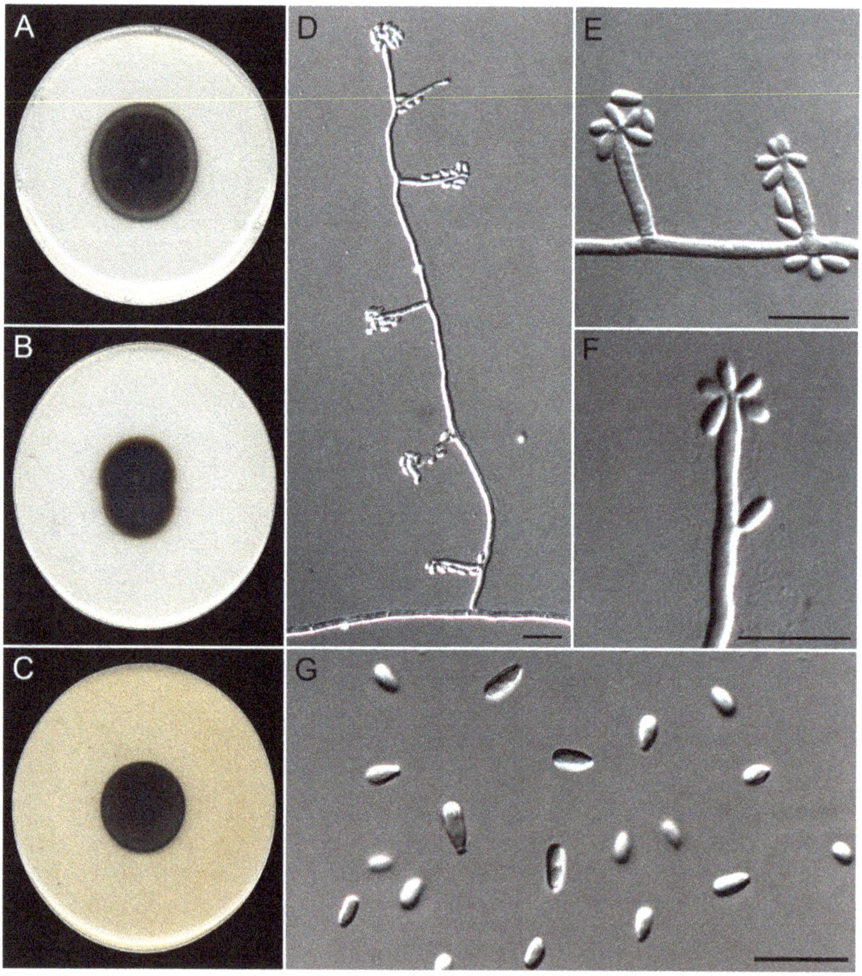


**Figure 9.** *Exophiala littoralis* (ex-type FMR 19606). (**A–C**) Colonies on PDA, MEA, and OA at 25 °C after 14 d. (**D**) Micronematous conidiophore. (**E**) Conidiogenous cells and conidia are borne directly from hyphae. (**F**). Detail of the apical annellated zone of the conidiogenous cells. (**G**) Conidia. Scale bars: 10 µm.

*Additional specimens were examined.* Spain, Catalonia, Mediterranean coast, Tarragona, Platja del Miracle, 41°6′19″ N, 1°15′37″ E, from sediments at 6 m depth, June 2021, *G. Quiroga-Jofre* and *D. Guerra-Mateo* (FMR 19607).

*Notes*. The genus *Exophiala* represents a polyphyletic group in the family *Herpotrichiellaceae*, where *E. littoralis* is phylogenetically placed in the so-called *jeanselmei*-clade [45,60]. This species is morphologically similar to other species of this clade, such as *E. hongkongensis*, *E. jeanselmei*, *E. nishimurae*, *E. oligosperma*, and *E. pseudooligosperma*, by the production of cylindrical conidiogenous cells (Figure 5; Refs. [61–66]). However, the production of claviform conidia with thick and truncate bases, the absence of budding cells, and the phylogenetic distance with the ITS/LSU barcodes were clear features to distinguish *E. littoralis* from the rest of the members of the *jeanselmei*-clade.

***Nigrocephalum paracollariferum*** Guerra-Mateo, Cano & Gené, sp. nov. Figure 10.

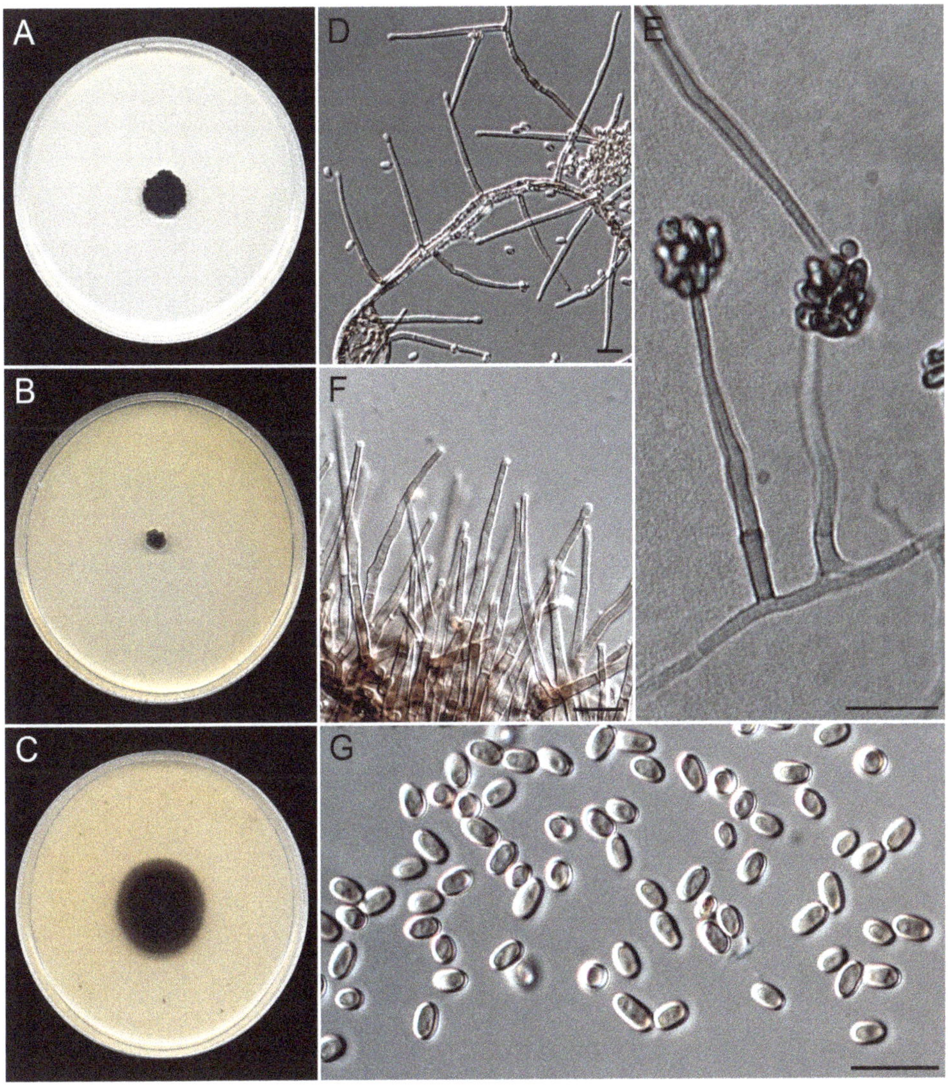


**Figure 10.** *Nigrocephalum paracollariferum* (ex-type FMR 20069). (**A–C**) Colonies on PDA, MEA, and OA at 25 °C after 14 d. (**D**) Micronematous conidiophores. (**E**) Conidiogenous cells with slimy heads of conidia. (**F**) Detail of the funnel-shaped collarettes in the apex of the conidiogenous cells. (**G**) Conidia. Scale bars: 10 μm.

MycoBank 852017

*Etymology*. The name refers to the morphological similarity and phylogenetic relationship with the type species of the genus *N. collariferum*.

*Type*. Spain, Catalonia, Mediterranean coast, Tarragona, Platja de la Arrabassada, 41°6′53″ N, 1°16′48″ E, from sediments at 20 m depth, March 2022, *G. Quiroga-Jofre* and *D. Guerra-Mateo* (holotype CBS H-25352, ex-type FMR 20069, CBS 151313).

*Classification*. *Sordariomycetes*, *Glomerellales*, *Plectosphaerellaceae*.

*Saprobic* on marine sediments at 6, 13, and 20 m depth. *Asexual morph* on OA at 25 °C. *Mycelium* is composed of septate, branched, finely to roughly warted, thick-walled, pale olive to brown, 2–3 µm wide hyphae. *Conidiophores* micronematous, usually consisting of single phialides on slightly darker subtending cells, occasionally basitonously branched, arising in acute angles from vegetative hyphae, smooth-walled, pale olive-brown to dark brown. *Phialides* are mostly lateral, subcylindrical to subulate, wavy at the apex, with funnel-shaped collarettes and conspicuous periclinal thickening, pale olive-brown, 30–46 × 1.5–2.5 µm. *Conidia* are one-celled, thick- and smooth-walled, pale brown, ellipsoidal, 3.8–4.5 × 2.3–3 µm, arranged in slimy dark heads. *Chlamydospores* were not observed. *Sexual morph* is unknown.

*Culture characteristics* (after 14 days at 25 °C). Colonies on PDA, reaching 10–12 mm diam., crateriform, radially sulcate, velvety, brownish gray (5E2), margin irregularly lobated; reverse light gray (5D1). On MEA, reaching 4–5 mm diam., flat, gray (5F1) at the center to brownish gray (5E2) and yellowish white (4A2) at the periphery, margin entire, slightly lobated; reverse concolorous. On OA, attaining 22–24 mm diam., flat, brownish gray (5E2) to white (1A1) at the periphery, margin diffusing in the medium. The diffusible pigment was not observed on any medium.

*Cardinal temperatures for growth*. Minimum 10 °C (4 mm), optimum 25 °C (10–12 mm), maximum 35 °C (2 mm).

*Additional specimens were examined*. Spain, Catalonia, Mediterranean coast, Tarragona, Platja de la Arrabassada, 41°7′3″ N, 1°16′42″ E, from sediments at 6 m depth, March 2022, *G. Quiroga-Jofre* and *D. Guerra-Mateo* (FMR 19852); ibid., N 41°6′57″ N, E 1°16′45″ E, from sediments at 13 m depth, June 2022, *G. Quiroga-Jofre* and *D. Guerra-Mateo* (FMR 20174).

*Notes*. The genus *Nigrocephalum* was erected to accommodate a dark-pigmented *Acremonium* species, *A. collariferum*, within the family *Plectosphaerellaceae* [46]. The genus is monotypic, with *N. collariferum* isolated from human toenails. *Nigrocephalum paracollariferum* is morphologically similar to the type species; it differs only in the completely ellipsoidal conidia and the absence of chlamydospores in any of the culture media tested, instead of the reniform in lateral view conidia and intercalary chlamydospores of the type species [67].

***Queenslandipenidiella verrucosa*** Guerra-Mateo, Cano & Gené, sp. nov. Figure 11.

Mycobank MB 848254

*Etymology*. The name refers to the verrucose ornamentation of the conidiophore.

*Type*. Spain, Catalonia, Mediterranean coast, Tarragona, Platja del Miracle, 41°6′19″ N, 1°15′37″ E, from sediments at 13 m depth, June 2021, *G. Quiroga-Jofre* and *D. Guerra-Mateo* (holotype CBS H-25258, ex-type FMR 19477, CBS 149939).

*Classification*. *Dothideomycetes*, *Mycosphaerellales*, *Teratosphaeriaceae*.

*Saprobic* on marine sediments at 6, 13, and 20 m depth. *Asexual morph* on OA at 25 °C. *Mycelium* is composed of septate, branched, smooth to verrucose, thick-walled, brown, 2–3 µm wide hyphae. *Conidiophores* semi-macronematous to macronematous, arising directly from superficial mycelium, erect, unbranched, occasionally branched, brown, verrucose, thick-walled, up to 110 µm long; stipe up to 5-septate, cells 10–25 × 2–3 µm. *Conidiogenous cells* are integrated, terminal or intercalary, verrucose, thick-walled, brown, cylindrical to subcylindrical, 10–25 × 2–4.5 µm, often slightly inflated terminally, polyblastic, proliferating sympodially at the apex, showing 2–4 conidiogenous loci, slightly denticulate and pigmented, giving rise to at least two sets of ramoconidia. *Ramoconidia* with up to 6 distal conidiogenous loci, finely roughened, thick-walled, brown, subcylindrical to cylindrical-oblong, 3.5–7(–8) × 3–4 µm, become sinuous and tortuous with age. *Conidia* solitary or arranged in short chains, smooth- to finely rough-walled, brown, subglobose to obpyriform, 3–4.5 × 2–4 μm. *Sexual morph* is not observed.

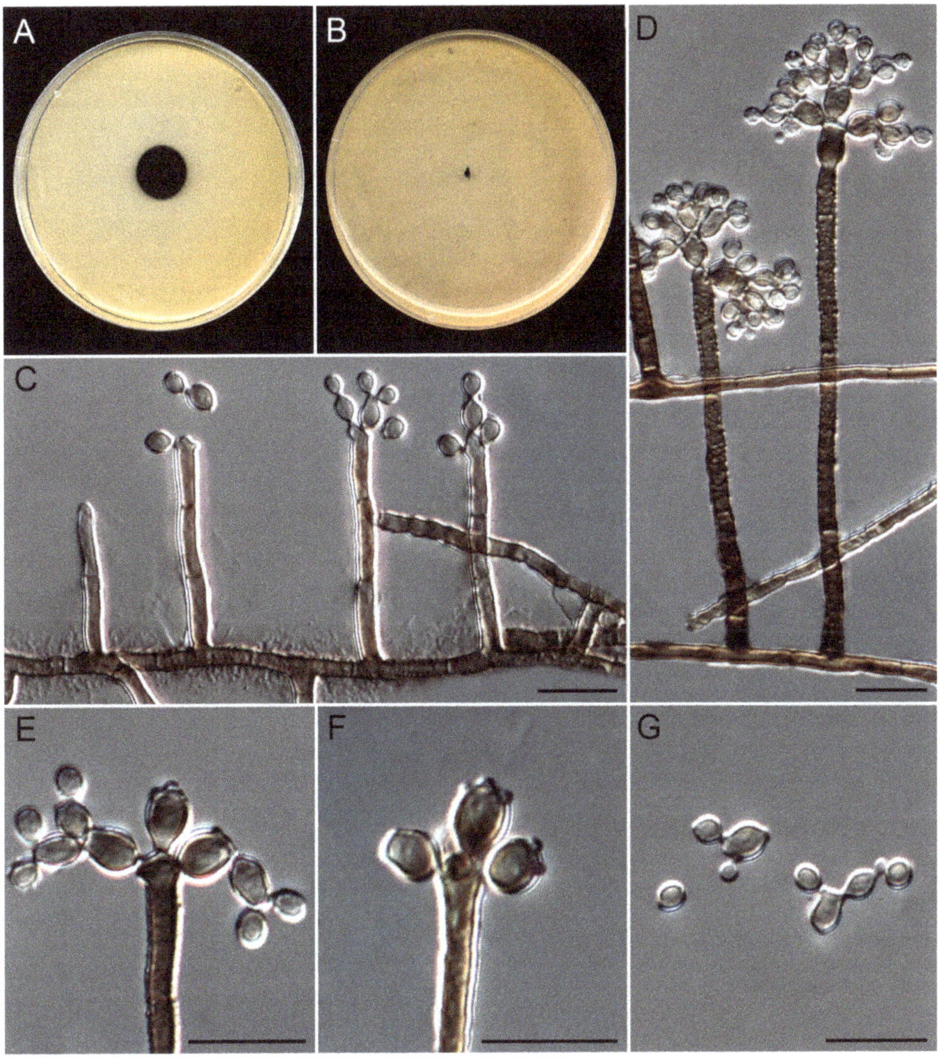


**Figure 11.** *Queenslandipenidiella verrucosa* (ex-type FMR 19477). (**A**,**B**) Colonies on MEA and OA at 25 °C after 14 d. (**C**) Development of the conidiophores. (**D**) Mature conidiophores: the conidiophore on the right shows an inflated conidiogenous cell. (**E**) Detail of the conidiogenous cell and two sets of ramoconidia. (**F**) Detail of the conidiogenous loci in ramoconidia. (**G**) Ramoconidia and conidia. Scale bars: 10 μm.

*Culture characteristics* (after 14 days at 25 °C). Colonies on MEA 2%, attaining 8–10 mm diam. (16–19 mm after 28 days), circular, felty, slightly raised, greenish gray (30D2) to grayish green (30D6) and dark green (30F8) at the periphery, margin regular and feathery; reverse gray (30F1) to dark green (30F8) at the periphery. On OA, attaining 2 mm diam. (3–4 mm after 28 days), raised, irregular, with abundant aerial mycelium, margin feathery, pastel gray (30D1); reverse gray (30F1). Sporulation is abundant in all culture media. The diffusible pigment was not observed on any medium.

*Cardinal temperatures for growth.* Minimum 15 °C (3 mm), optimum 25 °C (10 mm), maximum 35 °C (2 mm).

*Additional specimens examined*: Spain, Catalonia, Mediterranean coast, Tarragona, Platja del Miracle, 41°6′19″ N, 1°15′37″ E, from sediments at 6 m depth, June 2021, *G. Quiroga-Jofre* and *D. Guerra-Mateo* (FMR 19481); ibid., from sediments at 13 m depth (FMR 19478, FMR 19479, 19480); ibid., from sediments at 20 m depth (FMR 19473, FMR 19474, FMR 19475, FMR 19476).

*Notes.* The genus *Queenslandipenidiella* was erected to accommodate *Penidiella kurandae* based on the phylogenetic distance and distinct morphological features regarding *P. columbiana*, the type species of the genus *Penidiella* [44,68]. *Queenslandipenidiella* verrucosa can be primarily distinguished from *Q. kurandae* based on conidiophore morphological traits, which are verrucose and commonly unbranched in the former species and smooth and penicillate-branched in *Q. kurandae* [44,68]. It is worth mentioning that *Q. verrucosa* shows a considerable phylogenetic distance with respect to *Q. kurandae*. However, the phylogenetic distance between these two species for each gene marker is within the range of interspecific variability accepted in *Teratosphaeriaceae* (ITS 90%, LSU 97%, *tef*1-α 89%, *tub*2 88%; Ref. [44]). This classification, however, may be subject to taxonomical changes with further studies on the biodiversity and phylogeny of the family.

***Schizochlamydosporiella*** Guerra-Mateo, Gené & Cano, gen. nov.

MycoBank 852018

*Etymology.* The name refers to the production of chlamydospores released by schizolythic secession over time.

*Classification. Sordariomycetes, Sordariales, Schizotheciaceae.*

*Mycelium* is superficial and immersed, septate, torulose, smooth- and thin-walled, hyaline to pale brown, with hyphae with intercalary chlamydospore-like cells. *Chlamydospore-like cells* are single or in chains, sometimes with several branching loci, one- or two-celled, smooth- and thick-walled, brown, subglobose to subcylindrical, ellipsoidal, barrel-shaped, or irregularly shaped, with terminal, dark pigmented, thick-walled scars, released from the chain by schizolythic secession. *Sexual morph* is unknown.

*Type species. Schizochlamydosporiella marina* Guerra-Mateo, Gené & Cano.

***Schizochlamydosporiella marina*** Guerra-Mateo, Gené & Cano, sp. nov. Figure 12.

MycoBank 852019

*Etymology.* The name refers to the habitat where the type strain was isolated, marine sediments.

*Type.* Spain, Catalonia, Mediterranean coast, Tarragona, Platja de la Arrabassada, 41°6′45″ N, 1°16′51″ E, from sediments at 27 m depth, March 2022, *G. Quiroga-Jofre* and *D. Guerra-Mateo* (holotype CBS H-25353, ex-type FMR 20114, CBS 151315).

*Saprobic* on marine sediments at 27 m of depth. *Asexual morph* on OA at 25 °C. *Mycelium is* composed of superficial and immersed, septate, torulose, smooth- and thin-walled, hyaline to pale brown, 1–2 µm wide hyphae with intercalary chlamydospore-like cells. *Chlamydospore-like cells* are single or in chains; single chlamydospore cells, one-celled, rarely two-celled and constricted at septum, smooth- and thick-walled, brown, subglobose to subcylindrical, ellipsoidal, barrel-shaped or irregularly shaped, (8–)10–15(–17) × 3–6 µm, with terminal, dark pigmented and thick-walled scars; chains reaching up to 190 µm in length, occasionally branched, with up to three branching loci, cells released from the chain by schizolythic secession in old cultures. *Sexual morph* is unknown.

*Culture characteristics* (after 28 days at 25 °C). Colonies on MEA reaching 32–36 mm diam., slightly raised, radially sulcate underneath the aerial mycelium, cottony, golden gray (4C2) to dark gray (1F1) and yellowish gray (4B2) at periphery, margin entire and lobated; reverse dark gray (1F1) to golden gray (4C2). On OA, reaching 85 mm diam., velvety with short aerial mycelium that becomes prominent towards the margin, brownish gray (4D2) at center with scattered patches of dark gray (1F1), margin entire; reverse dark gray (1F1). On CMA, reaching 56–60 mm diam., slightly raised, radially sulcate, velvety, golden gray (4C2) at the center to dark gray (1F1) and olive brown (4D3) at the periphery, margin entire and lobated; reverse dark gray (1F1). On PCA, reaching 65–70 mm diam., predominantly submerged with short and sparse aerial mycelium, dark gray (1D1) to brownish gray towards the margin (4D2), margin feathery; reverse dark gray (1F1).

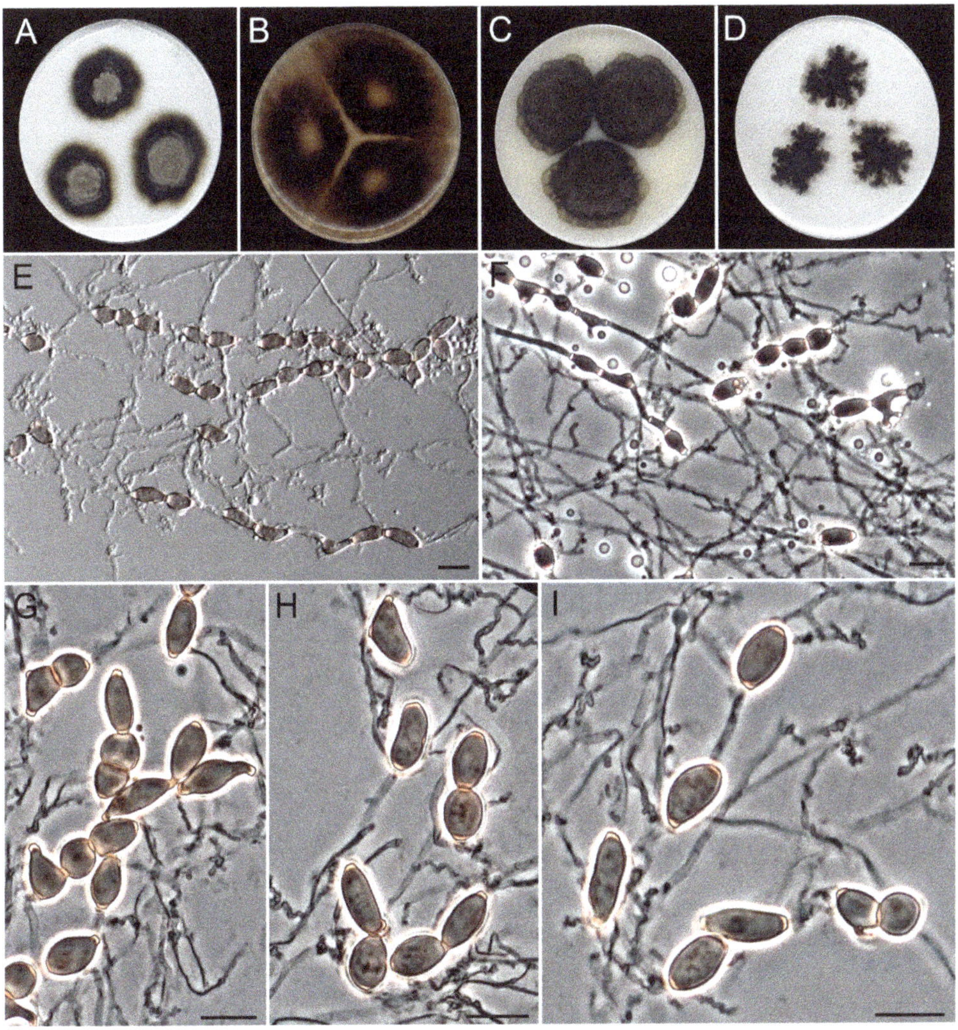

**Figure 12.** *Schizochlamydosporiella marina* (ex-type FMR 20114). (**A–D**) Colonies on MEA, OA, CMA, and PCA at 25 °C after 28 d. (**E**) Intercalary chains of chlamydospore-like cells. (**F**) Chains of chlamydospore-like cells under phase contrast microscopy (PCM). (**G**) Chains are composed of 0- and 1-septate chlamydospore-like cells (PCM). (**H**,**I**) Chlamydospore-like cells released by schizolythic secession show terminal, thick-walled scars (PCM). Scale bars: 10 µm.

*Cardinal temperatures for growth*. Minimum 5 °C (2 mm), optimum 30 °C (56 mm), maximum 40 °C (2 mm).

*Notes*. *Schizochlamydosporiella marina* represents an independent lineage within the *Schizotheciaceae* (Figure 8 and Figure S12 in Supplementary Materials). This family was erected to resolve an independent lineage in *Lasiosphaeriaceae s.l.* that comprises the genus *Schizothecium*, together with many species whose identification needs a taxonomical revaluation [47,48]. The *Schizotheciaceae* were circumscribed based on the morphological characters of the sexual morph, which was not produced by our species despite testing different culture media, even mixed with plant material (leaves of *Chamaerops humilis* and *Cymodocea nodosa*), and up to two months of incubation. *Schizochlamydosporiella marina* only produces brown chlamydospore-like cells that are released by schizolythic secession. By definition, chlamydospores represent resting spores that are not dispersed [69]. Thus, the cells produced by *S. marina* suggest a kind of dispersal asexual structure like conidia, but we could not observe in vitro germination of the cells. Of note is that this particular type of asexual morph has not been reported before in the family. The asexual morphs in *Schizotheciaceae* are commonly displayed as either phialidic conidiogenous cells that produce ovoid to clavate, hyaline to pale brown conidia, or holoblastic conidia in a chrysosporium-like fashion, like in *A. microcarpa* [47,70]. We propose *S. marina* as a type species of the novel genus *Schizochlamydosporiella* based on the great phylogenetic distance with its closest species, *A. microcarpa* (CBS 692.82) and unique morphology in comparison to other members of the family. However, the morphological description of our fungus is clearly susceptible to emendation when new isolates of the species become available.

## 4. Discussion

This work aimed to detect the diversity of ascomycetous taxa from marine sediments in the Miracle and Arrabassada Beaches, western Mediterranean Sea. For this, we combined the use of direct plating and an SMF pre-treatment with three culture media and two incubation temperatures. This methodology enabled the detection of 94 species, which represent a diverse fungal community that exceeds the number of species typically detected in culture-dependent studies in the marine environment [71–74]. The combination of both approaches improved our ability to detect fungi in marine sediments, including rare and novel species. In fact, although most of the species were detected through direct plating, up to 20 species were exclusively recovered with SMF (Figure 4).

### *4.1. Effect of In Vitro Conditions for Isolation of Marine Fungi*

Conventional methods for studying marine fungi include dilution plating, membrane filtering, baiting, single spore isolation, and the deployment of wood or man-made panels in the sea [14,17]. This study represents the first time that SMF has been employed for the detection of marine fungi.

Our results show that, independently of the media and temperatures used for isolation, we detected different communities with direct plating and SMF. It is noteworthy that the SMF-pre-treated sediment was stored at $-20$ °C, and despite this, fungal propagules were still viable and recovered in synthetic media. This suggests that SMF could be used as a method to store fungi over time. Although the extent of fungal viability under this treatment should be tested experimentally in future research, this opens the door for a more efficient study of the hard-to-reach marine samples. For these reasons, we recommend the use of SMF in surveys of marine fungi. Particularly in combination with direct plating, enabling the in-depth study of the sample.

Culture media restricts the diversity of detectable fungi. In this regard, we used three isolation media, but SWMEA3% proved to be the most successful one (Figure 4), consistent with the results obtained by other authors [31]. However, to the extent of our knowledge, this medium has never been used in diversity studies. In contrast, DRBC represents a typical medium for the isolation of marine fungi [11,13,73,74]. This medium enabled the isolation of another great portion of species and, most notably, the detection of six *Talaromyces* species that were not obtained with any other medium (Figure 4). In addition, we used a medium supplemented with cycloheximide (PDA+C). Although its use in diversity studies is rare due to the toxicity of this antimicrobial drug to eukaryotes, many marine fungi are resistant to this protein synthesis inhibitor [32]. Therefore, this medium provided the smallest pool of diversity. However, it enabled the isolation of some onygenalean fungi that are likely undescribed ascomycetes (e.g., *Gymnoascoideus* sp., *Malbranchea* sp. 1–3, among others).

In addition, the combination of different temperatures to improve the detection of culturable fungi has been successful in marine habitats [71]. In this case, we combined two incubation temperatures, 25 and 15 °C. We recovered more species from 15 °C; however, this result may be biased by our ability to detect fungal colonies in primary plates. In general, fungi grow more slowly on plates incubated at 15 °C than at 25 °C, thus enabling better detection of single colonies before fast-growing fungi colonize the surface of the plate.

### 4.2. Diversity and Insights of the Recovered Species

In both the Miracle and Arrabassada Beaches, the diversity and the number of ascomycetes that we recovered increased proportionately with the depth of the collection. This trend was particularly evident below 13 m, reaching the maximum diversity at 27–30 m of depth (Figure 2). However, this observation is biased by the small scale of our work in terms of depth. Depth seems to be one of the main factors determining the geographical structure of marine fungi, resulting in an inverse relationship with fungal diversity [75–77]. This raises the question: what explains an increase in fungal diversity around 30 m of depth? The answer may be a combination of different environmental factors like temperature, hydrostatic pressure, light availability and its effect on primary production, and the flux of organic matter from the euphotic zone to the deep sediment [71,75–77].

The fungal community recovered in this study was dominated by members of the genera *Aspergillus* and *Penicillium*. This trend seems to be constant in surveys of marine fungal diversity worldwide [78,79] and in the Mediterranean Sea [12]. Arrabassada Beach was characterized by the prevalence of the genus *Scedosporium*, with the species *S. apiospermum* and *S. boydii*. Although little is known about this genus in the marine environment, novel species are being described from marine sediments [80]. Other ascomycetes that are typically recovered in sediments of the Mediterranean Sea comprise the genera *Cladosporium* and *Emericellopsis* [11–13]. Although in our case only *Cladosporium* (*C.*) *cladosporioides* was isolated, four species of *Emericellopsis* (*Em.*) were recovered (i.e., *Em. maritima*, *Em. microspora*, *Em. minima*, and *Em. salmosynnemata*), with *Em. Maritima* and *Em. Salmosynnemata* detected at 6 and 20 m depth in both sampled areas. In addition, we recovered some rare species that represent the first reports for the marine environment, such as *Byssoonygena ceratinophyla*, for which only the ex-type strain is known [81]. We identified the strain FMR 19986 as *Chantransiopsis* (*Ch.*) *c.f. decumbens* based on its morphological resemblance to the protologue of *Ch. decumbens* [82], but with slightly smaller conidia. This species belongs to the *Laboulbeniales* and is supposed to represent an obligate pathogen of arthropods [83], but the marine strain grows perfectly in vitro. Of note is that only SSU sequences for this fungus were available for comparison, and although our strain showed a similarity of 99% with *Ch. decumbens* (LG589), its molecular identification could not be confirmed. Further phylogenetic studies are needed to resolve the taxonomy of the genus *Chantransiopsis* and allied genera.

On the other hand, it is noteworthy the high prevalence of the novel species *Q. verrucosa* in marine sediments. In the sampled areas described in this study, we have isolated multiple strains in sediments ranging in depth from 6 to 20 m. However, we have also detected this species in other areas along Tarragona's golden coast and the Ebro River Delta (unpublished data), indicating that it is a well-adapted species to the marine habitat. The genus *Queenslandipenidiella* is classified in *Teratosphaeriaceae*, a family of melanized fungi that predominantly inhabit plant material, while others are saprobes that can colonize extreme environments like rock surfaces and even marine habitats [44,84]. The rest of the novel species reported in this study were recovered in less abundance. The ecological knowledge of the genus *Nigrocephalum* is particularly limited. Regarding the new genus *Schizochlamydosporiella*, it is surprising the broad temperature range for growth (5–40 °C) of the new species *S. marina* found at 27 m of depth, although this property is coherent with certain fungi in *Sordariales* [85].

There is evidence of fungal activity throughout the marine environment [86]. In marine food webs, fungi can appear as saprobes, mutualists, endosymbionts, and parasites [87]. Ascomycetes represent active contributors in carbon cycling, degrading pelagic carbohydrates, and reallocating carbon to sediments [15,78,86]. Among the ascomycetes that we have detected in this area, the genera *Aspergillus* and *Penicillium*, together with species like *C. cladosporioides*, *Em. minima*, and *Fusarium solani*, have been reported to degrade different types of plastic polymers [88]. Another species, *Amphichorda* (*A.*) *littoralis*, has been recovered from both sediments and floating rubber, suggesting a role in plastic degradation [89]. Some marine fungi are known to degrade environmental pollutants, like crude oil [90–94]. We have detected the species *E. xenobiotica* and the novel species *E. littoralis* exclusively in sediments from Miracle Beach, adjacent to the port of Tarragona (Figure 1). Some species in *Exophiala* represent extremotolerant colonizers of polluted habitats and nutrient-poor substrates like hydrocarbons [95–97]. Moreover, these two species are members of the *Exophiala jeanselmei*-clade, which includes agents of chromoblastomycosis and mycetoma in humans, and like *E. littoralis*, its members can grow at 37 °C [60,98–100]. These two *Exophiala* species, as well as *A. littoralis*, represent interesting candidates for future research on their metabolic activities.

Overall, the fungal community detected in this work seems to represent indigenous taxa of the marine environment. This idea is supported by the proximity to a commercial port and the high prevalence of microplastics and other pollutants in submerged sediments from the coast of Tarragona city (Miracle and Arrabassada Beaches; Ref. [25]).

## 5. Conclusions

SMF was successfully employed for the first time to detect fungi in marine sediments and to store them over time. The combination of direct plating and SMF pretreatment, along with three media and two isolation temperatures, resulted in the detection of 94 species of filamentous ascomycetes. These species represent indigenous taxa of the marine environment, including novel species such as *E. littoralis*, *N. paracollariferum*, *Q. verrucosa*, and *S. marina*.

**Supplementary Materials:** The following supporting information can be downloaded at: https://www.mdpi.com/article/10.3390/jof10040281/s1, Table S1: Collection details of the strains detected in marine sediments and GenBank accession numbers of the barcodes used for identification, Table S2: Accession numbers of the strains included in the phylogenetic analyses, Figure S1: RaxML phylogenetic tree representing the individual ITS alignment of the *Exophiala jeanselmei* clade, Figure S2: RaxML phylogenetic tree representing the individual LSU alignment of the *Exophiala jeanselmei* clade, Figure S3: RaxML phylogenetic tree representing the individual ITS alignment of *Nigrocephalum* and representative species of *Plectosphaerellaceae*, Figure S4: RaxML phylogenetic tree representing the individual LSU alignment of *Nigrocephalum* and representative species of *Plectosphaerellaceae*, Figure S5: RaxML phylogenetic tree representing the individual *tef*1-α alignment of *Nigrocephalum* and representative species of *Plectosphaerellaceae*, Figure S6: RaxML phylogenetic tree representing the individual *rpb*2 alignment of *Nigrocephalum* and representative species of *Plectosphaerellaceae*, Figure S7: RaxML phylogenetic tree representing the individual ITS alignment of *Queenslandipenidiella* and representative species of *Teratosphaeriaceae*, Figure S8: RaxML phylogenetic tree representing the individual LSU alignment of *Queenslandipenidiella* and representative species of *Teratosphaeriaceae*, Figure S9: RaxML phylogenetic tree representing the individual ITS alignment of *Schizotheciaceae*, Figure S10: RaxML phylogenetic tree representing the individual LSU alignment of *Schizotheciaceae*, Figure S11: RaxML phylogenetic tree representing the individual *rpb*2 alignment of *Schizotheciaceae*, Figure S12: Phylogenetic tree inferred from a concatenated alignment of ITS, LSU, and *rpb*2 sequences of 28 strains representing *Schizotheciaceae*. Refs. [101–138] are cited exclusively in Table S2.

**Author Contributions:** Conceptualization, D.G.-M. and J.G.; methodology, D.G.-M., J.G., J.F.C.-L. and A.F.-B.; software, D.G.-M. and J.F.C.-L.; validation, J.F.C.-L. and J.G.; formal analysis, D.G.-M., J.F.C.-L. and J.G.; investigation, D.G.-M., J.G. and J.F.C.-L.; resources, J.F.C.-L. and J.G.; data curation, J.F.C.-L. and J.G.; writing—original draft preparation, D.G.-M. and J.G.; writing—review and editing, D.G.-M., J.G., J.F.C.-L. and A.F.-B.; visualization, D.G.-M. and J.G.; supervision, D.G.-M., J.F.C.-L. and

J.G.; project administration, J.G.; funding acquisition, J.G. All authors have read and agreed to the published version of the manuscript.

**Funding:** This study was supported by the grant PID2021-128068NB-100 funded by MCIU/AEI/10.13039/501100011033/ and by "ERDF A way of making Europe".

**Institutional Review Board Statement:** Not applicable.

**Informed Consent Statement:** Not applicable.

**Data Availability Statement:** Data are contained within the article and Supplementary Materials.

**Acknowledgments:** The authors thank Vladimir Baulin (Universitat Rovira i Virgili) and Gabriel Quiroga-Jofre (Tarraco Diving Center) for their help and services in the collection of the samples. D.G.-M. is grateful to University Rovira i Virgili and Diputación de Tarragona for a Martí-Franqués doctoral grant.

**Conflicts of Interest:** The authors declare no conflicts of interest.

## References

1. Niskanen, T.; Lücking, R.; Dahlberg, A.; Gaya, E.; Suz, L.M.; Mikryukov, V.; Liimatainen, K.; Druzhinina, I.; Westrip, J.R.S.; Mueller, G.M.; et al. Pushing the Frontiers of Biodiversity Research: Unveiling the Global Diversity, Distribution, and Conservation of Fungi. *Annu. Rev. Environ. Resour.* **2023**, *48*, 149–176. [CrossRef]
2. Antonelli, A.; Smith, R.J.; Fry, C.; Simmonds, M.S.J.; Kersey, P.J.; Pritchard, H.W.; Abbo, M.S.; Acedo, C.; Adams, J.; Ainsworth, A.M.; et al. *State of the World's Plants and Fungi*; Royal Botanic Gardens (Kew); Sfumato Foundation: London, UK, 2020.
3. Jones, E.B.G. History of Marine Mycology—A Personal Perspective. *Bot. Mar.* **2023**, *66*, 453–470. [CrossRef]
4. Calabon, M.S.; Jones, E.B.G.; Pang, K.-L.; Abdel-Wahab, M.A.; Jin, J.; Devadatha, B.; Sadaba, R.B.; Apurillo, C.C.; Hyde, K.D. Updates on the Classification and Numbers of Marine Fungi. *Bot. Mar.* **2023**, *66*, 213–238. [CrossRef]
5. Debeljak, P.; Baltar, F. Fungal Diversity and Community Composition across Ecosystems. *J. Fungi* **2023**, *9*, 510. [CrossRef] [PubMed]
6. Jones, E.B.G.; Pang, K.-L.; Abdel-Wahab, M.A.; Scholz, B.; Hyde, K.D.; Boekhout, T.; Ebel, R.; Rateb, M.E.; Henderson, L.; Sakayaroj, J.; et al. An Online Resource for Marine Fungi. *Fungal Divers.* **2019**, *96*, 347–433. [CrossRef]
7. Coll, M.; Piroddi, C.; Steenbeek, J.; Kaschner, K.; Lasram, F.B.R.; Aguzzi, J.; Ballesteros, E.; Bianchi, C.N.; Corbera, J.; Dailianis, T.; et al. The Biodiversity of the Mediterranean Sea: Estimates, Patterns, and Threats. *PLoS ONE* **2010**, *5*, e11842. [CrossRef] [PubMed]
8. Scheffers, B.R.; Joppa, L.N.; Pimm, S.L.; Laurance, W.F. What We Know and Don't Know about Earth's Missing Biodiversity. *Trends Ecol. Evol.* **2012**, *27*, 501–510. [CrossRef]
9. Bovio, E.; Gnavi, G.; Prigione, V.; Spina, F.; Denaro, R.; Yakimov, M.; Calogero, R.; Crisafi, F.; Varese, G.C. The Culturable Mycobiota of a Mediterranean Marine Site after an Oil Spill: Isolation, Identification and Potential Application in Bioremediation. *Sci. Total Environ.* **2017**, *576*, 310–318. [CrossRef] [PubMed]
10. Garzoli, L.; Gnavi, G.; Tamma, F.; Tosi, S.; Varese, G.C.; Picco, A.M. Sink or Swim: Updated Knowledge on Marine Fungi Associated with Wood Substrates in the Mediterranean Sea and Hints about Their Potential to Remediate Hydrocarbons. *Prog. Oceanogr.* **2015**, *137*, 140–148. [CrossRef]
11. Cecchi, G.; Cutroneo, L.; Di Piazza, S.; Capello, M.; Zotti, M. Culturable Fungi from Dredged and Marine Sediments from Six Ports Studied in the Framework of the SEDITERRA Project. *J. Soils Sediments* **2021**, *21*, 1563–1573. [CrossRef]
12. Marchese, P.; Garzoli, L.; Gnavi, G.; O'Connell, E.; Bouraoui, A.; Mehiri, M.; Murphy, J.M.; Varese, G.C. Diversity and Bioactivity of Fungi Associated with the Marine Sea Cucumber *Holothuria Poli*: Disclosing the Strains Potential for Biomedical Applications. *J. Appl. Microbiol.* **2020**, *129*, 612–625. [CrossRef] [PubMed]
13. Greco, G.; Cecchi, G.; Di Piazza, S.; Cutroneo, L.; Capello, M.; Zotti, M. Fungal Characterisation of a Contaminated Marine Environment: The Case of the Port of Genoa (North-Western Italy). *Webbia* **2018**, *73*, 97–106. [CrossRef]
14. Hassett, B.T.; Vonnahme, T.R.; Peng, X.; Jones, E.B.G.; Heuzé, C. Global Diversity and Geography of Planktonic Marine Fungi. *Bot. Mar.* **2020**, *63*, 121–139. [CrossRef]
15. Barone, G.; Rastelli, E.; Corinaldesi, C.; Tangherlini, M.; Danovaro, R.; Dell'Anno, A. Benthic Deep-Sea Fungi in Submarine Canyons of the Mediterranean Sea. *Prog. Oceanogr.* **2018**, *168*, 57–64. [CrossRef]
16. Phukhamsakda, C.; Nilsson, R.H.; Bhunjun, C.S.; de Farias, A.R.G.; Sun, Y.-R.; Wijesinghe, S.N.; Raza, M.; Bao, D.-F.; Lu, L.; Tibpromma, S.; et al. The Numbers of Fungi: Contributions from Traditional Taxonomic Studies and Challenges of Metabarcoding. *Fungal Divers.* **2022**, *114*, 327–386. [CrossRef]
17. Zhang, Z.-F.; Liu, F.; Liu, L.-R.; Li, M.; Cai, L.; Liu, S.; Mao, J. Culturing the Uncultured Marine Fungi in the Omics Age: Opportunities and Challenges. *Fungal Biol. Rev.* **2024**, *48*, 100353. [CrossRef]
18. Calgua, B.; Mengewein, A.; Grunert, A.; Bofill-Mas, S.; Clemente-Casares, P.; Hundesa, A.; Wyn-Jones, A.P.; López-Pila, J.M.; Girones, R. Development and Application of a One-Step Low Cost Procedure to Concentrate Viruses from Seawater Samples. *J. Virol. Methods* **2008**, *153*, 79–83. [CrossRef]

19. Calgua, B.; Rodriguez-Manzano, J.; Hundesa, A.; Suñen, E.; Calvo, M.; Bofill-Mas, S.; Girones, R. New Methods for the Concentration of Viruses from Urban Sewage Using Quantitative PCR. *J. Virol. Methods* **2013**, *187*, 215–221. [CrossRef]
20. Calgua, B.; Fumian, T.; Rusiñol, M.; Rodriguez-Manzano, J.; Mbayed, V.A.; Bofill-Mas, S.; Miagostovich, M.; Girones, R. Detection and Quantification of Classic and Emerging Viruses by Skimmed-Milk Flocculation and PCR in River Water from Two Geographical Areas. *Water Res.* **2013**, *47*, 2797–2810. [CrossRef]
21. Rusiñol, M.; Hundesa, A.; Cárdenas-Youngs, Y.; Fernández-Bravo, A.; Pérez-Cataluña, A.; Moreno-Mesonero, L.; Moreno, Y.; Calvo, M.; Alonso, J.L.; Figueras, M.J.; et al. Microbiological Contamination of Conventional and Reclaimed Irrigation Water: Evaluation and Management Measures. *Sci. Total Environ.* **2020**, *710*, 136298. [CrossRef]
22. Bofill-Mas, S.; Hundesa, A.; Calgua, B.; Rusiñol, M.; de Motes, C.M.; Girones, R. Cost-Effective Method for Microbial Source Tracking Using Specific Human and Animal Viruses. *JoVE* **2011**, *58*, e2820. [CrossRef]
23. Gonzales-Gustavson, E.; Cárdenas-Youngs, Y.; Calvo, M.; da Silva, M.F.M.; Hundesa, A.; Amorós, I.; Moreno, Y.; Moreno-Mesonero, L.; Rosell, R.; Ganges, L.; et al. Characterization of the Efficiency and Uncertainty of Skimmed Milk Flocculation for the Simultaneous Concentration and Quantification of Water-Borne Viruses, Bacteria and Protozoa. *J. Microbiol. Methods* **2017**, *134*, 46–53. [CrossRef] [PubMed]
24. Garcia-Lozano, C.; Pintó, J. Current Status and Future Restoration of Coastal Dune Systems on the Catalan Shoreline (Spain, NW Mediterranean Sea). *J. Coast Conserv.* **2018**, *22*, 519–532. [CrossRef]
25. Expósito, N.; Rovira, J.; Sierra, J.; Folch, J.; Schuhmacher, M. Microplastics Levels, Size, Morphology and Composition in Marine Water, Sediments and Sand Beaches. Case Study of Tarragona Coast (Western Mediterranean). *Sci. Total Environ.* **2021**, *786*, 147453. [CrossRef]
26. Folch, J. *La Mar de Tarragona: Una Aproximació Naturalista al Litoral Més Desconegut*; Publicacions Universitat Rovira i Virgili: Tarragona, Spain, 2021; pp. 1–200.
27. Ruiz-Fernández, J.M.; Guillén, J.E.; Ramos-Segura, A.; Otero, M.; Tello-Antón, M.O.; Mateo, M.A.; Bernardeau-Esteller, J.; Rueda, J.L.; Urra, J.; Mateo, A.; et al. *Atlas de las Praderas Marinas de España*; IEO: Murcia, Spain; IEL: Alicante, Spain; UICN: Málaga, Spain, 2016; pp. 1–681.
28. Folch, J.; Pascual, X. Inventario de Especies. Parc Subaquàtic. Tarragona. Societat d'Exploracions Submarines—Autoridad Portuaria de Tarragona. Available online: https://www.porttarragona.cat/es/catalogo-de-publicaciones/item/inventario-de-especies (accessed on 14 February 2024).
29. Capdevila, M.; Folch, J. Fauna Malacològica Del Parc Subaquàtic de Tarragona (El Tarragonès, Catalunya, Espanya). *Spira* **2009**, *3*, 33–51.
30. QGIS Development Team. QGIS Geographic Information System. 2024. Available online: https://www.qgis.org (accessed on 20 January 2024).
31. Kossuga, M.H.; Romminger, S.; Xavier, C.; Milanetto, M.C.; do Valle, M.Z.; Pimenta, E.F.; Morais, R.P.; de Carvalho, E.; Mizuno, C.M.; Coradello, L.F.C.; et al. Evaluating Methods for the Isolation of Marine-Derived Fungal Strains and Production of Bioactive Secondary Metabolites. *Rev. Bras. Farm.* **2012**, *22*, 257–267. [CrossRef]
32. Wingfield, B.D.; Wingfield, M.J.; Duong, T.A. Molecular Basis of Cycloheximide Resistance in the Ophiostomatales Revealed. *Curr. Genet.* **2022**, *68*, 505–514. [CrossRef]
33. Gannibal, P.B. Polyphasic Approach to Fungal Taxonomy. *Biol. Bull. Rev.* **2022**, *12*, 18–28. [CrossRef]
34. Kornerup, A.; Wanscher, J.H. *Methuen Handbook of Colour*, 3rd ed.; Methuen: London, UK, 1978; p. 256.
35. Müller, F.M.C.; Werner, K.E.; Kasai, M.; Francesconi, A.; Chanock, S.J.; Walsh, T.J. Rapid Extraction of Genomic DNA from Medically Important Yeasts and Filamentous Fungi by High-Speed Cell Disruption. *J. Clin. Microbiol.* **1998**, *36*, 1625–1629. [CrossRef] [PubMed]
36. White, T.J.; Bruns, T.; Lee, S.; Taylor, J.W. Amplification and Direct Sequencing of Fungal Ribosomal RNA Genes for Phylogenetics. In *PCR Protocols: A Guide to Methods and Applications*; Mycological Society of America: New York, NY, USA, 1990; pp. 315–322.
37. Vilgalys, R.; Hester, M. Rapid Genetic Identification and Mapping of Enzymatically Amplified Ribosomal DNA from Several Cryptococcus Species. *J. Bacteriol.* **1990**, *172*, 4238–4246. [CrossRef] [PubMed]
38. Glass, N.L.; Donaldson, G.C. Development of Primer Sets Designed for Use with the PCR to Amplify Conserved Genes from Filamentous Ascomycetes. *Appl. Environ. Microbiol.* **1995**, *61*, 1323–1330. [CrossRef] [PubMed]
39. Woudenberg, J.H.C.; Aveskamp, M.M.; de Gruyter, J.; Spiers, A.G.; Crous, P.W. Multiple Didymella Teleomorphs Are Linked to the Phoma Clematidina Morphotype. *Persoonia Mol. Phylogeny Evol. Fungi* **2009**, *22*, 56–62. [CrossRef] [PubMed]
40. O'Donnell, K.; Cigelnik, E. Two Divergent Intragenomic RDNA ITS2 Types within a Monophyletic Lineage of the Fungusfusarium are Nonorthologous. *Mol. Phylogenet. Evol.* **1997**, *7*, 103–116. [CrossRef]
41. Stielow, J.B.; Lévesque, C.A.; Seifert, K.A.; Meyer, W.; Irinyi, L.; Smits, D.; Renfurm, R.; Verkley, G.J.M.; Groenewald, M.; Chaduli, D.; et al. One Fungus, Which Genes? Development and Assessment of Universal Primers for Potential Secondary Fungal DNA Barcodes. *Persoonia Mol. Phylogeny Evol. Fungi* **2015**, *35*, 242–263. [CrossRef] [PubMed]
42. Carbone, I.; Kohn, L.M. A Method for Designing Primer Sets for Speciation Studies in Filamentous Ascomycetes. *Mycologia* **1999**, *91*, 553–556. [CrossRef]
43. Liu, Y.J.; Whelen, S.; Hall, B.D. Phylogenetic Relationships among Ascomycetes: Evidence from an RNA Polymerse II Subunit. *Mol. Biol. Evol.* **1999**, *16*, 1799–1808. [CrossRef] [PubMed]

44. Quaedvlieg, W.; Binder, M.; Groenewald, J.Z.; Summerell, B.A.; Carnegie, A.J.; Burgess, T.I.; Crous, P.W. Introducing the Consolidated Species Concept to Resolve Species in the *Teratosphaeriaceae*. *Persoonia Mol. Phylogeny Evol. Fungi* **2014**, *33*, 1–40. [CrossRef]
45. Torres-Garcia, D.; García, D.; Réblová, M.; Jurjević, Ž.; Hubka, V.; Gené, J. Diversity and Novel Lineages of Black Yeasts in Chaetothyriales from Freshwater Sediments in Spain. *Persoonia Mol. Phylogeny Evol. Fungi* **2023**, *51*, 194–228. [CrossRef]
46. Giraldo, A.; Crous, P.W. Inside Plectosphaerellaceae. *Stud. Mycol.* **2019**, *92*, 227–286. [CrossRef]
47. Marin-Felix, Y.; Miller, A.N.; Cano-Lira, J.F.; Guarro, J.; García, D.; Stadler, M.; Huhndorf, S.M.; Stchigel, A.M. Re-Evaluation of the Order *Sordariales*: Delimitation of *Lasiosphaeriaceae s. str.*, and Introduction of the New Families *Diplogelasinosporaceae*, *Naviculisporaceae*, and *Schizotheciaceae*. *Microorganisms* **2020**, *8*, 1430. [CrossRef]
48. Marin-Felix, Y.; Miller, A.N. Corrections to Recent Changes in the Taxonomy of the *Sordariales*. *Mycol. Progr.* **2022**, *21*, 69. [CrossRef]
49. Thompson, J.D.; Higgins, D.G.; Gibson, T.J. CLUSTAL W: Improving the Sensitivity of Progressive Multiple Sequence Alignment through Sequence Weighting, Position-Specific Gap Penalties and Weight Matrix Choice. *Nucleic Acids Res.* **1994**, *22*, 4673–4680. [CrossRef]
50. Tamura, K.; Stecher, G.; Peterson, D.; Filipski, A.; Kumar, S. MEGA6: Molecular Evolutionary Genetics Analysis Version 6.0. *Mol. Biol. Evol.* **2013**, *30*, 2725–2729. [CrossRef]
51. Edgar, R.C. MUSCLE: Multiple Sequence Alignment with High Accuracy and High Throughput. *Nucleic Acids Res.* **2004**, *32*, 1792–1797. [CrossRef]
52. Miller, M.A.; Pfeiffer, W.; Schwartz, T. The CIPRES Science Gateway: Enabling High-Impact Science for Phylogenetics Researchers with Limited Resources. In Proceedings of the 1st Conference of the Extreme Science and Engineering Discovery Environment: Bridging from the eXtreme to the Campus and Beyond, New York, NY, USA, 16 July 2012; Association for Computing Machinery: New York, NY, USA, 2012; pp. 1–8.
53. Stamatakis, A. RAxML Version 8: A Tool for Phylogenetic Analysis and Post-Analysis of Large Phylogenies. *Bioinformatics* **2014**, *30*, 1312–1313. [CrossRef] [PubMed]
54. Hillis, D.M.; Bull, J.J. An Empirical Test of Bootstrapping as a Method for Assessing Confidence in Phylogenetic Analysis. *Syst. Biol.* **1993**, *42*, 182–192. [CrossRef]
55. Ronquist, F.; Teslenko, M.; van der Mark, P.; Ayres, D.L.; Darling, A.; Höhna, S.; Larget, B.; Liu, L.; Suchard, M.A.; Huelsenbeck, J.P. MrBayes 3.2: Efficient Bayesian Phylogenetic Inference and Model Choice Across a Large Model Space. *Syst. Biol.* **2012**, *61*, 539–542. [CrossRef]
56. Darriba, D.; Taboada, G.L.; Doallo, R.; Posada, D. JModelTest 2: More Models, New Heuristics and Parallel Computing. *Nat. Methods* **2012**, *9*, 772. [CrossRef]
57. Guindon, S.; Gascuel, O. A Simple, Fast, and Accurate Algorithm to Estimate Large Phylogenies by Maximum Likelihood. *Syst. Biol.* **2003**, *52*, 696–704. [CrossRef]
58. Hespanhol, L.; Vallio, C.S.; Costa, L.M.; Saragiotto, B.T. Understanding and Interpreting Confidence and Credible Intervals around Effect Estimates. *Braz. J. Phys. Ther.* **2019**, *23*, 290–301. [CrossRef] [PubMed]
59. Shannon, P.; Markiel, A.; Ozier, O.; Baliga, N.S.; Wang, J.T.; Ramage, D.; Amin, N.; Schwikowski, B.; Ideker, T. Cytoscape: A Software Environment for Integrated Models of Biomolecular Interaction Networks. *Genome Res.* **2003**, *13*, 2498–2504. [CrossRef] [PubMed]
60. de Hoog, G.S.; Vicente, V.A.; Najafzadeh, M.J.; Harrak, M.J.; Badali, H.; Seyedmousavi, S. Waterborne *Exophiala* Species Causing Disease in Cold-Blooded Animals. *Persoonia Mol. Phylogeny Evol. Fungi* **2011**, *27*, 46–72. [CrossRef] [PubMed]
61. McGinnis, M.R.; Padhye, A. A *Exophiala jeanselmei*, a New Combination for *Phialophora jeanselmei*. *Mycotaxon* **1977**, *5*, 341–352.
62. Vitale, R.G.; Hoog, G.S. de Molecular Diversity, New Species and Antifungal Susceptibilities in the *Exophiala spinifera* Clade. *Med. Mycol.* **2002**, *40*, 545–556. [CrossRef] [PubMed]
63. de Hoog, G.S.; Vicente, V.; Caligiorne, R.B.; Kantarcioglu, S.; Tintelnot, K.; Gerrits van den Ende, A.H.G.; Haase, G. Species Diversity and Polymorphism in the *Exophiala spinifera* Clade Containing Opportunistic Black Yeast-Like Fungi. *J. Clin. Microbiol.* **2003**, *41*, 4767–4778. [CrossRef]
64. Yang, X.; Feng, M.; Yu, Z. *Exophiala pseudooligosperma* sp. nov., a Novel Black Yeast from Soil in Southern China. *Int. J. Syst. Evol. Microbiol.* **2021**, *71*, 5116. [CrossRef] [PubMed]
65. Woo, P.C.Y.; Ngan, A.H.Y.; Tsang, C.C.C.; Ling, I.W.H.; Chan, J.F.W.; Leung, S.-Y.; Yuen, K.-Y.; Lau, S.K.P. Clinical Spectrum of *Exophiala* Infections and a Novel *Exophiala* Species, *Exophiala hongkongensis*. *J. Clin. Microbiol.* **2020**, *51*, 260–267. [CrossRef] [PubMed]
66. De Hoog, G.S.; Guarro, J.; Gené, J.; Ahmed, S.; Al-Hatmi, A.M.S.; Figueras, M.J.; Vitale, R.G. *Atlas of Clinical Fungi*, 4th ed.; Foundation Atlas of Clinical Fungi: Hilversum, The Netherlands, 2020.
67. Weisenborn, J.L.; Kirschner, R.; Piepenbring, M.A. New Darkly Pigmented and Keratinolytic Species of *Acremonium* (Hyphomycetes) with Relationship to the *Plectosphaerellaceae* from Human Skin and Nail Lesions in Panama. *Nova Hedwig.* **2010**, *90*, 457–468. [CrossRef]
68. Crous, P.W.; Groenewald, J.Z.; Stone, J.K. *Penidiella kurandae*. *Fungal Planet* **2007**, *16*.

69. Taxonomy of Fungi Imperfecti: Proceedings of the First International Specialists' Workshop Conference on Criteria and Terminology in the Classification of Fungi Imperfecti, Kananaskis, Alberta, Canada. In *Taxonomy of Fungi Imperfecti*; University of Toronto Press: Toronto, ON, Canada, 1971; ISBN 978-1-4875-8916-5.
70. Udagawa, S.I.; Muroi, T. Notes on some Japanese ascomycetes XVI. *Nippon Kingakukai Kaiho* **1981**, *22*, 11–26.
71. Rédou, V.; Navarri, M.; Meslet-Cladière, L.; Barbier, G.; Burgaud, G. Species Richness and Adaptation of Marine Fungi from Deep-Subseafloor Sediments. *Appl. Environ. Microbiol.* **2015**, *81*, 3571–3583. [CrossRef] [PubMed]
72. Gonçalves, M.F.M.; Esteves, A.C.; Alves, A. Revealing the Hidden Diversity of Marine Fungi in Portugal with the Description of Two Novel Species, Neoascochyta Fuci Sp. Nov. and Paraconiothyrium Salinum Sp. Nov. *Int. J. Syst. Evol. Microbiol.* **2020**, *70*, 5337–5354. [CrossRef] [PubMed]
73. Luo, Y.; Xu, W.; Luo, Z.-H.; Pang, K.-L. Diversity and Temperature Adaptability of Cultivable Fungi in Marine Sediments from the Chukchi Sea. *Bot. Mar.* **2020**, *63*, 197–207. [CrossRef]
74. Garmendia, G.; Alvarez, A.; Villarreal, R.; Martinez Silveira, A.; Wisniewski, M.; Vero, S. Fungal Diversity in the Coastal Waters of King George Island (Maritime Antarctica). *World J. Microbiol. Biotechnol.* **2021**, *37*, 142. [CrossRef]
75. Morales, S.E.; Biswas, A.; Herndl, G.J.; Baltar, F. Global Structuring of Phylogenetic and Functional Diversity of Pelagic Fungi by Depth and Temperature. *Front. Mar. Sci.* **2019**, *6*, 131. [CrossRef]
76. Tisthammer, K.H.; Cobian, G.M.; Amend, A.S. Global Biogeography of Marine Fungi Is Shaped by the Environment. *Fungal Ecol.* **2016**, *19*, 39–46. [CrossRef]
77. Li, W.; Wang, M.; Burgaud, G.; Yu, H.; Cai, L. Fungal Community Composition and Potential Depth-Related Driving Factors Impacting Distribution Pattern and Trophic Modes from Epi- to Abyssopelagic Zones of the Western Pacific Ocean. *Microb. Ecol.* **2019**, *78*, 820–831. [CrossRef] [PubMed]
78. Baltar, F.; Zhao, Z.; Herndl, G.J. Potential and Expression of Carbohydrate Utilization by Marine Fungi in the Global Ocean. *Microbiome* **2021**, *9*, 106. [CrossRef]
79. Nicoletti, R.; Trincone, A. Bioactive Compounds Produced by Strains of *Penicillium* and *Talaromyces* of Marine Origin. *Mar. Drugs* **2016**, *14*, 37. [CrossRef]
80. Wang, M.-M.; Yang, S.-Y.; Li, Q.; Zheng, Y.; Ma, H.-H.; Tu, Y.-H.; Li, W.; Cai, L. *Microascaceae* from the Marine Environment, with Descriptions of Six New Species. *J. Fungi* **2024**, *10*, 45. [CrossRef] [PubMed]
81. Guarro, J.; Punsola, L.; Cano, J. *Byssoonygena ceratinophila*, gen. et sp. nov. a New Keratinophilic Fungus from Spain. *Mycopathologia* **1987**, *100*, 159–161. [CrossRef]
82. Thaxter, R. On Certain Peculiar Fungus-Parasites of Living Insects. *Bot. Gaz.* **1914**, *58*, 235–253. [CrossRef]
83. Blackwell, M.; Haelewaters, D.; Pfister, D.H. Laboulbeniomycetes: Evolution, Natural History, and Thaxter's Final Word. *Mycologia* **2020**, *112*, 1048–1059. [CrossRef] [PubMed]
84. Wei, X.; Guo, S.; Gong, L.-F.; He, G.; Pang, K.-L.; Luo, Z.-H. Cultivable Fungal Diversity in Deep-Sea Sediment of the East Pacific Ocean. *Geomicrobiol. J.* **2018**, *35*, 790–797. [CrossRef]
85. Wang, X.W.; Han, P.J.; Bai, F.Y.; Luo, A.; Bensch, K.; Meijer, M.; Kraak, B.; Han, D.Y.; Sun, B.D.; Crous, P.W.; et al. Taxonomy, Phylogeny and Identification of *Chaetomiaceae* with Emphasis on Thermophilic Species. *Stud. Mycol.* **2022**, *101*, 121–243. [CrossRef] [PubMed]
86. Gladfelter, A.S.; James, T.Y.; Amend, A.S. Marine Fungi. *Curr. Biol.* **2019**, *29*, R191–R195. [CrossRef]
87. Gonçalves, M.F.M.; Esteves, A.C.; Alves, A. Marine Fungi: Opportunities and Challenges. *Encyclopedia* **2022**, *2*, 559–577. [CrossRef]
88. Ekanayaka, A.H.; Tibpromma, S.; Dai, D.; Xu, R.; Suwannarach, N.; Stephenson, S.L.; Dao, C.; Karunarathna, S.C. A Review of the Fungi That Degrade Plastic. *J. Fungi* **2022**, *8*, 772. [CrossRef]
89. Guerra-Mateo, D.; Gené, J.; Baulin, V.; Cano-Lira, J.F. Phylogeny and Taxonomy of the Genus *Amphichorda* (*Bionectriaceae*): An Update on Beauveria-like Strains and Description of a Novel Species from Marine Sediments. *Diversity* **2023**, *15*, 795. [CrossRef]
90. Bonugli-Santos, R.C.; dos Santos Vasconcelos, M.R.; Passarini, M.R.Z.; Vieira, G.A.L.; Lopes, V.C.P.; Mainardi, P.H.; dos Santos, J.A.; de Azevedo Duarte, L.; Otero, I.V.R.; da Silva Yoshida, A.M.; et al. Marine-Derived Fungi: Diversity of Enzymes and Biotechnological Applications. *Front. Microbiol.* **2015**, *6*, 269. [CrossRef] [PubMed]
91. Kumar, V.; Sarma, V.V.; Thambugala, K.M.; Huang, J.-J.; Li, X.-Y.; Hao, G.-F. Ecology and Evolution of Marine Fungi with Their Adaptation to Climate Change. *Front. Microbiol.* **2021**, *12*, 719000. [CrossRef] [PubMed]
92. Zeghal, E.; Vaksmaa, A.; Vielfaure, H.; Boekhout, T.; Niemann, H. The Potential Role of Marine Fungi in Plastic Degradation—A Review. *Front. Mar. Sci.* **2021**, *8*, 738877. [CrossRef]
93. Barnes, N.M.; Damare, S.R.; Bhatawadekar, V.C.; Garg, A.; Lotlikar, N.P. Degradation of Crude Oil-Associated Polycyclic Aromatic Hydrocarbons by Marine-Derived Fungi. *3 Biotech.* **2023**, *13*, 335. [CrossRef] [PubMed]
94. Vaksmaa, A.; Polerecky, L.; Dombrowski, N.; Kienhuis, M.V.M.; Posthuma, I.; Gerritse, J.; Boekhout, T.; Niemann, H. Polyethylene Degradation and Assimilation by the Marine Yeast *Rhodotorula mucilaginosa*. *ISME Commun.* **2023**, *3*, 68. [CrossRef]
95. De Hoog, G.S.; Zeng, J.S.; Harrak, M.J.; Sutton, D.A. Exophiala Xenobiotica Sp. Nov., an Opportunistic Black Yeast Inhabiting Environments Rich in Hydrocarbons. *Antonie Van Leeuwenhoek* **2006**, *90*, 257–268. [CrossRef] [PubMed]
96. Isola, D.; Zucconi, L.; Onofri, S.; Caneva, G.; de Hoog, G.S.; Selbmann, L. Extremotolerant Rock Inhabiting Black Fungi from Italian Monumental Sites. *Fungal Divers.* **2016**, *76*, 75–96. [CrossRef]

97. Madrid, H.; Hernández-Restrepo, M.; Gené, J.; Cano, J.; Guarro, J.; Silva, V. New and Interesting Chaetothyrialean Fungi from Spain. *Mycol. Progr.* **2016**, *15*, 1179–1201. [CrossRef]
98. Zeng, J.S.; Sutton, D.A.; Fothergill, A.W.; Rinaldi, M.G.; Harrak, M.J.; de Hoog, G.S. Spectrum of Clinically Relevant Exophiala Species in the United States. *J. Clini. Microbiol.* **2007**, *45*, 3713–3720. [CrossRef]
99. Zeng, J.; Feng, P.; van den Ende, A.H.G.G.; Xi, L.; Harrak, M.J.; de Hoog, G.S. Multilocus Analysis of the *Exophiala jeanselmei* Clade Containing Black Yeasts Involved in Opportunistic Disease in Humans. *Fungal Divers.* **2014**, *65*, 3–16. [CrossRef]
100. Seyedmousavi, S.; Guillot, J.; de Hoog, G.S. Phaeohyphomycoses, Emerging Opportunistic Diseases in Animals. *Clin. Microbiol. Rev.* **2013**, *26*, 19–35. [CrossRef]
101. Thitla, T.; Kumla, J.; Khuna, S.; Lumyong, S.; Suwannarach, N. Species Diversity, Distribution, and Phylogeny of *Exophiala* with the Addition of Four New Species from Thailand. *J. Fungi* **2022**, *8*, 766. [CrossRef] [PubMed]
102. Schoch, C.L.; Robbertse, B.; Robert, V.; Vu, D.; Cardinali, G.; Irinyi, L.; Meyer, W.; Nilsson, R.H.; Hughes, K.; Miller, A.N.; et al. Finding Needles in Haystacks: Linking Scientific Names, Reference Specimens and Molecular Data for Fungi. *Database* **2014**, *2014*, bau061. [CrossRef] [PubMed]
103. Borman, A.M.; Fraser, M.; Szekely, A.; Larcombe, D.E.; Johnson, E.M. Rapid Identification of Clinically Relevant Members of the Genus *Exophiala* by Matrix-Assisted Laser Desorption Ionization–Time of Flight Mass Spectrometry and Description of Two Novel Species, *Exophiala campbellii* and *Exophiala lavatrina*. *J. Clin. Microbiol.* **2017**, *55*, 1162–1176. [CrossRef] [PubMed]
104. Borman, A.M.; Fraser, M.; Schilling, W.; Jones, G.; Pearl, R.; Linton, C.J.; Johnson, E.M. *Exophiala campbellii* Causing a Subcutaneous Palmar Cyst in an Otherwise Healthy UK Resident. *Med. Mycol. Case Rep.* **2020**, *29*, 43–45. [CrossRef] [PubMed]
105. Singh, S.; Rudramurthy, S.M.; Padhye, A.A.; Hemashetter, B.M.; Iyer, R.; Hallur, V.; Sharma, A.; Agnihotri, S.; Gupta, S.; Ghosh, A.; et al. Clinical Spectrum, Molecular Characterization, Antifungal Susceptibility Testing of *Exophiala* spp. From India and Description of a Novel *Exophiala* Species, *E. arunalokei* sp. Nov. *Front. Cell. Infect. Microbiol.* **2021**, *11*, 686120. [CrossRef] [PubMed]
106. Tibpromma, S.; Hyde, K.D.; McKenzie, E.H.C.; Bhat, D.J.; Phillips, A.J.L.; Wanasinghe, D.N.; Samarakoon, M.C.; Jayawardena, R.S.; Dissanayake, A.J.; Tennakoon, D.S.; et al. Fungal Diversity Notes 840–928: Micro-Fungi Associated with *Pandanaceae*. *Fungal Divers.* **2018**, *93*, 1–160. [CrossRef]
107. Yong, L.K.; Wiederhold, N.P.; Sutton, D.A.; Sandoval-Denis, M.; Lindner, J.R.; Fan, H.; Sanders, C.; Guarro, J. Morphological and Molecular Characterization of *Exophiala polymorpha* Sp. Nov. Isolated from Sporotrichoid Lymphocutaneous Lesions in a Patient with Myasthenia Gravis. *J. Clin. Microbiol.* **2015**, *53*, 2816–2822. [CrossRef]
108. Teixeira, M.M.; Moreno, L.F.; Stielow, B.J.; Muszewska, A.; Hainaut, M.; Gonzaga, L.; Abouelleil, A.; Patané, J.S.L.; Priest, M.; Souza, R.; et al. Exploring the Genomic Diversity of Black Yeasts and Relatives (*Chaetothyriales*, *Ascomycota*). *Stud. Mycol.* **2017**, *86*, 1–28. [CrossRef] [PubMed]
109. Crous, P.W.; Cowan, D.A.; Maggs-Kölling, G.; Yilmaz, N.; Thangavel, R.; Wingfield, M.J.; Noordeloos, M.E.; Dima, B.; Brandrud, T.E.; Jansen, G.M.; et al. Fungal Planet Description Sheets: 1182–1283. *Persoonia Mol. Phylogeny Evol. Fungi* **2021**, *46*, 313–528. [CrossRef]
110. Vu, D.; Groenewald, M.; de Vries, M.; Gehrmann, T.; Stielow, B.; Eberhardt, U.; Al-Hatmi, A.; Groenewald, J.Z.; Cardinali, G.; Houbraken, J.; et al. Large-Scale Generation and Analysis of Filamentous Fungal DNA Barcodes Boosts Coverage for Kingdom Fungi and Reveals Thresholds for Fungal Species and Higher Taxon Delimitation. *Stud. Mycol.* **2019**, *92*, 135–154. [CrossRef] [PubMed]
111. Gueidan, C.; Villaseñor, C.R.; de Hoog, G.S.; Gorbushina, A.A.; Untereiner, W.A.; Lutzoni, F. A Rock-Inhabiting Ancestor for Mutualistic and Pathogen-Rich Fungal Lineages. *Stud. Mycol.* **2008**, *61*, 111–119. [CrossRef] [PubMed]
112. Irinyi, L.; Serena, C.; Garcia-Hermoso, D.; Arabatzis, M.; Desnos-Ollivier, M.; Vu, D.; Cardinali, G.; Arthur, I.; Normand, A.-C.; Giraldo, A.; et al. International Society of Human and Animal Mycology (ISHAM)-ITS Reference DNA Barcoding Database—The Quality Controlled Standard Tool for Routine Identification of Human and Animal Pathogenic Fungi. *Med. Mycol.* **2015**, *53*, 313–337. [CrossRef] [PubMed]
113. Arzanlou, M.; Groenewald, J.Z.; Gams, W.; Braun, U.; Shin, H.-D.; Crous, P.W. Phylogenetic and Morphotaxonomic Revision of *Ramichloridium* and Allied Genera. *Stud. Mycol.* **2007**, *58*, 57–93. [CrossRef] [PubMed]
114. Tsang, C.-C.; Tang, J.Y.M.; Chan, K.-F.; Lee, C.-Y.; Chan, J.F.W.; Ngan, A.H.Y.; Cheung, M.; Lau, E.C.L.; Li, X.; Ng, R.H.Y.; et al. Diversity of Phenotypically Non-Dermatophyte, Non-*Aspergillus* Filamentous Fungi Causing Nail Infections: Importance of Accurate Identification and Antifungal Susceptibility Testing. *Emerg. Microbes Infect.* **2019**, *8*, 531–541. [CrossRef] [PubMed]
115. Sun, W.; Su, L.; Yang, S.; Sun, J.; Liu, B.; Fu, R.; Wu, B.; Liu, X.; Cai, L.; Guo, L.; et al. Unveiling the Hidden Diversity of Rock-Inhabiting Fungi: *Chaetothyriales* from China. *J. Fungi* **2020**, *6*, 187. [CrossRef]
116. Moussa, T.A.A.; Al-Zahrani, H.S.; Kadasa, N.M.S.; Moreno, L.F.; Gerrits van den Ende, A.H.G.; de Hoog, G.S.; Al-Hatmi, A.M.S. Nomenclatural Notes on *Nadsoniella* and the Human Opportunist Black Yeast Genus *Exophiala*. *Mycoses* **2017**, *60*, 358–365. [CrossRef] [PubMed]
117. Nascimento, M.M.F.; Vicente, V.A.; Bittencourt, J.V.M.; Gelinski, J.M.L.; Prenafeta-Boldú, F.X.; Romero-Güiza, M.; Fornari, G.; Gomes, R.R.; Santos, G.D.; Gerrits Van Den Ende, A.H.G.; et al. Diversity of Opportunistic Black Fungi on Babassu Coconut Shells, a Rich Source of Esters and Hydrocarbons. *Fungal Biol.* **2017**, *121*, 488–500. [CrossRef] [PubMed]
118. Crous, P.W.; Wingfield, M.J.; Guarro, J.; Cheewangkoon, R.; van der Bank, M.; Swart, W.J.; Stchigel, A.M.; Cano-Lira, J.F.; Roux, J.; Madrid, H.; et al. Fungal Planet Description Sheets: 154–213. *Persoonia Mol. Phylogeny Evol. Fungi* **2013**, *31*, 188–296. [CrossRef] [PubMed]

119. Kiyuna, T.; An, K.-D.; Kigawa, R.; Sano, C.; Miura, S.; Sugiyama, J. Molecular Assessment of Fungi in "Black Spots" That Deface Murals in the Takamatsuzuka and Kitora Tumuli in Japan: *Acremonium* Sect. *Gliomastix Including Acremonium tumulicola Sp. Nov. and Acremonium felinum Comb. Nov. Mycoscience* **2011**, *52*, 1–17. [CrossRef]
120. Réblová, M.; Hernández-Restrepo, M.; Fournier, J.; Nekvindová, J. New Insights into the Systematics of *Bactrodesmium* and Its Allies and Introducing New Genera, Species and Morphological Patterns in the *Pleurotheciales* and *Savoryellales* (Sordariomycetes). *Stud. Mycol.* **2020**, *95*, 415–466. [CrossRef]
121. Crous, P.W.; Braun, U.; Groenewald, J.Z. *Mycosphaerella* Is Polyphyletic. *Stud. Mycol.* **2007**, *58*, 1–32. [CrossRef] [PubMed]
122. Egidi, E.; de Hoog, G.S.; Isola, D.; Onofri, S.; Quaedvlieg, W.; de Vries, M.; Verkley, G.J.M.; Stielow, J.B.; Zucconi, L.; Selbmann, L. Phylogeny and Taxonomy of Meristematic Rock-Inhabiting Black Fungi in the Dothideomycetes Based on Multi-Locus Phylogenies. *Fungal Divers.* **2014**, *65*, 127–165. [CrossRef]
123. Crous, P.W.; Groenewald, J.Z. Why Everlastings Don't Last. *Persoonia Mol. Phylogeny Evol. Fungi* **2011**, *26*, 70–84. [CrossRef] [PubMed]
124. Scott, J.A.; Ewaze, J.O.; Summerbell, R.C.; Arocha-Rosete, Y.; Maharaj, A.; Guardiola, Y.; Saleh, M.; Wong, B.; Bogale, M.; O'Hara, M.J.; et al. Multilocus DnA Sequencing of the Whiskey Fungus Reveals a Continental-Scale Speciation Pattern. *Persoonia Mol. Phylogeny Evol. Fungi* **2016**, *37*, 13–20. [CrossRef] [PubMed]
125. Crous, P.W.; Carnegie, A.J.; Wingfield, M.J.; Sharma, R.; Mughini, G.; Noordeloos, M.E.; Santini, A.; Shouche, Y.S.; Bezerra, J.D.P.; Dima, B.; et al. Fungal Planet Description Sheets: 868–950. *Persoonia Mol. Phylogeny Evol. Fungi* **2019**, *42*, 291–473. [CrossRef] [PubMed]
126. Hunter, G.C.; Wingfield, B.D.; Crous, P.W.; Wingfield, M.J. A Multi-Gene Phylogeny for Species of Mycosphaerella Occurring on Eucalyptus Leaves. *Stud. Mycol.* **2006**, *55*, 147–161. [CrossRef]
127. Crous, P.W.; Summerell, B.A.; Carnegie, A.J.; Wingfield, M.J.; Hunter, G.C.; Burgess, T.I.; Andjic, V.; Barber, P.A.; Groenewald, J.Z. Unravelling *Mycosphaerella*: Do You Believe in Genera? *Persoonia Mol. Phylogeny Evol. Fungi* **2009**, *23*, 99–118. [CrossRef]
128. Wang, X.W.; Yang, F.Y.; Meijer, M.; Kraak, B.; Sun, B.D.; Jiang, Y.L.; Wu, Y.M.; Bai, F.Y.; Seifert, K.A.; Crous, P.W.; et al. Redefining *Humicola* Sensu Stricto and Related Genera in the *Chaetomiaceae*. *Stud. Mycol.* **2019**, *93*, 65–153. [CrossRef]
129. Miller, A.N.; Huhndorf, S.M. Multi-Gene Phylogenies Indicate Ascomal Wall Morphology Is a Better Predictor of Phylogenetic Relationships than Ascospore Morphology in the *Sordariales* (*Ascomycota*, Fungi). *Mol. Phylogen. Evol.* **2005**, *35*, 60–75. [CrossRef]
130. Dai, D.-Q.; Phookamsak, R.; Wijayawardene, N.; Li, W.; Bhat, D.J.; Xu, J.; Taylor, J.; Hyde, K.; Chukeatirote, E. Bambusicolous Fungi. *Fungal Divers.* **2016**, *82*, 1–105. [CrossRef]
131. Cai, L.; Jeewon, R.; Hyde, K.D. Phylogenetic Investigations of *Sordariaceae* Based on Multiple Gene Sequences and Morphology. *Mycol. Res.* **2006**, *110*, 137–150. [CrossRef]
132. Cai, L.; Jeewon, R.; Hyde, K.D. Phylogenetic Evaluation and Taxonomic Revision of *Schizothecium* Based on Ribosomal DNA and Protein Coding Genes. *Fungal Divers.* **2005**, *19*, 1–21.
133. Kruys, Å.; Huhndorf, S.M.; Miller, A.N. Coprophilous Contributions to the Phylogeny of *Lasiosphaeriaceae* and Allied Taxa within *Sordariales* (*Ascomycota*, Fungi). *Fungal Divers.* **2015**, *70*, 101–113. [CrossRef]
134. Miller, A.N.; Huhndorf, S.M. Using Phylogenetic Species Recognition to Delimit Species Boundaries within *Lasiosphaeria*. *Mycologia* **2004**, *96*, 1106–1127. [CrossRef] [PubMed]
135. Wang, X.W.; Houbraken, J.; Groenewald, J.Z.; Meijer, M.; Andersen, B.; Nielsen, K.F.; Crous, P.W.; Samson, R.A. Diversity and Taxonomy of *Chaetomium* and Chaetomium-like Fungi from Indoor Environments. *Stud. Mycol.* **2016**, *84*, 145–224. [CrossRef]
136. Huhndorf, S.M.; Miller, A.N.; Fernández, F.A. Molecular Systematics of the *Sordariales*: The Order and the Family *Lasiosphaeriaceae* Redefined. *Mycologia* **2004**, *96*, 368–387. [CrossRef] [PubMed]
137. Bell, A.; Mahoney, D.; Debuchy, R. Podospora Bullata, a New Homothallic Ascomycete from Kangaroo Dung in Australia. Available online: https://ascomycete.org/Journal/Article/art-0180 (accessed on 14 February 2024).
138. Harms, K.; Milic, A.; Stchigel, A.M.; Stadler, M.; Surup, F.; Marin-Felix, Y. Three New Derivatives of Zopfinol from *Pseudorhypophila mangenotii* Gen. et Comb. Nov. *J. Fungi* **2021**, *7*, 181. [CrossRef]

Journal of *Fungi*

MDPI

*Article*

# Diversity of Rock-Inhabiting Fungi in Tarragona Province, Spain

**Angie Paola Sastoque, José Francisco Cano-Lira * and Alberto Miguel Stchigel**

Mycology Unit, Medical School, Universitat Rovira i Virgili, C/Sant Llorenç 21, 43201 Reus, Tarragona, Spain; angiepaola.sastoque@urv.cat (A.P.S.); albertomiguel.stchigel@urv.cat (A.M.S.)
* Correspondence: jose.cano@urv.cat

**Abstract:** Rock-inhabiting fungi (RIF) are usually extremely tolerant or extremophilic, as they can survive on natural and artificial rocks despite being exposed to stressful conditions. RIF have serious negative effects on the appearance and cohesion of rocky substrates, causing the alteration and decomposition of building materials, but also on human and animal health, as they can act as opportunistic pathogens. Their identification is therefore of great importance, especially in urban areas. In the present study, culturing techniques *for* isolating fungi, and a polyphasic taxonomic approach to their identification, were used to assess the diversity of micromycetes that darken the surfaces of buildings in various villages and cities in Tarragona Province (Spain). Sixty-four species of RIF belonging to forty-one genera were identified, including a new genus (*Coccodomyces*) and the following six new fungal species: *Coccodomyces pleiosporus*, *Exophiala caementiphila*, *Exophiala multiformis*, *Neocatenulostroma spinulosum*, *Neodevriesia longicatenispora*, and *Paradevriesia holothallica*. Thus, we have established that building materials are ecological niches where a high biodiversity of RIF can develop.

**Keywords:** *Ascomycota*; building materials; dark discoloration; extremophiles fungi; opportunistic pathogens

## 1. Introduction

Within the wide variety of stressful environments, rocky surfaces (whether natural or artificial) are extreme for most microorganisms because of the limited availability of nutrients, the wide variation in temperatures throughout the seasons, the extremely drying effect of wind, and the harmful effects of solar radiation on microbial life through the action of ultraviolet radiation. However, genetic damage also can be caused by the emission of alpha particles generated by uranium decay, elements present in certain types of rocks used in construction (such as granite), exposure to environmental pollution, etc., as well as combinations of these factors [1–4]. Despite the extreme environmental conditions, rock substrates are inhabited by a high diversity of microorganisms, mostly bacteria and fungi. In fact, the proliferation of fungi is the main cause of the blackening (dark discoloration) of the outdoor surfaces of buildings and monuments. Previous studies have focused on the characterization of rock-inhabiting microbial communities and the nature and physiology of fungi growing on natural or artificial rocky substrates [5–13]. The fungi inhabiting monuments and urban buildings made of marble, stone, or concrete have drawn attention because of their negative impact on the cohesion and aesthetics of the construction materials. They play an important role in biodeterioration and influence public health by increasing the risk of contracting opportunistic mycoses and fungal allergies [1,7,13–23]. The free-living, oligotrophic, dematiaceous, and non-lichenized microfungi reported in these studies have been grouped and named according to their natural habitat as rock-inhabiting fungi (RIF), although they have been isolated even from non-rocky surfaces of urban buildings and living plants, and have been reported as opportunistic animal or human pathogens [4,24–27]. RIF, as mentioned above, possess peculiar physiological and structural features related to stress tolerance, such as their oligotrophic metabolism, which allows them to survive in substrates with a limited amount and diversity of nutrients but with a wide fluctuation of available water. This would cause the fungi to assimilate a wide spectrum of nitrogen and carbon sources, including recalcitrant carbon sources and even aliphatic and aromatic hydrocarbons [24,28–33]. In addition, RIF have thick, highly-melanized cell walls, a feature that protects the fungal cell against the damaging effects of a wide range of electromagnetic radiation (i.e., non-ionizing UV, ionizing X-rays, γ- and ß-radiation), extreme temperatures, and osmotic shock, and allows for mechanical penetration into hard inorganic materials and tissues of host plants or animals, including humans [4,24,34–36].

RIF can be divided into groups according to their morphological traits. De Hoog and Hermanides-Nijhof [37] introduced the term "**meristematic fungi**" to describe slow-growing fungi that form cauliflower-like colonies and aggregates of thick-walled melanized cells. They are reproduced by isodiametric enlargement and subdividing cells, releasing propagules by disarticulation or by endogenous conidiogenesis. Sometimes meristematic fungi produce few blastic conidia and/or budding cells. The authors also introduced the term "**black yeasts**" to describe a group of fungi that have in common melanized cell walls and the production of daughter yeast-like cells by monopolar or multilateral budding, which may be embedded in an extracellular polymeric molecule's matrix. Most of these fungi can also produce hyphae and conidia from phialides or anellides [26,37]. In 1982, the term "**microcolonial fungi**" (MCF) was introduced by Staley et al. [38] to refer to the appearance and in situ growth of fungal colonies that are cauliflower-like, fairly small (up to 1 mm diameter), spherical, smooth, brown or black, that can be found growing on rocky substrates, glass, or metal surfaces, and whose micromorphology is characterized by meristematic or yeast-like growth with densely aggregated thick-walled cells [4,12,26,38,39]. Lastly, "**dematiaceous mycelial fungi**" are those that are stress-tolerant, have dark-colored cell walls, and expansively growing melanized colonies that do not display the morphological characteristics of the previous groups [3,4,26]. Even though the aforementioned groups have their own morphological characteristics, it is often difficult to classify each fungus within only one group, because they have common traits. For instance, some meristematic fungi can also be classified morphologically as black yeasts and vice versa, and some microcolonial fungi can be considered meristematic at the same time [4,26]. This is due to the great growth flexibility of RIF, being able to shift from one growth mode to another according to the needs and stress conditions of the environment which they inhabit [11,26,40]. Several studies have shown that meristematic growth facilitates survival at extreme temperatures and desiccation, and can lead to the formation of microcolonies that reduce energy consumption [3,4,11,26,41]. Instead, black yeasts contribute to the formation of biofilms through their extracellular polymeric substance matrices, which protect them from osmotic stress and make them thermotolerant and highly resistant to antifungal agents. The meristematic growth of RIF is the most common form of chromoblastomycosis in animals and humans (producing muriform propagules in infected tissue), while the production of conidia by budding in the so-called black yeasts is frequently associated with superficial infections, such as *tinea nigra*, and systemic mycoses [24,26,27,37,42].

Moreover, the grouping of RIF under different names according to morphological characteristics is artificial from a taxonomic and evolutionary point of view, as previous phylogenetic studies have shown that they belong to evolutionarily distant orders, namely Capnodiales, Chaetotyriales, Dothideales, and Eurotiales [4,43–48]. Recently, a culture-independent molecular technique, i.e., high-throughput amplicon sequencing, has helped to expand our knowledge of the taxonomy of RIF and their role in microbial communities evolving on rocky substrates. However, their application still has some drawbacks, such as contamination with exogenous DNA [49] and limited identification due to the use of short amplicons [50–52]. Longer amplicon sequencing, as well as shotgun sequencing, will certainly be two important tools for the study of RIF communities at the species level [52,53].

The main objective of our study was to evaluate, through the use of culture-dependent techniques and a taxonomic polyphasic approach, the biodiversity of the RIF that alter (darken) the facades of buildings in different localities of the Province of Tarragona (Spain), and to determine whether some of them present typical characteristics compatible with extremophilic fungi. In addition, the taxonomic identity of the RIF found can alert us about the potential health risk for the inhabitants of these municipalities if they inhale their spores, as they are taxa that frequently trigger hypersensitivity reactions or cause opportunistic mycoses, especially in immunocompromised individuals.

## 2. Materials and Methods

### 2.1. Sampling Sites

The sampling sites were located in four towns in the Tarragona Province (Catalonia community, Spain): Calafell, Els Pallaresos, Montbrió del Camp, and Reus, with a Csa (Mediterranean hot summer) subtype climate described in the Köppen-Geiger classification [54] and settled on calcareous soils. All locations are surrounded by different sorts of Mediterranean crops (carob trees, hazelnuts, cereals, olive trees, vines, etc.), pine forests, and bushes (https://atlasnacional.ign.es/wane/Suelos (accessed on 12 June 2023)). The town of Reus also has an ornamental flora that comprises a large number of exotic plants, a legacy of nineteenth-century culture. The main meteorological data of the sampled sites are given in Table 1.

**Table 1.** Main climatological data for the sampling sites [1].

| Parameters | Calafell | Els Pallaresos | Montbrió del Camp | Reus |
|---|---|---|---|---|
| Lineal distance to the shore (km) | 0 | 5.7 | 7.4 | 9.0 |
| Average annual temperature (°C) | 15.9 | 16 | 15.6 | 15.0 |
| Minimum average annual temperature (°C) | 8.4 | 8 | 7.7 | 6.3 |
| Maximum average annual temperature (°C) | 24.2 | 24 | 24.3 | 24 |
| Average annual rainfall (mm) | 602 | 513 | 550 | 525 |
| Relative humidity: lower value (%) | 64.87 | 67.8 | 61.57 | 57.27 |
| Relative humidity: highest value (%) | 74.60 | 72.6 | 75.41 | 75.69 |
| Prevailing wind direction | S/W | S/W | S/W | S/W |

[1] data from https://es.climate-data.org/ (accessed on 12 June 2023) and https://es.weatherspark.com/ (accessed on 12 June 2023) (S = South; W = Western).

### 2.2. Sampling Design

In November 2020, the samples were taken from the rocky exteriors of urban buildings in four towns in the Province of Tarragona. Five samples were taken at five points at each location, except for Els Pallaresos, where ten samples were taken (five in the urban area and five in the industrial one). The sort of samples collected, and their geographical location, are given in Table 2.

The samples were collected with sterile swabs soaked in Ringer's solution (an isotonic solution that contains NaCl (0.86%), KCl (0.03%), $CaCl_2$ (0.01%) and $NaHCO_3$ (0.02%) [55], by rubbing an area of 20 $cm^2$ (approx.) and then placing the swabs into sterile transport tubes containing 1 mL of Ringer's solution. In the laboratory, the samples were stored at 4 °C until processed.

### 2.3. Fungal Isolation

Samples were vortexed three times at 1200 RPM for two minutes (leaving five minutes between each operation) and inoculated directly with the swab moistened with the sample, using the technique of loop depletion by duplicate, since the samples were incubated at two different temperatures, 15 °C and 25 °C, in the dark on different culture media.

Subsequently, the culture media were placed in 90 mm diameter sterile disposable Petri dishes: potato dextrose agar (PDA; Pronadisa, Madrid, Spain; [56]); potato carrot agar (PCA; 20 g/L potato, 20 g/L carrot, 20 g/L bacteriological agar, 0.2 g/L chloramphenicol; [57]); tap water agar (TWA; 15 g/L bacteriological agar, 0.2 g/L chloramphenicol; [58]); and dichloran-Rose Bengal-chloramphenicol agar (DRBC; Condalab, Madrid, Spain; [59,60]). Fungal colony growth was followed weekly under a stereoscope for up to three months. To obtain the fungal strains, vegetative and reproductive structures were collected using sterile needles (tuberculin/insulin type) and transferred onto PDA and onto oatmeal agar (OA; 30 g/L filtered oatmeal flakes after boiling for one hour, 20 g/L bacteriologic agar, 0.2 g/L chloramphenicol; [61]) in 50 mm diameter sterile disposable Petri dishes, which were incubated at the same temperatures as described earlier.

**Table 2.** Sampling points, geographical location, and sort of sampled sources.

| Town | Sample Name | Coordinates (UTM) * | Source |
|---|---|---|---|
| Calafell | C1 | 31T 79,177.10 61,802.90 | Darkened concrete fence of a garden house |
| | C2 | 31T 79,134.60 61,766.60 | Darkened concrete wall |
| | C3 | 31T 79,796.60 61,222.00 | Darkened concrete wall |
| | C4 | 31T 78,841.90 61,117.50 | Blackened block wall |
| | C5 | 31T 78,584.00 61,023.10 | Darkened concrete wall |
| Montbrió del Camp | M1 | 31T 32,720.20 54,291.00 | Darkened concrete wall |
| | M2 | 31T 32,658.70 54,252.30 | Darkened brick wall |
| | M3 | 31T 32,414.20 54,091.30 | Darkened concrete wall |
| | M4 | 31T 32,413.80 54,090.90 | Darkened concrete wall |
| | M5 | 31T 32,378.60 54,109.80 | Darkened clay wall |
| Els Pallaresos (Urban area) | P1 | 31T 54,937.40 59,758.10 | Darkened concrete wall |
| | P2 | 31T 54,954.60 59,694.90 | Darkened brick wall |
| | P3 | 31T 54,952.50 59,638.90 | Darkened concrete wall |
| | P4 | 31T 55,523.50 59,652.30 | Darkened concrete fence of a home garden |
| | P5 | 31T 55,545.50 59,571.40 | Darkened metal railing of a small natural park |
| Els Pallaresos (Industrial area) | S1 | 31T 54,978.50 59,782.70 | Darkened concrete wall |
| | S2 | 31T 54,945.50 59,754.60 | Darkened metal fence near a tree |
| | S3 | 31T 54,916.20 59,764.70 | Blackened cement blocks wall |
| | S4 | 31T 54,899.30 59,817.50 | Darkened PVC pipe for pluvial drain |
| | S5 | 31T 54,946.80 59,861.60 | Darkened concrete wall |
| Reus | R1 | 31T 49,684.80 57,714.70 | Darkened concrete wall |
| | R2 | 31T 49,916.90 57,650.50 | Darkened concrete wall |
| | R3 | 31T 42,102.90 57,713.20 | Darkened concrete wall |
| | R4 | 31T 46,377.40 57,475.40 | Darkened brick wall |
| | R5 | 31T 41,270.60 58,092.00 | Blackened block wall |

* UTM = Universal Transverse Mercator.

### 2.4. Phenotypic Characterization

Vegetative and reproductive fungal structures were collected from mature colonies of the fungal strains using sterile hypodermic needles and syringes (tuberculin/insulin type) and deposited onto a drop of 65% lactic acid (as the mounting medium), between slide and cover slide. Then, these structures were observed and measured using an Olympus BH-2 bright field microscope with an eye scale (Olympus Corporation, Tokyo, Japan). For those strains of special interest, the morphology was studied in detail using the slide culture technique according to Egidi et al. [48], which consists of inoculating blocks of 1 $cm^2$ of OA and PCA into wet chambers and incubating at 25 °C for up to 12 weeks. Images of the fungal structures were acquired with a DeltaPix Infinity X coupled to the Zeiss Axio Imager M1 microscope (Oberkochen, Germany), using Nomarski (interference contrast) and phase contrast condensers. These images were edited via Adobe Photoshop CS6 v. 13.0 (Adobe Systems, San Jose, CA, USA).

The culture characteristics were documented, growing the strains on malt extract agar (MEA; 30 g/L malt extract, 5 g/L peptone, 15 g/L bacteriologic agar; [61,62]) or MEA 2% (20 g/L malt extract, 15 g/L bacteriologic agar; [63]), OA, PCA, and PDA in 90 mm diameter disposable Petri dishes and incubating at 25 °C for two to four weeks. The colors were described according to Kornerup and Wanscher (1978) [64]. In addition, when necessary, culturing was carried out by inoculating the fungal strain onto sterile plant material (such as oak and palm leaves, filter paper, and pine needles) and into TWA according to Smith et al. [65]. Optimal, minimum, and maximum growth temperatures were determined by growing the strains on PDA at 5, 12, 15, 20, 25, 30, 35, 37, 40, and 45 °C for two weeks.

### 2.5. DNA Extraction, Amplification, and Sequencing

Fungal strains were grown on PDA for one or two weeks at 25 °C. After that, the mycelium was removed by scraping using a sterile scalpel to extract the DNA according to the FastDNA kit protocol (Bio; Vista, CA, USA) plus 50 mg of 425–600 µm size-fractionated glass beads, acid-washed [Sigma] with a FastPrep-24™ instrument (Thermo Savant, Holbrook, NY, USA). The DNA was quantified using a NanoDrop 2000 instrument (Thermo-Scientific, Madrid, Spain). We amplified a fragment of the 28S nrRNA gene (LSU), of the second largest subunit of the RNA polymerase II (*rpb*2), of the β-tubulin (*tub*2), and of the translation elongation factor-1α (*tef*1), and the entire rDNA internal transcribed spacer region (ITS). The primers used are listed in Table 3.

**Table 3.** List of primers and annealing temperatures used for amplification of gene targets.

| Locus | Primer | Sequence (5′→3′) * | Orientation | Annealing Temperature (°C) | Reference |
|---|---|---|---|---|---|
| ITS/LSU | ITS5 | GGAAGTAAAAGTCGTAACAAGG | Forward | 53–55 | [66] |
| | LR5 | ATCCTGAGGGAAACTTC | Reverse | 53–55 | [67] |
| LSU | NL1 | GCATATCAATAAGCGGAGGAAAAG | Forward | 53–55 | [68] |
| | NL4b | GGTCCGTGTTTCAAGACGG | Reverse | 53–55 | [68] |
| *rpb2* | fRpb2-5F | GGGGWGGAYCAGAAGAAG | Forward | 55–60 | [69] |
| | fRpb2-7R | CCCATRGCTTGYTTRCCCAT | Reverse | 55–60 | [69] |
| *tub2* | T10 | ACGATAGGTTCACCTCCAGAC | Forward | 55–57 | [70] |
| | Bt2a | GGTAACCAAATCGGTGCTGCTTTC | Forward | 55–57 | [71] |
| | Bt2b | ACCCTCAGTGTAGTGACCCTTGGC | Reverse | 55–57 | [71] |
| *tef1* | EF-1H | ATGGGTAAGGARGACAAGAC | Forward | 57 | [72] |
| | EF-2T | GGAAGTACCAGTGATCATGTT | Reverse | 57 | [72] |
| | EF1-983F | GCYCCYGGHCAYCGTGAYTTYAT | Forward | 57 | [73] |
| | EF1-2218R | ATGACACCRACRGCRACRGTYTG | Reverse | 57 | [73] |
| | EF1-728F | CATCGAGAAGTTCGAGAAGG | Forward | 57 | [74] |
| | EF1-986R | TACTTGAAGGAACCCTTACC | Reverse | 57 | [74] |

* H = A, C or T; R = A or G; W = A or T; Y = C or T.

The PCR reactions were performed using EmeraldAmp® GT PCR Master Mix (Takara Bio Inc., Saint-Germain-en-Laye, France) according to the manufacturer's manual for 25 µL of reaction [75]. In each 25 µL reaction tube, 5 pmol of each primer and 50 ng of template DNA were added. The amplification was carried out using a MyCycler ™Thermal Cycler (Bio-Rad, Feldkirchen, Germany) under the following conditions: initial denaturation temperature of 94 °C for 5 min, 35 cycles of denaturation temperature of 95 °C for 30 s, annealing of the primer at the temperature stipulated in Table 3 for 45 s, primer extension at 72 °C for 120 s, and a final extension step at 72 °C for 7 min. The amplicons were sequenced in both directions with the same primer pair used for amplification at Macrogen Spain (Macrogen Inc., Madrid, Spain). The consensus sequences were obtained using the SeqMan software version 7.0.0 (DNAStar Lasergene, Madison, WI, USA), and then deposited at the European Nucleotide Archive (ENA) (Table S1 ).

### 2.6. Fungal Identification and Phylogenetic Analyses

The nucleotide sequence of each *locus* generated in this study was subjected to a comparison with that at the National Center for Biotechnology Information (NCBI) database using the Basic Local Alignment Search Tool [76] (BLAST; https://blast.ncbi.nlm.nih.gov/Blast.cgi (accessed on 12 December 2023)). Fungal strains were identified at the species level when the ITS sequences displayed a level of identity ≥ 98% with those of ex-type and/or reference strains in the database.

Phylogenetic analyses were performed using the phylogenetic markers LSU (for general analysis), *rpb2* (specifically for the genus *Neocatenulostroma*), and ITS (for the rest of the genera). The nucleotide sequences of the markers of the most phylogenetically related taxa were retrieved from GenBank (https://www.ncbi.nlm.nih.gov (accessed on 14 May 2023); see Table S2). The nucleotide sequences were aligned separately using the ClustalW algorithm [77] in the MEGA software v. 7.0 [78] and manually adjusted using the same software. Phylogenetic reconstructions were made for each phylogenetic marker by maximum likelihood (ML) and Bayesian inference (BI) with RAxML [79] in CIPRES web (https://www.phylo.org/ (accessed on 14 May 2023)) [80] and MrBayes 3.2.6 [81], respectively. The best substitution model for each gene matrix was estimated using MrModelTest v. 2.3.25 [82]. For ML analyses, the nearest-neighbour interchange was used as the heuristic method for tree inference. Support for internal branches was assessed with 1000 ML bootstrapped pseudo-replicates. A bootstrap support (BS) of ≥70 was considered significant. For BI analyses, Markov chain Monte Carlo (MCMC) [83] sampling was carried out with four million generations, with samples taken every 1000 generations. The 50% majority rule consensus trees and posterior probability values (PP) were calculated after removing the first 25% of the resulting trees for burn-in. A PP value of ≥0.95 was considered significant.

### 2.7. Physiological Characterization of the Strains of Interest

Eleven strains, mostly producers of yeast-like cells, were grown in MEA or PDA and incubated at 25 °C for 47 days, depending on their growth rates. Then, inocula were prepared by aseptically adding small colony portions by a loop in sterile water into 16 mm diameter glass tubes with a screw cap until a final concentration of $5 \times 10^5$ CFU/mL was reached, or an Optical Density (OD) of $OD_{520}$ 0.20–0.25, filtering through sterile gauze to remove hyphae if necessary. The evaluation of the nutritional abilities and tolerance tests were based on the methods described by van der Walt and Yarrow [84] and adapted for black yeasts by de Hoog et al. [85] and Wollenzien et al. [86]. The pH tolerance was checked according to Uribe [87]. The assimilation of carbon sources was carried out according to Schwarz et al. [88] and Alvarez et al. [89], using the commercial kit API 50 CH (bioMérieux, Marcy, l'Etoile, France), the inoculum was prepared in a Yeast Nitrogen Base (YNB; 6.7 g/L [BD Difco, Madrid, Spain], 0.5 g/L L-chloramphenicol, 0.05% *w/v* bacteriologic agar), and the results were read every week. All assimilation assays (×triplicate) were incubated for up to 2 or 3 weeks at 25 °C and were either stationary or horizontally shaken, as indicated by the procedure.

## 3. Results

### 3.1. Phenotypic and Molecular Identification of the Fungal Strains

Using a polyphasic approach, we identified a total of 224 fungal strains (32 from Calafell, 63 [urban area] and 58 [industrial area] from Els Pallaresos, 14 from Montbrió del Camp, and 57 from Reus) belonging to 41 genera and 64 species. However, not all the strains were molecularly identified (Table S1).

The strains belonged to the orders previously reported as RIF with Pleosporales being the most abundant and recovered from all sampled sites. In addition, we recovered representatives of the orders Botryosphaeriales, Capnodiales, Cladosporiales, Chaetotyriales, Coniochaetales, Dothideales, Hypocreales, Lichenostigmatales, Mucorales, Sordariales, and Xylariales [1].

Figure 1 shows the relative abundance (based on the number of strains) of the genera recovered from the samples. *Alternaria* (45) was the most abundant, followed by *Cladosporium* 35), *Penicillium* (18), *Fusarium* (15), *Xenodidymella* (12), *Aureobasidium* (10), and *Epicoccum* (10). Several genera had only one strain, i.e., *Angustimassarina*, *Apiospora*, *Beauveria*, *Coniochaeta*, *Cosmospora*, *Dothiorella*, *Juxtiphoma*, *Neodidymelliopsis*, *Paraconiothyrium*, *Paraphoma*, and *Pseudoseptoria*. Additionally, we isolated several previously unknown species belonging to the genera *Exophiala*, *Neocatenulostroma*, *Neodevriesia*, *Paradevriesia*, and *Phaeococcomyces* ([90]) (Figure 1, written in red), as well as the new genus *Coccodomyces*; all of them were isolated from Els Pallaresos and are suggested here as new taxa.

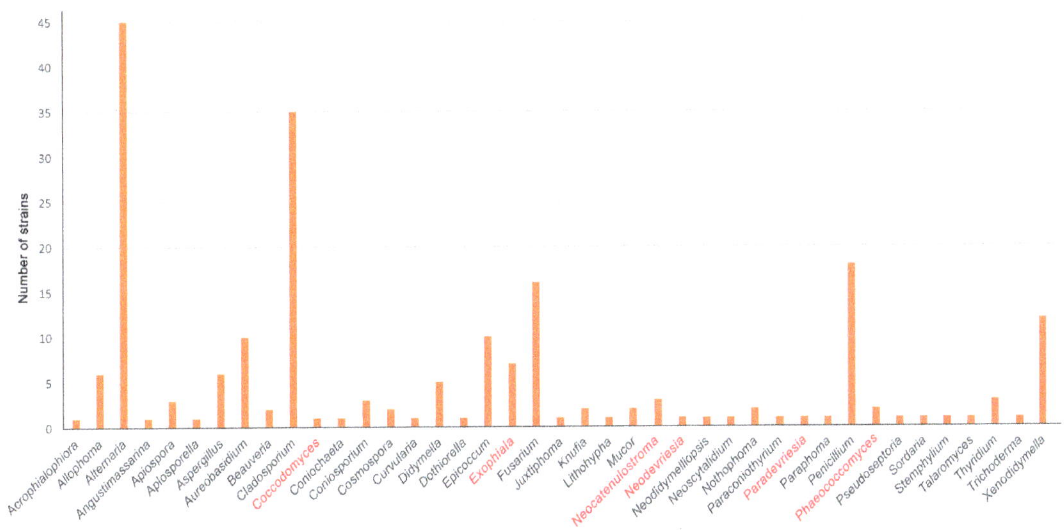


**Figure 1.** Relative abundance of the genera found in the samples analyzed (red color indicates the genera where new species are reported).

The relative abundance of fungal genera was also compared among the sampled sites, and the results are shown as a heatmap (Figure 2). Thus, Montbrió del Camp and Calafell had the least fungal diversity, while Els Pallaresos had the highest biodiversity, followed by Reus. The most abundant genera, *Alternaria*, *Cladosporium*, and *Penicillium*, were isolated from all locations, though many of the genera were isolated from only one location (Figure 2). Regarding the preference for growing on certain substrates, it was found that *Acrophialophora*, *Angustimassarina*, *Aplosporella*, *Beauveria*, *Cosmospora*, *Curvularia*, *Dothiorella*, *Juxtiphoma*, *Lythohypha*, *Mucor*, *Neodidymelliopsis*, *Neoscytalidium*, *Nothophoma*, *Paradevriesia*, *Paraphoma*, *Pseudoseptoria*, *Sordaria*, *Stemphylium*, *Talaromyces*, and *Thyridium* only grew on concrete, while *Coccodomyces*, *Neodevriesia*, *Paraconiothyrium*, and *Phaeococcomyces* were found on metal structures, and *Necatenulostroma* and *Trichoderma* on a PVC pipe.

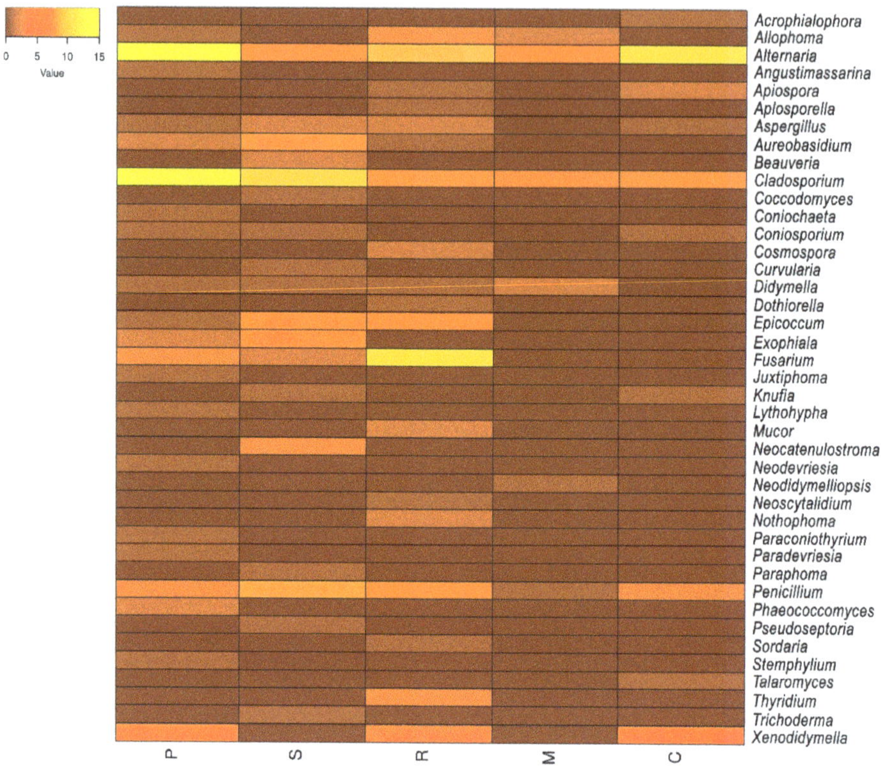


**Figure 2.** Heatmap showing the relative abundance and distribution of fungal genera among studied sites. (P, Els Pallaresos, urban area; S, Els Pallaresos, industrial area; R, Reus; M, Montbrió del Camp; C, Calafell).

### 3.2. Phylogeny

The accession numbers of the nucleotide sequences obtained in this study are listed in Table S1, while those retrieved from NCBI databases to build the phylogenetic trees are listed in Table S2.

It is important to mention that the *rpb*2 *locus* was not always correctly amplified. The general LSU phylogenetic tree (Figure 3), based on 899 positions including gaps, confirmed that our strains FMR 18793, FMR 18795, and FMR 18825 belonged to the order Capnodiales, while FMR 18977 and FMR 18809 were located within the order Chaetotyriales, and FMR 18827 in the order Pleosporales, all of these being considered as potential new taxa.

#### 3.2.1. Order Dothideales

For the strain FMR 18827, the ITS alignment included 12 ingroups and two outgroups (*Dothidea sambuci* CBS198.58 and *Stylodothis pucciniodes* CBS 193.58) with a total of 575 characters (including gaps). The BI and ML showed similar topology and congruent results. In the phylogenetic tree (Figure 4) FMR 18827 was located within the order Dothideales, in the same linage (fully supported) as *Rhizosphaera* spp., *Hormonema merioides* CBS 906.85, and *Phaeocryptopus nudus* CBS 268.37, and in the same terminal clade (0.70 PP/71% BS) as *Gonatobotryum apiculatum* CBS 182.68, *Dothiora mahoniae* CBS 264.92, and *Scleroconidioma sphagnicola* UAMH 9731.

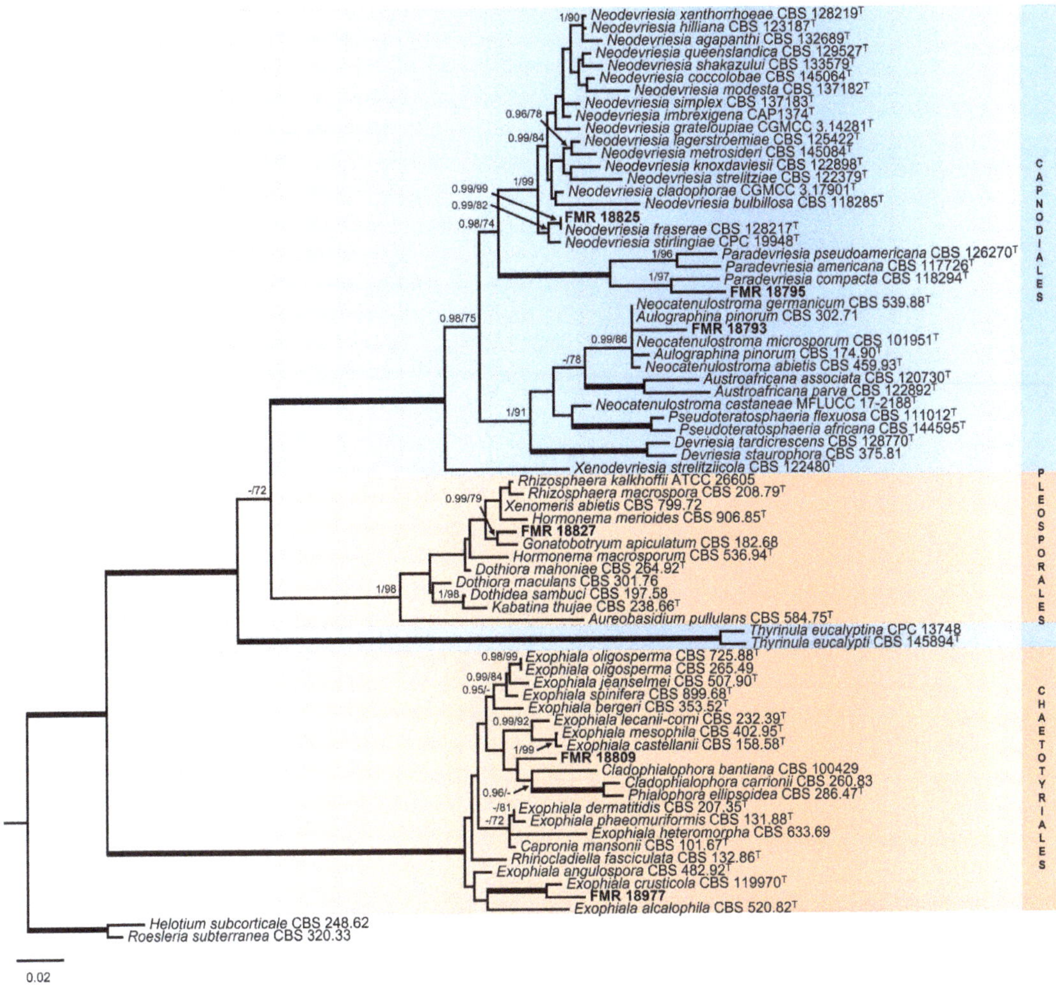


**Figure 3.** Maximum likelihood tree based on LSU alignment (899 pb) of the sequences from our strains and retrieved from the GenBank. Bayesian posterior probabilities (PP) equal to or above 0.95 and the RAxML bootstrap support values (BS) $\geq 70\%$ are presented at the nodes (PP/BS). Thickened branches indicate full support (PP = 1 and BS = 100%). *Roesleria subterranea* (CBS 320.33) and *Helotium subcorticale* (CBS 248.62) were used as outgroups. Strains corresponding to the potential new species are indicated in **bold**. $^{T}$ represents the ex-type strain of the species.

### 3.2.2. Genus *Exophiala*

For *Exophiala* spp., the ITS alignment comprised 41 ingroups and three outgroups (*Cladophialophora bantiana* CBS 101158 and CBS100429, and *Cladophialophora carrioni* CBS 260.83) with 586 characters including gaps. The BI and ML showed similar topology and congruent results. In the phylogenetic tree (Figure 5) the strains FMR 18794, FMR 18810, and FMR 19066 were placed in a well-supported (0.98 PP/92% BS) terminal clade corresponding to *E. xenobiotica*. On the other hand, the strain FMR 18977 was placed in a well-supported (0.99 PP/100% BS) terminal clade together with the type strain of *E. crusticola* (CBS 119970), but as a different species (BLAST Id = 88%), while the strain FMR 18809 was located in other well-supported (1 PP/100% BS) terminal clade together the type strain of *E. asiatica* (CBS 122847), but as a different species (BLAST Id = 88%).

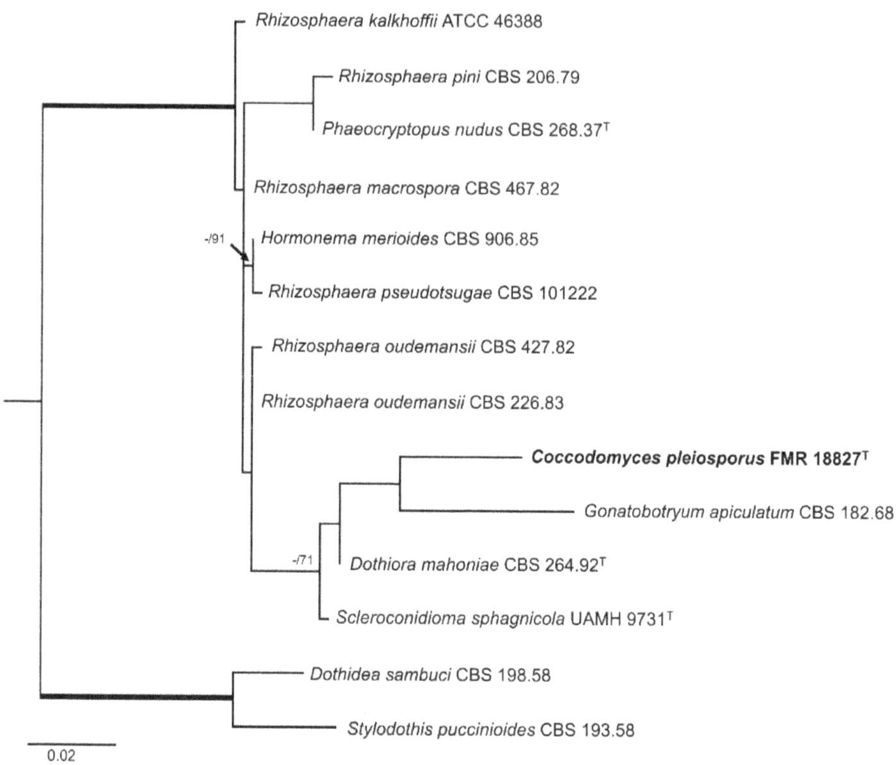


**Figure 4.** Maximum likelihood tree based on ITS alignment (575 bp) of the sequences from our strain FMR 18827 and those retrieved from the GenBank. Bayesian posterior probabilities (PP) equal to or above 0.95 and the RAxML bootstrap support values (BS) $\geq$ 70% are presented at the nodes (PP/BS). Thickened branches indicate full support (PP = 1 and BS = 100%). *Dothidea sambuci* (CBS 198.58) and *Stylodothis pucciniodes* (CBS 193.58) were used as outgroups. The new genus is indicated in **bold**. $^{T}$ represents the ex-type strain of the species.

### 3.2.3. Order Capnodiales

For the genera *Neodevriesia* and *Paradevriesia*, the ITS sequence alignment included 25 ingroups and two outgroups (*Amycosphaerella africana* CBS 116154 and *Brunneosphaerella jonkershoekensis* CBS 130594) with 514 characters including gaps. The BI analysis showed a similar tree topology and congruent results compared to those obtained in ML. The phylogenetic inference (Figure 6) shows that our strain FMR 18825 forms a well-supported (1 PP/99% BS) terminal clade, together with *N. fraserae* (CBS 128217) (BLAST Id = 99%) and *N. stirlingiae* (CBS 133581) (BLAST Id = 98%), within a main fully supported clade corresponding to all species of *Neodevriesia*. Moreover, the BLAST search for the strain FMR 18825 against *N. fraserae* and *N. stirlingiae* indicated a low similarity for *tub2* (92% and 90%, respectively) and for *rpb2* (94% and 93%, respectively). The main clade corresponding to the species of *Paradevriesia* (0.99 PP/85% BS) included our strain FMR 18795 (BLAST Id = 92%), which was placed as a different species in a branch (0.99 PP/89% BS) with the type strain of *P. compacta*.

Regarding the genus *Neocatenulostroma*, the LSU phylogenetic tree (Figure 3) revealed that all species were placed in a well-supported clade (0.99 PP/86% BS), which also included *Aulographina pinorum* (CBS 174.90 and CBS 302.71) and our strain FMR 18793, with the genus *Austroafricana* (*A. parva* and *A. associata*) as a sister clade, and placed *N. castenae* (MFLUCC 17-2188) [91] outside these clades. In order to solve the genera and species boundaries, a *rpb2* phylogenetic tree was built. The *rpb2* alignment comprised eight ingroups with a total of 910 characters, including gaps, and with *Thyrinula eucalyptina* (CPC 13748) and *Thyrinula eucalypti* (CBS 145894) as outgroups. The BI and ML showed similar topology and congruent results. The *rpb2* phylogenetic tree (Figure 7) confirmed what was observed in the LSU phylogeny: On the one hand, FMR 18793 was placed as a new species of the genus, with *N. germanicum* as its phylogenetically closest species and together forming a sister clade with *N. microsporum*. On the other hand, *N. pinorum* was included in the genus *Neocatenulostroma*. Because *N. castaneae* lacks the *rpb2* nucleotide sequence, it could not, unfortunately, be included in such phylogenetic analysis.

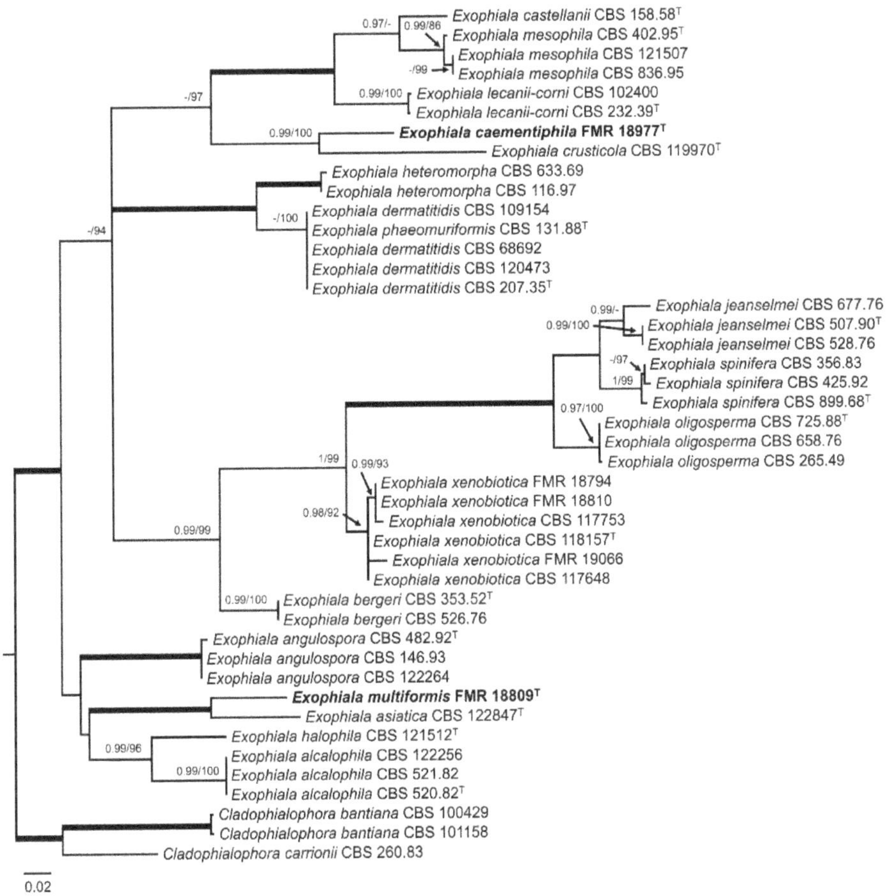


**Figure 5.** Maximum likelihood tree based on ITS alignment (586 bp) of the sequences of our strains (FMR) and those retrieved from the GenBank (CBS). Bayesian posterior probabilities (PP) equal to or above 0.95 and the RAxML bootstrap support values (BS) $\geq$ 70% are presented at the nodes (PP/BS). Thickened branches indicate full support (PP = 1 and BS = 100%). *Cladophialophora bantiana* (CBS 101158 and CBS 100429) and *Cl. carrioni* (CBS 260.83) were used as outgroups. The new species are indicated in **bold**. $^{T}$ represents the ex-type strain of the species.

### 3.3. Taxonomy

*Coccodomyces* Sastoque, Cano and Stchigel, gen. nov. Mycobank MB841929.

*Etymology*: From Latin *coccum-*, grain, granule, berry, seed, and Latin < Greek *-μύκης*, mushroom, fungus.

*Classification*—*Incertae sedis*, Dothideales, Pezizomycotina, Ascomycota.

*Colonies* spreading, flat, crateriform or umbonated, margins filiform due to the submerged mycelium, becoming black and leathery with age. *Mycelium* is composed of septate, hyaline, and thin-walled hyphae when young, becoming dark brown, thick-walled, and torulose with age, frequently ending in long terminal non-septate segments. *Hyphae* hyaline to copper-brown, composed of cylindrical, sub-cylindrical to globose cells, thick-walled when pigmented, increasing the degree of constriction and the number of transversal septa with age, forming also additional longitudinal and oblique septa; additionally, there are globose cells produced by the blowing-out of the pre-existent cells, giving a granulose look to the hyphae. *Conidiophores* absent. *Conidiogenous cells* integrated to the hyphae, mono- to polyblastic, bearing lateral denticles (conidiogenous *loci*). *Conidia* holoblastic, one-celled, produced asynchronously in slimy masses, smooth- and thin- to thick-walled, ellipsoidal to globose, hyaline at first, becoming dark brown with age, developing in new conidiogenous cells producing holoblastic conidia by budding, but also forming two-celled conidia by the development of a medial septum which is always constricted due to the swelling of both cells, which also forms in new conidiogenous cells producing inflated, one- or two-celled dark brown conidia in short chains, which can also develop additional longitudinal and oblique septa. *Sexual morph* unknown.

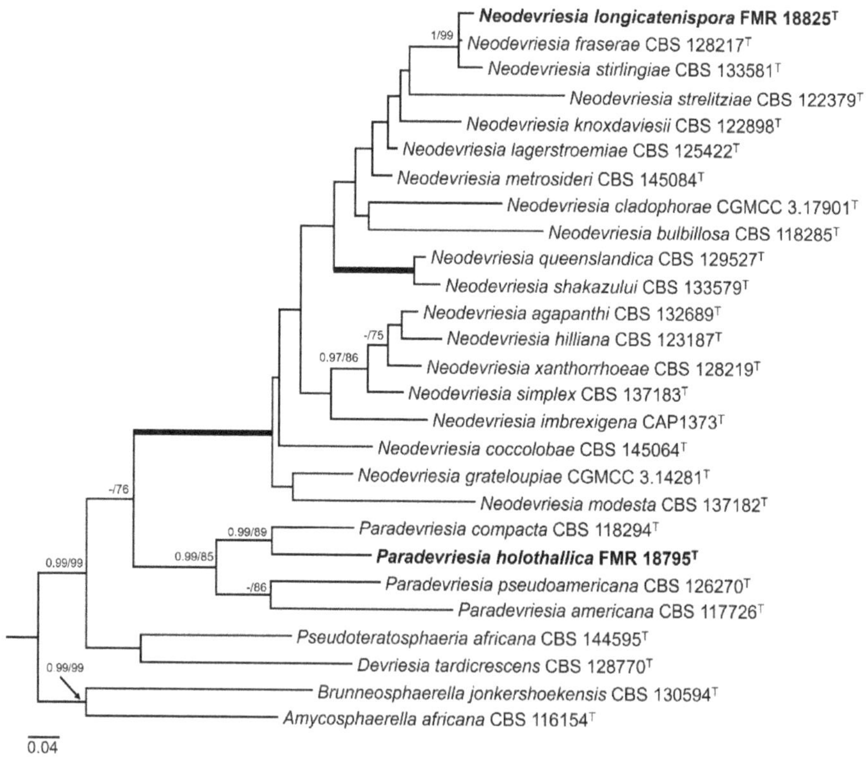


**Figure 6.** Maximum likelihood tree based on ITS alignment (514 pb) of the sequences of our strains (FMR) and those retrieved from the GenBank. Bayesian posterior probabilities (PP) equal to or above 0.95 and the RAxML bootstrap support values (BS) $\geq$ 70% are presented at the nodes (PP/BS). Thickened branches indicate full support (PP = 1 and BS = 100%). *Amycosphaerella africana* (CBS 116154) and *Brunneosphaerella jonkershoekensis* (CBS 130594) were used as outgroups. The new species proposed in this study are indicated in **bold**. $^{T}$ represents the ex-type strain of the species.

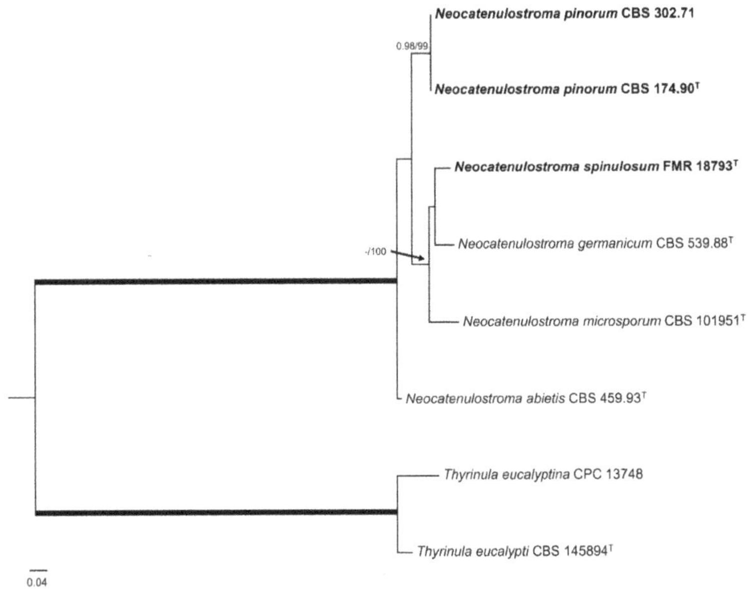


**Figure 7.** Maximum likelihood tree based on *rpb2* alignment (910 pb) of the sequences of our strain FMR 18793 and those retrieved from the GenBank. Bayesian posterior probabilities (PP) equal to or above 0.95 and the RAxML bootstrap support values (BS) ≥ 70% are presented at the nodes (PP/BS). Thickened branches indicate full support (PP = 1 and BS = 100%). *Thyrinula eucalyptina* (CPC 13748) and *Thyrinula eucalypti* (CBS 145894) were used as outgroups. The new species/combinations are indicated in **bold**. T represents the ex-type strain of the species.

*Coccodomyces pleiosporus* Sastoque, Cano and Stchigel, sp. nov. MycoBank MB841950. Figure 8.

*Etymology*. From Greek *πλειο-*, more, *-σπόρος*, spore, due to the production of conidia which are variable in shape and number of cells.

*Description*—On potato dextrose agar after two weeks at 25 °C—*Mycelium* composed of septate, hyaline, and thin-walled branching hyphae when young, becoming dark brown, thick-walled, and torulose with age, 4–8 μm wide (Figure 8E–G), ending in terminal aseptate segments of up to 300 μm long (Figure 8F). *Hyphae* hyaline to copper-brown, guttulate when young, thick-walled when pigmented, composed of cylindrical, subcylindrical to globose cells, 5–20 × 3–7 μm, increasing in the degree of constriction at and the number of transversal septa with age, and then forming additional longitudinal and oblique septa; additional globose cells produced by the blowing-out of the pre-existent cell, giving a granulose look to the hyphae (Figure 8G). *Conidiophores* absent. *Conidiogenous cells* integrated to hyphae, smooth or nearly so, thin-walled, prismatic, barrel-shaped, or globose, 5–20 × 3–10 μm, occasionally arising as globose lateral cells, mono- to polyblastic, bearing lateral denticles of 1–3 × 0.5–2 μm (Figure 8H,I). *Conidia* holoblastic, one-celled, hyaline at first, becoming dark brown with age, produced asynchronously in slimy masses, smooth- and thin- to thick-walled, ellipsoidal to globose, 6–13 × 2–7 μm, covered with a pigmented mucous layer with age and attaching the conidial mass to the hyphae; conidia also developing in new conidiogenous cells producing holoblastic conidia by budding but also forming two-celled conidia by the development of a medial septum which is always constricted due to the swelling of both cells, which also develop in new conidiogenous cells producing inflated, one- or two-celled dark brown conidia, 11–22 × 6–10 μm, disposed in short chains, which can also develop additional longitudinal and oblique septa (Figure 8H–L). *Chlamydospores*, *endoconidia*, and *sexual morph* not seen.

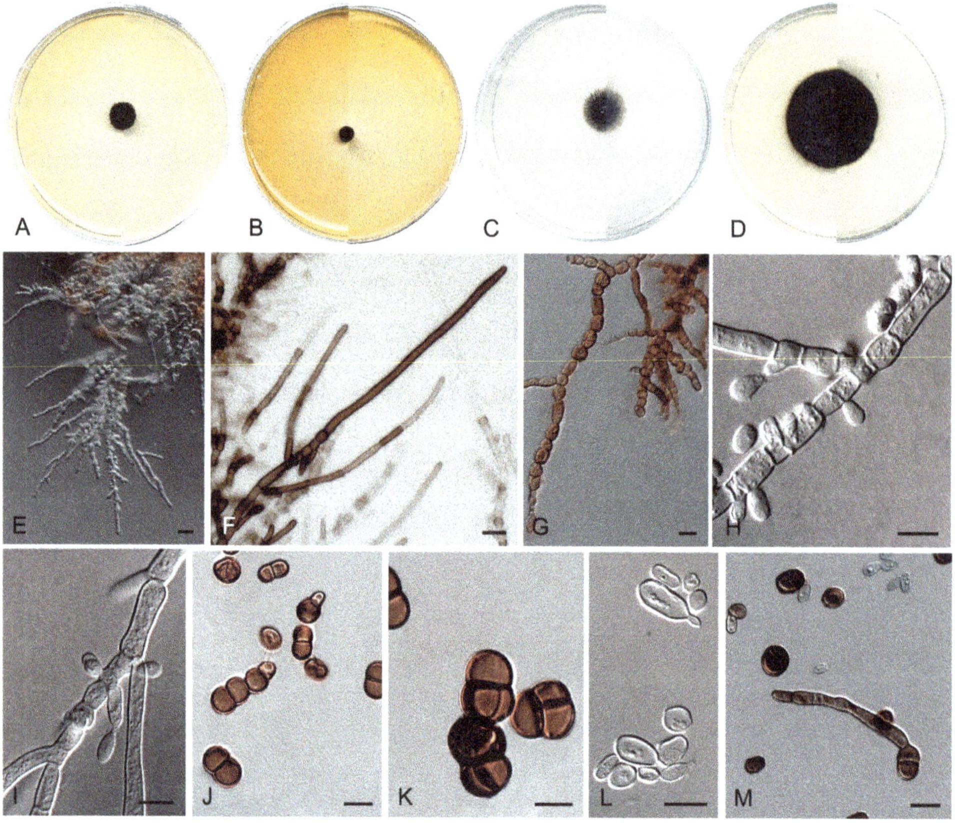


**Figure 8.** *Coccodomyces pleiosporus* FMR 18827. (**A–D**) Colony on MEA, OA, PCA, PDA (after 2 wks at 25 ± 1 °C; surface, left; reverse, right). (**E–G**) Isodiametric and torulose hyphae, hyaline to copper-brown colored, with terminal aseptate segments. (**H**,**I**) Conidiogenous cells and holoblastic conidia. (**J–L**) Formation of new conidia by septation of pre-existent conidia (**J**,**K**) or by budding cells (**L**,**M**) Conidial germination. DIC Nomarski. Scale bars = 10 µm.

*Culture characteristics*—(After 14 days at 25 °C, Figure 8A–D) Colonies reaching 11 mm diameter on MEA, 6 mm on OA, 24 mm on PCA, and 37 mm on PDA, spreading, flat, glistening, dry, and slightly umbonated or crateriform, with filiform margins due to the presence of submerged hyphae mostly on PCA, without aerial mycelium, becoming coriaceous with age, surface black (6F3) (according to Kornerup and Wanscher [64]) on all culture media tested, with pale yellow (2A3) margins (surface and reverse) on MEA and PDA; reverse, grey (29F1) and black (6F3) on MEA, black (6F3) on OA and PCA, and nickel-green (27F3) and black (6F3) on PDA. Minimum, optimum, and maximum temperature of growth: 5 °C, 25 °C, and 30 °C, respectively.

*Type*—SPAIN, Tarragona Province, Els Pallaresos, isolated from a blackened metal fence of an industrial warehouse N 41°10′34.2″ E 1°16′14.4″, 20 November 2020, J. F. Cano-Lira and A. M. Stchigel, isol. A. P. Sastoque (holotype CBS H-24941, ex-type FMR 18827 = CBS 149014; ITS and LSU sequences GenBank OW273979 and OW370575, respectively).

*Diagnosis*—Morphologically, *Coccodomyces pleiosporus* differs from *Gonatobotryum apiculatum*, the closest species, in its diameter and the shape of the colony on PDA, being filiform and coriaceous. *C. pleiosporus* has torulose and hyaline to copper-brown hyphae with terminal aseptate segments up to 300 µm long, an absence of conidiophores, conidiogenous cells integrated into the hyphae, conidia holoblastic, but also producing holoblastic conidia by budding or forming two-celled conidia by the development of a medial septum that is always constricted due to the swelling of both cells. By contrast, *G. apiculatum* presents conidiogenous ampullae with echinulate cicatrized scars, its conidiophores are erect or flexuous, unbranched, and nodose, and its conidia are catenate and aseptate [92]. The sexual morph of *C. pleiosporus* was not observed.

*Notes*—Based on a mega BLAST search of NCBIs GenBank nucleotide database, the closest hit using the **ITS** sequence was *Scleroconidioma sphagnicola* (strain UAMH 9731, GenBank NR_121294; identities = 490/507 (96.65%), gaps 2/507 (0%)), and the closest hit using the **LSU** sequence was *Plowrightia abietis* (strain ATCC 24339, GenBank EF114703; Identities = 819/828 (98.91%), gaps 0/828 (0%)).

*Exophiala caementiphila* Sastoque, Stchigel, and Cano, sp. nov. Mycobank MB 849258. Figure 9.

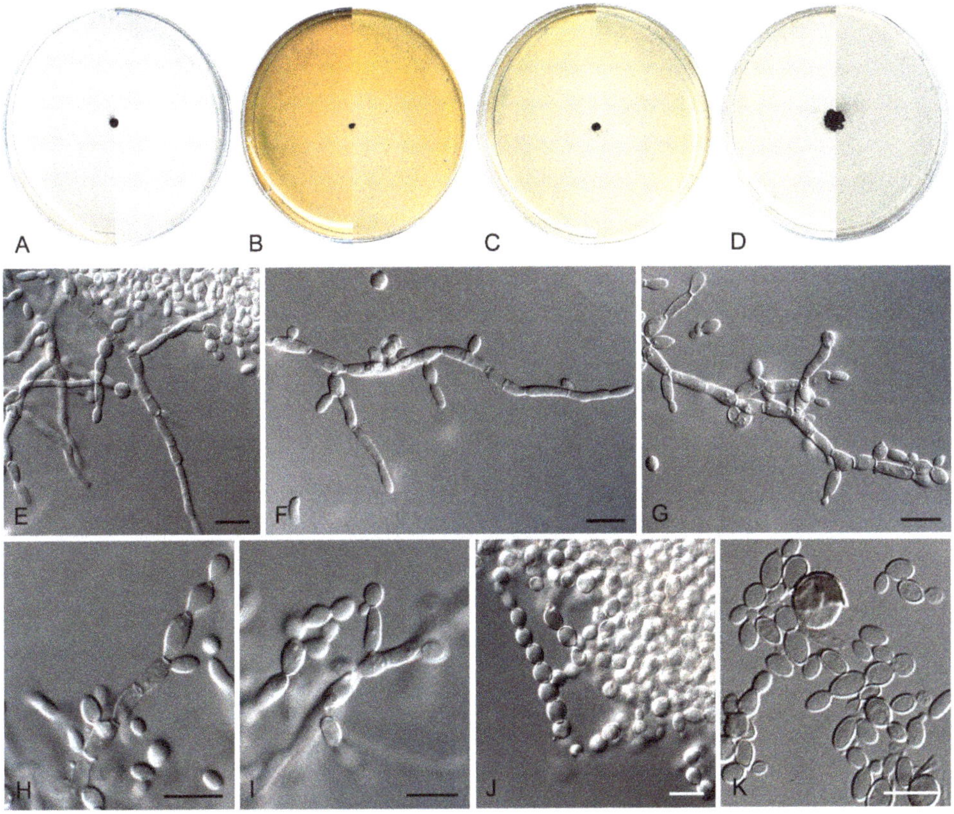

**Figure 9.** *Exophiala caementiphila* FMR 18977. (**A–D**) Colony on MEA, OA, PCA, PDA (after 2 wks at 25 ± 1 °C; surface, left; reverse, right). (**E–G**) Mycelium and anelidic conidiogenous cells (**H,I**) Enteroblastic conidia. (**J**) Conidia chains. (**K**) Yeast-like cells. DIC Nomarski. Scale bars = 10 µm.

*Etymology*. From Latin *caementum-*, cement, and from Latin < Greek -*φιλία*, friendship, because the substrate on the fungus develops.

*Classification*—Herpotrichiellaceae, Chaetothyriales, Chaetothyriomycetidae, Eurotiomycetes, Pezizomycotina, Ascomycota.

*Description*—On potato dextrose agar after two weeks at 25 °C—*Mycelium* scarce, submerged, composed of pale olivaceous-brown to olivaceous-brown, septate, branching, smooth- and thin-walled hyphae, consisting of cylindrical or torulose cells, 2–4 µm wide

(Figure 9E). *Conidiogenous cells* annellidic, mono- to polyblastic, integrated to the hyphae or discrete, then laterally disposed, pale olivaceous-brown to olivaceous-brown, cylindrical, sub-cylindrical, ellipsoidal to broadly ovoid, 4–12 × 2–4 µm (Figure 9F–H). *Conidia* enteroblastic, non-septate, subhyaline to pale olivaceous-brown, smooth- and thin-walled, cylindrical with rounded ends, ellipsoidal, ovoid to nearly globose, 2–8 × 2–4 µm, often forming chains. *Yeast-like cells* very abundant, non-septate, subhyaline to olivaceous brown, smooth- and thin-walled, ellipsoidal to ovoid or nearly globose, 2–7 × 2–5 µm, developing secondary conidia to form long chains (Figure 9I–K). *Chlamydospores* and *sexual morph* not seen.

*Culture characteristics*—(After 14 days at 25 °C, Figure 9A–D) Colonies reaching 3.0 mm diameter on MEA and PCA, 1.5 mm diameter on OA, and 5.0 mm diameter on PDA, circular, margins entire and regular, convex to slightly pulvinate, mucoid at first, soon becoming smooth to verrucose, dry and glistening, cerebriform with age on MEA, OA, and PDA, but mucoid on PCA. Surface and reverse black (6F3) on MEA, OA, and PDA, and black (6F3) with dark brown (6F4) margins on PCA. Minimum, optimum, and maximum temperature of growth: 5 °C, 15–25 °C, and 25 °C, respectively.

*Type*—Spain, Tarragona Province, Els Pallaresos, isolated from a blackened wall of an industrial warehouse N 41°10′34.8″ E 1°16′13.4″, 20 November 2020, coll. J. F. Cano-Lira and A. M. Stchigel, isol. A. P. Sastoque (holotype CBS H-25346, ex-type FMR 18977 = CBS 150902; ITS, LSU, *tef* and *tub2* sequences GenBank OX380503, OX380504, OX380501, and OX380502, respectively).

*Diagnosis*—*Exophiala caementiphila* exhibits the most important features of the genus, such as annelidic conidiogenous cells, the presence of budding cells and torulose hyphae, as well as the formation of chains of conidia (its cladophialophora-like synanamorph). However, *E. caementiphila* differs from *E. crusticola*, the phylogenetically closest species, in having scarce mycelium, integrated or discrete conidiogenous cells, and conidia forming chains [93]. Physiologically, *E. crusticola*, in contrast to *E. caementiphila*, is able to assimilate arginine and ornithine, but not galactose, inulin, maltose, raffinose, and sucrose.

*Notes*—Based on a mega BLAST search of NCBIs GenBank nucleotide database, the closest hit using the **ITS** sequence was *E. crusticola* (strain CBS 119970, GenBank NR_165997; Identities = 493/569 (86.24%), gaps = 41/569 (7%)), and using the **LSU** sequence it was the same strain (GenBank MH874623; Identities = 424/434 (97.70%), gaps = 0/434 (0%)).

*Exophiala multiformis* Sastoque, Cano, and Stchigel, sp. nov. MycoBank MB842313. Figure 10.

*Etymology*. From Latin *multi-*, many, and *-formis*, shapes, because of the diversity of reproductive structures produced.

*Classification*—Herpotrichiellaceae, Chaetothyriales, Chaetothyriomycetidae, Eurotiomycetes, Pezizomycotina, Ascomycota

*Description*—On potato dextrose agar after two weeks at 25 °C—*Mycelium* abundant, composed of olivaceous-brown, smooth- and thin-walled septate hyphae, 1–3 µm wide, surrounded by a partially soluble dark pigment (Figure 10L). *Conidiophores* of three kinds: micronematous, reduced to an olivaceous brown to brown annellidic conidiogenous cell integrated to the vegetative hyphae or discrete, if discrete sub-cylindrical, cylindrical to flask-shaped, 4–8 × 2–4 µm; semi-micronematous, arising as short lateral branches from the vegetative hyphae, one-celled, mostly septate at the base, cylindrical but constrained at the septum, 5–10 × 2–4 µm, truncated at both ends, bearing one, or rarely two, annellidic conidiogenous cells which can also proliferate percurrently to form long chains of additional conidiogenous cells; macronematous, erect, straight, unbranched, smooth- and thin-walled, brown but becoming paler towards the apex, continuous to 2-septate, up to 40 µm long, bearing a terminal integrated conidiogenous cell proliferating sympodially, resulting in one or two conspicuous conidiogenous *loci*, cylindrical or nearly so, 6–15 × 3–4 µm, scars not seen (Figure 10F–J). *Conidia* non-septate, subhyaline to pale olivaceous-brown, smooth- and thin-walled, ellipsoidal to ovoid, 2–6 × 1–4 µm. *Budding cells* barely present, morphologically like those conidia from annellidic conidiogenous cells, 3–6 x 2.5–4.5 µm (Figure 10K). *Chlamydospores* and *sexual morph* not seen.

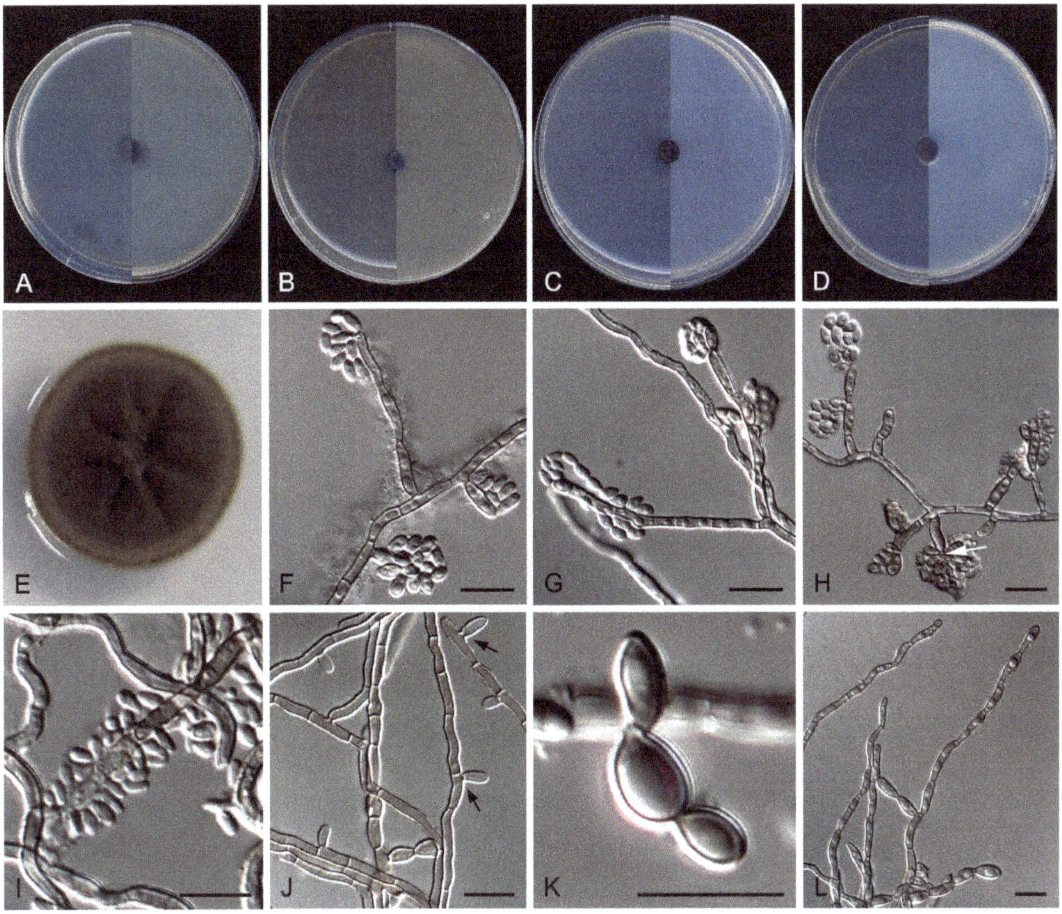


**Figure 10.** *Exophiala multiformis* FMR 18809. (**A**–**D**) Colony on MEA, OA, PCA, PDA (after 2 wks at 25 ± 1 °C; surface, left; reverse, right). (**E**) Colony appearance on PDA at 25 ± 1 °C. (**F**–**H**) Macro, semi-micro, and micronemotous conidiphores and annelidic conidiogenous cells (white arrow). (**I**) Semi-micronematous conidiophore with sympodial proliferation. (**J**) Conidiogenous cells integrated into the hyphae (black arrows). (**K**) Yeast-like cells. (**L**) Cylindrical and torulose hyphae. DIC Nomarski. Scale bars = 10 µm.

*Culture characteristics*—(After 14 d at 25 °C, Figure 10A–E) Colonies reaching 8.5 mm diameter on MEA, 7 mm on OA, 8 mm on PCA, and 9 mm on PDA, restricted, circular, margins entire, flat to slightly convex or elevated at the center, velvety on MEA and PDA, but brightening (due to the production of yeast-like cells) at the center on OA and PCA. Surface color on all culture media tested dull green (29E4) with light green margins (30E6), and reverse bronze green (30F3). Minimum, optimum, and maximum temperature of growth: 15 °C, 25 °C, and 30 °C, respectively.

*Type*—Spain, Tarragona Province, Els Pallaresos, isolated from a blackened metal fence of an industrial warehouse N 41°10′34.2″ E 1°16′14.4″, 20 November 2020, coll. J. F. Cano-Lira and A. M. Stchigel, isol. A. P. Sastoque (holotype CBS H24940, ex-type FMR 18809 = CBS 149013; ITS and LSU sequences GenBank OU624180 and OU624179, respectively.

*Diagnosis—Exophiala multiformis* differs from *E. asiatica*, the phylogenetically closest species, by its larger and thinner annellidic conidiogenous cells (4–8 × 2–4 μm in *E. multiformis* vs. 4.5–6.0 × 4–5 μm in *E. asiatica*), the presence of torulose hyphae (nearly absent in *E. asiatica)* and the bigger annelloconidia (2–6 × 1–4 μm in *E. multiformis* vs. 3–4.5 × 1–2 μm in *E. asiatica*), and by the presence of well-developed conidiophores whose apices proliferate sympodially (absents in *E. asiatica*). Moreover, *E. asiatica* can grow at up to 40 °C, while *E. multiformis* only grows at up to 30 °C [94].

*Notes*—Based on a Mega BLAST search of NCBIs GenBank nucleotide database, the closest hit using the **ITS** sequence was *E. asiatica* (strain CBS 122847, GenBank MH863242; Identities = 458/522 (87.74%), 19/522 (3%)), and using the **LSU** sequence it was *Fonsecaea brasiliensis* (strain CBS 127815, GenBank MH877954; Identities = 748/766 (97.70%), 1/766 (0%)).

*Neocatenulostroma* Quaedvlieg and Crous.

*Type species—Neocatenulostroma abietis* (Butin and Pehl) Quaedvl. and Crous. Figure 11.

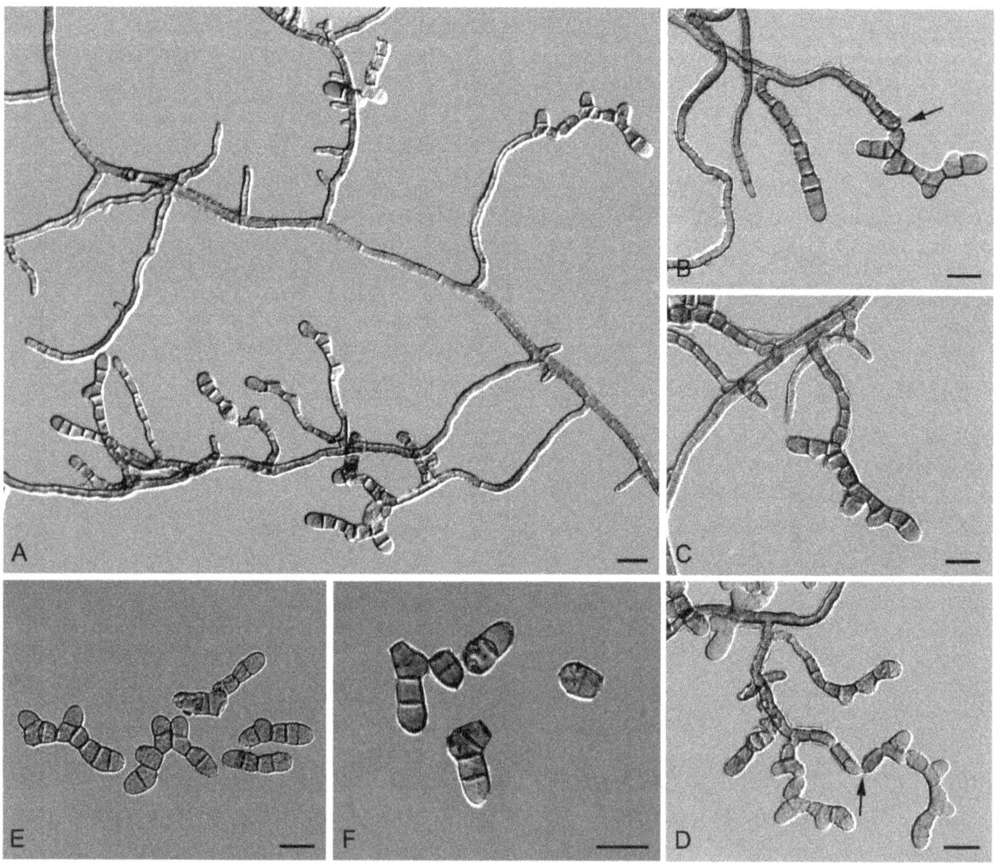


**Figure 11.** *Neocatenulostroma abietis* CBS 459.93. (**A**) Overview of conidiophores arising laterally from the mycelium. (**B–D**) Macronematous conidiophores with catenated thallic–arthric conidia disarticulated by schizolytic secession (arrows). (**E**,**F**) Thallic–arthric multi-septate conidia, with transverse and occasionally oblique septa, catenate, variously shaped, straight, or curved. DIC Nomarski. Scale bars = 10 μm.

*Classification*—Teratosphaeriaceae, Capnodiales, Dothideomycetes, Pezizomycotina, Ascomycota.

*Emended description*—*Mycelium* composed of pale brown to brown, septate, branched, smooth- and thin-walled hyphae (Figure 11A). *Asexual state* consisting of macronematous, mainly straight, caespitose, short, smooth-walled, olivaceous-brown, closely packed conidiophores, emerging laterally from the mycelium through a stoma or forming sporodochia; *conidia* thallic–arthric (Figure 11B,D), olivaceous to red-brown, multi-septate, with transverse and occasionally oblique septa (Figure 11E,F), catenate, in branched chains, with secondary meristematic development, variously shaped (ellipsoidal, cylindrical, Y-shaped, or irregularly-shaped), straight or curved, with truncated to rounded ends, secession schizolytic (Figure 11B–F). *Sexual state* consisting of amphigenous, immersed, substomatal, subepidermal ascomata, with a small papilla or not papillated, globose to subglobose, with a periphysate central ostiole, peridium comprising in two layers, outer layer thick, brown, with *textura angularis*, inner layer thin and hyaline; *asci* 8-spored, bitunicate, obclavate to globose; *ascospores*, medially 1-septate, hyaline to pale brown, broadly fusiform with obtuse apices, eguttulate.

*Neocatenulostroma pinorum* (Arx and E. Müll.) Sastoque, Stchigel, and Cano, comb. nov. Mycobank MB326815. Figure 12.

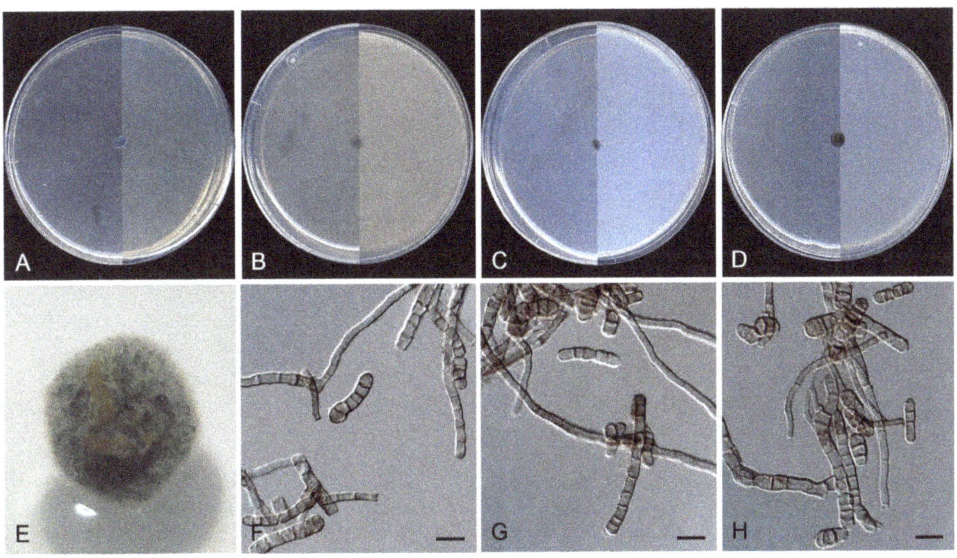


**Figure 12.** *Neocatenulostroma pinorum* CBS 174.90. (**A–D**) Colony on MEA, OA, PCA, PDA (after 2 wks at 25 ± 1 °C; surface, left; reverse, right). (**E**) Colony appearance on PDA. (**F–H**) Fertile hyphae from which arthroconidia are released by schizolytic secession. Conidia smooth, 0–4 septa, straight to slightly curved. DIC Nomarski. Scale bars = 10 μm.

*Basionym*—*Aulographina pinorum* (Desm.) Arx and E. Müll., Sydowia 14: 332 (1960)
≡ *Aulographum pinorum* Desm., Annls Sci. Nat., Bot., sér 2 10: 314 (1838)

*Description*—On oatmeal agar after two weeks at 25 °C—*Mycelium* consisting of septate, branching, olivaceous brown to brown, smooth- and thin-walled, 2–4 μm wide hyphae (Figure 12F–H). *Asexual state* (produced in vitro) consisting of fertile hyphae arising laterally from the mycelium, forming intercalary arthroconidia of thickened walls; *arthroconidia* 0–4-septate, olivaceous brown to brown, smooth- and thick-walled, sub-cylindrical to cylindrical, (6–)9–20(–26) × (2.5–)3–5(–6) μm, straight or slightly curved, rounded or truncated at both ends, slightly constricted at the septa, disposed in long branched or unbranched chains, secession schizolytic (Figure 12F–H). *Sexual state* (produced in vivo on leaves of *Pi-*

*nus silvestris*, *P. maritima*, and *P. nigra*) consisting of superficial, scattered to gregarious, dark brown to black, carbonaceous thyriothecia-like ascomata, 100–200 μm high × 78–94 μm diameter ($\overline{X}$ = 85 × 146 μm; *n* = 5), opening by a longitudinal or Y-shaped sunken slit; peridium 10–23.7 μm wide, thinner at the base, comprising of condensed hyphae; hamathecium comprising numerous, 1.1–1.9 μm wide, filiform, flexuous pseudoparaphyses; *asci* 8-spored, bitunicate, cylindrical-clavate to clavate, 37–43 × 7.9–11.5 μm ($\overline{X}$ = 41.3 × 9.6 μm; *n* = 10), with a short, broad pedicel, apically rounded, with a distinct ocular chamber; *ascospores* 2–3-seriate, partially overlapping, hyaline, smooth-walled to verrucose, ellipsoid to fusiform or ovoid with rounded ends, laterally compressed, 10–12 × 4.5–6 μm ($\overline{X}$ = 10.8 × 5 μm; *n* = 5), slightly constricted at the septum, with a mucilaginous appendage at each end.

*Culture characteristics*—(After 14 days at 25 °C) Colonies on MEA 2% 3–5 mm diameter, umbonate, circular, entire, tough and mostly bright, margins lobulate and slightly filamentous, floccose, ivy-green (1F3) to olive (2F4) and margins dark-grey (1F1); reverse dark-gray (1F1) (Figure 12A). Colonies on OA 5–6 mm diameter, circular, flat with a mound at the center and opaque, margins entire and regular, dark-green (30F4), velvety mostly at the central area with celadon green (30D3) hyphae, and dull-green (30E4) margins; reverse dark-green (30F4) with dull-green (30E4) margins (Figure 12B). Colonies on PCA 3.5–4 mm diameter, convex to pulvinate, wrinkled, circular, entire, tough, and mostly bright, margins lobulate and slightly filamentous, sparse velvety, greenish-grey (30F2) and with dull-green (30E3) hyphae; reverse bronze-green (30F3) (Figure 12C). Colonies on PDA 4.5–6 mm diameter, convex, and wrinkled, circular, entire, tough and opaque, margins entire and regular, velvety, olive (1E3) to olive-brown (4E5) at the center, with spherical, black (6F3), slightly velvety, 27–50 × 23–43 stromata; reverse dark-gray (1F1) and margins olive (1E5) (Figure 12D,E).

*Specimen examined*—France, Trédarzec, on needles of *Pinus insignis* (Pineaceae), Coll. Desmaziere (CBS 174.90).

*Diagnosis*—The genus currently contains five species, *N. abietis* (the type species), *N. castaneae*, *N. germanicum*, *N. microsporus*, and *N. spinulosum*. *Neocatenulostroma pinorum* presents most of the traits of the genus, showing thallic conidiogenesis similar to *N. abietis* and *N. spinulosum*, but differs in the conidial size: 8–24 × 5–7 in *N. abietis*, and (5–)7–20(–22) × (2.5–)3–4(–5.5) μm in *N. spinulosum*. The colonies form pseudostromata on TWA with needles of *Pinus* sp. after four weeks at 25 °C, composed of quite compact hyphae, but not differentiated cells. *N. pinorum* is the only known species of the genus to produce a sexual morph in vivo.

*Notes*—Based on a BLAST search of the NCBI GenBank nucleotide database, the closest hit using the **ITS** sequence was *N. abietis* (strain CPC 14996, GenBank FJ372387.1; identities = 511/512 (99.80%) gaps 0/512 (0%)). Using the **LSU** sequence, it was *N. microsporus* (strain CBS 110890, GenBank EU019255.2; Identities = 1266/1289 (98.22%), gaps 23/1289 (1%)). For the ***rpb2*** sequence, it was *N. abietis* (strain CBS 459.93, GenBank OX431256; Identities = 793/833 (95.20%), gaps 0/833 (0%)). Our *rpb2* phylogenetic tree corroborates the placement of our strain as a species of the genus *Neocatenulostroma*. Due to the recent migration of the species to other genera [95], the genus *Aulographina* is currently invalidated.

*Neocatenulostroma spinulosum* Sastoque, Cano and Stchigel, sp. nov. Mycobank MB847923. Figure 13.

*Etymology*—From Latin *spinulosus*, having small spines, because of the ornamentation of the hyphae and conidia with age.

*Description*—On oatmeal agar after two weeks at 25 °C—*Mycelium* consisting of septate, branching, pale brown to brown, smooth- and thin-walled hyphae, 2–4.5 μm wide, from which arise laterally the fertile hyphae (Figure 13F–H,L). *Arthroconidia* 0–5-septate, pale brown to brown, smooth to asperulate, thick-walled, sub-cylindrical to cylindrical, (5–)7–20(–22) × (2.5–)3–4(–5.5) μm, straight or slightly curved, rounded or truncated at the ends, slightly constricted at the septa, disposed in long unbranched or branched chains, secession schizolytic (Figure 13I–K). *Sexual morph* not observed.

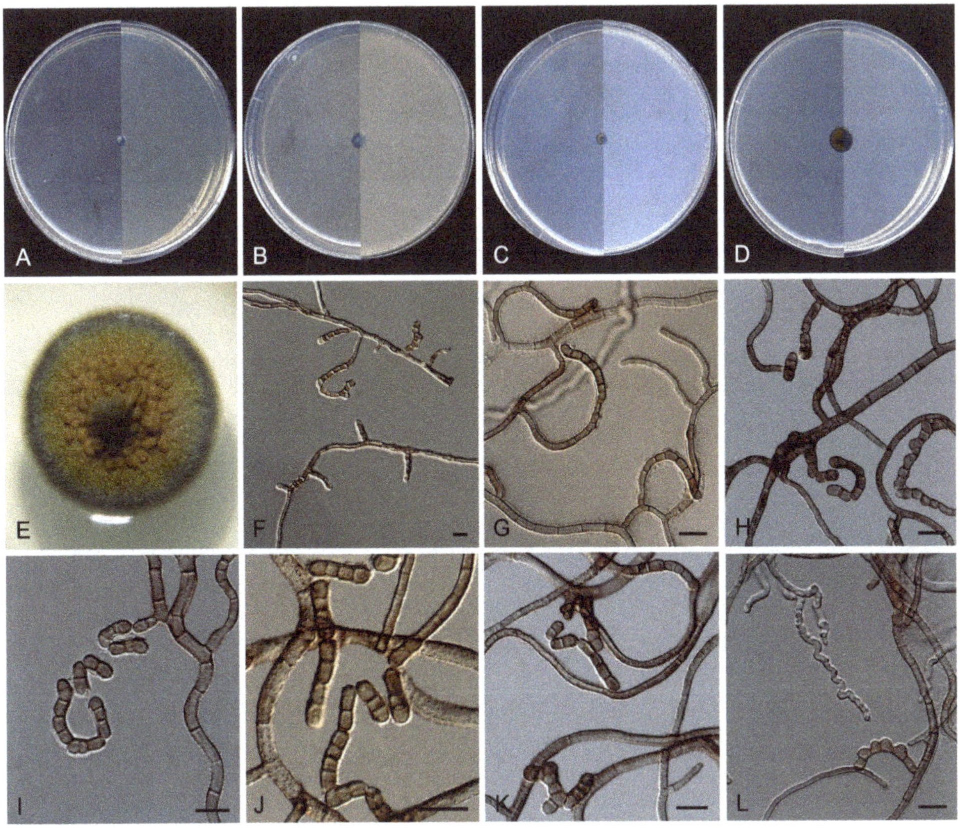


**Figure 13.** *Neocatenulostroma spinulosum* FMR 18793. (**A–D**) Colony on MEA, OA, PCA, PDA (after 2 wks at 25 ± 1 °C; surface, left; reverse, right). (**E**) Colony appearance on PDA. (**F–K**) Mycelium and fertile hyphae from which arthroconidia are released by schizolytic secession. (**L**) Asperulate conidia, 0–5 septa. DIC Nomarski. Scale bars = 10 µm.

*Culture characteristics*—(After 14 days at 25 °C) Colonies on MEA 8 mm diameter, umbonate and acuminate, circular, entire, restricted, tough and bright, margins entire and flat, black (6F3), with sparse groups of olive (1F5) aerial hyphae, margins olive (1F4); reverse dark-gray (1F1) and margins olive (1F4) (Figure 13A). Colonies on OA 6 mm diameter, circular, flat with a mound in the center and bright, margins entire, bronze-green (30F3) with sparse greenish-grey (30E3) aerial mycelium in the central area, and grass-green (30E7) margins; reverse bronze green (30F3) with olive (30E7) margins (Figure 13B). Colonies on PCA 8–9 mm diameter, pulvinate and wrinkled at the top, circular, entire, tough and mostly opaque, margins entire and submerged on the culture medium, velvety with yellow-green (3F3) hyphae, with patches bright and dark-grey (1F1) in the central area and margins dark-green (1F3) without aerial hyphae; reverse dark-gray (1F1), margins dark-green (1F3) (Figure 13C). Colonies on PDA 7.5–9 mm diameter, circular, entire, tough, opaque, crater-shaped with smooth, bright and spherical bumps at the center, margins entire and slightly buried on the culture medium, greyish-green (2E6) with velvety hyphae, bright and dark-grey (1F1) with olive-brown (4E5) bumps in the central area, margins olive-green (2F6) without aerial hyphae; reverse dark-gray (1F1) and margins olive-green (2F6) (Figure 13D,E). Minimum, optimum, and maximum temperature of growth: 5 °C, 20 °C and, 25 °C, respectively.

*Type*—Spain, Tarragona Province, Els Pallaresos, isolated from a darkened PVC pipe for pluvial drain at an industrial warehouse N 41°10′36.2″ E 1°16′12.4″, 11 November 2020, coll. J. F. Cano-Lira and A. M. Stchigel, isol. A. P. Sastoque (holotype CBS H-25262, ex-type FMR 18793 = CBS 150899; ITS, LSU, *rpb2*, and *tef1* sequences GenBank OX628944, OX628945, OX628946 and OX628947, respectively).

*Diagnosis*—*Neocatenulostroma spinulosum* differs from *N. germanicum*, its phylogenetically closest species, by the conidiogenesis, which is holoblastic in *N. germanicum*, as well as in the size and the septation of the conidia, which are (8–)10–15(–20) × 4–5(–6) µm in size and obliquely septate in *N. germanicum* [96].

*Notes*—Based on a mega BLAST search of NCBIs GenBank nucleotide database, the closest hit using the **ITS** sequence was *A. pinorum* (strain CBS 302.71 (Type material), GenBank GU214622.1; identities = 468/470 (99.57%) gaps 0/470 (0%)). Using the **LSU** sequence, it was *N. abietis* (strain CBS 290.90, GenBank MH873896.1; Identities = 473/471 (98.34%), gaps 0/481 (0%)). For the ***rpb2*** sequence, it was *N. microsporum* (strain CBS 101951, GenBank OX431256; Identities = 830/873 (95.07%), gaps 2/873 (0%)). Regarding the ***tef1*** sequence, it was *N. abietis* (isolate AFTOL-ID1789, GenBank DQ677933.1; identities = 774/791 (97.85%) gaps 0/791 (0%)).

*Neodevriesia longicatenispora* Sastoque, Stchigel and Cano, sp. nov. Mycobank MB 847920. Figure 14.

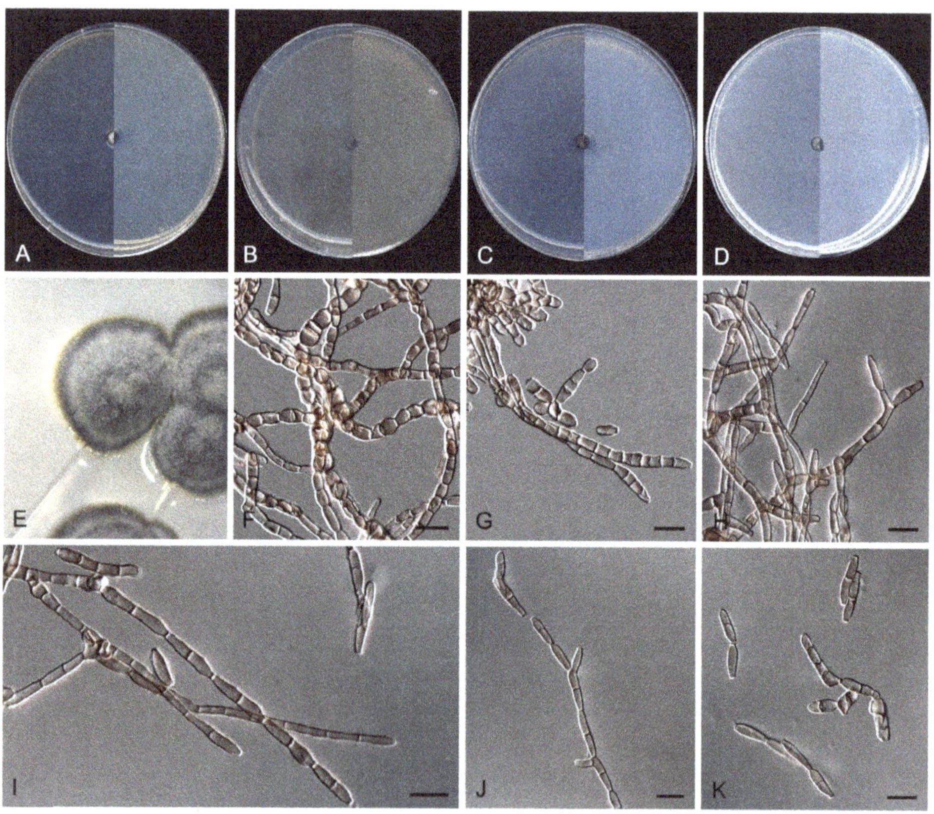

**Figure 14.** *Neodevriesia longicatenispora* FMR 18825. (**A–D**) Colony on MEA, OA, PCA, PDA (after 12 wks at 25 ± 1 °C; surface, left; reverse, right). (**E**) Colony appearance on PDA at 25 ± 1 °C. (**F**) Torulose hyphae. (**G–I**) Conidiophores and conidia. (**J**) Ramoconidia. (**K**) Free conidia or in chains. DIC Nomarski. Scale bars = 10 µm.

*Etymology*. From Latin *longus-*, long, *-catena-*, chain, and *-sporis*, spores, because of the production of conidia in long chains.

*Classification*—Neodevriesiaceae, Capnodiales, Dothideomycetes, Pezizomycotina, Ascomycota.

*Description*—On oatmeal agar after twelve weeks at 25 °C—*Mycelium* consisting of pale brown to olivaceous brown, smooth- and thick-walled, septate, branching, isodiametric to slightly torulouse hyphae, 2–3 μm wide (Figure 14F). *Conidiophores* arising from the hyphae, mononematous, olivaceous brown, smooth- and thick-walled, straight to flexuous, unbranched, 1–12-septate, sub-cylindrical, 10–80 × 2–4 μm (Figure 14G–I). *Conidiogenous* cells sympodially proliferating, integrated or terminal, septate or non-septate, brown, sub-cylindrical, 5–14 × 3–4 μm, with flattened scars,1–3 μm wide, darkened along the rim, neither thickened nor refractive. *Ramoconidia* 0–1(–2)-septate, olivaceous brown, smooth- and thick-walled, constricted at both ends and slightly constricted at septa, sub-cylindrical, 6–17 × 2.5–4 μm (Figure 14J). *Conidia* 0–1-septate, olivaceous-brown to light brown, smooth- and thick-walled, disposed in persistent, long, branching chains, tapering towards both ends, sub-cylindrical to narrowly fusoid, 6–18 × 2–4 μm, slightly truncate at septum level, scars flattened, somewhat darkened and thickened, 1–3 μm wide (Figure 14K). *Chlamydospores* absent.

*Culture characteristics*—(After 14 days at 25 °C, Figure 14A–E) Colonies reaching 2 mm diameter on MEA and PCA, 3 mm on OA, and 3.5 mm on PDA, circular, entire, and regular margins. Pulvinate on MEA, OA, and PCA, crater-like on PDA, compact, dry, and velvety due to the aerial mycelium. Greyish green (29E5) on the surface, with white aerial mycelium (29A1) on MEA, OA, and PDA, but dull green (30E4) with bronze margins (30F3) on PCA and dark grey (1F1) with margins dark green to greyish green (29F5-30E5) reverse on all culture media tested. Minimum, optimum, and maximum temperature of growth: 15 °C, 25 °C, and 25 °C, respectively.

*Type*—Spain, Tarragona Province, Els Pallaresos, isolated from a blackened metal railing of an industrial warehouse N 41°10′28.6″ E 1°16′40.3″, 11 November 2020, coll. J. F. Cano-Lira and A. M. Stchigel, isol. A. P. Sastoque (holotype CBS H-25246, ex-type FMR 18825 = CBS 149963; ITS, LSU, *tub2* and *rpb2* sequences GenBank OX342400, OX342401, OX342226 and OX342225).

*Diagnosis*—*Neodevriesia longicatenispora* differs from *N. stirlingiae*, its phylogenetic closest species, by the size of conidiophores, conidiogenous cells, scars, ramoconidia, conidia, and hila, which are 10–50 × 4–5 μm, 8–15 × 3–4 μm, 1–2 μm diameter, 15–30 × 4–5 μm, (7–)12–16(–20) × (3–)4(–5) μm and 1–2 μm diameter, respectively, in *N. stirlingiae*. The number of septa in ramoconidia and conidia is also different among them, being 1–3 septate and 0–3 septate in *N. stirlingiae*, respectively. Moreover, *N. stirlingiae* does not form long chains of conidia and produces chlamydospores (absent in *N. longicatenispora*) [97].

*Notes*—Based on a mega BLAST search of NCBIs GenBank nucleotide database, the closest hit using the **ITS** sequence was *N. fraserae* (strain CBS 128217 (Type material), GenBank NR_144961.1; identities = 521/527 (98.86%) gaps 1/527 (0%)). Using the **LSU** sequence, it was *N. stirlingiae* (strain CPC 19948, GenBank NG_042755.1; Identities = 848/854 (99.30%), gaps 0/854 (0%)) and *N. fraserae* CBS 128217 (GenBank OX346373; Identities = 575/583 (98.63%), gaps 0/583 (0%)). For the ***tub2*** sequence, it was *N. fraserae* (strain CBS 128217, GenBank OX346373; Identities = 317/340 (93.24%), gaps 3/340 (0%)) and *N. stirlingiae* (strain CPC 19948, GenBank OX346410; Identities = 386/428 (90.19%), gaps 7/428 (1%)). Regarding the ***rpb2*** sequence, it was *N. fraserae* (strain CBS 128217, GenBank OX346372; Identities = 777/830 (93.61%), gaps 0/830 (0%)) and *N. stirlingiae* (strain CPC 19948, GenBank OX346315; Identities = 685/736 (93.07%), gaps 0/736 (0%)).

*Paradevriesia holothallica* Sastoque, Cano and Stchigel, sp. nov. MycoBank MB 842105. Figure 15.

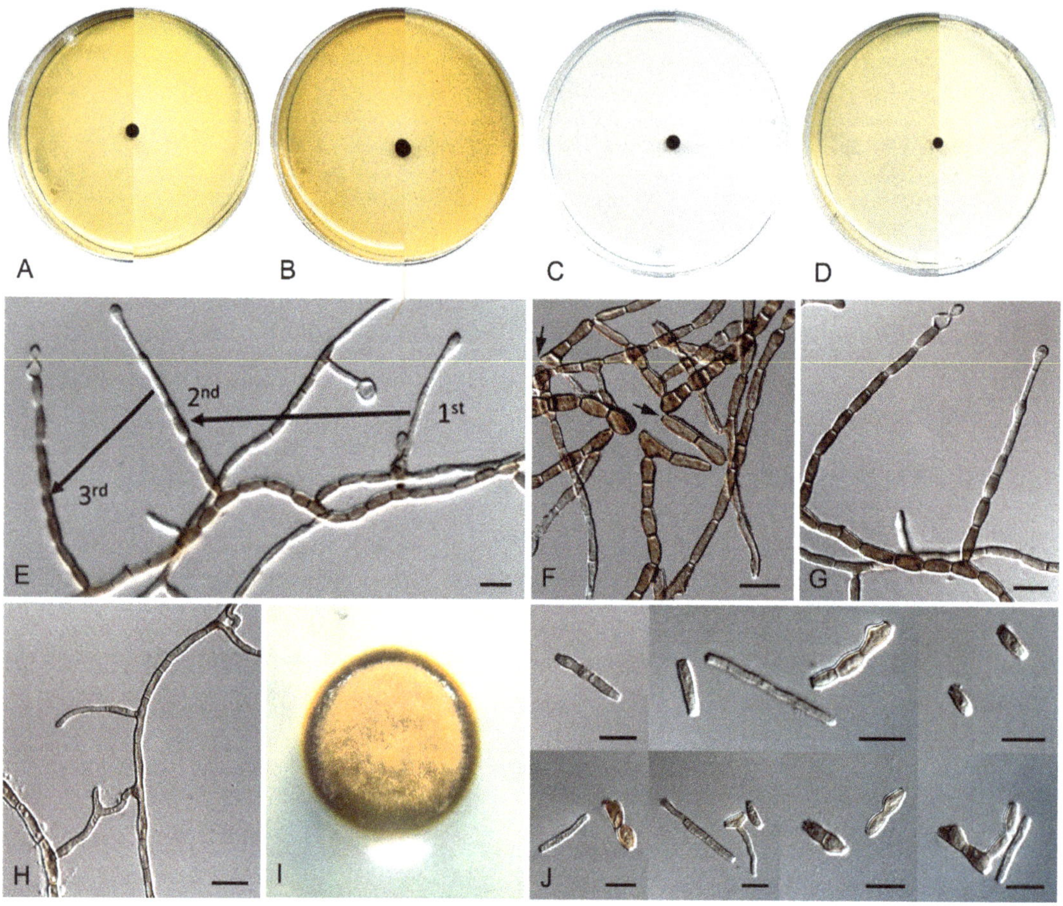


**Figure 15.** *Paradevriesia holothallica* FMR 18795. (**A–D**) Colony on MEA, OA, PCA, PDA (after 2 wks at 25 ± 1 °C; surface, left; reverse, right). (**E**) Conidiogenesis: hyphae become torulose and then disarticulate with schizolitic secession. (**F**) Disarticulation of conidia with schizolytic secession (black arrows). (**G**) Hyphae with swollen terminal cells. (**H**) Anastomosis. (**I**) Colony appearance on PDA at 25 ± 1 °C. (**J**) Holothallic conidia. DIC Nomarski. Scale bars = 10 μm.

*Etymology*. From Greek *όλος-*, whole, and *-θαλλός*, sprout, in reference to the sort of the conidial ontogeny (holothallic).

*Classification*—Paradevriesiaceae, Capnodiales, Dothideomycetes, Pezizomycotina, Ascomycota.

*Description*—On potato dextrose agar after two weeks at 25 °C—*Mycelium* composed of septate, brown, guttulate, smooth to slightly verrucose, moderately thick-walled, branching hyphae, 1.5–3 μm wide; anastomosis frequently present and terminal part of the hyphae may swell (Figure 15G,H). *Conidiophores* absent. *Conidia* holothallic, 0–2 septate, brown, moderately thick-walled, smooth to slightly verruculose, guttulate, mostly prismatic, subcylindrical or barrel-shaped, occasionally T-shaped or ellipsoidal, almost flattened at both ends and without scars, 5–15 × 3–5 μm (Figure 15J), formed by remodeling and disarticulation of pre-existing hyphae sections by schizolytic secession (Figure 15E,F). After 12 wks on PDA at 25 °C, hyphae become moniliform, and the conidia are mostly barrel-shaped to ellipsoidal.

*Culture characteristics*—(After 14 days at 25 °C, Figure 15A–D,I) Colonies reaching 5 mm diameter on MEA, 6 mm on OA and PCA, and 3 mm on PDA, circular with wide, glistening, and regular margins. Flat to raise on OA and PCA, convex to pulvinate on MEA and PDA, compact, granulose, dry, velvety due to the aerial mycelium on PDA, and slightly velvety and glistening on MEA, OA, and PCA. Greenish black (30F4) on the surface, with black margins (6F3) and a black (6F3) reverse on all culture media tested. Minimum, optimum, and maximum temperature of growth: 15 °C, 25 °C, and 30 °C, respectively.

*Type*—SPAIN, Tarragona Province, Els Pallaresos, isolated from the darkened surface of a wall surrounding a garden N 41°10′31.3″ E 1°16′39.2″, 20 November 2020, coll. J. F. Cano-Lira and A. M. Stchigel, isol. A. P. Sastoque (holotype CBS H-24939, ex-type FMR 18795 = CBS 149012; ITS, LSU, and *rpb2* sequences GenBank OX031242, OX031243 and OX031309).

*Diagnosis—Paradevriesia holothallica* presents most of the traits of the genus but lacks conidiophores, the conidia have a holothallic ontogeny, the *hila* are absent, size and shape of conidia are very variable and have a schizolytic secession.

*Notes*—Based on a mega BLAST search of NCBIs GenBank nucleotide database, the closest hit using the **ITS** sequence was *P. compacta* (strain CBS 118294 (Type material), GenBank NR_144955.1; identities = 422/458 (92.14%) gaps 8/458 (1%)); using the **LSU** sequence it was *P. compacta* (strain CBS 118294, GenBank NG_059089.1; Identities = 760/784 (96.94%), gaps 2/784 (0%)) and for ***rpb2*** it was *Capnodiales* sp. (strain CBS 118294, GenBank GU371751.1; identities = 350/397 (88.16%) gaps 0/397 (0%)).

### 3.4. Physiology

The results of the carbon sources assimilation test for our strains are given in Table S3. *Exophiala* strains, *P. kinklidomatophilus*, *K. perfecta*, and *K. epidermidis* displayed the lowest spectrum of organic molecules assimilated. The strains of *E. xenobiotica* (FMR 19066 and CBS 118157) presented the same assimilation profile, while FMR 18810 was quite different. Among all *Exophiala* strains tested, *E. multiformis* showed the broadest range of carbon source assimilation. *Aureobasidium pullulans* and *A. microstrictum* were the species that showed greater differences than other *Aureobasidium* spp. in the assimilation pattern of organic compounds.

The results of the nitrogen sources assimilation test, fermentation, and production of acid from glucose, urease, and DNase for our strains are shown in Table S4. Most of the species assimilated all nitrogen sources tested, except *P. kinklidomatophilus* and *K. perfecta* which were not able to assimilate any of them; the assimilation pattern of *C. pleiosporus* (FMR 18827) was different from that observed in the other strains tested, assimilating only six of the eleven nitrogen sources used. The strains *A. pullulans*, *A. microstictum*, and *E. xenobiotica* (CBS 118157 and FMR 19066) were osmotolerant, whereas *C. pleiosporus* was weak. *K. epidermidis* and *E. xenobiotica* (CBS 118157 and FMR 19066) were tolerant to cycloheximide, and *P. kinklidomatophilus*, *K. perfecta*, and *C. pleiosporus* were negative for urease. All strains were negative for DNase. *A. pullulans*, *A. microstictum*, and *C. pleiosporus* produced acid from glucose. Notably, *P. kinklidomatophilus* and *K. perfecta* (FMR 18715) did not ferment any of the tested compounds, while *K. epidermidis* fermented almost all of them (except for galactose, lactose, and inulin). The strains belonging to the genera *Aureobasidium* and *Exophiala* showed a good fermentation capacity. Tolerance to NaCl was variable, *P. kinklidomatophilus*, *K. perfecta*, *E. caementiphila*, and *A. microstictum* were non-tolerant, while for $MgCl_2$ and $CaCl_2$ only *P. kinklidomatophilus* and *K. perfecta* were a little tolerant or intolerant (Table S5). Practically all strains grew from 5 °C to 30 °C, and between pH 3 and 11, but the strains *K. perfecta* and *E. xenobiotica* (FMR 19066) were able to grow in up to pH 12 (Table S5). The production of gelatinase was positive for *P. kinklidomatophilus*, *A. pullulans*, *C. pleiosporus*, and *A. microstictum* (Table S5).

## 4. Discussion

In our study of RIF involved in the alteration (darkening) of the surfaces of various urban buildings in different localities of the Province of Tarragona (Spain), 41 genera of fungi were found. Among them, *Alternaria*, *Cladosporium*, and *Penicillium* (present in all localities), and *Aspergillus*, *Aureobasidium*, *Beauveria*, *Curvularia*, *Epicoccum*, *Fusarium*, and *Trichoderma* (present only in Els Pallaresos and Reus) have been previously described as allergenic to humans [11,16,22,98,99]. Within these genera, 64 species have been identified, including *Alternaria infectoria*, *Aureobasidium pullulans*, *Didymella glomerata*, *Didymella microchlamydospora*, *Exophiala xenobiotica*, *Knufia epidermdis*, *Neoscytalidium dimidiatum*, and *Stemphylium vesicarium* (all found in Els Pallaresos, Montbrió del Camp and Reus), which are considered opportunistic pathogens for humans (Table S1) [27,36,42,100–104].

In addition, a new genus and six new species have been identified, all from darkened surfaces of the exterior walls of various buildings in the village of Els Pallaresos. *Coccodomyces pleiosporus* (Figure 8) was isolated from a metal fence and found to produce holoblastic conidia, cells with meristematic growth, and yeast-like cells. *Exophiala caementiphila* (FMR 18977) (Figure 9) was isolated from concrete and *E. multiformis* (FMR 18809) (Figure 10) from a metal fence; both species can be easily distinguished morphologically, the former produces abundant yeast-like cells while the latter is predominantly mycelial. *Neodevriesia longicatenispora*, isolated from the blackened metal railing of an industrial warehouse, produced cladosporium-like conidiophores (Figure 14), and *Paradevriesia holothallica*, isolated from the darkened surface of a wall surrounding a garden, showed holothallic conidiogenesis (Figure 15), making it the only species in the genus with this type of conidiogenesis. In addition, *Neocatenulostroma spinulosum* (Figure 13), the only fungus isolated from a PVC pipe, exhibits thallic–arthric conidiogenesis. This prompted us to review the conidia of other species [35,86] and to modify the description of the genus.

It was difficult to establish whether the meteorological conditions had a direct impact on the distribution and diversity of RIF in the sampled areas because no significant differences were found among the fungal communities isolated (Table 1). However, the fungal taxa we found were comparable to those reported in previous studies with similar meteorological data [5,7,8,10,21,39,48]. In addition, some factors that can favor the establishment and development of a rich RIF community are the presence of different plants as well as the composition of the soils surrounding the rocky substrates [1,4,8,11,26,105]. The local vegetation in our study consisted of Mediterranean crops, pine forests (Els Pallaresos), as well as ornamental plants (in Reus). This can influence in two ways, either because many RIF species have a life cycle linked to plants, animals, and humans, thus enabling them to reach these rocky substrates and colonize these niches successfully, or because nearby plants can contribute with decomposing materials (leaves, seeds, pollen, exudates, etc.) to enrich the rock surfaces with different types of compounds necessary for the germination of fungal propagules and the growth of RIF. This influence could be confirmed both by the species that have been considered as the first reports mentioned above, some of them better known as plant-associated, and by the frequency at which the plant-associated fungi have been found growing on rocky substrates [7,15]. On the other hand, the soil that surrounds them will also influence the fungal species that might reach the rocky substrates, i.e., by spreading the spores through the air [8,15,22,41]. In this sense, a substantial number of the strains isolated from all sampled sites (Figure 1) belonged to typical soil-borne fungal genera (i.e., *Aspergillus*, *Fusarium*, *Mucor*, and *Penicillium*,) or to fungi frequently observed in the soil (*Alternaria* and *Cladosporium*). Therefore, the influence of these factors could help explain the differences in RIF communities as well as their abundance among the points sampled in this study.

Regarding substrates, RIF can grow not only on different types of rocks (marble, stone, concrete, granite, sandstone, limestone, basalt, or travertine) but also on the surface of various hard materials, including glass, plastic, roof tiles, solar panels, steamers, humidifiers, and concrete dishwashers [2,4,6,9,21,26,39], as observed in our study (Table 2). These findings confirm the great flexibility and adaptability of RIF to different sorts of materials, suggesting that the chemical composition of the substrates has a minimal effect on fungal metabolism because, perhaps, the greatest part of its source of nutrients is external [8]. On the other hand, a potential positive association has been suggested between substrates with a high capacity for water retention and a diverse RIF fungal community [106]. As in previous reports [26,33,107], we found *A. pullulans* and *E. xenobiotica* on metallic structures, but also *P. brasiliense* and *Con. Leucoplaca*, as well as some of the new taxa, *C. pleiosporus*, *E. multiformis*, *N. longicatenulospora*, and *P. kinklidomatophilus* [90], which represent new reports for this material. On the other hand, only *N. spinulosum* was found on a PVC pipe, indicating its particular ability to colonize and develop on this polymer substrate, probably because *N. spinulosum* is an extremely oligotrophic fungus. The limited amount of nutrients in these materials would favor the development of RIF, with an oligotrophic metabolism, compared to cosmopolitan microorganisms, whose development is highly dependent on the nutrients available in such substrates [8].

The physiological characteristics of our tested RIF strains do not allow them to be considered as strictly extremophilic organisms (e.g., the tested strains do not show optimal growth at "high" or "low" temperatures, nor at a "low" humidity or "extreme pH"), as one might expect from microorganisms adapted to colonize the surface of various types of building materials whose chemical composition, in addition to their exposure to unfavorable weather conditions, makes them uninhabitable for most microorganisms.

## 5. Conclusions

Our results showed a high biodiversity of RIF in different sites and building materials sampled in the Province of Tarragona. Although this was comparable to the previous studies in urban areas of Europe, some of the RIF we found were first reports and also new species. The urban and industrial areas of the village of Els Pallaresos had the highest RIF biodiversity, followed by Reus. A large number of representatives of the order Pleosporales were found at all sampling sites, but representatives of the orders Capnodiales, Chaetothyriales, and Dothideales were also found. The present study also allowed us to discover a new genus (*Coccodomyces*) and six new species (*Coccodomyces pleiosporus*, *Exophiala caementiphila*, *Exophiala multiformis*, *Neocatenulostroma spinulosum*, *Neodevriesia longicatenispora*, and *Paradevriesia holothallica*) as well as to determine the correct phylogenetic placement of *Aulographina pinorum* (transferred to the genus *Neocatenulostroma*). Regarding the physiology of the RIF species, all tested strains exhibited growth within the temperature range of 5 to 30 °C and at pH levels between 3 and 11. However, osmotolerance was observed only in strains with the potential characteristics of extremophiles. We had expected that most strains would have a maximum growth temperature above 35 °C, considering their prolonged exposure to high temperatures in late spring and summer in the Mediterranean areas studied, but no thermophilic or highly thermotolerant fungi were found. Likewise, no xerophilic species of RIF were found. The resistance of the isolated RIF species to desiccation and temperature shocks must, therefore, be attributed to a thick melanized cell wall, the extracellular production of melanin, and the formation of biofilms [4,34].

**Supplementary Materials:** The following supporting information can be downloaded at: https://www.mdpi.com/article/10.3390/jof10030170/s1, Table S1: Geographical origin and sources, and loci sequenced and accession numbers (GenBank/EMBL) for their nucleotide sequences of the fungi isolated in this study [108–114]. Table S2: Geographical origin and sources, loci and accession numbers of nucleotide sequences of fungal strains included in the phylogenetic analysis. Table S3: Carbon source assimilation; Table S4: Nitrogen source assimilation, osmotolerance, cycloheximide resistance, urease and DNAse production, acid from glucose and sugar fermentation for the fungal strains tested; Table S5: Halotolerance, thermotolerance, pH tolerance, and gelatinase production for the fungal strains tested.

**Author Contributions:** Conceptualization, A.M.S. and J.F.C.-L.; methodology, A.P.S.; software, A.P.S. and J.F.C.-L.; validation, A.M.S. and J.F.C.-L.; formal analysis, A.P.S., A.M.S. and J.F.C.-L.; investigation, A.P.S.; resources, J.F.C.-L.; data curation, A.M.S. and J.F.C.-L.; writing—original draft preparation, A.P.S.; writing—review and editing, A.P.S., A.M.S. and J.F.C.-L.; visualization, A.M.S. and J.F.C.-L.; supervision, A.M.S. and J.F.C.-L.; project administration, J.F.C.-L.; funding acquisition, J.F.C.-L. All authors have read and agreed to the published version of the manuscript.

**Funding:** This research was supported by the Spanish *Ministerio de Economía y Competitividad*, grant CGL2017-88094-P, and by the student grant 2020 FI SDUR 00212.

**Institutional Review Board Statement:** Not applicable.

**Informed Consent Statement:** Not applicable.

**Data Availability Statement:** Data are contained within the article and Supplementary Materials.

**Acknowledgments:** We are grateful to Nuria Pilas López lab technician of the Mycology Unit for technical support.

**Conflicts of Interest:** The authors declare no conflicts of interest.

## References

1. Scheerer, S.; Ortega-Morales, O.; Gaylarde, C. Microbial deterioration of stone monuments-an updated overview. In *Advances in Applied Microbiology*, 1st ed.; Elesvier Inc.: Amsterdam, The Netherlands, 2009; pp. 97–139. [CrossRef]
2. Sterflinger, K.; Krumbein, W.E. Multiple stress factors affecting growth of rock-inhabiting black fungi. *Bot. Acta* **1995**, *108*, 490–496. [CrossRef]
3. Zak, J.C.; Wildman, H.G. Fungi in stressful environments. In *Biodiversity of Fungi Inventory and Monitoring Methods*; Mueller, G.M., Bills, G.F., Foster, M.S., Eds.; Elsevier Academic Press: Boston, MA, USA, 2004; pp. 513–531. [CrossRef]
4. Liu, B.; Fu, R.; Wu, B.; Liu, X.; Xiang, M. Rock-inhabiting fungi: Terminology, diversity, evolution and adaptation mechanisms. *Mycology* **2021**, *13*, 1–31. [CrossRef]
5. Berti, L.; Marvasi, M.; Perito, B. Characterization of the community of black meristematic fungi inhabiting the external white marble of the Florence Cathedral. *J. Fungi* **2023**, *9*, 665. [CrossRef]
6. Choe, Y.-H.; Kim, M.; Woo, J.; Lee, M.J.; Lee, J.I.; Lee, E.J.; Lee, Y.K. Comparing rock-inhabiting microbial communities in different rock types from a high arctic polar desert. *FEMS Microbiol. Ecol.* **2018**, *9*, 1–13. [CrossRef] [PubMed]
7. Isola, D.; Zucconi, L.; Onofri, S.; Caneva, G.; De Hoog, G.S.; Selbmann, L. Extremotolerant rock inhabiting black fungi from Italian monumental sites. *Fungal Divers.* **2016**, *76*, 75–96. [CrossRef]
8. Ruibal, C.; Platas, G.; Bills, G. High diversity and morphological convergence among melanised fungi from rock formations in the Central Mountain System of Spain. *Persoonia* **2008**, *21*, 93–110. [CrossRef]
9. Ruibal, C.; Selbmann, L.; Avci, S.; Martin-Sanchez, P.M.; Gorbushina, A.A. Roof-inhabiting cousins of rock-inhabiting fungi: Novel melanized microcolonial fungal species from photocatalytically reactive subaerial surfaces. *Life* **2018**, *8*, 30. [CrossRef]
10. Santo, A.P.; Cuzman, O.A.; Petrocchi, D.; Pinna, D.; Salvatici, T.; Perito, B. Black on white: Microbial growth darkens the external marble of florence cathedral. *Appl. Sci.* **2021**, *11*, 6163. [CrossRef]
11. Tiquia-Arashiro, S.; Grube, M. *Fungi in Extreme Environments: Ecological Role and Biotechnological Significance*; Tiquia-Arashiro, S.M., Grube, M., Eds.; Springer Nature Switzerland AG: Gewerbestrasse, Switzerland, 2019; pp. 39–57. [CrossRef]
12. Wollenzien, U.; de Hoog, G.; Krumbein, W.; Urzí, C. On the isolation of microcolonial fungi occurring on and in marble and other calcareous rocks. *Sci. Total. Environ.* **1995**, *167*, 287–294. [CrossRef]
13. Sterflinger, K.; De Baere, R.; de Hoog, G.; De Wachter, R.; Krumbein, W.E.; Haase, G. Coniosporium perforans and C. apollinis, two new rock-inhabiting fungi isolated from marble in the Sanctuary of Delos (Cyclades, Greece). *Antonie van Leeuwenhoek* **1997**, *72*, 349–363. [CrossRef]
14. Cappitelli, F.; Nosanchuk, J.D.; Casadevall, A.; Toniolo, L.; Brusetti, L.; Florio, S.; Principi, P.; Borin, S.; Sorlini, C. Synthetic consolidants attacked by melanin-producing fungi: Case study of the biodeterioration of Milan (Italy) cathedral marble treated with acrylics. *Appl. Environ. Microbiol.* **2007**, *73*, 271–277. [CrossRef] [PubMed]
15. Gorbushina, A.A.; Lyalikova, N.N.; Vlasov, D.Y.; Khizhnyak, T.V. Microbial communities on the monuments of Moscow and St. Petersburg: Biodiversity and trophic relations. *Microbiology* **2002**, *71*, 350–356. [CrossRef]
16. Marfenina, O.E.; Makarova, N.V.; Ivanova, A.E. Opportunistic fungi in soils and surface air of a megalopolis (for the Tushino Region, Moscow). *Microbiology* **2011**, *80*, 870–876. [CrossRef]
17. Viles, H.A.; Gorbushina, A.A. Soiling and microbial colonisation on urban roadside limestone: A three year study in Oxford, England. *Build. Environ.* **2003**, *38*, 1217–1224. [CrossRef]
18. May, E. Microbes on building stone—For good or ill? *Culture* **2003**, *24*, 5–8.
19. Warscheid, T.; Braams, J. Biodeterioration of stone: A review. *Int. Biodeterior. Biodegrad.* **2000**, *46*, 343–368. [CrossRef]
20. Kurup, V.P.; Shen, H.-D.; Banerjee, B. Respiratory fungal allergy. *Microbes Infect.* **2000**, *2*, 1101–1110. [CrossRef]

21. Marvasi, M.; Donnarumma, F.; Frandi, A.; Mastromei, G.; Sterflinger, K.; Tiano, P.; Perito, B. Black microcolonial fungi as deteriogens of two famous marble statues in Florence, Italy. *Int. Biodeterior. Biodegrad.* **2012**, *68*, 36–44. [CrossRef]
22. Páramo-Aguilera, L.; Ortega-Morales, B.O.; Narváez-Zapata, J.A. Culturable fungi associated with urban stone surfaces in Mexico City. *Electron. J. Biotechnol.* **2012**, *15*, 1–17. [CrossRef]
23. Suihko, M.-L.; Alakomi, H.-L.; Gorbushina, A.; Fortune, I.; Marquardt, J.; Saarela, M. Characterization of aerobic bacterial and fungal microbiota on surfaces of historic Scottish monuments. *Syst. Appl. Microbiol.* **2007**, *30*, 494–508. [CrossRef]
24. Ametrano, C.G.; Muggia, L.; Grube, M. Extremotolerant black fungi from rocks and lichens. In *Fungi in Extreme Environments: Ecological Role and Biotechnological Significance*; Tiquia-Arashiro, S.M., Grube, M., Eds.; Springer Nature Switzerland AG: Cham, Switzerland, 2019; pp. 39–57. [CrossRef]
25. de Hoog, G.; Beguin, H.; Batenburg-van de Vegte, W.H. *Phaeotheca triangularis*, a new meristematic black yeast from a humidifier. *Antonie van Leeuwenhoek* **1997**, *71*, 289–295. [CrossRef]
26. Sterflinger, K. Black yeasts and meristematic fungi: Ecology, diversity and identification. In *Biodiversity and Ecophysiology of Yeasts the Yeast Handbook*; Péter, G., Rosa, C., Eds.; Springer: Berlin/Heidelberg, Germany, 2006. [CrossRef]
27. Vicente, V.; Attili-Angelis, D.; Pie, M.; Queiroz-Telles, F.; Cruz, L.; Najafzadeh, M.; de Hoog, G.; Zhao, J.; Pizzirani-Kleiner, A. Environmental isolation of black yeast-like fungi involved in human infection. *Stud. Mycol.* **2008**, *61*, 137–144. [CrossRef] [PubMed]
28. Prenafeta-Boldú, F.X.; Summerbell, R.; De Hoog, G.S. Fungi growing on aromatic hydrocarbons: Biotechnology's unexpected encounter with biohazard? *FEMS Microbiol. Rev.* **2006**, *30*, 109–130. [CrossRef] [PubMed]
29. de Hoog, G.S.; Ende AHGG van den Uijthof, J.M.J.; Untereiner, W.A. Nutritional physiology of type isolates of currently accepted species of *Exophiala* and *Phaeococcomyces*. *Antonie van Leeuwenhoek* **1995**, *68*, 43–49. [CrossRef] [PubMed]
30. De Hoog, G.S.; Zeng, J.S.; Harrak, M.J.; Sutton, D.A. *Exophiala xenobiotica* sp. nov., an opportunistic black yeast inhabiting environments rich in hydrocarbons. *Antonie van Leeuwenhoek* **2006**, *90*, 257–268. [CrossRef] [PubMed]
31. Prenafeta-Boldú, F.X.; Kuhn, A.; Luykx, D.M.; Anke, H.; van Groenestijn, J.W.; de Bont, J.A. Isolation and characterization of fungi growing on volatile aromatic hydrocarbons as their sole carbon and energy source. *Mycol. Res.* **2001**, *105*, 477–484. [CrossRef]
32. Isola, D.; Scano, A.; Orrù, G.; Prenafeta-Boldú, F.X.; Zucconi, L. Hydrocarbon-contaminated sites: Is there something more than *Exophiala xenobiotica*? New insights into black fungal diversity using the long cold incubation method. *J. Fungi* **2021**, *7*, 817. [CrossRef] [PubMed]
33. de Hoog, G.S.; Yurlova, N.A. Conidiogenesis, nutritional physiology and taxonomy of *Aureobasidium* and *Hormonema*. *Antonie van Leeuwenhoek* **1994**, *65*, 41–54. [CrossRef] [PubMed]
34. Cordero, R.J.; Casadevall, A. Functions of fungal melanin beyond virulence. *Fungal Biol. Rev.* **2017**, *31*, 99–112. [CrossRef]
35. Sterflinger, K.; de Hoog, G.S.; Haase, G. Phylogeny and ecology of meristematic ascomycetes. *Stud. Mycol.* **1999**, *43*, 5–22.
36. Wheeler, M.H.; Bell, A.A. Melanins and their importance in pathogenic fungi. *Curr. Top Med. Mycol.* **1988**, *2*, 338–387. [CrossRef]
37. de Hoog, G.S.; Hermanides-Nijhof, E.J. The black yeasts and allied hyphomycetes. *Stud. Mycol.* **1977**, *15*, 151–222.
38. Staley, J.T.; Palmer, F.; Adams, J.B. Microcolonial fungi: Common inhabitants on desert rocks? *Science* **1982**, *215*, 1093–1095. [CrossRef] [PubMed]
39. Gorbushina, A.A.; Krumbein, W.E.; Hamman, C.H.; Panina, L.; Soukharjevski, S.; Wollenzien, U. Role of black fungi in color change and biodeterioration of antique marbles. *Geomicrobiol. J.* **1993**, *11*, 205–221. [CrossRef]
40. Slepecky, R.A.; Starmer, W.T. Phenotypic plasticity in fungi: A review with observations on *Aureobasidium pullulans*. *Mycologia* **2009**, *101*, 823–832. [CrossRef] [PubMed]
41. Gorbushina, A.A. Life on the rocks. *Environ. Microbiol.* **2007**, *9*, 1613–1631. [CrossRef] [PubMed]
42. Chowdhary, A.; Perfect, J.; de Hoog, G.S. Black Molds and Melanized Yeasts Pathogenic to Humans. *Cold Spring Harb. Perspect. Med.* **2015**, *5*, a019570. [CrossRef] [PubMed]
43. Teixeira, M.M.; Moreno, L.F.; Stielow, B.; Muszewska, A.; Hainaut, M.; Gonzaga, L.; Abouelleil, A.; Patané, J.S.L.; Priest, M.; Souza, R.; et al. Exploring the genomic diversity of black yeasts and relatives (Chaetothyriales, Ascomycota). *Stud. Mycol.* **2017**, *86*, 1–28. [CrossRef] [PubMed]
44. Abdollahzadeh, J.; Groenewald, J.; Coetzee, M.; Wingfield, M.; Crous, P. Evolution of lifestyles in Capnodiales. *Stud. Mycol.* **2020**, *95*, 381–414. [CrossRef]
45. Crous, P.; Schoch, C.; Hyde, K.; Wood, A.; Gueidan, C.; de Hoog, G.; Groenewald, J. Phylogenetic lineages in the Capnodiales. *Stud. Mycol.* **2009**, *64*, 17–47. [CrossRef]
46. Schoch, C.; Crous, P.; Groenewald, J.; Boehm, E.; Burgess, T.; de Gruyter, J.; de Hoog, G.; Dixon, L.; Grube, M.; Gueidan, C.; et al. A class-wide phylogenetic assessment of Dothideomycetes. *Stud. Mycol.* **2009**, *64*, 1–15. [CrossRef]
47. Sun, W.; Su, L.; Yang, S.; Sun, J.; Liu, B.; Fu, R.; Wu, B.; Liu, X.; Cai, L.; Guo, L.; et al. Unveiling the hidden diversity of rock-inhabiting fungi: Chaetothyriales from China. *J. Fungi* **2020**, *6*, 187. [CrossRef]
48. Egidi, E.; de Hoog, G.S.; Isola, D.; Onofri, S.; Quaedvlieg, W.; de Vries, M.; Verkley, G.J.M.; Stielow, J.B.; Zucconi, L.; Selbmann, L. Phylogeny and taxonomy of meristematic rock-inhabiting black fungi in the Dothideomycetes based on multi-locus phylogenies. *Fungal Divers.* **2014**, *65*, 127–165. [CrossRef]
49. Cuscó, A.; Catozzi, C.; Viñes, J.; Sanchez, A.; Francino, O. Microbiota profiling with long amplicons using Nanopore sequencing: Full-length 16S rRNA gene and the 16S-ITS-23S of the rrn operon. *F1000Research* **2019**, *7*, 1755. [CrossRef]

50. Nagano, Y.; Miura, T.; Tsubouchi, T.; Lima, A.O.; Kawato, M.; Fujiwara, Y.; Fujikura, K. Cryptic fungal diversity revealed in deep-sea sediments associated with whale-fall chemosynthetic ecosystems. *Mycology* **2020**, *11*, 263–278. [CrossRef]
51. Luo, Y.; Wei, X.; Yang, S.; Gao, Y.-H.; Luo, Z.-H. Fungal diversity in deep-sea sediments from the Magellan seamounts as revealed by a metabarcoding approach targeting the ITS2 regions. *Mycology* **2020**, *11*, 214–229. [CrossRef]
52. Zhang, Z.-F.; Liu, Y.; Li, M. Pacific Biosciences Single-Molecule Real-Time (SMRT) Sequencing Reveals High Diversity of Basal Fungal Lineages and Stochastic Processes Controlled Fungal Community Assembly in Mangrove Sediments. 2020. Available online: https://www.researchsquare.com/article/rs-97364/v1 (accessed on 4 January 2024). [CrossRef]
53. Wu, B.; Hussain, M.; Zhang, W.; Stadler, M.; Liu, X.; Xiang, M. Current insights into fungal species diversity and perspective on naming the environmental DNA sequences of fungi. *Mycology* **2019**, *10*, 127–140. [CrossRef] [PubMed]
54. Beck, H.E.; Zimmermann, N.E.; McVicar, T.R.; Vergopolan, N.; Berg, A.; Wood, E.F. Present and future köppen-geiger climate classification maps at 1-km resolution. *Sci. Data* **2018**, *5*, 180214. [CrossRef]
55. Ringer, S. Regarding the action of hydrate of soda, hydrate of ammonia, and hydrate of potash on the ventricle of the frog's heart. *J. Physiol.* **1882**, *3*, 195–202. [CrossRef] [PubMed]
56. Hawksworth, D.; Kirk, P.; Sutton, B.; Pegler, D. *Ainsworth & Bisby's Dictionary of the Fungi*, 8th ed.; Prensa de la Universidad de Oxford, Ed.; CAB International: Wallingford, UK, 1996.
57. Onions, A.; Pitt, J. Appendix: Media. In *Filamentous Fungi*; Hawksworth, D., Kirsop, B., Eds.; Cambridge University Press: Cambridge, UK, 1988; Available online: https://onlinelibrary.wiley.com/doi/abs/10.1002/jobm.3620300217 (accessed on 15 May 2023).
58. Atlas, R.M. *Handbook of Microbiological Media*, 4th ed.; CRC Press Taylor & Francis Group: Washington, DC, USA, 2010; 2036p. [CrossRef]
59. Jarvis, B. Comparison of an improved rose bengal-chlortetracycline agar with other media for the selective isolation and enumeration of moulds and yeasts in foods. *J. Appl. Bacteriol.* **1973**, *36*, 723–727. [CrossRef]
60. King, A.D.; Hocking, A.D.; Pitt, J.I. Dichloran-rose bengal medium for enumeration and isolation of molds from foods. *Appl. Environ. Microbiol.* **1979**, *37*, 959–964. [CrossRef] [PubMed]
61. Samson, R.A.; Houbraken, J.; Thrane, U.; Frisvad, J.C.; Andersen, B. *Food and Indoor Fungi*, 2nd ed.; CBS-KNAW Fungal Biodiversity Centre: Utrecht, The Netherlands, 2010.
62. Reiss, J. Ein selektives kulturmedium für der Nachweiss von *Aspergillus flavus*. *Zbl. Bokt. Hyg. I. Abt. Orig.* **1972**, *220*, 564–566.
63. Crous, P.W.; Verkley, G.J.M.; Groenewald, J.Z.; Samson, R.A. *Fungal Biodiversity*; CBS Laboratory Manual Series; The Westerdijk Fungal Biodiversity Institute: Utrecht, The Netherlands, 2009.
64. Kornerup, A.; Wanscher, J. *Methuen Handbook of Colour*, 3rd ed.; Eyre Methuen: London, UK, 1978; 252p.
65. Smith, H.; Wingfield, M.J.; Crous, P.W.; Coutinho, I.A. Sphaeropsis sapinea and Botryosphaeria dothidea endophytic in Pinus spp. and Eucalyptus spp. in South Africa. *S. Afr. J. Bot.* **1996**, *62*, 86–88. [CrossRef]
66. White, T.J.; Bruns, T.; Lee, S.; Taylor, J. Amplification and direct sequencing of fungal ribosomal RNA genes for phylogenetics. In *PCR Protocols: A Guide to Methods and Applications*, 2nd ed.; Innis, M., Gelfand, D., Sninsky, J., White, T., Eds.; Academic Press, Inc.: San Diego, CA, USA, 1990; pp. 315–322. [CrossRef]
67. Vilgalys, R.; Hester, M. Rapid genetic identification and mapping of enzymatically amplified ribosomal DNA from several Cryptococcus species. *J. Bacteriol.* **1990**, *172*, 4238–4246. [CrossRef] [PubMed]
68. Kurtzman, C.P.; Robnett, C.J. Identification of clinically important ascomycetous yeasts based on nucleotide divergence in the 5′ end of the large-subunit (26S) ribosomal DNA gene. *J. Clin. Microbiol.* **1997**, *35*, 1216–1223. [CrossRef] [PubMed]
69. Liu, Y.J.; Whelen, S.; Hall, B.D. Phylogenetic relationships among ascomycetes: Evidence from an RNA polymerse II subunit. *Mol. Biol. Evol.* **1999**, *16*, 1799–1808. [CrossRef] [PubMed]
70. O'Donnell, K.; Cigelnik, E. Two divergent intragenomic rDNA ITS2 types within a monophyletic lineage of the fungus. *Mol. Phylogenet. Evol.* **1997**, *7*, 103–116. [CrossRef]
71. Glass, N.L.; Donaldson, G.C. Development of primer sets designed for use with the PCR to amplify conserved genes from filamentous ascomycetes. *Appl. Environ. Microbiol.* **1995**, *61*, 1323–1330. [CrossRef]
72. O'Donnell, K.; Kistler, H.C.; Cigelnik, E.; Ploetz, R.C. Multiple evolutionary origins of the fungus causing panama disease of banana: Concordant evidence from nuclear and mitochondrial gene genealogies. *Proc. Natl. Acad. Sci. USA* **1998**, *9*, 2044–2049. [CrossRef]
73. Rehner, S.A.; Buckley, E. A *Beauveria* phylogeny inferred from nuclear ITS and EF1-α sequences: Evidence for cryptic diversification and links to *Cordyceps* teleomorphs. *Mycologia* **2005**, *97*, 84–98. [CrossRef]
74. Cohen, S.D. A protocol for direct sequencing of multiple gene specific PCR products from *Discula umbrinella*, a fungal endophyte, utilizing bufferless precast electrophoresis. *J. Microbiol. Methods* **2005**, *61*, 131–135. [CrossRef]
75. Takara Bio Inc. *EmeraldAmp®GT PCR Master Mix*; Takara Bio Europe SAS: Saint-Germain-en-Laye, France, 2012; p. 1.
76. Altschul, S.F.; Gish, W.; Miller, W.; Myers, E.W.; Lipman, D.J. Basic local alignment search tool. *J. Mol. Biol.* **1990**, *215*, 403–410. [CrossRef]
77. Thompson, J.D.; Higgins, D.G.; Gibson, T.J. CLUSTAL W: Improving the sensitivity of progressive multiple sequence alignment through sequence weighting, position-specific gap penalties and weight matrix choice. *Nucleic Acids Res.* **1994**, *22*, 4673–4680. [CrossRef]

78. Kumar, S.; Stecher, G.; Tamura, K. MEGA7: Molecular Evolutionary Genetics Analysis Version 7.0 for Bigger Datasets. *Mol. Biol. Evol.* **2016**, *33*, 1870–1874. [CrossRef]
79. Stamatakis, A. RAxML version 8: A tool for phylogenetic analysis and post-analysis of large phylogenies. *Bioinformatics* **2014**, *30*, 1312–1313. [CrossRef] [PubMed]
80. Miller, M.A.; Pfeiffer, W.; Schwartz, T. The CIPRES science gateway: Enabling high-impact science for phylogenetics researchers with limited resources. In Proceedings of the 1st Conference of the Extreme Science and Engineering Discovery Environment: Bridging from the Extreme to the Campus and Beyond, Chicago, IL, USA, 16–20 July 2012; Available online: https://dl.acm.org/doi/abs/10.1145/2335755.2335836 (accessed on 14 May 2023).
81. Ronquist, F.; Teslenko, M.; van der Mark, P.; Ayres, D.L.; Darling, A.; Höhna, S.; Larget, B.; Liu, L.; Suchard, M.A.; Huelsenbeck, J.P. Mrbayes 3.2: Efficient bayesian phylogenetic inference and model choice across a large model space. *Syst. Biol.* **2012**, *61*, 539–542. [CrossRef] [PubMed]
82. Nylander, J.A.A. *MrModeltest Version 2. Program Distributed by the Author*; Evolutionary Biology Centre, Uppsala University: Uppsala, Sweden, 2004.
83. Metropolis, N.; Ulam, S. The monte carlo method. *J. Am. Stat. Assoc.* **1949**, *44*, 335–341. Available online: https://www.jstor.org/stable/2280232 (accessed on 15 May 2023). [CrossRef]
84. van der Walt, J.P.; Yarrow, D. Methods for the isolation, maintenance, classification and identification of yeasts. In *The Yeasts*; Elsevier Science Publishers B.V.: Amsterdam, The Netherlands, 1984; pp. 45–104. [CrossRef]
85. de Hoog, G.S.; Guého, E.; Masclaux, F.; Gerrits van den Ende, A.H.G.; Kwon-Chung, K.J.; Mcginnis, M.R. Nutritional physiology and taxonomy of human-pathogenic *Cladosporium-Xylohypha* species. *Med. Mycol.* **1995**, *33*, 339–347. [CrossRef]
86. Wollenzien, U.; de Hoog, G.S.; Krumbein, W.; Uijthof, J.M.J. *Sarcinomyces petricola*, a new microcolonial fungus from marble in the Mediterranean basin. *Antonie van Leeuwenhoek* **1997**, *71*, 281–288. [CrossRef]
87. Uribe Gutiérrez, L.A. *Caracterización Fisiológica de Levaduras Aisladas de la Filósfera de Mora*; Pontificia Universidad Javeriana: Bogotá, Columbia, 2007; Available online: https://repository.javeriana.edu.co/bitstream/handle/10554/8298/tesis276.pdf;jsessionid=C4CC63BB7E5E435DF74B4DCD756156A5?sequence=1 (accessed on 11 January 2023).
88. Schwarz, P.; Lortholary, O.; Dromer, F.; Dannaoui, E. Carbon assimilation profiles as a tool for identification of Zygomycetes. *J. Clin. Microbiol.* **2007**, *45*, 1433–1439. [CrossRef] [PubMed]
89. Alvarez, E.; Stchigel, A.M.; Cano, J.; Sutton, D.A.; Fothergill, A.W.; Chander, J.; Salas, V.; Rinaldi, M.G.; Guarro, J. Diversidad filogenética del hongo mucoral emergente *Apophysomyces*: Propuesta de tres nuevas especies. *Rev. Iberoam. Micol.* **2010**, *27*, 80–89. [CrossRef]
90. Crous, P.; Osieck, E.; Jurjevi, Ž.; Boers, J.; Van Iperen, A.; Starink-Willemse, M.; Dima, B.; Balashov, S.; Bulgakov, T.; Johnston, P.; et al. Fungal Planet description sheets: 1284–1382. *Persoonia* **2021**, *47*, 178–374. [CrossRef]
91. Phukhamsakda, C.; Nilsson, R.H.; Bhunjun, C.S.; de Farias, A.R.G.; Sun, Y.-R.; Wijesinghe, S.N.; Raza, M.; Bao, D.-F.; Lu, L.; Tibpromma, S.; et al. The numbers of fungi: Contributions from traditional taxonomic studies and challenges of metabarcoding. *Fungal Divers.* **2022**, *114*, 327–386. [CrossRef]
92. Gao, Y.; Liu, H.F.; Song, Z.X.; Du, X.Y.; Deng, J.X. Identification and characterization of *Gonatobotryum apiculatum* causing leaf spot and blight on *Sinowilsonia henryi*. *Mycobiology* **2020**, *48*, 70–74. [CrossRef]
93. Bates, S.T.; Reddy, G.S.N.; Garcia-Pichel, F. *Exophiala crusticola* anam. nov. (affinity Herpotrichiellaceae), a novel black yeast from biological soil crust in the Western United States. *Int. J. Syst. Evol. Microbiol.* **2006**, *56*, 2697–2702. [CrossRef]
94. Li, D.M.; Li, R.Y.; De Hoog, G.; Wang, Y.X.; Wang, D.L. *Exophiala asiatica*, a new species from a fatal case in China. *Med. Mycol.* **2009**, *47*, 101–109. [CrossRef] [PubMed]
95. Crous, P.; Wingfield, M.; Cheewangkoon, R.; Carnegie, A.; Burgess, T.; Summerell, B.; Edwards, J.; Taylor, P.; Groenewald, J. Foliar pathogens of eucalypts. *Stud. Mycol.* **2019**, *94*, 125–298. [CrossRef] [PubMed]
96. Crous, P.; Braun, U.; Groenewald, J. Mycosphaerella is polyphyletic. *Stud. Mycol.* **2007**, *58*, 1–32. [CrossRef] [PubMed]
97. Crous, P.W.; Shivas, R.G.; Wingfield, M.J.; Summerell, B.A.; Rossman, A.Y.; Alves, J.L.; Adams, G.C.; Barreto, R.W.; Bell, A.; Coutinho, M.L.; et al. Fungal Planet description sheets: 128–153. *Persoonia* **2012**, *29*, 146–201. [CrossRef] [PubMed]
98. Kasprzyk, I.; Grinn-Gofroń, A.; Ćwik, A.; Kluska, K.; Cariñanos, P.; Wójcik, T. Allergenic fungal spores in the air of urban parks. *Aerobiologia* **2021**, *37*, 39–51. [CrossRef]
99. Egbuta, M.A.; Mwanza, M.M.; Babalola, O.O. Health risks associated with exposure to filamentous fungi. *Int. J. Environ. Res. Public Health* **2017**, *14*, 719. [CrossRef]
100. Dionne, B.; Neff, L.; Lee, S.A.; Sutton, D.A.; Wiederhold, N.P.; Lindner, J.; Fan, H.; Jakeman, B. Pulmonary fungal infection caused by *Neoscytalidium dimidiatum*. *J. Clin. Microbiol.* **2015**, *53*, 2381–2384. [CrossRef]
101. Salehi, M.; Zibafar, E.; Mahmoudi, S.; Hashemi, S.; Gatmiri, S.; Shoar, M.G.; Manshadi, S.D.; Jahanbin, B.; Alizadeh, R.; Hosseinpour, L.; et al. First report of invasive pulmonary infection by *Didymella microchlamydospora* and successful treatment with voriconazole. *Clin. Microbiol. Infect.* **2019**, *25*, 392–393. [CrossRef] [PubMed]
102. Stricker, S.M.; Gossen, B.D.; McDonald, M.R. Risk assessment of secondary metabolites produced by fungi in the genus *Stemphylium*. *Can. J. Microbiol.* **2021**, *67*, 445–450. [CrossRef] [PubMed]
103. Gaona-Álvarez, C.; González-Velasco, C.; Morais-Foruria, F.; Alastruey-Izquierdo, A. Unusual aetiology of keratitis in a patient with bullous keratopathy. *Enfermedades Infecc. Y Microbiol. Clin.* **2020**, *38*, 84–85. [CrossRef] [PubMed]

104. Garcia-Hermoso, D.; Valenzuela-Lopez, N.; Rivero-Menendez, O.; Alastruey-Izquierdo, A.; Guarro, J.; Cano-Lira, J.F.; Stchigel, A.M. Diversity of coelomycetous fungi in human infections: A 10-y experience of two European reference centres. *Fungal Biol.* **2019**, *123*, 341–349. [CrossRef]
105. Ruibal, C.; Gueidan, C.; Selbmann, L.; Gorbushina, A.; Crous, P.; Groenewald, J.; Muggia, L.; Grube, M.; Isola, D.; Schoch, C.; et al. Phylogeny of rock-inhabiting fungi related to Dothideomycetes. *Stud. Mycol.* **2009**, *64*, 123–133. [CrossRef] [PubMed]
106. Coleine, C.; Stajich, J.E.; de Los Ríos, A.; Selbmann, L. Beyond the extremes: Rocks as ultimate refuge for fungi in drylands. *Mycologia* **2021**, *113*, 108–133. [CrossRef]
107. Ramanauskas, R.; Juzeliûnas, E.; Buèinskienë, D.; Lugauskas, A.; Peèiulytë, D. Investigation of microbiologically influenced corrosion 1. Characterization of natural outdoor conditions in Lithuania. *Chemija* **2005**, *16*, 25–34.
108. De Leo, F.; Urzì, C.; de Hoog, G.S. Two Coniosporium species from rock surfaces. *Stud. Mycol.* **1999**, *1999*, 70–79.
109. Crous, P.; Luangsa-Ard, J.; Wingfield, M.; Carnegie, A.; Hernández-Restrepo, M.; Lombard, L.; Roux, J.; Barreto, R.; Baseia, I.; Cano-Lira, J.; et al. Fungal Planet description sheets: 785–867. *Persoonia* **2018**, *41*, 238–417. [CrossRef]
110. Hughes, S.J. Conidiophores, conidia, and classification. *Can. J. Bot.* **1953**, *31*, 577–659. [CrossRef]
111. Quaedvlieg, W.; Binder, M.; Groenewald, J.Z.; Summerell, B.A.; Carnegie, A.J.; Burgess, T.I.; Crous, P.W. Introducing the Consolidated Species Concept to resolve species in the Teratosphaeriaceae. *Persoonia* **2014**, *33*, 1–40. [CrossRef] [PubMed]
112. von Arx, J.A.; Müller, E. Über die neue Ascomycetengattung Aulographina. *Sydowia* **1960**, *14*, 330–333.
113. Wang, M.-M.; Shenoy, B.D.; Li, W.; Cai, L. Molecular phylogeny of *Neodevriesia*, with two new species and several new combinations. *Mycologia* **2017**, *109*, 965–974. [CrossRef]
114. Crous, P.W.; Schumacher, R.K.; Wingfield, M.J.; Lombard, L.; Giraldo, A.; Christensen, M.; Gardiennet, A.; Nakashima, C.; Pereira, O.L.; Smith, A.; et al. Fungal systematics and evolution: FUSE 1. *Sydowia* **2015**, *67*, 118. [CrossRef]

Journal of *Fungi*

MDPI

*Article*

# A Deep Insight into the Diversity of Microfungal Communities in Arctic and Antarctic Lakes

**Alessia Marchetta [1,†], Maria Papale [2,†], Alessandro Ciro Rappazzo [2], Carmen Rizzo [2,3], Antonio Camacho [4], Carlos Rochera [4], Maurizio Azzaro [2], Clara Urzì [1], Angelina Lo Giudice [2] and Filomena De Leo [1,*]**

1 Department of Chemical, Biological, Pharmaceutical and Environmental Sciences, University of Messina, Viale F. Stagno d'Alcontres, 31, 98166 Messina, Italy

2 Institute of Polar Sciences, National Research Council (CNR-ISP), Spianata S. Raineri 86, 98122 Messina, Italy; angelina.logiudice@cnr.it (A.L.G.)

3 Stazione Zoologica Anton Dohrn, National Institute of Biology, Sicily Marine Centre, Department Ecosustainable Marine Biotechnology, Villa Pace, Contrada Porticatello 29, 98167 Messina, Italy

4 Cavanilles Institute of Biodiversity and Evolutionary Biology, University of Valencia, C/Catédratico José Beltrán, 2, E46980 Paterna, Spain

* Correspondence: fdeleo@unime.it; Tel.: +39-0906765201

† These authors contributed equally to this work.

**Abstract:** We assessed fungal diversity in water and sediment samples obtained from five Arctic lakes in Ny-Ålesund (Svalbard Islands, High Arctic) and five Antarctic lakes on Livingston and Deception Islands (South Shetland Islands), using DNA metabarcoding. A total of 1,639,074 fungal DNA reads were detected and assigned to 5980 ASVs amplicon sequence variants (ASVs), with only 102 (1.7%) that were shared between the two Polar regions. For Arctic lakes, unknown fungal taxa dominated the sequence assemblages, suggesting the dominance of possibly undescribed fungi. The phylum *Chytridiomycota* was the most represented in the majority of Arctic and Antarctic samples, followed by *Rozellomycota*, *Ascomycota*, *Basidiomycota*, and the less frequent *Monoblepharomycota*, *Aphelidiomycota*, *Mortierellomycota*, *Mucoromycota*, and *Neocallimastigomycota*. At the genus level, the most abundant genera included psychrotolerant and cosmopolitan cold-adapted fungi including *Alternaria*, *Cladosporium*, *Cadophora*, *Ulvella* (Ascomycota), *Leucosporidium*, *Vishniacozyma* (*Basidiomycota*), and *Betamyces* (*Chytridiomycota*). The assemblages displayed high diversity and richness. The assigned diversity was composed mainly of taxa recognized as saprophytic fungi, followed by pathogenic and symbiotic fungi.

**Keywords:** Antarctica; Arctic; polar lakes; extreme environments; fungal community; fungal diversity; fungal ecology; next generation sequencing

## 1. Introduction

The Polar regions (both Arctic and Antarctic) share the common characteristics of extremely harsh climate and offer us a unique and irreplaceable platform to discover and study extremophilic organisms [1,2]. A major feature of Arctic landscapes is the large number of lakes and ponds, which in some regions can cover up to 90% of the total surface area [3]. They contribute significantly to Arctic biodiversity, offering a diverse range of habitats for aquatic organisms, from microorganisms to animals and plants, and providing food and freshwater to migratory nesting birds, resident animals, and humans [4]. Also, the Antarctic continent and sub-Antarctic islands represent some of the most diverse and interesting lake districts on the planet [5]. Apart from subglacial lakes, here, lake ecosystems are found in the limited ice-free areas or oases, where microbial communities of viruses, archaea, bacteria, microalgae, and fungi represent the largest reservoir of biodiversity [6].

Polar regions host an extraordinary diversity of lake types, ranging from freshwater to hypersaline, from highly acidic to alkaline and from perennially ice-covered waters to concentrated brines that never freeze. Many polar lakes are ultra-oligotrophic or extremely unproductive but lakes highly enriched by animal or human activities can also be found [7].

High latitude lakes have broad global significance, acting as residences of unique species and communities and as early detectors of environmental change. Multiple stressors associated with local and global human impact such as contaminant influxes, increased exposures to ultraviolet radiation and climate change have a striking impact on these aquatic ecosystems. In fact, even small changes in physical, chemical or biological characteristics can be amplified into major shifts in limnological properties and in lake ecosystem structure and functioning [8–11]. The combination of the above-mentioned features makes polar lakes interesting and unique environments to study the taxonomy and ecology of microbial communities living under extreme conditions.

Among the different microbial groups present in polar lakes, fungi are the biggest osmotrophic specialists, producing a plethora of secretory enzymes and obtaining nutrients through extracellular digestion and endocytosis. Thanks to diverse metabolic strategies and high morphological diversity, fungi have conquered numerous ecological niches and have shared various interactions with other living organisms and inorganic surfaces. In fact, fungi are found virtually in all environments throughout the globe, including extreme environments, such as torrid and polar deserts [12,13], hypersaline salterns [14], and deep-sea [15].

The global diversity of fungi was first estimated by Hawksworth [16] to be 1.5 million species. However, the increasing development of DNA sequencing technologies which occurred in the last ten years led to expanding the estimates of fungal species numbers to 2.2–13.2 millions. Despite this high estimate diversity, to date, only around 150,000 fungal species have been described [17]. Fungi constitute a well-founded component of terrestrial ecology due to more than 100 years of research that has highlighted their role in biogeochemical cycling and promoting biodiversity [18], while aquatic ecosystems, in contrast, were long overlooked as fungal habitats. However, fungal diversity, quantitative abundance, ecological functions and, in particular, their interactions with other microorganisms remain mainly speculative, unexplored and missing from current general concepts in aquatic ecology and biogeochemistry [19].

Most previous studies of polar lake fungal communities have used traditional culture-dependent methods [1,10,20–24] which do not reveal the full complexity of the resident fungal diversity. Recent applications of metabarcoding approaches have focused especially on sediment of lakes of maritime Antarctica [25–27] and there is only one study on water samples from lakes of continental Antarctica [28]. Very few metagenomic studies have been conducted on Arctic lakes, exclusively from water samples [2,29,30].

In this paper, we used a metabarcoding approach to study the microfungal diversity both in water and sediment sampled from Arctic and Antarctic lakes, to study the fungal community composition of these two different lake matrices in both Polar regions deeper, and to better understand their ecological roles in such extreme environments.

## 2. Materials and Methods

### 2.1. Study Sites and Samples Description

Two sampling campaigns were conducted between 5 and 18 August 2021 in the area of Ny-Ålesund (Svalbard Archipelago, High-Arctic Norway) and between 25 January and 1 February 2022 in Livingston and Deception Islands (South Shetland Islands, Antarctica), respectively. Water and sediment samples were collected from the littoral zone of five Arctic lakes which include Solvannet (L1), Glacier (L2), Knudsenheia (L3), Storvatnet (L4), Tvillingvatnet (L5), and five Antarctic lakes including Argentina (LA), Sofia (LS), Balleneros (LB), Telefon (LT), and Zapatilla (LZ). At each sampling point, the physical–chemical parameters were also measured, particularly temperature, pH, Oxygen ($O_2$%), and conductivity (uS/cm), using manual field probes (a CyberScan PC 300 probe was used for Conductivity, pH, and temperature, while oxygen concentration was measured by a

Hanna HI9143 probe). Sampling locations and physical-chemical data for each sampling point are shown in Table 1 and Figure 1.

**Table 1.** Geographical and physical–chemical data for each sampling site.

| Area | Lake | Sample ID | Coordinates | | Physical-Chemical Parameters | | | |
|---|---|---|---|---|---|---|---|---|
| | | | | | Water Temperature(°C) | pH | $O_2$ % | Cond (uS/cm) |
| Ny-Ålesund(Arctic) | Solvannet | L1 | 78°55′33.121″ N | 11°56′19.618″ E | 6.53 | 8.19 | 100.77 | 398 |
| | Glacier | L2 | 78°55′2.64″ N | 11°47′26.52″ E | 8.83 | 7.85 | 98.67 | 150 |
| | Knudsenheia | L3 | 78°56′40.801″ N | 11°51′34.74″ E | 9.00 | 8.40 | 106.50 | 2660 |
| | Storvatnet | L4 | 78°55′27.181″ N | 11°52′43.68″ E | 7.90 | 8.09 | 102.67 | 241 |
| | Tvillingvatnet | L5 | 78°55′3.4788″ N | 11°51′55321″ E | 8.13 | 7.82 | 102.47 | 222 |
| Livingston Island(Antarctica) | Argentina | LA | 62°40′22.39″ S | 60°24′18.12″ W | 0.20 | 5.56 | 66.00 | 65 |
| | Sofia | LS | 62°40′12.19″ S | 60°23′17.90″ W | 0.33 | 5.47 | 84.48 | 20 |
| Deception Island(Antarctica) | Balleneros | LB | 62°58′51.1″ S | 60°34′27.1″ W | 6.04 | 3.60 | 83.80 | 423 |
| | Telefon | LT | 62°55′39.9″ S | 60°41′21.3″ W | 7.37 | 6.03 | 86.12 | 467 |
| | Zapatilla | LZ | 62°59′00.24″ S | 60°40′29.07″ W | 6.80 | 5.30 | 84.65 | 81 |

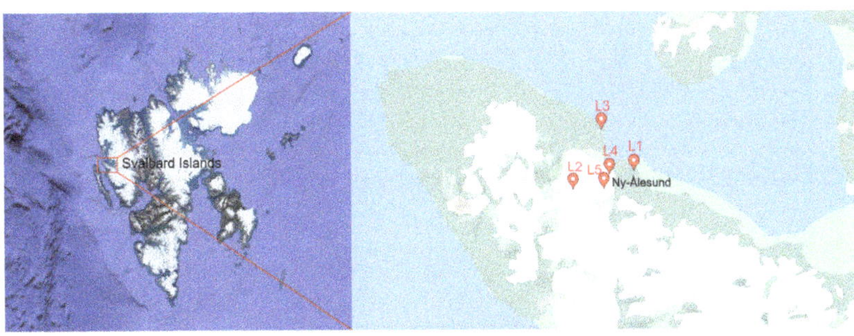


(a)

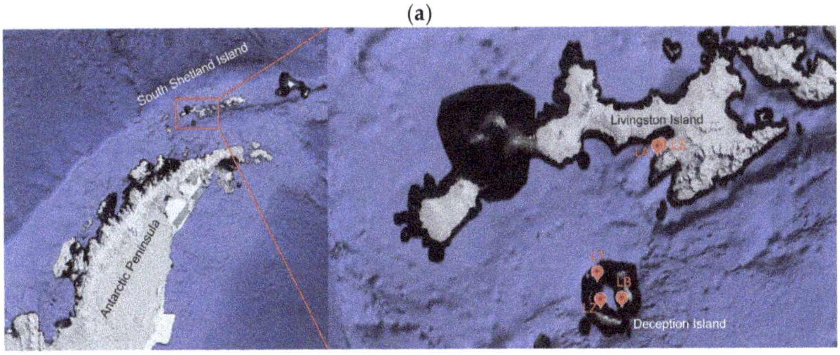


(b)

**Figure 1.** Maps showing the location of the sampling sites in (**a**) Ny-Ålesund area (Svalbard Islands) and in (**b**) Livingston and Deception Islands (Antarctica). L1 = Lake Solvannet, L2 = Lake Glacier, L3 = Lake Knudsenheia, L4 = Lake Storvatnet, L5 = Lake Tvillingvatnet, LA = Lake Argentina, LS = Lake Sofia, LB = Lake Balleneros; LT = Lake Telefon, LZ = Lake Zapatilla.

Water samples (n = 10, identified with "w") were manually collected using a presterilized 2 L-plastic bottle and immediately transported to the labs of the research station to be processed. Here, 1 L of water samples were filtered on polycarbonate membranes (diameter 47 mm; 0.22 μm pore size), in triplicate, and immediately frozen at −20 °C for transport until sample processing in the laboratory at CNR-ISP of Messina, Italy. Surface sediment samples (n = 10, identified with "s") were collected at the interface water sediments (water depth of 30–60 cm). The first 10 cm of the surface sediment were sampled using a precleaned scoop and pre-sterilized plastic containers, and immediately transported to the labs of the research station where they were directly stored at −20 °C until sample processing. Samples were than processed in the laboratory at CNR-ISPof Messina, Italy.

### 2.2. Total DNA Extraction, Bioinformatic Analyses and Fungal Identification

Total DNA was extracted from membranes and from 1 g of sediment using the DNeasy® PowerSoil® Pro Kit (Qiagen, Hilden, Germany) according to the manufacturer's instructions. Before the DNA extraction, the samples were gradually thawed at 4 °C. In all DNA extraction steps, we proceeded under strict control conditions within a laminar flow hood to recover the fungal DNA and avoid contaminations. DNA concentrations and purity were quantified by a NanoDrop ND-1000 UV–Vis spectrophotometer. Extracted DNA was used as a template for generating PCR amplicons. The internal transcribed spacer 2 region (ITS2) of the nuclear ribosomal DNA was used as a DNA barcode for molecular species identification. Fungal ITS2 was amplified using the following primers: IlluAdp_ITS31_NeXTf 5′-CATCGATGAAGAACGCAG-3′ and IlluAdp_ITS4_NeXTr5′-TCCTSCGCTTAT TGATATGC-3′ [31]. Sequencing was performed using the Illumina MiSeq platforms, in paired-end form, following the standard protocols of the company EurofinsEurope Services (Germany). FastQC was used to check the quality of raw sequences [32]. Sequences were preprocessed, quality filtered, trimmed, de-noised, merged, modeled, and analyzed by R package DADA2 [33] to infer amplicon sequence variants (ASVs), i.e., biologically relevant variants rather than an arbitrarily clustered group of similar sequences. Particularly we filtered reads by length (minimum length between 150 and 140 bp), by ambiguous bases (no reads with N base were maintained in the analysis), and all sequences were trimmed at the ends after quality control (trimLeft = 17, trimRight = 15). During the analysis, filters for reducing replicate, length, and chimera errors were also applied. Fungal taxonomy annotation was performed using the ITS fungal database, UNITE—Unified system for the DNA based fungal species linked to the classification [34], formatted for DADA2, offering an updated framework for annotating fungal taxonomy (unified system for the DNA based fungal species linked to the classification, identity, similarity used cutoff 95%). Finally, a manual inspection was done, and sequences with an abundance of below 0.1% were considered together in the minor groups of retrieved fungi. All sequences have been submitted to the National Center for Biotechnology Information (NCBI) under the BioProject PRJNA1000778.

### 2.3. Fungal Diversity, Distribution and Predictive Functional Profiling

The numbers of reads obtained for each water and sediment sample were used to quantify taxon alpha diversity, richness and dominance, using the following indices: Fisher, Shannon, Chao1, ACE, Simpson and InvSimpson. Venn diagrams were prepared using the retrieved ASVs using InteractiVenn online tool [35] to compare the fungal assemblages present in the lake samples. Functional assignments of fungal ASVs at species and genus levels were analysed using tool FUNGuild [36]. FUNGuild v1.1 is a flat database hosted by GitHub (https://github.com/UMNFuN/FUNGuild (accessed on 20 May 2023)), accessible for use and annotation by any interested party under GNU General Public License.

### 2.4. Statistical Analyses

To compare the fungal community compositions across groups of samples, the Bray–Curtis similarity analysis was performed and similarity matrices were used to obtain dendrograms using R base packages. Principal component analyses (PCAs) were performed using the factoextra R package, on data from selected physical and chemical properties of sediments and waters, and the relative abundance of significant fungal groups. The environmental variables used in these analyses were as follows: oxygen ($O_2$%), temperature (°C), water electrical conductivity (Cond uS/cm), and pH.

## 3. Results

### 3.1. Influence of Environmental Paramenters in Lake Clustering

In the principal component analysis of the physico-chemical parameters, it was possible to observe that lakes were completely separated by environmental factors. In fact, the Arctic and Antarctic samples were completely distinct and generated two different groupings, with the only exception represented by L3, which was strictly related to the conductivity, due to its closeness to the sea cost and therefore it is considerably influenced by salt water. Results are shown in Figure 2.

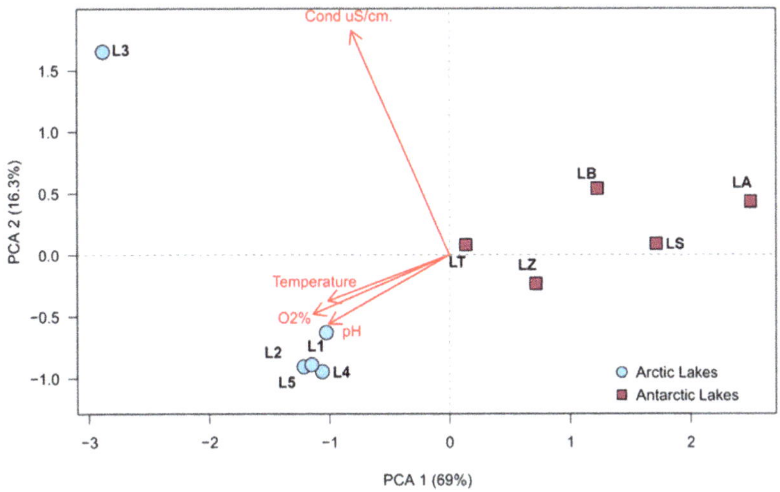


**Figure 2.** Principal component analysis obtained by recorded environmental parameters, made by factoextra R package.

### 3.2. Fungal Taxonomy

A total of 1,639,074 DNA merged reads of good quality were detected in the water and sediment samples from the ten lakes, with an average length of between 350 and 450 bp, representing 5980 ASVs. Unfortunately, two of the sequenced samples (i.e., L1-s and L2-s) did not produce good results in the first enrichment steps, with creation of a low-quality library and subsequently the impossibility to continue with the NGS sequencing. This was probably due to low the ITS DNA quantity, even though the total extracted DNA showed high concentration and good quality (Table S1). In the Arctic lakes, the analysis of phyla showed that almost in all samples an average of 50% ASVs were related to Fungi_phy_Incertae_sedis, consisting of ASVs whose taxonomical relationships and positions are unknown or not defined. The dominant phylum was represented by *Chytridiomycota* in all samples examined, except for L4-s, where the phylum *Rozellomycota* was most represented. The highest value of ASVs assigned to *Chytridiomycota* was observed in L4-w (46.2%) and the lowest value in L5-w (4.9%). *Rozellomycota* was the second most represented phylum, but with an uneven distribution: the highest abundance value was 20.5% (L4-s) and the lowest was 0.16% (L1-w). The phyla *Ascomycota* and *Monoblepharomycota* were retrieved with percentages higher than 1% only in L5-s (5.49%) and L3-s (2.41%), respectively. Finally, the phylum *Basidiomycota* was found with percentages higher than 1% in L2-w (1.52%) and L5-s (3.76%). The fungal community structure in water and sediment samples of the Arctic lakes is shown in Figure 3.

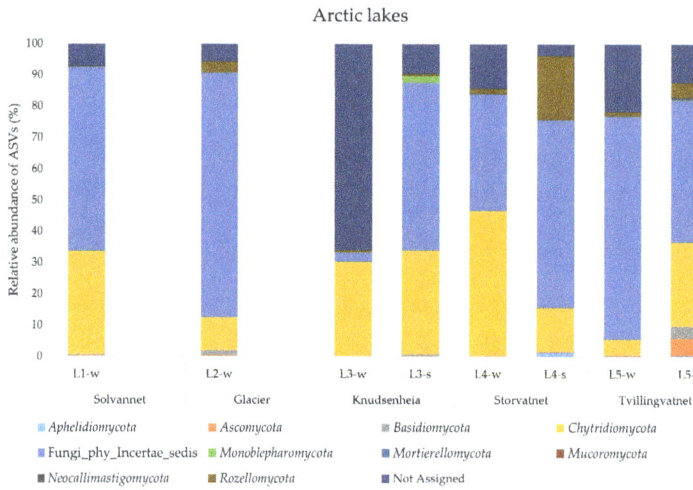


**Figure 3.** Fungal community structure at the phylum level in the water and sediment samples from Arctic lakes.

Overall, the same phyla were retrieved in Antarctic samples, but with some crucial differences. The results are shown in Figure 4. First, Fungi_phy_Incertae_sedis showed an average percentage of around 26% and were about completely absent (0.04%) in LA-w, underlining a fungal community composition related mostly to known phyla. In addition, in Antarctic samples *Chytridiomycota* were the most abundant phylum and ubiquitously distributed in all analyzed samples. Their highest value (68.5%) was retrieved of LS-w and the lowest value was obtained for LT-w (12.85%). The phylum *Ascomycota* in Antarctic lakes was retrieved with a higher value than the Arctic samples. It was found in all lakes with a value comprised between 65.8% (water of LA) and 0.45% (LZ-w). Phyla *Rozellomycota* and *Monoblepharomycota* were retrieved with percentages higher than 1% only in LA-s (40.3%) and LZ-w (3.16%), respectively. Finally, the phylum *Basidiomycota* was retrieved in all Antarctic samples with an average percentage of 4.5%. Their highest value was found in LB-w (23.9%) and the lowest in LZ-w (0.06%).

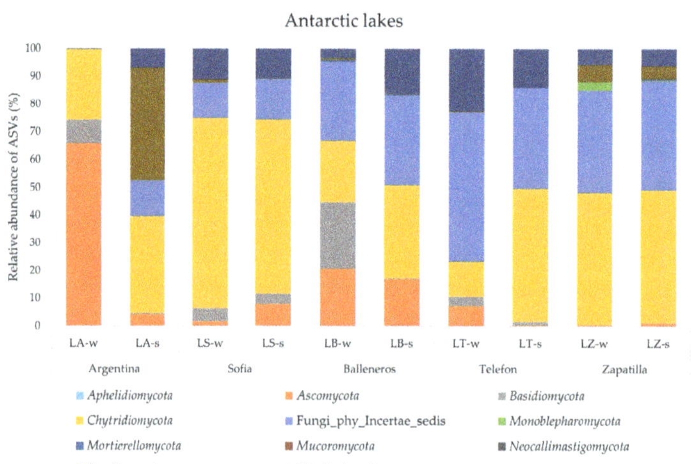


**Figure 4.** Fungal community structure at the phylum level in the water and sediment samples from Antarctic lakes.

Based on all the retrieved phyla, a principal component analysis (PCA) was performed, and the results were used for selecting fungal taxa that had high variance (*Ascomycota*, *Basidiomycota*, *Chytridiomycota*, *Rozellomycota* and Fungi_phy_Incertae_sedis). With that, then was performed the construction of the PCA to identify groups of samples with similar community compositions (Figure 5).

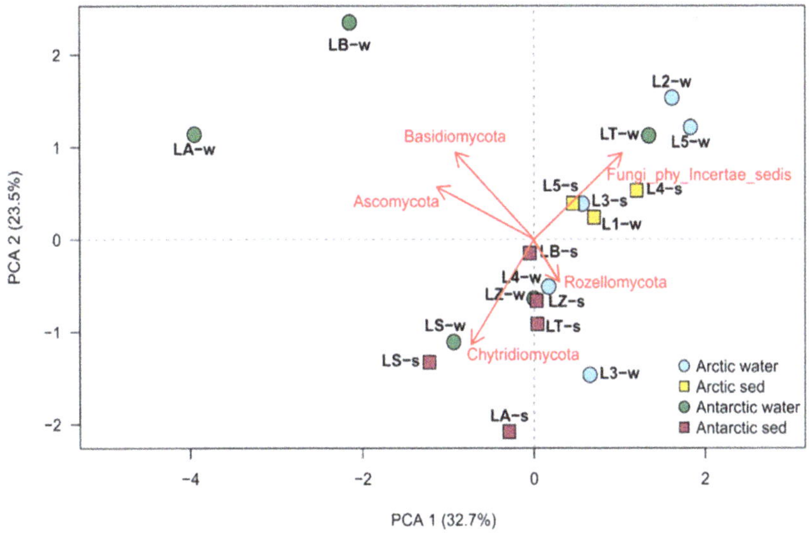


**Figure 5.** Principal component analysis obtained with the retrieved phyla information parameters, made by factoextra R package.

The PCA showed a distinction between the Arctic and Antarctic samples, in fact six of the eight Arctic samples clustered together and their separation was driven by Fungi_phy_Incertae_sedis. Two exceptions in the Arctic cluster were water from both L4 and L3 which were found to be related to most Antarctic samples and related to the presence of *Rozellomycota*. Finally, two water samples of Antarctic region (LA and LB) were completely separated by other groups and noticeably related to the phyla *Ascomycota* and *Basidiomycota*.

The most abundant genera (percentage above 1% for at least one sample) were summarized in a heat map (Table 2). Overall, a total of 17 and 50 genera were detected in lakes from the Arctic and Antarctic regions (Table 2), respectively. In the Arctic, L5-s showed most of the genus-wide affiliated sequences. *Betamyces* was the most abundant genus in the Arctic lakes, and it was retrieved in all analysed samples. In particular, the highest abundance was retrieved in L5-s (16.6%). The genera *Pseudeurotium* and *Amylocorticiellum* were retrieved with more than 1% of abundance (2.1% and 2.6%, respectively) in L5-s, together with the genus *Leucosporidium* (1.4%) in L2-w. The genus *Pseudeurotium* was present in almost all samples, the genus *Leucosporidium* was present in half of the Arctic samples, instead *Amylocorticiellum* and was only found in L5-s. For Antarctic samples, LB-w showed most of the genus-wide affiliated sequences. *Cladosporium* represented the most abundant genus retrieved, with the highest value that was observed in LA-w (39.5%). Similarly, the genus *Cadophora* was mainly retrieved in LA-w with a percentage of 19.5%, followed by the genera *Malassezia* (8.38%) and *Alternaria* (5.9%), with the latter that was found exclusively in this sample. *Ulvella* was the most abundant genus in LB-w and LB-s (14.2% and 13.3%, respectively), followed by *Vishniacozyma*, which was retrieved with a percentage of 7.5% in LB-w. The genus *Betamyces* was found in all studied samples from Antarctic lakes, exception for LT and from LB-s, with the highest abundance in LZ-s (4.32%). In LS, *Coleophoma* was the most represented genus, both in sediment (6.51%) and water

(1.05%). The ascomycetous genus *Metschnikowia* was the most abundant (3%) in LT-w. Interesting to note is that even though Arctic samples showed very few numbers of genera, *Amylocorticiellum, Helicodendron, Iodophanus, Scolecolachnum* (all found in L5-s), *Haptocillium* (L4-s), *Knufia* (L3-w), and *Zygophlyctis* (L1-w) were exclusively present in the Artic and completely absent in Antarctica.

### 3.3. Fungal Diversity and Distribution

A total of 3334 ASVs were obtained from the Arctic lake samples (both water and sediment) and, of them, only 3 ASVs (0.09%) were shared between the two matrices (Table S2). Overall, lakes L4 vs. L5 and L1 vs. L3 shared the highest percentages of ASVs (7.1%and 6.9%, respectively), while L1 vs. L5 shared the lowest percentages of ASVs (2.8%). Similarly, the difference in ASV distribution between water and sediment samples was evaluated, trying to understand if lakes showed similar populations, or water and sediment have different inhabitants. Considering the two different matrices, 1732 ASVs and 1881 ASVs were obtained from Arctic water and sediment samples, respectively, with 279 ASVs (8.36%) being in common. Comparing the ASVs obtained exclusively from water samples from each Arctic lake, L1-w vs. L3-w shared the highest percentage of ASVs, followed by L2-w vs. L5-w and L4-w vs. L5-w (16.5%, 9.3% and 8.2%, respectively), while the lowest percentages of shared ASVs was observed for samples L3-w vs. L4-w (3.2%) (Table S2). Lower percentages of shared ASVs were obtained from sediment samples L3-s vs. L4-s, L3-s vs. L5-s, and L4-s vs. L5-s (1.6%, 3.4% and 3.8%, respectively) (Table S2). Antarctic lake samples (both water and sediment) gave a total of 2748 ASVs and, of them, 4 ASVs (0.15%) were shared among samples (Table S2). The highest percentages of ASVs were shared between LA vs. LS and LB vs. LT (11.6% and 10.2%, respectively), while LT vs. LZ shared the lowest percentages of ASVs (1.9%). In particular, 1270 ASVs and 1843 ASVs were obtained from Antarctic water and sediment samples, respectively, and a total of 365 ASVs (13.28%) were shared between the two matrices (Table S2). With regard to ASVs obtained exclusively from water samples from each Antarctic lake, LB-w vs. LT-w shared the highest percentage of ASVs (12.5%), while the lowest percentages of shared ASVs was observed for samples LA-w vs. LT-w (1.3%) (Table S2). For ASVs obtained exclusively from sediment samples of each Antartic lake, LA-s vs. LS-s shared the highest percentages of ASVs, followed by LB-s vs. LT-s (13.4 and 7.2%, respectively). Only 0.5% of ASVs were shared between LT-s and LZ-s (Table S2). Considering the two Polar regions, of the total 5980 ASVs retrieved, only 1.7% (102 ASVs) were shared between the Arctic and Antarctic samples (Table S2). The percentage of common ASVs between water and sediment samples were higher than that underlined between the Arctic and Antarctic samples, showing a value of 10.87% (Table S2). Finally, following the above results, the ASVs were examined separated by region (Arctic and Antarctic) and by matrix (water and sediment). The results are shown in Figure 6. Arctic vs. Antarctic water and Arctic vs. Antarctic sediment showed really low percentage of common ASVs (1.79% and 1.16%, respectively). Put differently, water vs. sediment from Arctic and water vs. sediment from Antarctic region, shared higher number of ASVs (8.36% and 13.28%, respectively).

**Table 2.** Genera retrieved at a percentage above 1% in Arctic and Antarctic regions.

| Phylum | Genus | Arctic | | | | | | | | Antarctic | | | | | | | | | |
|---|---|---|---|---|---|---|---|---|---|---|---|---|---|---|---|---|---|---|---|
| | | Solvann | Glacier | Knudsenheia | | Storvatnet | | Tvillingvatnet | | Argentina | | Sofia | | Balleneros | | Telefon | | Zapatilla | |
| | | L1-w | L2-w | L3-w | L3-s | L4-w | L4-s | L5-w | L5-s | LA-w | LA-s | LS-w | LS-s | LB-w | LB-s | LT-w | LT-s | LZ-w | LZ-s |
| Ascomycota | *Alatospora* | | | | | | | | | | | | | | | | | | |
| | *Alfaria* | | | | | | | | | | | | | | | | | | |
| | *Alternaria* | | | | | | | | | | | | | | | | | | |
| | *Antarctolichenia* | | | | | | | | | | | | | | | | | | |
| | *Arthroderma* | | | | | | | | | | | | | | | | | | |
| | *Aspergillus* | | | | | | | | | | | | | | | | | | |
| | *Beauveria* | | | | | | | | | | | | | | | | | | |
| | *Cadophora* | | | | | | | | | | | | | | | | | | |
| | *Candida* | | | | | | | | | | | | | | | | | | |
| | *Cladosporium* | | | | | | | | | | | | | | | | | | |
| | *Coleophoma* | | | | | | | | | | | | | | | | | | |
| | *Collophora* | | | | | | | | | | | | | | | | | | |
| | *Coniosporium* | | | | | | | | | | | | | | | | | | |
| | *Debaryomyces* | | | | | | | | | | | | | | | | | | |
| | *Geomyces* | | | | | | | | | | | | | | | | | | |
| | *Haptocillium* | | | | | | | | | | | | | | | | | | |
| | *Helicodendron* | | | | | | | | | | | | | | | | | | |
| | *Heydenia* | | | | | | | | | | | | | | | | | | |
| | *Hyaloscypha* | | | | | | | | | | | | | | | | | | |
| | *Iodophanus* | | | | | | | | | | | | | | | | | | |
| | *Knufia* | | | | | | | | | | | | | | | | | | |
| | *Mastodia* | | | | | | | | | | | | | | | | | | |
| | *Metschnikowia* | | | | | | | | | | | | | | | | | | |
| | *Nectriopsis* | | | | | | | | | | | | | | | | | | |
| | *Neobulgaria* | | | | | | | | | | | | | | | | | | |
| | *Penicillium* | | | | | | | | | | | | | | | | | | |
| | *Polyphilus* | | | | | | | | | | | | | | | | | | |
| | *Pseudeurotium* | | | | | | | | | | | | | | | | | | |
| | *Pseudogymnoascus* | | | | | | | | | | | | | | | | | | |
| | *Scolecolachnum* | | | | | | | | | | | | | | | | | | |
| | *Tetracladium* | | | | | | | | | | | | | | | | | | |
| | *Thelebolus* | | | | | | | | | | | | | | | | | | |

**Table 2.** *Cont.*

| Phylum | Genus | Arctic | | | | | | | | Antarctic | | | | | | | | | |
|---|---|---|---|---|---|---|---|---|---|---|---|---|---|---|---|---|---|---|---|
| | | Solvannet | Glacier | Knudsenheia | | Storvatnet | | Tvillingvatnet | | Argentina | | Sofia | | Balleneros | | Telefon | | Zapatilla | |
| | | L1-w | L2-w | L3-w | L3-s | L4-w | L4-s | L5-w | L5-s | LA-w | LA-s | LS-w | LS-s | LB-w | LB-s | LT-w | LT-s | LZ-w | LZ-s |
| *Ascomycota* | *Ulvella* | | | | | | | | | | | | | | | | | | |
| *Basidiomycota* | *Amylocorticiellum* | | | | | | | | | | | | | | | | | | |
| | *Camptobasidium* | | | | | | | | | | | | | | | | | | |
| | *Coriolopsis* | | | | | | | | | | | | | | | | | | |
| | *Cryolevonia* | | | | | | | | | | | | | | | | | | |
| | *Cryptococcus* | | | | | | | | | | | | | | | | | | |
| | *Cutaneotrichosporon* | | | | | | | | | | | | | | | | | | |
| | *Cystofilobasidium* | | | | | | | | | | | | | | | | | | |
| | *Glaciozyma* | | | | | | | | | | | | | | | | | | |
| | *Hyphodermella* | | | | | | | | | | | | | | | | | | |
| | *Leucosporidium* | | | | | | | | | | | | | | | | | | |
| | *Malassezia* | | | | | | | | | | | | | | | | | | |
| | *Mrakia* | | | | | | | | | | | | | | | | | | |
| | *Naganishia* | | | | | | | | | | | | | | | | | | |
| | *Phenoliferia* | | | | | | | | | | | | | | | | | | |
| | *Pseudobensingtonia* | | | | | | | | | | | | | | | | | | |
| | *Scopuloides* | | | | | | | | | | | | | | | | | | |
| | *Sidera* | | | | | | | | | | | | | | | | | | |
| | *Vishniacozyma* | | | | | | | | | | | | | | | | | | |
| *Chytridiomycota* | *Betamyces* | | | | | | | | | | | | | | | | | | |
| | *Entophlyctis* | | | | | | | | | | | | | | | | | | |
| | *Lobulomyces* | | | | | | | | | | | | | | | | | | |
| | *Zygophlyctis* | | | | | | | | | | | | | | | | | | |
| *Mortierellomycota* | *Entomortierella* | | | | | | | | | | | | | | | | | | |
| | *Mortierella* | | | | | | | | | | | | | | | | | | |

Abundance range in percentage value (%). 0 – 40

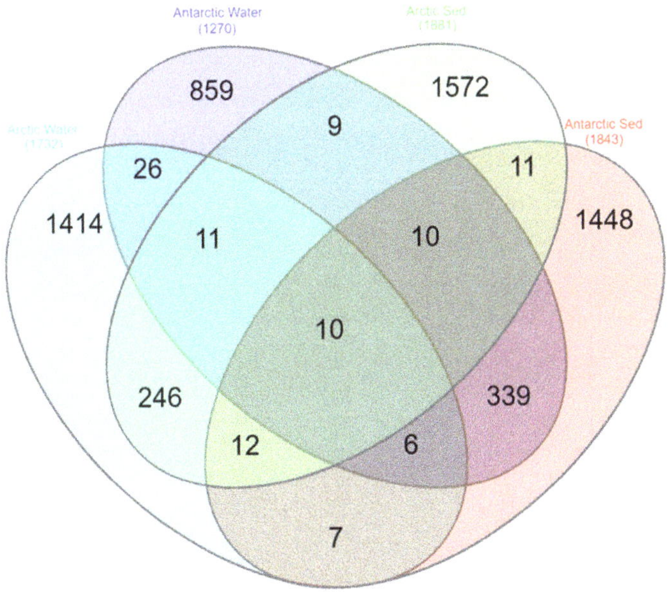


**Figure 6.** Venn diagram generated using all the retrieved ASVs separated by region (Arctic and Antarctic) and matrix of origin (water and sediment). The diagram was made by InteractiVenn online tool [35].

The diversity indices were calculated for each water and sediment sample, based on final ASVs obtained after bioinformatics analyses (Table 3). Sediment showed slightly higher diversity values (mean of indices value: Chao1 512.9; ACE 513.3; Shannon 3.9; Simpson 0.9; InvSimpson 18.5; Fisher 72.5) if compared with water (mean of indices value: Chao1 354.1; ACE 354.1; Shannon 3.4; Simpson 0.8; InvSimpson 17.7; Fisher 48.2). The highest value of Shannon index was found in water sample of LS (4.82) (Livingston Island, Antarctica) and a comparable value was found also in the sediment of the same lake (4.49). Instead, the lowest Shannon diversity value of was retrieved in the water sample of L3 (Svalbard Island, High Arctic), and also sediment of the same lake showed a value lower than the average sediment Shannon index value (3.17).

### 3.4. Predicitive Functional Profiling of Fungal Communities

In total, a function was assigned to 103 ASVs (accounted for 1.7% of total ASVs). Genera with confidence level of "possible" were classified as "uncertained" and excluded from the functional analyses in this study. Retrieved function were assigned to saprotroph (57%), pathotroph (30.9%) and symbiotroph (12.1%) (Table S3). For the Arctic samples, the functionality was assigned to a minor number of ASVs, and pathotroph and saprotroph showed 3.03% and 6.7% in L2 and, 12.7% and 18.8% in L5, respectively. Particularly, the saprotrophic function was underlined for the ASV 24 affiliated to the taxon *Betamyces* (L5). Interestingly, it was also observed the assignment of symbiotroph function to a great number of sequences retrieved in LA and related to the taxon *Cadophora*. In general, a predictive ecological function could be assigned to a very low percentage of ASVs. This was probably due to the fact that our samples come from scarcely studied environments and they have communities composed of a great number of organisms not yet identified.

**Table 3.** Diversity indices calculated using the total retrieved ASVs.

| Polar Region | Matrice | Sample | Observed | Chao1 | ACE | Shannon (H′) | Simpson | InvSimpson | Fisher |
|---|---|---|---|---|---|---|---|---|---|
| Arctic | Water | L1-w | 274 | 274.38 | 275.00 | 2.89 | 0.88 | 8.38 | 37.17 |
| | | L2-w | 563 | 563.40 | 564.21 | 4.80 | 0.98 | 53.50 | 80.69 |
| | | L3-w | 184 | 185.75 | 186.92 | 1.72 | 0.57 | 2.34 | 22.62 |
| | | L4-w | 585 | 587.02 | 589.17 | 4.10 | 0.95 | 19.78 | 86.61 |
| | | L5-w | 470 | 471.49 | 473.39 | 2.99 | 0.86 | 7.40 | 66.10 |
| | | mean | 415.2 | 416.41 | 417.74 | 3.3 | 0.85 | 18.28 | 58.64 |
| | Sediment | L3-s | 461 | 461.63 | 461.93 | 3.17 | 0.83 | 5.80 | 64.12 |
| | | L4-s | 645 | 645.32 | 645.81 | 4.41 | 0.96 | 26.24 | 95.50 |
| | | L5-s | 885 | 885.68 | 886.39 | 4.69 | 0.97 | 29.46 | 132.98 |
| | | mean | 663.7 | 664.21 | 664.71 | 4.09 | 0.92 | 20.5 | 97.54 |
| Antarctica | Water | LA-w | 102 | 111.55 | 113.20 | 2.45 | 0.85 | 6.58 | 10.86 |
| | | LB-w | 489 | 489.08 | 489.32 | 4.01 | 0.95 | 19.44 | 66.15 |
| | | LS-w | 390 | 390.75 | 390.92 | 4.82 | 0.98 | 43.99 | 54.75 |
| | | LT-w | 222 | 222.00 | 222.00 | 3.34 | 0.89 | 9.44 | 26.57 |
| | | LZ-w | 243 | 244.91 | 244.85 | 2.55 | 0.83 | 5.83 | 30.79 |
| | | mean | 289.2 | 291.66 | 292.06 | 3.43 | 0.9 | 17.06 | 37.83 |
| | Sediment | LA-s | 636 | 636.53 | 637.06 | 3.98 | 0.92 | 13.33 | 89.92 |
| | | LB-s | 409 | 409.06 | 409.39 | 3.87 | 0.95 | 20.38 | 55.38 |
| | | LS-s | 481 | 481.33 | 481.97 | 4.49 | 0.97 | 30.32 | 66.56 |
| | | LT-s | 186 | 186.00 | 186.00 | 2.79 | 0.85 | 6.68 | 21.76 |
| | | LZ-s | 398 | 398.13 | 398.37 | 3.80 | 0.94 | 15.41 | 53.39 |
| | | mean | 422 | 422.21 | 422.56 | 3.79 | 0.93 | 17.22 | 57.4 |

## 4. Discussion

Microorganisms are the dominant life forms in the Arctic and Antarctic regions. Amongst the groups of microorganisms occurring in these regions, fungi are one of the most abundant and better distributed in the various environments, playing a crucial role in the micro- and macro food webs. Particularly fungi inhabiting polar lakes play a key role in biogeochemical cycles and the mineralization of organic matter, which are essential for the balance of micro- and macronutrients in lake systems. Many fungal species display multiple stress tolerance capabilities, surviving the combination of low temperatures, high salinity, pH variation, seasonally high UV radiation and low nutrient availability experienced in different polar lakes [1,6,9]. However, despite their importance, the availability of studies of fungal diversity in polar lakes has increased only in recent years, but it still remains scant and fragmentary. To date, this work represents the first study of fungal communities in water and sediment of both Arctic and Antarctic lakes.

### 4.1. Fungal Diversity

The total fungal community detected through metabarcoding showed, within the analyzed lakes, comparable values of diversity for both Poles (Table 3), underlining no overall difference between the Arctic and Antarctic lakes, although in general, a slightly greater diversity for sediment than for water was observed. Comparable results were obtained by Perini et al. [30], who calculated fungal diversity in waters of a lake from Ny-Ålesund with a Shannon index H′ = 3.27, corresponding to the mean value obtained in this study (H′ = 3.3), while greater values (comprised between H′ = 3.83 and 5.24) were obtained by Zhang et al. [2]. The diversity data of fungal sequence assemblages detected in the sediment of Antarctic lakes LA, LB, LS, LT, and LZ studied here were greater than those reported in previous culture-based studies [1,10,24], and comparable with those reported in DNA metabarcode study by Ogaki et al. [25], de Souza et al. [26], Rosa et al. [27], and Gonçalves et al. [37] for other Antarctic lakes. However, results obtained by de Souza et al. [26] in sediment of Soto Lake, located in Deception Island (Antarctic Peninsula), which hosts three of the lakes examined in the present study (i.e., Balleneros, Telefon and Zapatilla), showed lower diversity indices (Fischer = 10.27). Although comparable diversity was observed between the studied lakes in the two regions (Arctic and Antarctic), a considerable difference was observed in terms of community composition. In fact, only 102 ASVs (out of 5980 ASVs; 1.7%) were shared between the Arctic and Antarctic samples. This result suggests that fungal distribution varies between the lakes in the two Polar regions and each counterpart hosts specific fungal taxa. Not only geographical distance, but also physical-chemical parameters that separate lakes of the two regions (as it is shown in Figure 2) and different sampling time could contribute to shaping the composition of the fungal community in lakes belonging to the two different regions.

### 4.2. Fungal Phyla

In Arctic lakes, unknown fungi dominated the sequence assemblages, with almost half of the obtained ASVs being assigned to Fungi_phy_Incertae_sedis. This assignation suggests the dominance of possibly undescribed fungi, or that these taxa provide examples of sequences not currently included in publicly accessible databases. The problem arises from the scarcity of metabarcode and metagenome studies of fungal communities in these polar ecosystems, in particular for Arctic lakes. This fact is also corroborated by the results obtained by the PCA of each fungal group for each sample (Figure 5), where almost all fungal assemblages in Arctic water and sediment correlated with the group of Fungi_phy_Incertae_sedis. The most represented identified phyla were, in order, *Chytridiomycota, Rozellomycota, Ascomycota, Basidiomycota,* which were commonly reported by Comeau et al. [29], Zhang et al. [2] and Perini et al. [30] in Arctic lakes. Less frequent, instead, were *Monoblepharomycota, Aphelidiomycota, Mortierellomycota, Neocallimastigomycota* and *Mucoromycota* which were never reported in Arctic lakes, but (excluding the phylum *Neocallimastigomycota*) were previously detected in Antarctic lakes [25–27,37].

In Antarctic samples, the percentage of ASVs assigned to unknown fungi was lower (average of 26.70%) than the ASVs obtained from Arctic samples, probably due to the higher number of studies of fungal communities in Antarctic lacustrine systems that increased considerably in recent years. *Chytridiomycota* results as the most represented phylum, followed in the order by *Ascomycota, Rozellomycota, Basidiomycota, Monoblepharomycota, Mortierellomycota* and *Aphelidiomycota.* Different studies showed that members of the phyla *Chytridiomycota* and *Cryptomycota* (*Rozellomycota*) dominated the fungal community composition in European freshwater lakes [38,39]. Similar results were also obtained in marine and polar freshwater environments [2,29,40] and the recent use of DNA metabarcoding approaches has revealed the presence of *Chytridiomycota* and *Rozellomycota* assigned sequences also in Antarctica, with reports from soil [41], air [42], mosses [43], permafrost [44], marine sediments [45], snow [46] and, recently, in lake sediments [25–27] and lake water [28]. The phyla *Rozellomycota* and *Chytridiomycota* have some physiological advantages for inhabiting aquatic ecosystems, including their mobility and capacity to parasitize numerous phytoplankton species such as diatoms, green algae, dinoflagellates and cyanobacteria [47,48]. Furthermore, the *Chytridiomycota* are implicated in a variety of ecological processes, such as the transfer of organic matter from phytoplankton into zooplankton via saprophytic and parasitic activity [49]. Taxa in this group are thought to mediate the transfer of organic matter from phytoplankton to zooplankton via saprophytic and parasitic activity described as the "mycoloop" [49]. The phylum *Ascomycota* was the most represented in water samples of LAwith a percentage of 65.87%. This dominance could be related to the origin of this lake, which derives from the ice-melting of a close glacier. In a recent DNA metabarcoding study [46], it was reported that *Ascomycota* represents the dominant phylum in Antarctic snow. So, by the water supply from the glacier, this phylum could enrich the fungal composition of the lake. In our study the infrequent phylum *Neocallimastigomycota* was found exclusively in Arctic lakes, in water of L1 and in sediment of L3, L4 and L5. This phylum was never reported in studies of polar lakes. The members of *Neocallimastigomycota* are anaerobic-flagellate fungi residing in the rumen and alimentary tract of larger mammalian and some reptilian, marsupial and avian herbivores, where they play an important role in the degradation of plant material [50]. The detection of this phylum in Arctic lakes analysed in this study could be due to the presence of birds or reindeers which visit these lakes.

### 4.3. Fungal Genera

Generally, it is very difficult to reach the genus level for fungi by metabarcoding analyses, and in our case particularly due to the scarcity of information and deposited sequences of fungi in the environments under consideration. In our study, the most represented genera found in Arctic lakes were *Betamyces* (*Chytridiomycota*), *Pseuderotium* (*Ascomycota*), *Amylocorticiellum* and *Leucosporidium* (*Basidiomycota*), while *Cladosporium, Cadophora, Ulvella, Alternaria, Coleophoma, Metschnikowia* for *Ascomycota* and *Vishniacozyma, Malassezia* for *Basidiomycota* were the most represented genera in Antarctic lakes, some of which have previously been reported from different environment in the Polar regions. The genus *Betamyces* (*Chytridiomycota*) was retrieved in freshwater ecosystem in Ny-Ålesund (Arctic) [2] and in sediment of Antarctic lakes, where *Betamyces* spp. dominated the assemblages [37]. The genus includes only one known species, *Betamyces americae-meridionalis*, which was isolated from pollen baits at the Paraná River (Buenos Aires, Argentina) and in soil in Costa Rica [51]. The genus *Pseuderotium* (*Ascomycota*) was previously reported in Arctic aquatic environments (streams, ponds, melting ice water, and estuaries) in the Ny-Ålesund Region using 454 pyrosequencing [2]. In a metabarcode study of a sediment core of Lake Boeckella (Antarctic peninsula) the genus *Pseuderotium*, with the species *P. hygrophilum*, represented one of the most abundant taxa [27]. The genus was isolated from different Arctic and Antarctic environments and substrata, such as lakes [1,10,30], active layer of the ice-free oases in continental Antarctica [52], sponges [53], and soil [54]. The genus *Leucosporidium* was described in the course of investigation of Antarctic heterobasidiomycetes [55]. It comprises 12 species, mostly phychrotolerant or psychrophilic, which occur in plant materials, soils, and marine environments of high and moderate latitudes [56–58]. Among *Basidiomycota*, *Leucosporidium* was the most represented genus in fungal assemblages of Arctic freshwater [29] and its isolation from Antarctic lakes was previously reported [10,20,21,26,59,60]. The basidiomycetous genus *Amylocorticiellus* consists of four species with a widespread distibution. The species *Amylocorticiellus mollis* (known in 'building mycology' as *Leucogyrophana mollis*) resulted as the dominant wood-decaying fungus in samples taken from wooden historic constructions in Svalbard [61]. *Cladosporium, Cadophora* and *Alternaria* are melanized ascomycetous fungi distributed worldwide and occupy various ecological niches. They are known to be able to resist harsh environmental conditions such as high temperatures, scarcity of water, and high UV radiation [62]. Most species of this genus are plant pathogens or endophytes [63,64], wood destroyers [61], and soil inhabitants [65]. Some species of the genus *Cadophora* and *Cladosporium* are psychrotrophs [66,67] and the presence of these genera are well documented both in Antarctica [6,54] and in the Arctic [2,30,68]. *Malassezia* is a lipophilous basidiomycete yeast genus typically associated with vertebrate animals, but culture-independent studies revealed their presence in diverse acquatic and terrestrial ecosystems. In particular, *Malassezia* is reported among the fungal genera with widest distributions across various polar niches [69], having been reported several times by Arctic and Antarctic culture-independent studies [27,30,70–72]. The literature about the genus *Ulvella* is very scarce. There is only one species, *U. chlorospila* synonym *Pyrenula chlorospila* Arnold. The genus *Pyrenula* is a group of crustose lichens typically growing on smooth, shaded bark. It comprises 745 named species with worldwide distribution, most represented in the tropics and Europe [73], but never reported from polar environments. *Vishniacozyma* is a cosmopolitan genus, and it has been reported in cold environments around the world, including subglacial ice samples from Svalbard Islands [30], and soil and wood in Antarctica [59,60,70,74]. The ascomycetous yeast genus *Metschnikowia* was frequently reported from polar habitats such as sea ice, invertebrates, macroalgae, marine sediment, sea water [6,75]. *M. australis*, which is considered an endemic Antarctic species, was isolated from biofilms sampled in Lake Kroner (Deception Island) [24]. The genus *Coleophoma* includes species reported as plant pathogens, saprophytes or endophytes for different plant species [76], and it was reported from sediment of Lake Wanda and from moss samples from King George Island, Antarctica [22,43].

### 4.4. Fungal Ecology

In our study, we deeply investigate fungal community composition additionally including the use of functional prediction. As a result, Arctic and Antarctic fungi displayed different ecological roles as saprophytes, mutualists, symbionts, and/or parasites. Saprophytic fungi dominated the assemblages detected in water and sediment of the Arctic and Antarctic lakes examined, followed by plant and animal pathogens and symbionts. Similar results were reported for sediment of lakes on Vega Island, Elephant Island, Deception Island, Jame Ross Island, and Trinity Peninsula [25–27,37], all in maritime Antarctica. The same functional ecological profiles were reported in metabarcoding studies of different Antarctic habitats, such as air and snow [42,77], soil [41], freshwater [78] and rock surfaces [79]. According to Schütte et al. [80], fungi present in polar environments display the capability to degrade organic matter at low temperatures, thereby releasing compounds containing carbon, nitrogen, and other elements to other organisms. The dominance of saprophytic fungi in water and sediment of the examined Arctic and Antarctic lakes might indicate, as suggest by de Souza et al. [26] the presence in these environments of a complex saprophytic fungal community that plays a vital role in the decomposition of organic matter under extreme conditions.

## 5. Conclusions

To date, studies on the fungal communities of lake ecosystems in Arctic and Antarctic regions are decidedly few and the majority of them have been based primarily on traditional culture-dependent approaches, thus underlining a still large gap in the understanding of these sensitive environments. This study is the first to compare fungal communities in water and sediment of Arctic and Antarctic lakes. Metabarcoding analysis revealed complex fungal communities in water and sediment of Polar lakes, which may be considered hotspots of fungal diversity, potentially including new and previously unreported species. The results obtained show clearly distinct communities between the analyzed environments, probably due to the different environmental factors and limnological differences between the analyzed lakes, however, showing some common threads. In particular, the most frequently found phyla are generally ubiquitous, even if for the first time the presence of *Neocallimastigomycota* is reported in Arctic samples. Furthermore, the lake water and sediment fungal assemblages were dominated by saprophytes, which may contribute to the decomposition of organic matter under extreme conditions. However, further investigation is necessary to better understand the ecological role of freshwater fungi in polar lakes, and in particular their roles in nutrient cycling.

**Supplementary Materials:** The following supporting information can be downloaded at: https://www.mdpi.com/article/10.3390/jof9111095/s1, Table S1: Concentration and quality of DNA extracted from water and sediment samples; Table S2: ASVs distribution in Arctic and Antarctic lakes; Table S3: Predictive functional profile of fungal community in Polar lakes.

**Author Contributions:** Conceptualization M.P., A.L.G. and F.D.L.; methodology A.M. and M.P.; software M.P.; investigation A.M., M.P., A.C.R., A.C. and C.R. (Carlos Rochera); resources A.L.G., C.U., C.R. (Carmen Rizzo), A.C. and M.A.; writing—original draft preparation A.M. and F.D.L.; writing—review and editing A.M., F.D.L., M.P., C.R. (Carlos Rochera) and A.L.G.; supervision F.D.L., M.P. and A.L.G. All authors have read and agreed to the published version of the manuscript.

**Funding:** This research was funded by the Italian National Antarctic Research Program (grant n. PNRA18_00194; project MicroPolArS 2020-2023).

**Institutional Review Board Statement:** Not applicable.

**Informed Consent Statement:** Not applicable.

**Data Availability Statement:** All sequences have been submitted to the National Center for Biotechnology Information (NCBI) and are associated to the BioProject PRJNA1000778.

**Conflicts of Interest:** The authors declare no conflict of interest.

## References


1. Gonçalves, V.N.; Vaz, A.B.; Rosa, C.A.; Rosa, L.H. Diversity and distribution of fungal communities in lakes of Antarctica. *FEMS Microbiol. Ecol.* **2012**, *82*, 459–471. [CrossRef]
2. Zhang, T.; Wang, N.F.; Zhang, Y.Q.; Liu, H.Y.; Yu, L.Y. Diversity and distribution of aquatic fungal communities in the Ny-Ålesund region, Svalbard (High Arctic). *Microb. Ecol.* **2016**, *71*, 543–554. [CrossRef] [PubMed]
3. Pienitz, R.; Doran, P.T.; Lamoureux, S.F. Origin and geomorphology of lakes in the polar regions. In *Polar Lakes and Rivers: Limnology of Arctic and Antarctic Aquatic Ecosystems*; Vincent, W.F., Laybourn-Parry, J., Eds.; Oxford University Press: New York, NY, USA, 2008; pp. 25–41.
4. Rautio, M.; Dufresne, F.; Laurion, I.; Bonilla, S.; Vincent, W.F.; Christoffersen, K.S. Shallow freshwater ecosystems of the circumpolar Arctic. *Écoscience* **2011**, *18*, 204–222. [CrossRef]
5. Hodgson, D.A. Antarctic lakes. In *Encyclopedia of Lakes and Reservoirs*; Bengtsson, L., Herschy, R.W., Fairbridge, R.W., Eds.; Springer: Dordrecht, The Netherlands, 2012; pp. 26–31. [CrossRef]
6. Ogaki, M.B.; Vieira, R.; Lírio, J.M.; Rosa, C.A.; Rosa, L.H. Diversity and ecology of fungal assemblages present in lakes of Antarctica. In *Fungi of Antarctica: Diversity, Ecology and Biotechnological Applications*; Rosa, L.H., Ed.; Springer: Cham, Switzerland, 2019; pp. 69–97. [CrossRef]
7. Vincent, W.F.; Hobbie, J.E.; Laybourn-Parry, J. Introduction to the limnology of high-latitude lake and river ecosystems. In *Polar Lakes and Rivers: Limnology of Arctic and Antarctic Aquatic Ecosystems*; Vincent, W.F., Laybourn-Parry, J., Eds.; Oxford University Press: Oxford, UK, 2008; pp. 1–24.
8. Quayle, W.C.; Convey, P.; Peck, L.S.; Ellis-Evans, J.C.; Butler, H.G.; Peat, H.J. Ecological responses of maritime Antarctic lakes to regional climate change. In *Antarctic Peninsula Climate Variability: Historical and Palaeoenvironmental Perspectives*; Domack, E., Burnett, A., Leventer, A., Convey, P., Kirby, M., Bindschadler, R., Eds.; American Geophysical Union: Washington, DC, USA, 2003; pp. 159–170. [CrossRef]
9. Rosa, L.H.; Zani, C.L.; Cantrell, C.L.; Duke, S.O.; van Dijck, P.; Desideri, A.; Rosa, C.A. Fungi in Antarctica: Diversity, ecology, effects of climate change, and bioprospection for bioactive compounds. In *Fungi of Antarctica: Diversity, Ecology and Biotechnological Applications*; Rosa, L.H., Ed.; Springer: Cham, Switzerland, 2019; pp. 1–17. [CrossRef]
10. Ogaki, M.B.; Vieira, R.; Muniz, M.C.; Zani, C.L.; Alves, T.; Junior, P.A.; Murt, S.M.F.; Barbosa, E.C.; Oliveira, J.G.; Ceravolo, I.P.; et al. Diversity, ecology, and bioprospecting of culturable fungi in lakes impacted by anthropogenic activities in Maritime Antarctica. *Extremophiles* **2020**, *24*, 637–655. [CrossRef] [PubMed]
11. Camacho, A.; Rochera, C.; Picazo, A. Effect of experimentally increased nutrient availability on the structure, metabolic activities, and potential microbial functions of a maritime Antarctic microbial mat. *Front. Microb.* **2022**, *13*, 900158. [CrossRef] [PubMed]
12. Selbmann, L.; de Hoog, G.S.; Mazzaglia, A.; Friedmann, E.I.; Onofri, S. Fungi at the edge of life: Cryptendolithic black fungi from Antarctic desert. *Stud. Mycol.* **2005**, *51*, 1–32.
13. Gonçalves, V.N.; Cantrell, C.L.; Wedge, D.E.; Ferreira, M.C.; Soares, M.A.; Jacob, M.R.; Oliveira, F.S.; Galante, D.; Rodrigues, F.; Alves, T.M.A.; et al. Fungi associated with rocks of the Atacama Desert: Taxonomy, distribution, diversity, ecology and bioprospection for bioactive compounds. *Environ. Microbiol.* **2016**, *18*, 232–245. [CrossRef]
14. Gunde-Cimerman, N.; Zalar, P.; de Hoog, G.S.; Plemenitaš, A. Hypersaline waters in salterns: Natural ecological niches for halophilic black yeasts. *FEMS Microbiol. Ecol.* **2000**, *32*, 235–340. [CrossRef]
15. Marchetta, A.; Gerrits van den Ende, B.; Al-Hatmi, A.M.S.; Hagen, F.; Zalar, P.; Sudhadham, M.; Gunde-Cimerman, N.; Urzì, C.; de Hoog, G.S.; De Leo, F. Global molecular diversity of the halotolerant fungus *Hortaea werneckii*. *Life* **2018**, *8*, 31. [CrossRef] [PubMed]
16. Hawksworth, D.L. The fungal dimension of biodiversity: Magnitude, significance, and conservation. *Mycol. Res.* **1991**, *95*, 641–655. [CrossRef]
17. Hyde, K.D. The numbers of fungi. *Fungal Divers.* **2022**, *114*, 1. [CrossRef]
18. Peay, K.G.; Kennedy, P.G.; Talbot, J.M. Dimensions of biodiversity in the Earth mycobiome. *Nat. Rev. Microbiol.* **2016**, *14*, 434–447. [CrossRef] [PubMed]
19. Grossart, H.P.; Van den Wyngaert, S.; Kagami, M.; Wurzbacher, C.; Cunliffe, M.; Rojas-Jimenez, K. Fungi in aquatic ecosystems. *Nat. Rev. Microbiol.* **2019**, *17*, 339–354. [CrossRef] [PubMed]
20. Ellis-Evans, J.C. Fungi from maritime Antarctic freshwater environments. *Br. Antarct. Surv. Bull.* **1985**, *68*, 37–45.
21. Brunati, M.; Rojas, J.L.; Sponga, F.; Ciciliato, I.; Losi, D.; Göttlich, E.; de Hoog, G.S.; Genilloud, O.E.; Marinelli, F. Diversity and pharmaceutical screening of fungi from benthic mats of Antarctic lakes. *Mar. Genom.* **2009**, *2*, 43–50. [CrossRef]
22. Ogaki, M.B.; Teixeira, D.R.; Vieira, R.; Lírio, J.M.; Felizardo, J.P.; Abuchacra, R.C.; Cardoso, R.P.; Zani, C.L.; Alves, T.M.A.; Junior, P.A.S.; et al. Diversity and bioprospecting of cultivable fungal assemblages in sediments of lakes in the Antarctic Peninsula. *Fungal Biol.* **2020**, *124*, 601–611. [CrossRef]

23. Connell, L.; Segee, B.; Redman, R.; Rodriguez, R.J.; Staudige, H. Biodiversity and abundance of cultured microfungi from the permanently ice-covered Lake Fryxell, Antarctica. *Life* **2018**, *8*, 37. [CrossRef]
24. de Souza, L.M.D.; Ogaki, M.B.; Texeira, E.A.A.; de Mendes, G.C.A.; Convey, P.; Rosa, C.A.; Rosa, L.H. Communities of culturable freshwater fungi present in Antarctic lakes and detection of their low-temperature-active enzymes. *Braz. J. Microbiol.* **2022**. [CrossRef]
25. Ogaki, M.B.; Camara, P.E.A.S.; Pinto, O.H.B.; Lirio, J.M.; Coria, S.H.; Vieira, R.; Carvalho-Silva, M.; Convey, P.; Rosa, C.A.; Rosa, L.H. Diversity of fungal DNA in lake sediments on Vega Island, north-east Antarctic Peninsula assessed using DNA metabarcoding. *Extremophiles* **2021**, *25*, 257–265. [CrossRef]
26. de Souza, L.M.D.; Lirio, J.M.; Coria, S.H.; Lopes, F.A.C.; Convey, P.; Carvalho-Silva, M.; de Oliveira, F.S.; Rosa, C.A.; Câmara, P.E.A.S.; Rosa, L.H. Diversity, distribution and ecology of fungal communities present in Antarctic lake sediments uncovered by DNA metabarcoding. *Sci. Rep.* **2022**, *12*, 8407. [CrossRef]
27. Rosa, L.H.; Ogaki, M.B.; Lirio, J.M.; Vieira, R.; Coria, S.H.; Pinto, O.H.B.; Carvalho Silva, M.; Convey, P.; Rosa, C.A.; Camara, P.E.A.S. Fungal diversity in a sediment core from climate change impacted Boeckella Lake, Hope Bay, northeastern Antarctic Peninsula assessed using metabarcoding. *Extremophiles* **2022**, *26*, 1–10. [CrossRef]
28. Rojas-Jimenez, K.; Wurzbacher, C.; Bourne, E.C.; Chiuchiolo, A.; Priscu, J.C.; Grossart, H.P. Early diverging lineages within Cryptomycota and Chytridiomycota dominate the fungal communities in ice-covered lakes of the McMurdo Dry Valleys, Antarctica. *Sci. Rep.* **2017**, *7*, 15348. [CrossRef]
29. Comeau, A.M.; Vincent, W.F.; Bernier, L.; Lovejoy, C. Novel chytrid lineages dominate fungal sequences in diverse marine and freshwater habitats. *Sci. Rep.* **2016**, *6*, 30120. [CrossRef]
30. Perini, L.; Gostinčar, C.; Gunde-Cimerman, N. Fungal and bacterial diversity of Svalbard subglacial ice. *Sci. Rep.* **2019**, *27*, 20230. [CrossRef]
31. Guglielmin, M.; Azzaro, M.; Buzzini, P.; Battistel, D.; Roman, M.; Ponti, S.; Turchetti, B.; Sannino, C.; Borruso, L.; Papale, M.; et al. Possible unique ecosystem in the endoglacial hypersaline brines in Antarctica. *Sci. Rep.* **2023**, *13*, 177. [CrossRef] [PubMed]
32. Brown, J.; Pirrung, M.; McCue, L.A. FQC Dashboard: Integrates FastQC results into a web-based, interactive, and extensible FASTQ quality control tool. *Bioinformatics* **2017**, *33*, 3137–3139. [CrossRef] [PubMed]
33. Weißbecker, C.; Schnabel, B.; Heintz-Buschart, A. Dadasnakea Snakemake implementation of DADA2 to process amplicon sequencing data for microbial ecology. *GigaScience* **2020**, *9*, giaa135. [CrossRef] [PubMed]
34. Nilsson, R.H.; Larsson, K.H.; Taylor, A.F.S.; Bengtsson-Palme, J.; Jeppesen, T.S.; Schigel, D.; Kennedy, P.; Picard, K.; Glöckner, F.O.; Tedersoo, L.; et al. The UNITE database for molecular identification of fungi: Handling dark taxa and parallel taxonomic classifications. *Nucleic Acids Res.* **2019**, *47*, D259–D264. [CrossRef]
35. Heberle, H.; Meirelles, G.V.; da Silva, F.R.; Telles, G.P.; Minghim, R. InteractiVenn: A web-based tool for the analysis of sets through Venn diagrams. *BMC Bioinform.* **2015**, *16*, 169. [CrossRef] [PubMed]
36. Nguyen, N.H.; Song, Z.; Bates, S.T.; Branco, S.; Tedersoo, L.; Menke, J.; Schilling, J.S.; Kennedy, P.G. FUNGuild: An open annotation tool for parsing fungal community datasets by ecological guild. *Fungal Ecol.* **2016**, *20*, 241–248. [CrossRef]
37. Gonçalves, V.N.; de Souza, L.M.D.; Lirio, J.M.; Coria, S.H.; Lopes, F.A.C.; Convey, P.; Carvalho-Silva, M.; de Oliveira, F.S.; Camara, P.E.A.S.; Rosa, L.H. Diversity and ecology of fungal assemblages present in lake sediments at Clearwater Mesa, James Ross Island, Antarctica, assessed using metabarcoding of environmental DNA. *Fungal Biol.* **2022**, *126*, 640–647. [CrossRef]
38. Monchy, S.; Sanciu, G.; Jobard, M.; Rasconi, S.; Gerphagnon, M.; Chabé, M.; Cian, A.; Meloni, D.; Niquil, N.; Christaki, U.; et al. Exploring and quantifying fungal diversity in freshwater lake ecosystems using rDNA cloning/sequencing and SSU tag pyrosequencing. *Environ. Microbiol.* **2011**, *13*, 1433–1453. [CrossRef] [PubMed]
39. Wurzbacher, C.; Warthmann, N.; Bourne, E.; Attermeyer, K.; Allgaier, M.; Powell, J.R.; Detering, H.; Mbedi, S.; Grossart, H.P.; Monaghan, M.T. High habitat-specificity in fungal communities in oligo-mesotrophic, temperate Lake Stechlin (North-East Germany). *MycoKeys* **2016**, *16*, 17–44. [CrossRef]
40. Zhang, T.; Wang, N.F.; Zhang, Y.Q.; Liu, H.Y.; Yu, L.Y. Diversity and distribution of fungal communities in the marine sediments of Kongsfjorden, Svalbard (High Arctic). *Sci. Rep.* **2015**, *5*, 14524. [CrossRef] [PubMed]
41. Rosa, L.H.; da Silva, T.H.; Ogaki, M.B.; Pinto, O.H.B.; Stech, M.; Convey, P.; Caravalho-Silva, M.; Rosa, C.A.; Camara, P.E.A.S. DNA metabarcoding high-throughput sequencing uncovers cryptic fungal diversity in soils of protected and non-protected areas on Deception Island, Antarctica. *Sci. Rep.* **2020**, *10*, 21986. [CrossRef] [PubMed]
42. Rosa, L.H.; Pinto, O.H.B.; Convey, P.; Caravalho-Silva, M.; Rosa, C.A.; Camara, P.E.A.S. DNA metabarcoding to assess the diversity of airborne fungi present in air over Keller Peninsula, King George Island, Antarctica. *Microb. Ecol.* **2021**, *82*, 165–172. [CrossRef] [PubMed]
43. Rosa, L.H.; da Costa Coelho, L.; Pinto, O.H.B.; Carvalho-Silva, M.; Convey, P.; Rosa, C.A.; Camara, P.E.A.S. Ecological succession of fungal and bacterial communities in Antarctic mosses affected by a fairy ring disease. *Extremophiles* **2021**, *25*, 471–481. [CrossRef]
44. da Silva, T.H.; Camara, P.E.; Pinto, O.H.B.; Carvalho-Silva, M.; Oliveira, F.S.; Convey, P.; Rosa, C.A.; Rosa, L.H. Diversity of fungi present in permafrost in the south Shetland Islands, maritime Antarctic. *Microb. Ecol.* **2022**, *83*, 58–67. [CrossRef]
45. Ogaki, M.B.; Pinto, O.H.B.; Vieira, R.; Neto, A.A.; Convey, P.; Carvalho-Silva, M.; Rosa, C.A.; Camara, P.E.A.S.; Rosa, L.H. Fungi present in Antarctic deepsea sediments assessed using DNA metabarcoding. *Microb. Ecol.* **2021**, *82*, 157–164. [CrossRef]

46. Rosa, L.H.; de Menezes, G.C.A.; Pinto, O.H.B.; Convey, P.; Carvalho-Silva, M.; Simoes, J.C.; Rosa, C.A.; Camara, P.E.A.S. Fungal diversity in seasonal snow of Martel Inlet, King George Island, South Shetland Islands, assessed using DNA metabarcoding. *Polar Biol.* **2022**, *45*, 627–636. [CrossRef]
47. Kagami, M.; de Bruin, A.; Ibelings, B.W.; Van Donk, E. Parasitic chytrids: Their effects on phytoplankton communities and foodweb dynamics. *Hydrobiologia* **2007**, *578*, 113–129. [CrossRef]
48. Ishida, S.; Nozaki, D.; Grossart, H.P.; Kagami, M. Novel basal, fungal lineages from freshwater phytoplankton and lake samples. *Environ. Microbial.* **2015**, *7*, 435–441. [CrossRef]
49. Kagami, M.; Miki, T.; Takimoto, G. Mycoloop: Chytrids in aquatic food webs. *Front. Microbiol.* **2014**, *5*, 166. [CrossRef] [PubMed]
50. Hanafy, R.A.; Dagar, S.; Griffith, G.W.; Pratt, C.J.; Ypussef, N.H.; Elshahed, M.S. Taxonomy of the anaerobic gut fungi (Neocallimastigomycota): A review of classification criteria and description of currenttaxa. *Int. J. Syst. Evol. Microbiol.* **2022**, *72*, 005322. [CrossRef] [PubMed]
51. Letcher, P.M.; Velez, C.G.; Schultz, S.; Powell, M.J. New taxa are delineated in *Alphamycetaceae* (Rhizophydiales, Chytridiomycota). *Nova Hedwig.* **2012**, *94*, 9–29. [CrossRef]
52. Kochkina, G.A.; Ozerskaya, S.M.; Ivanushkina, N.E.; Chigineva, N.I.; Vasilenko, O.V.; Spirina, E.V.; Gilichinskii, D.A. Fungal diversity in the Antarctic active layer. *Microbiology* **2014**, *83*, 94–101. [CrossRef]
53. Henríquez, M.; Vergara, K.; Norambuena, J.; Beiza, A.; Maza, F.; Ubilla, P.; Araya, I.; Chávez, R.; San-Martín, A.; Darias, J.; et al. Diversity of cultivable fungi associated with Antarctic marine sponges and screening for their antimicrobial, antitumoral and antioxidant potential. *World J. Microbiol. Biotechnol.* **2014**, *30*, 65–76. [CrossRef]
54. Arenz, B.E.; Blanchette, R.A. Investigations of fungal diversity in wooden structures and soils at historic sites on the Antarctic Peninsula. *Can. J. Microbiol.* **2009**, *55*, 46–56. [CrossRef]
55. Fell, J.W.; Statzell, A.; Hunter, I.L.; Phaff, H.J. *Leucosporidium* gen. nov. the heterobasidiomycetous stage of several yeasts of the genus *Candida*. *Antonie van Leeuwenhoek* **1969**, *35*, 433–462. [CrossRef]
56. Yurkov, A.M.; Schäfer, A.M.; Begerow, D. *Leucosporidium drummii* sp. nov., a member of the *Microbotryomycetes* isolated from soil. *Int. J. Syst. Evol. Microbiol.* **2012**, *62*, 728–734. [CrossRef]
57. Laich, F.; Chávez, R.; Vaca, I. *Leucosporidium escuderoi* f. a., sp. nov., a basidiomycetous yeast associated with an Antarctic marine sponge. *Antonie van Leeuwenhoek* **2014**, *105*, 593–601. [CrossRef]
58. de García, V.; Coelho, M.A.; Maia, T.M.; Rosa, L.H.; Vaz, A.M.; Rosa, C.A.; Sampaio, J.P.; Gonçalves, P.; van Broock, M.; Libkind, D. Sex in the cold: Taxonomic reorganization of psychrotolerant yeasts in the order *Leucosporidiales*. *FEMS Yeast Res.* **2015**, *15*, fov019. [CrossRef] [PubMed]
59. Vaz, A.B.; Rosa, L.H.; Vieira, M.L.; Garcia, V.D.; Brandão, L.R.; Teixeira, L.C.; Rosa, C.A. The diversity, extracellular enzymatic activities and photoprotective compounds of yeasts isolated in Antarctica. *Braz. J. Microbiol.* **2011**, *42*, 937–947. [CrossRef]
60. Tsuji, M.; Fujiu, S.; Xiao, N.; Hanada, Y.; Kudoh, S.; Kondo, H.; Tsuda, S.; Hoshino, T. Cold adaptation of fungi obtained from soil and lake sediment in the Skarvsnes ice-free area, Antarctic. *FEMS Microbiol. Lett.* **2013**, *346*, 121–130. [CrossRef] [PubMed]
61. Mattsson, J.; Flyen, A.-C.; Nunez, M. Wood-decaying fungi in protected buildings and structures on Svalbard. *Agarica* **2010**, *29*, 5–14.
62. De Leo, F.; Marchetta, A.; Urzì, C. Black fungi on stone-built Heritage: Current knowledge and future outlook. *Appl. Sci.* **2022**, *12*, 3969. [CrossRef]
63. Walsh, E.; Duan, W.; Mehdi, M.; Naphri, K.; Khiste, S.; Scalera, A.; Zhang, N. *Cadophora meredithiae* and *C. interclivum*, new species from roots of sedge and spruce in a western Canada subalpine forest. *Mycologia* **2018**, *110*, 201–214. [CrossRef]
64. Zabouri, Y.; Cheriguene, A.; Chougrani, F.; Merzouk, Y.; Marchetta, A.; Urzi, C.; De Leo, F. Antifungal activity of lactic acid bacteria against phytopathogenic *Alternaria alternata* species and their molecular characterization. *J. Food Nutr. Res.* **2021**, *60*, 18–28.
65. Domsch, K.H.; Gams, W.; Anderson, T.-H. Compendium of soil fungi. *Eur. J. Soil Sci.* **2007**, *59*, 1007. [CrossRef]
66. Iliushin, V.A. First find of *Cadophora antarctica* Rodr. Andrade, Stchigel, Mac Cormack & Cano in the Arctic. *Czech Polar Rep.* **2020**, *10*, 147–152. [CrossRef]
67. Brück, S.A.; Contato, A.G.; Gamboa-Trujillo, P.; de Oliveira, T.B.; Cereia, M.; de Moraes Polizeli, M.L.T. Prospection of psychrotrophic filamentous fungi isolated from the High Andean Paramo Region of Northern Ecuador: Enzymatic activity and molecular identification. *Microorganisms* **2022**, *10*, 282. [CrossRef] [PubMed]
68. Kirtsideli, I.Y.; Vlasov, D.Y.; Barantsevich, E.P.; Krylenkov, V.A.; Sokolov, V.T. Microfungi from soil of polar island Izvestia Tsik (Kara Sea). *Mikol. I Fitopatol.* **2014**, *48*, 365–371.
69. Zalar, P.; Gunde-Cimerman, N. Cold-Adapted Yeasts in Arctic Habitats in Cold-Adapted Yeasts. In *Biodiversity, Adaptation Strategies and Biotechnological Significance*; Buzzini, P., Margesin, R., Eds.; Springer: Dordrecht, The Netherlands, 2014; pp. 49–74.
70. Arenz, B.E.; Held, B.W.; Jurgens, J.A.; Farrell, R.L.; Blanchette, R.A. Fungal diversity in soils and historic wood from the Ross Sea Region of Antarctica. *Soil. Biol. Biochem.* **2006**, *38*, 3057–3064. [CrossRef]
71. Bridge, P.D.; Spooner, B.M. Non-lichenized Antarctic fungi: Transient visitors or members of a cryptic ecosystem? *Fungal Ecol.* **2012**, *5*, 381–394. [CrossRef]
72. Connell, L.B.; Staudigel, H. Fungal diversity in a dark oligotrophic volcanic ecosystem (DOVE) on Mount Erebus, Antarctica. *Biology* **2013**, *2*, 798–809. [CrossRef]
73. Aptroot, A. A world key to the species of *Anthracothecium* and *Pyrenula*. *Lichenologist* **2012**, *44*, 5–53. [CrossRef]

74. Connell, L.; Redman, R.; Craig, S.; Scorzetti, G.; Iszard, M.; Rodriguez, R. Diversity of soil yeasts isolated from South Victoria Land, Antarctica. *Microb. Ecol.* **2008**, *56*, 448–459. [CrossRef]
75. Butinar, L.; Strmole, T.; Gunde-Cimerman, N. Relative incidence of ascomycetous yeasts in Arctic coastal environments. *Microb. Ecol.* **2011**, *61*, 832–843. [CrossRef] [PubMed]
76. Crous, P.W.; Groenewald, J.Z. They seldom occur alone. *Fungal Biol.* **2016**, *120*, 1392–1415. [CrossRef] [PubMed]
77. Rosa, L.H.; Pinto, O.H.B.; Šantl-Temkiv, T.; Convey, P.; Caravalho-Silva, M.; Rosa, C.A.; Camara, P.E.A.S. DNA metabarcoding high-throughput sequencing of fungal diversity in air and snow of Livingston Island, South Shetland Islands, Antarctica. *Sci. Rep.* **2020**, *10*, 21793. [CrossRef] [PubMed]
78. de Souza, L.M.D.; Ogaki, M.B.; Câmara, P.E.A.S.; Pinto, O.H.B.; Convey, P.; Caravalho-Silva, M.; Rosa, C.A.; Rosa, L.H. Assessment of fungal diversity present in lakes of Maritime Antarctica using DNA metabarcoding: A temporal microcosm experiment. *Extremophiles* **2021**, *25*, 77–84. [CrossRef] [PubMed]
79. de Menezes, G.C.A.; Câmara, P.E.A.S.; Pinto, O.H.B.; Caravalho-Silva, M.; Oliveira, F.S.; Souza, C.D.; Schaefer, C.E.G.R.; Convey, P.; Rosa, C.A.; Rosa, L.H. Fungal diversity present on rocks from a polar desert in continental Antarctica assessed using DNA metabarcoding. *Extremophiles* **2021**, *25*, 193–202. [CrossRef] [PubMed]
80. Schütte, U.M.; Henning, J.A.; Ye, Y.; Bowling, A.; Ford, J.; Genet, H.; Waldrop, M.P.; Turetsky, M.R.; White, J.R.; Bever, J.D. Effect of permafrost thaw on plant and soil fungal community in a boreal forest: Does fungal community change mediate plant productivity response? *J. Ecol.* **2019**, *107*, 1737–1752. [CrossRef]

Journal of
*Fungi*

MDPI

*Review*

# Exploring Fungal Diversity in Seagrass Ecosystems for Pharmaceutical and Ecological Insights

**Oshadi Rajakaruna [1,2,†], Nalin N. Wijayawardene [1,3,†], Susantha Udagedara [4], Prabath K. Jayasinghe [5], Sudheera S. Gunasekara [5], Nattawut Boonyuen [6], Thushara C. Bamunuarachchige [7,*] and Kahandawa G. S. U. Ariyawansa [2,*]**

1 Centre for Yunnan Plateau Biological Resources Protection and Utilization, College of Biological Resource and Food Engineering, Qujing Normal University, Qujing 655011, China; koshadirajakaruna@gmail.com (O.R.); nalinwijayawardene@yahoo.com (N.N.W.)
2 Department of Plant Sciences, Faculty of Science, University of Colombo, Colombo 00300, Sri Lanka
3 Tropical Microbiology Research Foundation, Pannipitiya 10230, Sri Lanka
4 Blue Resources Trust, Colombo 00700, Sri Lanka; susantha@blueresources.org
5 National Aquatic Resources Research and Development Agency (NARA), Crow Island, Colombo 01500, Sri Lanka; prabathj@nara.ac.lk (P.K.J.); sudheera@nara.ac.lk (S.S.G.)
6 National Center for Genetic Engineering and Biotechnology (BIOTEC), National Science and Technology Development Agency (NSTDA), 113 Thailand Science Park, Phahonyothin Road, Khlong Nueng, Khlong Luang, Pathum Thani 12120, Thailand; nattawut@biotec.or.th
7 Department of Bioprocess Technology, Faculty of Technology, Rajarata University of Sri Lanka, Mihintale 50300, Sri Lanka
* Correspondence: tcbamunu@tec.rjt.ac.lk (T.C.B.); sameera@pts.cmb.ac.lk (K.G.S.U.A.)
† These authors contributed equally to this work.

**Abstract:** Marine ecosystems are important in discovering novel fungi with interesting metabolites that have shown great potential in pharmaceutical and biotechnological industries. Seagrasses, the sole submerged marine angiosperm, host diverse fungal taxa with mostly unknown metabolic capabilities. They are considered to be one of the least studied marine fungal habitats in the world. This review gathers and analyzes data from studies related to seagrasses-associated fungi, including taxonomy and biogeography, and highlights existing research gaps. The significance of the seagrass–fungal associations remains largely unknown, and current understanding of fungal diversity is limited to specific geographical regions such as the Tropical Atlantic, Mediterranean, and Indo-Pacific. Our survey yielded 29 culture-dependent studies on seagrass-associated endophytic and epiphytic fungi, and 13 miscellaneous studies, as well as 11 meta-studies, with no pathogenic true fungi described. There is a significant opportunity to expand existing studies and conduct multidisciplinary research into novel species and their potential applications, especially from understudied geographical locations. Future research should prioritize high-throughput sequencing and mycobiome studies, utilizing both culture-dependent and -independent approaches to effectively identify novel seagrass-associated fungal taxa.

**Keywords:** biogeography; endophytes; epiphytes; marine ecosystems; pathogens; species diversity

## 1. Introduction

Fungi are a diverse group of organisms inhabiting a wide range of environments, playing significant functions in challenging ecosystems [1,2]. They can range from unicellular to filamentous taxa and exhibit various life modes, such as pathogens, saprobes, endophytes, and epiphytes [3]. All these different life modes of fungi contribute to the overall functioning of an ecosystem, though some may not always be beneficial, such as phytopathogenic fungi [4] and clinically important fungi [5]. Approximately, 156,000 species have been scientifically documented in Species Fungorum 2024 [6]. The global richness of fungi has been a popular topic among mycologists with diverse predictions made over the past three decades using various approaches [7–12]. Hawksworth and Lücking [10] estimated there to be between 2.2 to 3.8 million fungal species, but the most recent estimate by Niskanen et al. [11] predicts 2 to 3 million species to be more realistic, with over 90% of these species yet to be revealed. Marine fungi have been identified as a potential group for bridging gaps in missing species [11,13] and for discovering novel biological substances [14].

Microorganisms are considered ubiquitous in marine ecosystems, despite the environmental limitations posed by high salinity, low temperature, low water availability, and oligotrophic conditions [15]. Microorganisms represent 90% of the total oceanic biomass [16], making marine ecosystems a potential habitat for discovering novel microorganisms. This largely understudied habitat is rich in microorganisms producing bioactive natural products [17]. Marine fungi are one of the important groups of fungi that mycologists have continuously studied for over 50 years [18,19]. According to www.marinefungi.org (accession on 14 July 2024), 2041 species have been reported as marine fungi [19]. According to Calabon et al. [20] and Devadatha et al. [21], the majority of these taxa have been documented from mangroves and salt marshes. Additionally, substrates such as algae, silt, driftwood, and seagrasses have been researched globally [22]. While over 10,000 fungal species from different marine habitats have been predicted, less than 20% of them have been described [19]. Wijayawardene et al. [23] emphasized the need for investigating marine fungi from various microhabitats in understudied geographical regions and highlighted the role of multidisciplinary sciences in discovering novel fungi and understanding their potential uses.

For over 150 years, it has been known that plants and microorganisms have intimate relationships that affect overall plant fitness, growth, and survival [24]. Consequently, studying plant–microbe associations to uncover the elements of their interactions has become a topic of interest in modern times. As a component of the marine ecosystem, seagrasses host a diverse community of microorganisms, including bacteria, archaea, fungi, microalgae, and viruses [16]. These microbes have a fundamental impact on the physiology and well-being of seagrasses while playing a major role in controlling biogeochemical processes within entire seagrass meadows [16].

During our ongoing study on seagrass-associated fungi in the Puttalam lagoon, Sri Lanka, we isolated over 40 different fungal species from two seagrass species, namely, *Enhalus acaroides* (Linnaeus f.) Royle, 1839 and *Oceana serrulata* (R.Brown) Byng & Christenh. Concurrently, our literature survey revealed that a considerable number of studies on seagrass fungi (including fungus-like taxa) have been carried out, but the data (taxonomy, classification, applications, biogeography, and ecological data) are scattered. This review and opinion paper aims to compile the available data on seagrass-associated fungi (focusing on taxonomy, biogeography, and industrial applications) with the collaboration of fungal taxonomists, marine ecologists, and industry experts. Furthermore, we intend to highlight research gaps and emphasize the need for more research on seagrass-associated fungi to expand future studies in this field.

## 2. Marine Fungi

'Marine fungi' are defined in multiple ways in the literature. One of the earliest definitions is based on their requirement to grow in saline water [25]. The most quoted definition provided by Kohlmeyer and Kohlmeyer [26] restricts marine fungi to two ecological groups: obligate and facultative marine fungi. Obligate marine fungi exclusively grow and sporulate in marine environments, while facultative species sporulate in marine environments but can grow in freshwater or terrestrial environments [26,27]. Later, a broader definition was provided by Pang et al. [28], where marine fungi were defined as fungi that are repeatedly recovered from marine habitats and either: (1) grow and/or sporulate on substrates in marine environments; (2) form symbiotic relationships with other marine organisms; or (3) adapt and evolve at the genetic level or are metabolically active in marine environments. Recently, Pasqualetti et al. [15] defined fungi obtained from marine environments as "marine-derived" as they are isolated from marine sources. This definition includes obligates, facultatives, and fungi arising from dormant terrestrial or freshwater propagules. To address the confusion regarding the applicability of these terms, a revision of the taxonomy and a deeper understanding of the metabolomic basis of marine fungal life are necessary.

In general, marine fungi have been recovered from different habitats, including mangrove plants, macroalgae, coral reefs, drift and submerged wood, sponges, sea ice, sea garbage, seagrasses, deep-sea and benthic sediments, hydrothermal vents, oxygen-deficient environments, and the water column [29–31]. To provide a structured reference for marine fungal habitats, Wijayawardene et al. [23] defined these habitats as 'coastal terrestrial ecosystems', 'semi-coastal marine ecosystems', 'coastal marine ecosystems', and 'deep marine ecosystems'. Despite this categorization, fungal associations in many marine habitats, especially seagrasses, remain understudied, which is the primary focus of this review.

## 3. Seagrasses

Seagrasses are flowering plants (angiosperms) found in shallow marine waters [32] and are adapted for fully submerged conditions [33]. They can exist as monospecific or multispecific meadows, often extending over a larger surface area covering the seabed [34].

Despite their crucial role as foundation species in coastal ecosystems [35], seagrass populations are declining at an alarming rate. Studies indicate that the extent of seagrass beds has been declining by 110 km$^2$ per year since 1980, with 29% of their total extent lost since seagrass regions were first recorded in 1879 [36]. Consequently, there is an urgent necessity for focused worldwide conservation initiatives to protect the existing seagrass meadows [37].

### *Seagrass Distribution and Diversity*

Seagrasses are widely distributed along tropical and temperate shallow coastal waters spanning 159 countries on six continents, making them one of the most widespread coastal habitats on Earth [38,39]. They are often associated with important marine habitats such as corals, bivalve reefs, and mangroves in tropics and marshes, and kelp forests in temperate regions [40].

According to the estimates from the UNEP World Conservation Monitoring Centre (WCMC), the spatial distribution of seagrasses was approximated at 177,000 km$^2$ in 2001 [41]. However, a recent study by McKenzie et al. [42] revised this estimate, indicating a global seagrass distribution of 160,387 km$^2$, with a high to moderate level of confidence. Furthermore, UNEP [39] highlighted that the estimated global extent of seagrass ranges between 300,000 and 600,000 km$^2$. The depth limits of seagrasses are estimated to range from the intertidal zone to 90 m below mean sea level [43].

Seagrasses are currently divided into six families: *Zosteraceae*, *Hydrocharitaceae*, *Posidoniaceae*, *Cymodoceaceae*, *Ruppiaceae*, and *Zannichelliaceae* [44]. These six families contain twelve genera and 72 species worldwide [44]. The extinction risk assessment of the world's seagrass species reveals that three species are Endangered, seven Vulnerable, five Near Threatened, 48 are categorized as Least Concern, and nine species are Data-Deficient [44]. According to Rasheed and Unsworth [45], nearly 10% of all coastal seabeds are covered with seagrass and they exhibit low taxonomic diversity.

Amongst these, the tropical Indo-Pacific represents one of the most highly diverse seagrass bioregions in the world, accounting for around 35% of the total species. However, this region also faces significant data scarcity, particularly for population spatial distribution data, accounting for around 24% of the data deficiency [38].

## 4. Seagrass-Associated Fungi

Compared to the extensive studies on seagrass ecology, less is known about the diversity and ecological roles played by the fungal communities associated with seagrasses [35]. The fungal community associated with land plants plays a crucial role in influencing plant health and survival. Similarly, in seagrass ecosystems, the fungal community can significantly impact overall seagrass functioning. Understanding these associations can be a valuable tool for future seagrass restoration and conservation projects [46,47].

### 4.1. Seagrass-Associated Endophytic Fungi

In this section, our focus is on seagrass endophytic fungi recovered through culture-dependent approaches. Fungi derived from culture-independent studies will be discussed in the section on seagrass mycobiome studies.

The term 'endophyte' (endon = within, phyte = plant) is used to define any organism found within living plant tissues. In mycology, this term often specifically refers to fungi that reside within (inside) the plant tissues for at least part of their lifecycle without causing apparent harm or disease symptoms to the host [48]. The association between endophytes and their host plants can range from "mutualistic to opportunistically pathogenic" in nature [49]. According to Schulz and Boyle [50], endophytes are considered as virulent pathogens when they produce enzymes that damage host tissues to aid in colonization. However, a clear distinction between a pathogenic and non-pathogenic nature is often challenging. Recently, Hardoim et al. [51] suggested that the term endophyte should include all microorganisms that colonize internal plant tissues for all or part of their lifetime. Nevertheless, the distinction between a pathogen and an endophyte becomes apparent when they engage with the plant's defence system. A pathogen overcomes the plants' defences, inflicting damage on the host, whereas endophytes overcome plant defences by masking themselves without causing apparent damage or symptoms.

In general, endophytic fungi are considered to enhance the hosts' ability to tolerate environmental stresses, improve vigour, recycle nutrients, decrease susceptibility to pathogens and pests, and regulate the synthesis of phytohormones and metabolites [52–55]. In return, host plants provide them with organic nutrients, protection, and assurance of survival [56]. A recent study showed that two endophytic fungi, *Trichoderma* sp. and *Diaporthe* sp., isolated from a seagrass species, *Thalassia testudinum,* demonstrated significant bioactivity against pathogenic *Labyrinthula* infections, which have previously led to extensive seagrass die-offs in many parts of the world [57].

To the best of our knowledge, the first record of intercellular fungi in seagrass was reported by Kuo [58] in 1984, in the leaves of *Zostera muelleri*. Although they did not specifically use the term "endophytic", the symbiotic and intercellular nature of fungal filaments was demonstrated using microscopy. The first direct evidence of endophytic fungi isolated from living, healthy seagrass tissues was reported by Wilson [59] through the screening of the leaf tissues of *Thalassia testudinum, Halodule bennudensis,* and *Syringodium filiforme*. Since then, numerous investigations on the isolation and identification of seagrass endophytic fungi have been published, and all culture-based studies as well as isolated taxa reported thus far are summarized in Table S1.

Eurotialean fungi, specifically *Aspergillus* and *Penicillium,* are reported as the dominant endophytic taxa in most culture-dependent studies [60–64] (see Table S1). These two genera, *Penicillium* (100/572) and *Aspergillus* (63/572), account for more than 25% of the listed taxa (Figure 1A). Nevertheless, in most of these publications, fungal identification is restricted only to the generic level without support from DNA sequencing. Therefore, the taxonomic placements may not be very accurate.

Most of these dominant endophytic genera are reported to have a terrestrial origin [60,65–67]. While many studies report the absence of obligate marine fungi [66–68], some studies such as Cuomo et al. [69], Abdel-Wahab et al. [70], and Mata and Cebrián [61] mention the presence of obligate fungal species. According to Abdel-Wahab et al. [70], employing rigorous surface sterilization during endophytic isolations enhances the recovery of marine-derived fungal species, whereas a direct microscopic examination tends to identify obligate marine fungi. However, the validity of this statement is rather perplexing as the marine-derived, or obligatory, nature cannot be determined by morphological characters alone. In many studies, sterile forms are reported [61,63,66,67] and remain unidentified when morphological methods are used for identifications. In our ongoing study, over 15 endophytic fungi (out of a total of 40) isolated from two seagrass species, *E. acaroides* and *O. serrulata*, lacked any fruiting structures, necessitating molecular identification (Rajakaruna et al., unpublished). Gnavi et al. [71] highlight the need for molecular analyses to properly identify these sterile forms.

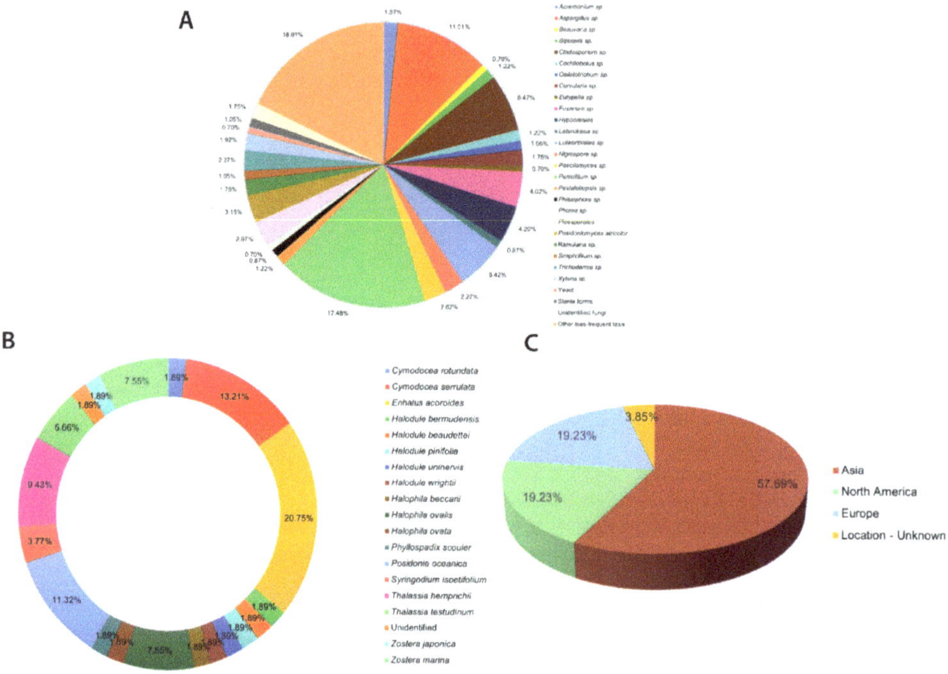


**Figure 1.** Summary of culture-dependent seagrass endophytic fungal studies listed in Table S1: (**A**) composition of endophytic fungi from different seagrass studies; (**B**) percentage number of studies carried out for each seagrass species; (**C**) continent-wide distribution of sampling sites for seagrass fungal endophytes in the world.

Seagrasses are reported to have a low diversity and density of fungal colonization compared to terrestrial plants [60,62,67,68]. However, Shoemaker and Wyllie-Echeverria [72] noted that the number of taxa isolated is "roughly similar" to that of land plants. While these comments appear contradictory, it is important to consider the diversity indices utilized by these authors, the climatic zones of the plants used for comparisons, and the total number of segments screened, before coming to a meaningful conclusion. Moreover, all the above predictions were based on culture-dependent techniques.

Nevertheless, it is stipulated that the low frequency of fungal colonization in seagrasses is attributed to multiple complex interactions between intrinsic and extrinsic factors. The endophytic fungal community in seagrasses is mainly influenced by intrinsic factors including tissue type ("district specificity") [73], age [60,62], morphological characters [59,60], and the phytochemical composition of tissues [62,67]. The antifungal metabolites produced by seagrasses are reported to limit internal fungal colonization [74]. Additionally, other groups of microbes associated with seagrasses, which produce antifungal compounds, can further reduce endophytic colonization [75].

External factors such as the nutrient content in the water column, water temperature, wind and wave actions, seasonal variations [59–61], and other physicochemical factors at the sampling sites may affect the fungal colonization within seagrasses. A recent study by Solé et al. [76] showed that human-generated noise significantly impacts the degradation of fungal symbionts in *Posidonia oceanica* roots, subsequently affecting the normal root functions. Moreover, the absence of mycorrhizal associations in seagrasses is said to be associated with limitations posed by high salinity and oligotrophic conditions in the marine sediments [77]. Although seagrasses are adapted for nutrient uptake through leaves, having root–fungal associations (similar to mycorrhizal associations in land plants) could be beneficial for absorbing nutrients from recalcitrant material under oligotrophic conditions [78]. A summary of these symbiotic root–fungal associations is given in Section 4.1.1.

Most of the seagrass endophytic fungal research has been conducted in the Asian, European, and North American regions (Figure 2). More than half of the sampling sites are concentrated on the Asian continent (Figures 1C and 2). The majority of studies are confined to a few seagrass species: *Enhalus acaroides, Cymodocea serrulata, Posidonia oceanica, Thalassia hemprichii,* and *Zostera* spp. (Figure 1B).

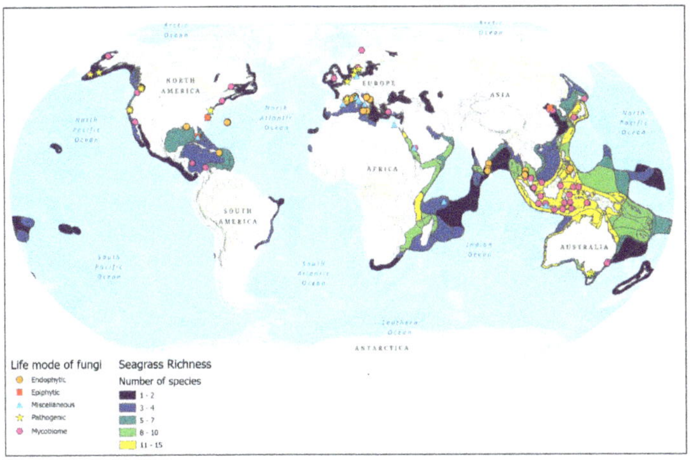


**Figure 2.** Biogeographic distribution of seagrass fungal studies (based on the data from Tables S1–S4).

#### 4.1.1. Root-Associated Endophytic Fungi

Some publications have explicitly investigated and uncovered interesting root–fungal relationships in seagrasses. Kuo et al. [79] reported the occurrence of fungi in the peripheral root tissues of two Australian seagrass species, *Posidonia australis* and *P. sinuosa*. To the best of our understanding, this is the first published investigation that specifically addresses the fungal inhabitants in seagrass roots.

Thereafter, for more than a decade, the topic remained unstudied. Nielsen et al. [77] attempted to find arbuscular mycorrhizal (AM) associations in the roots of *Zostera marina* and *Thalassia testudinum*. Since the majority of vascular plants contain these symbiotic associations, the authors aimed to address this gap in seagrass research. However, they were unable to observe AM associations in either seagrass species. The observed *Zostera marina* tissues showed unidentified fungal colonization that was not characteristic of AM. As they assumed, the lack of information on seagrass-associated root mycorrhizae in the 1900s is due to the inherent challenge of publishing negative results.

Torta et al. [80] conducted a study on root mycobiota of *Posidonia oceanica* and obtained a single species which they named *Lulwoana* sp. Without any supportive evidence, they claimed this species was a dark septate endophyte (DSE). Vohník and colleagues conducted numerous investigations to study fungi associated with seagrass roots but found no structures resembling mycorrhiza. However, in their study, Vohník et al. [81] described the presence of dark septate endophytes (DSE) with all typical structures, including extraradical and intraradical dark septate hyphae, dense melanized parenchymatous nets/hyphal sheaths on the root surface, and melanized intracellular microsclerotia for the first time in seagrasses. Subsequently, Vohník et al. [82] isolated three fungal species from the roots of the seagrass *Posidonia oceanica* with the dominant fungal species exhibiting characteristics of DSE. Sequencing data showed that this species belongs to the order *Pleosporales*, representing a new member in the *Aigialaceae*. A similar observation showing narrow endophytic diversity and *Pleosporales* dominance was reported from a culture-independent approach [83]. The mycobiont responsible for these dominant DSE associations in *Posidonia oceanica* was later described as *Posidoniomyces atricolor* [84].

However, the results of Vohník et al. [84] conflicted with the findings of Torta et al. [80] regarding the dominant DSE fungal groups, form, and distribution of fungal colonization. To resolve this, a comprehensive study was later conducted by Vohník et al. [78], at the same localities previously investigated by Torta et al. [80] and the dominant DSE associations in *Posidonia oceanica* was confirmed again as *Posidoniomyces atricolor* and not *Lulwoana* sp., as described by Torta et al. [80] previously. Lefebvre et al. [85] claim the colonization of *Posidoniomyces atricolor* in the degrading tissue of *Posidonia oceanica*, showing its saprobic nature, which extends beyond a typical plant endophytic association. Overall, the findings of these studies indicate that roots of seagrasses are colonized by DSE, but, until now, as already foreseen, no mycorrhizal association has been described in the roots of seagrasses.

In addition to these reports, Marina Carrasco-Acosta recently presented the isolation of an obligate marine fungus, *Cumulospora marina*, from the root tissues of *Cymodocea nodosa* at an IUMBM conference on extremophilic fungi [86]. Furthermore, Wang et al. [87] identified a novel lulworthioid fungus, *Halophilomyces hongkongensis*, colonizing the roots and rhizomes of the seagrass *Halophila ovalis*. However, the contribution of these fungal species on seagrass health is still unknown.

### 4.2. Seagrass Epiphytic Fungi

Epiphytes are spatially different from endophytes and are defined as organisms that live upon the plant surfaces [32]. Similar to endophytes, seagrasses provide excellent habitats for the epiphytic organisms [32,88]. The physicochemical changes in the surrounding water column provide environmental variability, while the host characteristics (such as leaf area) create structural variability for a variety of epiphytic organisms to colonize seagrass surfaces [89]. These variations shape the composition of the epiphytic community and their interactions with the host.

The epiphytic community of seagrasses is primarily composed of algae dominated by the *Rhodophyta*, while fungi remain largely unstudied [90]. In a previous study, fungi were recorded as "rarely found" in the epiphytic community of *Posidonia oceanica* leaves [91]. However, recent studies indicate that epiphytic fungi are abundant in seagrasses, and their true diversity and potential are yet to be fully discovered.

Typically, endophytic fungal isolates are obtained following surface sterilization with sodium hypochlorite and ethanol, whereas epiphytes are isolated using less stringent surface sterilization methods [49]. In several studies, surface sterilization methods are not explicitly specified. Thus, it is reasonable to believe that these studies have isolated both endophytic and epiphytic fungi. Table S2 provides a summary of these miscellaneous forms reported from seagrasses worldwide.

Genera such as *Cladosporium*, *Colletotrichum*, and *Penicillium*, and the order *Hypocreales*, which are often recorded as endophytes, have also been reported as epiphytes in a few studies [92,93]. However, the possibility of contamination, deficiencies in surface sterilization stringency, misidentification of fungi, and human error should also be considered when making inferences from these outcomes. Recently, a novel epiphytic root–fungus symbiosis has been reported to be associated with the roots of the Indo-Pacific seagrass *Thalassodendron ciliatum* [94].

In comparison, studies on the culture-dependent isolation of seagrass endophytic fungi are more common than those on epiphytic fungi. However, several studies conducted thus far have identified bioactive compounds from epiphytic fungi with potential industrial applications (Table 1). Nonetheless, there exists a notable disparity between these findings and the published records on the initial isolation of these fungi.

**Table 1.** Bioactivities of metabolites isolated from seagrass-derived fungi.

| Bioactivity | Bioactive Compound | Fungal Taxa with Bioactivity | Host Seagrass Species | Life Mode | Reference |
|---|---|---|---|---|---|
| Broad spectrum antimicrobial activity (antifungal and antibacterial activity against the tested isolates) | Not isolated | Not identified | *Zostera marina* | Epiphytic | [95] |
| Mild antibacterial activity against *S. aureus* ATCC25923 and methicillin-resistant *S. aureus* | Chrysophanol and Emodin | *Bipolaris* sp. | *Halophila ovalis* | Not specifically given (from leaves) | [96] |
| Antimicrobial activity against *Staphylococcus aureus* and anti-quorum sensing effect against *Pseudomonas aeruginosa* | Not isolated | *Penicillium crustosum* | *Posidonia oceanica* | Not specified | [97] |
| Mild antibacterial and antifungal activity | Zearalenone | *Fusarium* sp. | *Thalassia hemprichi* | Not specifically given (from leaves) | [98] |
| Antimicrobial activities (antibacterial and antifungal) | Not isolated | Few active isolates identified | *Cymodocea serrulata*, *Halophila ovalis* and *Thalassia hemprichii* | Endophytic | [65] |
| Antimicrobial activity against 10 human pathogens | Not isolated | Few active isolates identified | *Enhalus acoroides* | Endophytic (from leaves, roots, and rhizomes) | [62] |
| Antibacterial activity against *Micrococcus luteus* | Not identified | Fungi not identified | *Halophila ovalis*, *Thalassia hemprichii*. and *Syringodium isoetifolium* | Endophytic | [99] |
| Significant antifungal activity against *Penicillium griseofulvum* and *Aspergillus favus* | Terperstacin and 19-acetyl-4-hydroxydictyodiol | *Mariannaea humicola* | *Posidonia oceanica* | Epiphytic | [100] |
| Antimicrobial activity | Epicotripeptin, Cyclo (L-Pro-L-Val), Cyclo (L-Pro-L-Ile), Cyclo (L-Pro-L-Phe) Cyclo (L-Pro-L-Tyr) | *Epicoccum nigrum* | *Thalassia hemprichii* | Endophytic | [101] |
| Promising antimicrobial activity (against *Staphylococcus aureus*, *Pseudomonas aeruginosa*, and *Candida albicans*) | Not identified | *Epicoccum nigrum* | *Thalassia hemprichii* | Endophytic | [102] |
| Moderate antibacterial activity against *Vibrio alginolyticus* and V. *parahaemolyticus* | Not isolated | *Aspergillus versicolor* | *Enhalus acoroides* | Not specified | [103] |
| Strong antibacterial activity against *Vibrio alginolyticus* and V. *parahaemolyticus* | Not isolated | *Aspergillus unguis* | *Thalassia hemprichii* | Not specified | [104] |
| A bacteriostatic effect (doctoral dissertation) | Not identified | Not given | *Thalassia testudinum* | Endophytic | [105] |
| Antiviral activity | Halovir | *Scytalidium* sp. | *Halodule wrightii* | Not given | [106] |
| Antiplasmodial activity (however, significant activity is not detected from seagrass fungi) | N/A | N/A | N/A | Endophytic | [107] |
| Cytotoxic activity against P388 and HeLa cancer cells. | Malformin A1 | *Aspergillus tubingensis* | *Enhalus acoroides* | Not specified (should be endophytic) | [104,108] |

**Table 1.** *Cont.*

| Bioactivity | Bioactive Compound | Fungal Taxa with Bioactivity | Host Seagrass Species | Life Mode | Reference |
|---|---|---|---|---|---|
| In vitro cytotoxicity toward COLO 205 colon and SK-MEL-2 melanoma cancer cell lines. | Sansalvamide A | *Fusarium* sp. | *Halodule wrightii* | Surface fungi | [109] |
| Cytotoxicity against murine leukemia L1210 and human epidermoid carcinoma KB cells | Cladionol A (a new polyketide glycoside) | *Gliocladium* sp. | *Syringodium isoetifolium* | Not given | [110] |
| Cytotoxic activity against P388 and HeLa cancer cells | Not identified | *Penicillium thomii* | *Enhalus acoroides* | Endophytic | [111] |
| Anti-inflammatory activity | Thomimarine E | *Penicillium thomii* | *Zostera marina* | | [112] |
| Production of hydrophobins (useful as biosurfactants) | N/A | *Penicillium chrysogenum* | *Posidonia oceanica* | Not specified | [113] |
| Production of lignin-modifying enzymes | Not isolated | *Flavodon flavus* | *Thalassia hemprichii* | Saprobic fungi (from leaves) | [114] |
| Lignocellulosic enzyme activities | Not isolated | *Flavodon flavus* | *Thallasodendon ciliatum* | Saprobic fungi (from decaying leaves) | [115] |
| Removal of polycyclic aromatic hydrocarbons (for bioremediation) | Not isolated | *Flavodon flavus* | *Thalassia hemprichii* | Saprobic fungi (from leaves) | [114] |
| Production of chitin modifying enzymes | Not isolated | Many fungal species | Many seagrass species | N/A | [63] |
| Production of xylan degrading enzymes, useful in biofuel industry | Not specifically identified | N/A | *Thalassia* sp. *Syringodium* sp. (active isolates are from these species) | Endophytic | [116] |
| Production of ligninolytic enzymes and tannases | Many species | N/A | *Posidonia oceanica* | Not specified (can be endophytic or epiphytic) | [73] |
| Decolorized bleach plant effluent from paper and pulp mill, a range of synthetic dyes and molasses | N/A | *Flavodon flavus* | *Thalassia hemprichii* | Saprobic fungi (from decaying leaves) | [117] |
| Isolation of new dimeric chromanone and a phthalide (the applications of them are not tested as they are in small quantities) | Bipolarinone, Bipolarilide | *Bipolaris* sp. | *Halophila ovalis* | Not specifically given (from leaves) | [96] |
| Isolation of a new β-resorcylic macrolide (5′-hydroxyzearalenone) | N/A | *Fusarium* sp. | *Thalassia hemprichii* | Leaves | [98] |
| Four new eudesmane-type sesquiterpenes thomimarines A–D (1–4) | Thomimarines A–D (1–4) | *Penicillium thomii* | *Zostra marina* | Superficial mycobiota of the rhizome | [111] |
| A new azaphilone derivative, xylariphilone | Xylariphilone | *Xylariales* sp. | *Halophila ovalis* | Not given | [118] |
| Weak to potent antimicrobial activity against the plant pathogenic fungi and bacteria | Isolated, but reference not available | *Aspergillus alabamensis* | *Enhalus acoroides* | Pathogenic (from necrotic leaves) | [119] |
| Weak to moderate antifungal activity towards phytopathogenic test fungi | Isolated | *Aspergillus insuetus* | N/A | N/A | [120] |
| Alleviate salinity stress in crop plants | Not isolated | *Trichoderma longibrachiatum* | *Posidonia oceanica* | Root endophytes | [121] |

N/A = not applicable or not available.

### 4.3. Pathogenic Phytophthora and Halophytophthora Species of Seagrasses

Studies indicate that seagrass meadows are gradually thinning and diminishing due to the increased occurrence of diseases caused by pathogenic microorganisms. Seagrasses are infected by four main groups of pathogens: *Labyrinthula*, *Phytophthora*, *Halophytophthora*, and *Phytomyxea* [122,123]. Members of *Labyrinthula* (Labyrinthulids) and *Phytomyxea* (Plasmodiophorids) are classified under the Kingdom *Protista* [124,125] while *Phytophthora* and *Halophytophthora* are in the class *Oomycota*, Kingdom *Straminipila* [126].

Thus far, our literature review has revealed no direct records of any major diseases caused by fungal pathogens in seagrasses. Hu et al. [127] reported the bioactivities of *Aspergillus alabamensis*, a "phytopathogenic fungus" isolated from the seagrass *Enhalus acoroides*. However, it is a speculation, and its pathogenicity to seagrass has not been established.

In contrast, considerable scientific attention has been dedicated to *Labriyrnthula* pathogens responsible for the "seagrass wasting disease", which causes noticeable, extensive losses in many regions of the world. Little is known about the pathogenicity and disease ecology of other groups. Since the focus of this review is on fungal pathogens, we will only look at *Phytophthora* and *Halophytophthora* pathogens, which are considered as fungus-like Oomycetes.

Oomycetes are behaviourally similar but biologically distinct from other main groups belonging to Kingdom *Fungi* [126]. Historically, they were classified under 'Phycomycetes' or "lower fungi" [128]. Considering this early classification, a brief summary of *Phytophthora* and *Halophytophthora* species reported in seagrasses is given in Table S3. Relatively few studies have reported the discovery of pathogenic *Phytophthora* and *Halophytophthora* species in seagrasses. *Phytophthora* and *Halophytophthora* infections have been found to reduce the sexual reproduction of *Zostera marina* by sixfold [119]. Later, Govers et al. [129] demonstrated that treating *Z. marina* seeds with copper sulphate can effectively control *Phytophthora* and *Halophytophthora* infections, highlighting the effectiveness of this method in seed-based restoration projects.

Further, these pathogens are sensitive to annual and seasonal variations, and migratory bird species have been shown to impact disease dissemination [130]. Regardless of the significance of these pathogens in seagrass health, many studies are confined to *Zostera* spp. in the northern hemisphere. Comparable to *Zostera* spp., very little is known about the disease occurrence and pathogenicity of *Phytophthora* and *Halophytophthora* in other seagrass species.

### 4.4. Seagrass Mycobiome Studies

Conventional methods for investigating plant-associated fungi involve isolating and cultivating them on artificial media. Many investigations of seagrass-associated fungi have been undertaken using this standard methodology. However, culture-dependent approaches are often associated with inherent limitations, such as the inability of certain microorganisms to grow on culture media or under specific incubation conditions, as well as the masking of slow-growing microorganisms by fast growers [131]. Thus, these approaches often fail to capture the majority of microbial diversity within environmental samples [132].

As a result, culture-independent, high-throughput molecular methods have gained popularity in recent years to reveal the true diversity and abundance of microorganisms in complex environmental samples. A few meta-studies on seagrasses have been conducted focusing on the seagrass species such as *Zostera marina*, *Z. muelleri*, *Posidonia oceanica*, and *Halophilia* spp. Similar to the culture-dependent approaches, these studies report the dominance of *Eurotiales* fungi such as *Aspergillus* and *Penicillium* belonging to the phylum *Ascomycota*. However, unlike culture-dependent studies, the dominance of fungi belonging to the phylum *Chytridiomycota* is reported in a few studies [35,83,133,134]. This is the only phylum of true fungi that reproduces with zoospores (motile spores). Members of a relatively new order in *Chytridiomycota*, *Lobulomycetales*, have been reported in a few studies [35,83,133,134]. A concise summary of mycobiome studies reported thus far is given in Table S4. Designing new primer pairs and blocking oligonucleotides for fungal detection, especially for basal fungal lineages such as *Cryptomycota* and *Chytridiomycota*, can further refine these underrepresented groups associated with seagrasses [135,136].

## 5. Significance of Studying Seagrass-Associated Fungi

The significance of studying seagrass–fungal associations can be addressed from two perspectives. First, understanding seagrass-associated mycobionts can help to protect this vulnerable ecosystem. As previously noted, these relationships can have either a positive or negative impact on the overall health of seagrass beds. However, no studies have been conducted to investigate the molecular and biochemical mechanisms that govern the structure, activity, and function of these communities. A recent study demonstrated that endophytic fungi associated with seagrasses possess the ability to inhibit the growth of the devastative seagrass pathogen *Labyrinthula* spp. [57]. Therefore, comprehending these interactions and monitoring the compositional changes in fungal communities can serve as a crucial tool for seagrass transplant and restoration initiatives [46].

From a human-centric standpoint, seagrass-inhabiting fungi can be a novel reservoir of metabolites, useful in pharmaceutical and various other industrial applications. Since 'marine drugs' are becoming an appealing strategy for addressing antimicrobial resistance [95], seagrass-associated fungal communities can be studied for novel bioactive chemicals. However, previous reports on the screening and isolation of metabolites from fungi associated with seagrasses are limited, making it difficult to predict their real potential based on reliable scientific information. Table 1 lists some of the beneficial metabolites of seagrass-associated fungi and their bioactivities reported thus far. These bioactive metabolites are known to have antimicrobial, anticancer, anti-inflammatory, and antiviral properties. Some notable bioactive metabolites identified from seagrass mycoflora include Cladionol A [110], Sansalvamide [109], Malformin A1 [104], and Halovir A [106]. Peterson et al. [137] report the use of an omics-based high-throughput approach, a rather new approach for rapid bioactivity testing for seagrass-associated fungal metabolites.

Moreover, several publications report the production of lignocellulosic enzymes [114,115], xylan-degrading enzymes [116], and chitin-modifying enzymes [63] from seagrass-associated fungi. These enzymes have a wide range of biotechnological applications. For example, lignocellulosic and xylan-degrading enzymes are important in the biofuel industry, textile industry, paper and pulp industries, and in bioremediation [114,116]. However, to the best of our knowledge, no enzyme derived from a seagrass-associated fungus has reached mass-scale industrial production. This underscores the need for future studies in upscaling production following successful screening assays. Recently, a few studies have highlighted the potential of seagrass-associated fungi in agriculture, particularly as biocontrol agents and alleviating stress responses such as salinity stress [120,121,127].

## 6. Future Prospects

Currently, there is an emphasis on discovering new taxa from unexplored geographic regions that are quite promising and multifaceted. Nevertheless, it is essential that we incorporate the advances in mycology to expand fundamental studies (e.g., taxonomy based on polyphasic approaches and sequence-based nomenclature to name the Dark taxa). Hence, we recognize several key aspects to continue the research of fungi associated with seagrasses: 1. biodiversity exploration; 2. ecological role understanding; 3. biotechnological potential; and 4. use of emerging technologies.

### 6.1. Biodiversity Exploration

The exploration of biodiversity in the context of seagrass-associated fungi, particularly in uncharted regions such as parts of Southeast Asia and Northern and Southern America, presents an area rich with potential and complexity [22,46,135,138,139]. Seagrass ecosystems, known as biodiversity hotspots, are often under-researched regarding their fungal communities. These ecosystems may host unique fungal species adapted to specific seagrass environments, offering significant opportunities for discovering new fungal species and genera [80,138,140–142].

The geographical diversity of these regions suggests that the fungal diversity associated with seagrass beds could vary significantly, influenced by local environmental factors such as climate, water salinity, and the types of seagrasses present. This diversity is crucial for understanding the ecological roles these fungi play, from nutrient cycling and decomposition to aiding seagrasses in defence against pathogens and environmental stressors [16,93,143].

Advancements in molecular techniques, especially high-throughput DNA sequencing, have revolutionized the identification and cataloguing of fungal species. These methods are more efficient and accurate than traditional ones, capable of identifying even non-culturable fungi and uncovering cryptic species—different species previously thought to be the same due to similar appearances [70,84,133].

Exploring these ecosystems could also reveal new symbiotic relationships between fungi and seagrass or interactions with other microorganisms like bacteria and algae [35,134]. Additionally, understanding how these fungi have adapted to their specific environments can offer insights into their potential responses to global environmental changes, such as rising sea temperatures and ocean acidification [144,145].

Furthermore, each new fungal species discovered contributes to the global understanding of biodiversity, which is vital not just for academic knowledge but also for informing conservation strategies and understanding ecological balances. As seagrass meadows are among the most threatened ecosystems globally, knowing the full range of biodiversity within these areas is essential for their effective management and conservation. Overall, this exploration into the uncharted realms of seagrass-associated fungi holds the key to unlocking the complexities of these ecosystems and significantly contributes to our understanding of marine biodiversity and ecosystem health [22,146].

### 6.2. Ecological Role Understanding

Understanding the ecological roles of fungi in seagrass ecosystems is vital. This includes their roles in nutrient cycling, decomposition processes, and interactions with other microbial communities. As we uncover more about these fungi, we can better understand the overall health and functioning of seagrass ecosystems.

In marine ecology, the ecological roles of fungi within seagrass ecosystems are of paramount importance, encompassing a diverse array of functions that are crucial for the health and functionality of these underwater habitats. Fungi play a vital role in nutrient cycling, breaking down organic matter and releasing essential nutrients back into the environment, thereby maintaining the nutrient balance in these often nutrient-poor marine settings [73,147]. Additionally, they are key players in the decomposition processes, aiding in the breakdown of dead plant materials such as seagrass leaves and roots. This decomposition not only contributes to the recycling of organic matter but also supports the detritus-based food web that is central to the health of the seagrass ecosystem [148,149].

The interactions of fungi with other microbial communities within these ecosystems, including bacteria, viruses, and algae, are also significant. These interactions, which can range from symbiotic to competitive, affect the distribution and abundance of various microbial species and play a crucial role in the microbial dynamics of the ecosystem [150,151]. The secondary metabolites produced by these microbes impact the fungal colonization, which, in turn, can affect the general health of seagrass [75,152–154]. Interestingly, Nerva et al. [155] demonstrated the importance of mycoviruses, a group of viruses which affect fungi, in influencing the overall health of *Posidonia oceanica*.

Moreover, the health of seagrasses themselves can be directly influenced by the fungi associated with them. While some fungi provide protective benefits, helping seagrasses withstand environmental stressors or deter pathogens, others might be pathogenic and detrimental to seagrass health [16,147]. Conversely, different antifungal metabolites produced by seagrasses can affect internal fungal colonization [75,156,157].

Furthermore, fungi within seagrass ecosystems can act as bioindicators of environmental changes. Variations in fungal communities can signal alterations in environmental conditions, such as pollution or changes in water temperature and salinity, providing valuable insights into the health and stability of these ecosystems [158,159]. For instance, the Seagrass Microbiome Project launched in 2014 aims to identify seagrass–microbe interactions that can reveal important information about seagrass ecology, evolution, and function [160]. Additionally, the role of fungi in the decomposition process has implications for carbon sequestration in seagrass meadows, a significant factor in global carbon dynamics [161,162].

### 6.3. Biotechnological Potential

The biotechnological potential of marine fungi, particularly those associated with seagrass ecosystems, is a burgeoning field of research with significant implications for various industries, including medicine, agriculture, and industrial processes [163–165]. Marine fungi are a largely untapped resource, known for their unique ability to produce novel bioactive compounds [166,167]. These compounds, often not found in terrestrial fungi, arise from the adaptation of fungi to the challenging marine environment, characterized by high salt concentrations, varying pressure conditions, and intense competition for resources [168,169].

In medicine, the unique bioactive compounds derived from marine fungi have shown promise in the development of new pharmaceuticals. These compounds can have diverse biological activities, including antibacterial, antiviral, antifungal, anti-inflammatory, and anticancer properties. For example, some compounds might inhibit the growth of cancer cells or bacteria resistant to current antibiotics, offering new avenues for treatment where traditional medicines are failing [166,170].

In agriculture, these fungi could be a source of new biopesticides or growth enhancers. Given their origin in a highly competitive and harsh environment, these fungi may produce substances that are effective in controlling agricultural pests or diseases, potentially reducing the reliance on synthetic chemicals that can be harmful to the environment [164,171].

The industrial sector could also benefit from enzymes and other molecules produced by seagrass-associated fungi. These enzymes might be particularly useful in processes that require tolerance to saline conditions, such as in certain bioremediation applications or in the processing of marine-derived materials [172–174]. As reported by Panno et al. [171], the test fungal isolates recovered from seagrasses exhibit no degradative properties in their enzyme activity in high salt concentrations. This indicates that these enzymes could be valuable for future biotechnological applications that operate under extreme physiochemical conditions. Further, Raghukumar et al. [117] demonstrated the use of a seagrass-derived fungus to remove and detoxify wastewater from molasses-based alcohol distilleries, demonstrating their ability for bioremediation while enhancing the sustainability in biorefinery processes.

Moreover, the exploration of these fungi for biotechnological applications also contributes to the understanding of their ecological roles and potential for sustainable utilization. By identifying and harnessing these bioactive compounds, not only can new, potentially groundbreaking products be developed, but it also encourages the conservation of marine ecosystems like seagrass meadows, which are vital for the health of the marine environment [93,175,176].

### 6.4. Use of Emerging Technologies

The use of emerging technologies such as remote sensing, artificial intelligence (AI), machine learning in ecological studies, and nanotechnology can revolutionize our understanding of seagrass-associated fungi. These advanced technologies offer powerful tools for gaining new insights into the distribution, health, and ecological roles of these fungi in marine ecosystems [177–179].

Remote sensing technology, including satellite imaging and aerial photography, can monitor the health and extent of seagrass meadows over large areas and over time. This technology enables scientists to detect changes in seagrass coverage and condition, which can be indicative of the health of the associated fungal communities [180]. For instance, a decline in seagrass health might suggest issues such as disease or environmental stress, potentially linked to changes in fungal communities. Remote sensing also allows for the mapping of seagrass habitats, providing valuable data for conservation and management efforts [181,182].

Artificial intelligence and machine learning are rapidly becoming indispensable in ecological research. These technologies can process and analyze vast amounts of data much faster and more accurately than traditional methods [179]. In the context of seagrass-associated fungi, AI can be used to analyze complex datasets from remote sensing, genetic sequencing, and ecological surveys to identify patterns and relationships that might be invisible to the human eye. For example, machine-learning algorithms can help in predicting the distribution of fungal species based on environmental variables or in identifying changes in fungal communities in response to environmental stressors [183,184].

Furthermore, the integration of AI with genomic studies, such as metagenomics and high-throughput sequencing, is particularly promising. This integration allows for the rapid identification and classification of fungal species, even those that are rare or previously unknown. AI algorithms can analyze genetic data to uncover relationships between fungal species, their adaptation strategies, and their interactions with seagrasses and other marine organisms [185].

The production and utilization of nanoparticles using fungi, which is also known as myconanotechnology, is also an emerging topic of research [186]. These fungal-derived nanoparticles have a low toxicity and are eco-friendly, compared to the conventional nanoparticles synthesized by chemical or physical means [187]. As seagrass-associated fungi produce unique metabolites, they can hold unique biochemical mechanisms to generate nanoparticles with diverse chemical characteristics. Further, silver nanoparticles synthesized by this biological method can simplify the synthesis process by eliminating the extra steps required to prevent particle aggregation [188]. According to Abdelrahman et al. [189], it is possible to screen and optimize seagrass-associated fungi for nanoparticle synthesis, leading to nanomaterials with different bioactive properties.

The application of these emerging technologies in the study of seagrass-associated fungi not only enhances research capabilities but also contributes to more effective conservation strategies. By providing a more comprehensive and nuanced understanding of these ecosystems, technology can help in predicting the impacts of environmental changes and human activities, thereby informing management and restoration efforts [95,133]. More importantly, understanding the significance of fungal genetic resources through these new approaches can be used as a valuable tool for in situ conservation and support seagrass reforestation.

## 7. Conclusions

Seagrass beds are distinct ecosystems that serve as a reservoir for a diverse array of microbes, including fungi. These fungi can engage in different relationships with their host, spanning from mutualistic to potentially pathogenic. Culture-dependent and culture-independent methods are currently employed to study the diversity of seagrass-associated fungi.

The majority of culture-based methods concentrate on seagrass inhabiting endophytic fungi. Epiphytic fungi are often overlooked, or it could be argued that they are not easily identified directly, due to the challenges associated with handling samples during initial isolation. Thus far, *Eurotiales* species such as *Aspergillus* spp. and *Penicillium* spp. dominate the fungal community associated with seagrasses. There have been no reports of mycorrhizal associations in seagrass roots so far, although other distinct fungal associations have been discovered. Infections by *Phytophthora* and *Halophytophthora* have been documented in seagrasses, which were formerly classified as lower fungi. However, diseases caused by true fungi in seagrasses remain largely unidentified.

Overall, the majority of culture-based studies are confined to *Enhalus acaroides*, *Cymodocea serrulata*, *Posidonia oceanica*, *Thalassia hemprichii*, and *Zostera marina*. Meanwhile, mycobiome studies have been carried out on *Zostera* spp., *Halophilia* spp., and *Posidonia oceanica*. With 72 seagrass species worldwide, there is a significant opportunity to expand both culture-dependent and -independent research on a global scale. Regions with rich diversity, such as the Tropical Indo-Pacifics along the east coast of Africa, lack sufficient data.

The harsh environment in marine ecosystems may lead to the accumulation of metabolites in seagrass-associated fungi, which could be employed in extreme physiological conditions. These fungi could also be investigated as a source of new pharmaceutical lead compounds for development. Recently, a few attempts to employ these fungi to alleviate stress responses in plants have been reported. However, there is a lack of studies on their applications, and their full potential is still unknown.

Rapidly disappearing seagrass beds highlight the importance of advancing research to preserve their valuable genetic resources before they are lost without documentation. Exploring the ecological significance and relationships of these seagrass–fungal associations can aid in future seagrass restoration and transplantation initiatives. The application of modern technologies such as omics-based methods and AI can enhance research on seagrass fungi even further.

**Supplementary Materials:** The following supporting information can be downloaded at: https://www.mdpi.com/article/10.3390/jof10090627/s1, Table S1: Previous records on endophytic and epiphytic fungi isolated from seagrasses using culture-dependent approaches. Table S2: Previous records on miscellaneous fungi isolated from seagrasses using culture-dependent approaches. Table S3: Previous records on pathogenic fungi-like organisms isolated from seagrasses using culture-dependent approaches. Table S4: Summary of the culture-independent studies on seagrass-associated fungi. References [15,35,47,57,59–62,64–72,78,80,82–84,92,93,101,102,105,118,119,133–135,138,144,145,190–209] are cited in the supplementary tables.

**Author Contributions:** Conceptualization, N.N.W., O.R. and K.G.S.U.A.; software, O.R. and S.S.G.; data curation, O.R.; writing—original draft, O.R., N.N.W., S.U., P.K.J., S.S.G. and N.B.; writing—review and editing, O.R., N.N.W., K.G.S.U.A., S.U., P.K.J., S.S.G., N.B. and T.C.B.; visualization, O.R., S.S.G. and S.U.; supervision, N.N.W. and K.G.S.U.A.; project administration, K.G.S.U.A.; funding acquisition, T.C.B. All authors have read and agreed to the published version of the manuscript.

**Funding:** This work was supported by the Science and Technology Human Resource Development Project, Ministry of Higher Education, Sri Lanka, funded by the Asian Development Bank (Grant number R3RJ2) and Tropical Microbiology Research Foundation (TMRF).

**Institutional Review Board Statement:** Not applicable.

**Informed Consent Statement:** Not applicable.

**Data Availability Statement:** The original contributions presented in the study are included in the article/Supplementary Materials; further inquiries can be directed to the corresponding authors.

**Acknowledgments:** Oshadi Rajakaruna would like to thank W. M. Ayesha Sanahari, Department of Plant Sciences, University of Colombo for the help in preparing image plates using Photoshop. Nattawut Boonyuen expresses gratitude to BIOTEC-NSTDA for supporting the fungal research in Thailand. Nalin N. Wijayawardene thanks the High-Level Talent Recruitment Plan of Yunnan Provinces ("Young Talents" Program and "High-End Foreign Experts" Program).

**Conflicts of Interest:** The authors declare no conflicts of interest.

## References

1. Gladfelter, A.S.; James, T.Y.; Amend, A.S. Marine fungi. *Curr. Biol.* **2019**, *29*, R191–R195. [CrossRef] [PubMed]

2. Amend, A.; Burgaud, G.; Cunliffe, M.; Edgcomb, V.P.; Ettinger, C.L.; Gutiérrez, M.H.; Heitman, J.; Hom, E.F.Y.; Ianiri, G.; Jones, A.C.; et al. Fungi in the marine environment: Open questions and unsolved problems. *mBio* **2019**, *10*, e01189-18. [CrossRef] [PubMed]
3. Raghukumar, S. *Fungi in Coastal and Oceanic Marine Ecosystems*, 1st ed.; Springer International Publishing: New York, NY, USA, 2017; ISBN 978-3-319-54303-1.
4. Salvatore, M.M.; Andolfi, A. Phytopathogenic fungi and toxicity. *Toxins* **2021**, *13*, 689. [CrossRef]
5. Walsh, T.J.; Groll, A.; Hiemenz, J.; Fleming, R.; Roilides, E.; Anaissie, E. Infections due to emerging and uncommon medically important fungal pathogens. *Clin. Microbiol. Infect.* **2004**, *10*, 48–66. [CrossRef]
6. Species Fungorum. Available online: http://www.speciesfungorum.org (accessed on 11 July 2024).
7. Hawksworth, D.L. The Fungal dimension of biodiversity: Magnitude, significance, and conservation. *Mycol. Res.* **1991**, *95*, 641–655. [CrossRef]
8. Hawksworth, D.L. The Magnitude of fungal diversity: The 1·5 million species estimate revisited. *Mycol. Res.* **2001**, *105*, 1422–1432. [CrossRef]
9. Tedersoo, L.; Bahram, M.; Põlme, S.; Kõljalg, U.; Yorou, N.S.; Wijesundera, R.; Villarreal Ruiz, L.; Vasco-Palacios, A.M.; Quang Thu, P.; Smith, M.E.; et al. Global diversity and geography of soil fungi. *Science* **2014**, *346*, 1256688. [CrossRef]
10. Hawksworth, D.L.; Lücking, R. Fungal diversity revisited: 2.2 to 3.8 Million species. *Microbiol. Spectr.* **2017**, *5*, 79–95. [CrossRef]
11. Niskanen, T.; Lücking, R.; Dahlberg, A.; Gaya, E.; Suz, L.M.; Mikryukov, V.; Liimatainen, K.; Druzhinina, I.; Westrip, J.R.S.; Mueller, G.M.; et al. Pushing the frontiers of biodiversity research: Unveiling the global diversity, distribution, and conservation of Fungi. *Annu. Rev. Environ. Resour.* **2023**, *48*, 149–176. [CrossRef]
12. Blackwell, M. The Fungi: 1, 2, 3 ... 5.1 Million species? *Am. J. Bot.* **2011**, *98*, 426–438. [CrossRef]
13. Hawksworth, D.L.; Rossman, A.Y. Where are all the undescribed fungi? *Phytopathology* **1997**, *87*, 881–891. [CrossRef] [PubMed]
14. El-Bondkly, E.A.M.; El-Bondkly, A.A.M.; El-Bondkly, A.A.M. Marine endophytic fungal metabolites: A whole new world of pharmaceutical therapy exploration. *Heliyon* **2021**, *7*, e06362. [CrossRef]
15. Pasqualetti, M.; Giovannini, V.; Barghini, P.; Gorrasi, S.; Fenice, M. Diversity and ecology of culturable marine fungi associated with *Posidonia oceanica* leaves and their epiphytic algae *Dictyota dichotoma* and *Sphaerococcus coronopifolius*. *Fungal Ecol.* **2020**, *44*, 100906. [CrossRef]
16. Seymour, J.R.; Laverock, B.; Nielsen, D.A.; Trevathan-Tackett, S.M.; Macreadie, P.I. The microbiology of seagrasses. In *Seagrasses of Australia*; Larkum, A., Kendrick, G., Ralph, P., Eds.; Springer: Cham, Switzerland, 2018; pp. 343–392, ISBN 9783319713540.
17. Sarasan, M.; Puthumana, J.; Job, N.; Han, J.; Lee, J.S.; Philip, R. Marine algicolous endophytic fungi-a promising drug resource of the era. *J. Microbiol. Biotechnol.* **2017**, *27*, 1039–1052. [CrossRef]
18. Johnson, T.W.; Sparrow, F.K. *Fungi in Oceans and Estuaries*; J. Cramer: Weinheim, Germany, 1961.
19. Jones, E.B.G.; Pang, K.L.; Abdel-Wahab, M.A.; Scholz, B.; Hyde, K.D.; Boekhout, T.; Ebel, R.; Rateb, M.E.; Henderson, L.; Sakayaroj, J.; et al. An online resource for marine fungi. *Fungal Divers.* **2019**, *96*, 347–433. [CrossRef]
20. Calabon, M.S.; Jones, E.B.G.; Promputtha, I.; Hyde, K.D. Fungal biodiversity in salt marsh ecosystems. *J. Fungi* **2021**, *7*, 648. [CrossRef] [PubMed]
21. Devadatha, B.; Jones, E.B.G.; Pang, K.L.; Abdel-Wahab, M.A.; Hyde, K.D.; Sakayaroj, J.; Bahkali, A.H.; Calabon, M.S.; Sarma, V.V.; Sutreong, S.; et al. Occurrence and geographical distribution of mangrove fungi. *Fungal Divers.* **2021**, *106*, 137–227. [CrossRef]
22. Poli, A.; Varese, G.C.; Garzoli, L.; Prigione, V. Seagrasses, Seaweeds and Plant Debris: An extraordinary reservoir of fungal diversity in the Mediterranean Sea. *Fungal Ecol.* **2022**, *60*, 101156. [CrossRef]
23. Wijayawardene, N.N.; Dai, D.Q.; Jayasinghe, P.K.; Gunasekara, S.S.; Nagano, Y.; Tibpromma, S.; Suwannarach, N.; Boonyuen, N. Ecological and oceanographic perspectives in future marine fungal taxonomy. *J. Fungi* **2022**, *8*, 1141. [CrossRef]
24. Vandenkoornhuyse, P.; Quaiser, A.; Duhamel, M.; Le Van, A.; Dufresne, A. The importance of the Microbiome of the plant Holobiont. *New Phytol.* **2015**, *206*, 1196–1206. [CrossRef]
25. Jones, E.B.G.; Jennings, D.H. The effect of salinity on the growth of marine fungi in comparison with non-marine species. *Trans. Br. Mycol. Soc.* **1964**, *47*, 619–625. [CrossRef]
26. Kohlmeyer, J.; Kohlmeyer, E. *Marine Mycology: The Higher Fungi*; Academic Press: New York, NY, USA, 1979.
27. Bugni, T.S.; Ireland, C.M. Marine-Derived Fungi: A chemically and biologically diverse group of microorganisms. *Nat. Prod. Rep.* **2004**, *21*, 143–163. [CrossRef]
28. Pang, K.-L.; Overy, D.P.; Jones, E.B.G.; Calado, M.D.L.; Burgaud, G.; Walker, A.K.; Johnson, J.A.; Kerr, R.G.; Cha, H.-J.; Bills, G.F. 'Marine Fungi' and 'Marine-Derived Fungi' in natural product chemistry research: Toward a new consensual definition. *Fungal Biol. Rev.* **2016**, *30*, 163–175. [CrossRef]
29. Jones, E.B.G. Are there more marine fungi to be described? *Bot. Mar.* **2011**, *54*, 343–354. [CrossRef]
30. Gonçalves, M.F.M.; Esteves, A.C.; Alves, A. Marine fungi: Opportunities and challenges. *Encyclopedia* **2022**, *2*, 559–577. [CrossRef]
31. Manohar, C.S.; Raghukumar, C. Fungal diversity from various marine habitats deduced through culture-independent studies. *FEMS Microbiol. Lett.* **2013**, *341*, 69–78. [CrossRef] [PubMed]
32. Larkum, A.W.D.; Orth, R.J.; Duarte, C.M. *Seagrasses: Biology, Ecology and Conservation*, 1st ed.; Springer: Dordrecht, The Netherlands, 2006; ISBN 978-1-4020-2942-4.
33. Short, F.T.; Short, C.A.; Novak, A.B. Seagrasses. In *The Wetland Book*; Finlayson, C.M., Milton, G.R., Prentice, R.C., Davidson, N.C., Eds.; Springer: Dordrecht, The Netherlands, 2016; ISBN 9789400761735.

34. Duarte, C.M. Seagrasses. In *Encyclopedia of Biodiversity*, 2nd ed.; Academic Press: Cambridge, MA, USA, 2001; ISBN 9780128225622.
35. Ettinger, C.L.; Eisen, J.A. Characterization of the mycobiome of the seagrass, *Zostera marina*, reveals putative associations with marine chytrids. *Front. Microbiol.* **2019**, *10*, 2476. [CrossRef] [PubMed]
36. Waycott, M.; Duarte, C.M.; Carruthers, T.J.B.; Orth, R.J.; Dennison, W.C.; Olyarnik, S.; Calladine, A.; Fourqurean, J.W.; Heck, K.L.; Hughes, A.R.; et al. Accelerating loss of seagrasses across the globe threatens coastal ecosystems. *Proc. Natl. Aacad. Sci. USA* **2009**, *106*, 12377–12381. [CrossRef]
37. Orth, R.J.; Carruthers, T.J.B.; Dennison, W.C.; Duarte, C.M.; Fourqurean, J.W.; Heck, K.L.; Hughes, A.R.; Kendrick, G.A.; Kenworthy, W.J.; Olyarnik, S.; et al. A global crisis for seagrass ecosystems. *Bioscience* **2006**, *56*, 987–996. [CrossRef]
38. Short, F.; Carruthers, T.; Dennison, W.; Waycott, M. Global seagrass distribution and diversity: A bioregional model. *J. Exp. Mar. Biol. Ecol.* **2007**, *350*, 3–20. [CrossRef]
39. United Nations Environment Programme. *Out of the Blue: The Value of Seagrasses to the Environment and to People*; UNEP: Nairobi, Kenya, 2020.
40. Björk, M.; Short, F.T.; Mcleod, E.; Beer, S. *Managing Seagrasses for Resilience to Climate Change*; IUCN: Gland, Switzerland, 2008; ISBN 9782831710891.
41. Green, E.P.; Short, F.T. *World Atlas of Seagrasses*; University of California Press: Berkeley, CA, USA, 2003.
42. McKenzie, L.J.; Nordlund, L.M.; Jones, B.L.; Cullen-Unsworth, L.C.; Roelfsema, C.; Unsworth, R.K.F. The Global Distribution of Seagrass Meadows. *Environ. Res. Lett.* **2020**, *15*, 074041. [CrossRef]
43. Duarte, C.M. Seagrass Depth Limits. *Aquat. Bot.* **1991**, *40*, 363–377. [CrossRef]
44. Short, F.T.; Polidoro, B.; Livingstone, S.R.; Carpenter, K.E.; Bandeira, S.; Sidik Bujang, J.; Calumpong, H.P.; Carruthers, T.J.B.; Coles, R.G.; Dennison, W.C.; et al. Extinction risk assessment of the world's seagrass species. *Biol. Conserv.* **2011**, *144*, 1961–1971. [CrossRef]
45. Rasheed, M.A.; Unsworth, R.K.F. Long-term climate-associated dynamics of a tropical seagrass meadow: Implications for the future. *Mar. Ecol. Prog. Ser.* **2011**, *422*, 93–103. [CrossRef]
46. Wainwright, B.J.; Zahn, G.L.; Zushi, J.; Lee, N.L.Y.; Ooi, J.L.S.; Lee, J.N.; Huang, D. Seagrass-associated fungal communities show distance decay of similarity that has implications for seagrass management and restoration. *Ecol. Evol.* **2019**, *9*, 11288–11297. [CrossRef] [PubMed]
47. Quek, Z.B.R.; Zahn, G.; Lee, N.L.Y.; Ooi, J.L.S.; Lee, J.N.; Huang, D.; Wainwright, B.J. Biogeographic structure of fungal communities in seagrass *Halophilia ovalis* across the Malay Peninsula. *Environ. Microbiol. Rep.* **2021**, *13*, 871–877. [CrossRef] [PubMed]
48. Petrini, O. Fungal Endophytes of Tree Leaves. In *Microbial Ecology of Leaves*; Springer: Berlin/Heidelberg, Germany, 1991; pp. 179–197. [CrossRef]
49. Khiralla, A.; Spina, R.; Yagi, S.; Mohamed, I.; Laurain-Mattar, D. Endophytic fungi: Occurrence, classification, function and natural products. In *Endophytic Fungi: Diversity, Characterization and Biocontrol*; Nova Publishers: New York, NY, USA, 2016; ISBN 9781536103588.
50. Schulz, B.; Boyle, C. The endophytic continuum. *Mycol. Res.* **2005**, *109*, 661–686. [CrossRef]
51. Hardoim, P.R.; van Overbeek, L.S.; Berg, G.; Pirttilä, A.M.; Compant, S.; Campisano, A.; Döring, M.; Sessitsch, A. The Hidden World within Plants: Ecological and Evolutionary Considerations for Defining Functioning of Microbial Endophytes. *Microbiol. Mol. Biol. Rev.* **2015**, *79*, 293–320. [CrossRef]
52. Khare, E.; Mishra, J.; Arora, N.K. Multifaceted interactions between endophytes and plant: Developments and prospects. *Front. Microbiol.* **2018**, *9*, 2732. [CrossRef]
53. Yan, L.; Zhu, J.; Zhao, X.; Shi, J.; Jiang, C.; Shao, D. Beneficial effects of endophytic fungi colonization on plants. *Appl. Microbiol. Biotechnol.* **2019**, *103*, 3327–3340. [CrossRef]
54. Eid, A.M.; Salim, S.S.; Hassan, S.E.-D.; Ismail, M.A.; Fouda, A. Role of endophytes in plant health and abiotic stress management. In *Microbiome in Plant Health and Disease*; Springer: Singapore, 2019; pp. 119–144.
55. Sodhi, G.K.; Saxena, S. Role of Endophytic Fungi in Promoting Plant Health. In *Fungal Resources for Sustainable Economy*; Springer Nature: Singapore, 2023; pp. 319–345.
56. Rai, M.; Agarkar, G. Plant-Fungal Interactions: What triggers the fungi to switch among lifestyles? *Crit. Rev. Microbiol.* **2016**, *42*, 428–438. [CrossRef] [PubMed]
57. Ugarellia, K.; Jagels, A.; Choia, C.J.; Loesgenb, S.; Stingl, U. Fungal endophytes from *Thalassia testudinum* show bioactivity against seagrass pathogen *Labyrinthula* sp. *Front. Mar. Sci.* **2024**, *11*, 1359610.
58. Kuo, J. Structural Aspects of apoplast fungal hyphae in a marine angiosperm, *Zostera muelleri* Irmisch Ex Aschers. (Zosteraceae). *Protoplasma* **1984**, *121*, 1–7. [CrossRef]
59. Wilson, W.L. Isolation of Endophytes from Seagrasses from Bermuda. Master's Thesis, The University of New Brunswick, Saint John, NB, Canada, 1998.
60. Sakayaroj, J.; Preedanon, S.; Supaphon, O.; Jones, E.B.G.; Phongpaichit, S. Phylogenetic diversity of endophyte assemblages associated with the tropical seagrass *Enhalus acoroides* in Thailand. *Fungal Divers.* **2010**, *42*, 27–45. [CrossRef]
61. Mata, J.L.; Cebrián, J. Fungal endophytes of the seagrasses *Halodule wrightii* and *Thalassia testudinum* in the Northcentral Gulf of Mexico. *Bot. Mar.* **2013**, *56*, 541–545. [CrossRef]

62. Supaphon, P.; Phongpaichit, S.; Rukachaisirikul, V.; Sakayaroj, J. Diversity and antimicrobial activity of endophytic fungi isolated from the seagrass *Enhalus acoroides*. *Indian J. Geo-Mar. Sci.* **2014**, *43*, 785–797.
63. Venkatachalam, A.; Govinda Rajulu, M.; Thirunavukkarasu, N.; Suryanarayanan, T. Endophytic fungi of marine algae and seagrasses: A novel source of chitin modifying enzymes. *Mycosphere* **2015**, *6*, 345–355. [CrossRef]
64. Subrmaniyan, R.; Ponnambalam, S.; Thirunavukarassu, T. Inter species variations in cultivable endophytic fungal diversity among the tropical seagrasses. *Proc. Natl. Acad. Sci. India Sect B—Biol. Sci.* **2018**, *88*, 849–857. [CrossRef]
65. Supaphon, P.; Phongpaichit, S.; Rukachaisirikul, V.; Sakayaroj, J. Antimicrobial potential of endophytic fungi derived from three seagrass species: *Cymodocea serrulata*, *Halophila ovalis* and *Thalassia hemprichii*. *PLoS ONE* **2013**, *8*, e72520. [CrossRef]
66. Supaphon, P.; Phongpaichit, S.; Sakayaroj, J.; Rukachaisirikul, V.; Kobmoo, N.; Spatafora, J.W. Phylogenetic community structure of fungal endophytes in seagrass species. *Bot. Mar.* **2017**, *60*, 489–501. [CrossRef]
67. Devarajan, P.T.; Suryanarayanan, T.S.; Geetha, V. Endophytic fungi associated with the tropical seagrass *Halophila ovalis* (*Hydrocharitaceae*). *Indian J. Mar. Sci.* **2002**, *31*, 73–74.
68. Venkatachalam, A.; Thirunavukkarasu, N.; Suryanarayanan, T.S. Distribution and Diversity of Endophytes in Seagrasses. *Fungal Ecol.* **2015**, *13*, 60–65. [CrossRef]
69. Cuomo, V.; Vanzanella, F.; Fresi, E.; Cinelli, F.; Mazzella, L. Fungal flora of *Posidonia oceanica* and its ecological significance. *Trans. Br. Mycol. Soc.* **1985**, *84*, 35–40. [CrossRef]
70. Abdel-Wahab, M.A.; Bahkali, A.H.; Elgorban, A.M.; Jones, E.B.G. High-throughput amplicon sequencing of fungi and microbial eukaryotes associated with the seagrass *Halophila stipulacea* (Forssk.) Asch. from Al-Leith Mangroves, Saudi Arabia. *Mycol. Prog.* **2021**, *20*, 1365–1381. [CrossRef]
71. Gnavi, G.; Ercole, E.; Panno, L.; Vizzini, A.; Varese, G.C. *Dothideomycetes* and *Leotiomycetes* sterile mycelia isolated from the italian seagrass *Posidonia oceanica* based on RDNA Data. *SpringerPlus* **2014**, *3*, 508. [CrossRef]
72. Shoemaker, G.; Wyllie-Echeverria, S. Occurrence of rhizomal endophytes in three temperate northeast pacific seagrasses. *Aquat. Bot.* **2013**, *111*, 71–73. [CrossRef]
73. Panno, L.; Bruno, M.; Voyron, S.; Anastasi, A.; Gnavi, G.; Miserere, L.; Varese, G.C. Diversity, ecological role and potential biotechnological applications of marine fungi associated to the seagrass *Posidonia oceanica*. *New Biotechnol.* **2013**, *30*, 685–694. [CrossRef]
74. Ross, C.; Puglisi, M.P.; Paul, V.J. Antifungal defenses of seagrasses from the Indian River Lagoon, Florida. *Aquat. Bot.* **2008**, *88*, 134–141. [CrossRef]
75. Bibi, F.; Naseer, M.I.; Hassan, A.M.; Yasir, M.; Al-Ghamdi, A.A.K.; Azhar, E.I. Diversity and antagonistic potential of bacteria isolated from marine grass *Halodule uninervis*. *3 Biotech* **2018**, *8*, 48. [CrossRef]
76. Solé, M.; Lenoir, M.; Durfort, M.; Fortuño, J.M.; van der Schaar, M.; De Vreese, S.; André, M. Seagrass *Posidonia* is impaired by human-generated noise. *Commun. Biol.* **2021**, *4*, 743. [CrossRef]
77. Nielsen, S.L.; Thingstrup, I.; Wigand, C. Apparent Lack of Vesicular-Arbuscular Mycorrhiza (VAM) in the seagrasses *Zostera marina* L. and *Thalassia testudinum* Banks Ex Konig. *Aquat. Bot.* **1999**, *63*, 261–266. [CrossRef]
78. Vohník, M. Are Lulworthioid Fungi dark septate endophytes of the Dominant Mediterranean Seagrass *Posidonia oceanica*? *Plant Biol.* **2022**, *24*, 127–133. [CrossRef] [PubMed]
79. Kuo, J.; McComb, A.J.; Cambridge, M.L. Ultrastructure of the Seagrass Rhizosphere. *New Phytol.* **1981**, *89*, 139–143. [CrossRef]
80. Torta, L.; Lo Piccolo, S.; Piazza, G.; Burruano, S.; Colombo, P.; Ottonello, D.; Perrone, R.; Di Maida, G.; Pirrotta, M.; Tomasello, A.; et al. *Lulwoana* sp., a Dark Septate Endophyte in roots of *Posidonia oceanica* (L.) Delile seagrass. *Plant Biol.* **2015**, *17*, 505–511. [CrossRef] [PubMed]
81. Vohník, M.; Borovec, O.; Župan, I.; Vondrášek, D.; Petrtýl, M.; Sudová, R. Anatomically and morphologically unique dark septate endophytic association in the roots of the Mediterranean endemic seagrass *Posidonia oceanica*. *Mycorrhiza* **2015**, *25*, 663–672. [CrossRef]
82. Vohník, M.; Borovec, O.; Kolařík, M. Communities of Cultivable Root mycobionts of the seagrass *Posidonia oceanica* in the Northwest Mediterranean Sea are dominated by a hitherto undescribed Pleosporalean Dark Septate Endophyte. *Microb. Ecol.* **2016**, *71*, 442–451. [CrossRef] [PubMed]
83. Vohník, M.; Borovec, O.; Župan, I.; Kolařík, M.; Sudová, R. Fungal root symbionts of the seagrass Posidonia oceanica in the Central Adriatic Sea revealed by Microscopy, Culturing and 454-Pyrosequencing. *Mar. Ecol. Prog. Ser.* **2017**, *583*, 107–120. [CrossRef]
84. Vohník, M.; Borovec, O.; Kolaříková, Z.; Sudová, R.; Réblová, M. Extensive sampling and high-throughput sequencing reveal *Posidoniomyces atricolor* gen. Et sp. nov. (*Aigialaceae*, *Pleosporales*) as the dominant root mycobiont of the dominant Mediterranean seagrass *Posidonia oceanica*. *MycoKeys* **2019**, *55*, 59–86. [CrossRef]
85. Lefebvre, L.; Compère, P.; Gobert, S. The formation of aegagropiles from the Mediterranean seagrass *Posidonia oceanica* (L.) Delile (1813): Plant tissue sources and colonisation by melanised fungal mycelium. *Mar. Biol.* **2023**, *170*, 19. [CrossRef]
86. Carrasco-Acosta, M.; Poli, A.; Garcia-Jimenez, P.; Prigione, V.P.; Varese, G.C. The cultivable mycobiota associated with the seagrass *Cymodocea nodosa* for ecological and biotechnological purposes. In Proceedings of the IUBMB Focused Meeting on Extremophilic Fungi, Ljubljana, Slovenia, 19–22 September 2023; Biotechnical Faculty, University of Ljubljana: Ljubljana, Slovenia, 2023; p. 127.

87. Wang, X.; Pecoraro, L.; Chen, J.; Tang, Y.; Lee, S.; Chen, S.; Liu, H. *Halophilomyces hongkongensis*, a novel species and genus in the *Lulworthiaceae* with antibacterial potential, colonizing the roots and rhizomes of the seagrass *Halophila ovalis*. *J. Fungi* **2024**, *10*, 474. [CrossRef]
88. Michael, T.S.; Shin, H.W.; Hanna, R.; Spafford, D.C. A review of epiphyte community development: Surface interactions and settlement on seagrass. *J. Environ. Biol.* **2008**, *98*, 629–638.
89. Trevizan Segovia, B.; Sanders-Smith, R.; Adamczyk, E.M.; Forbes, C.; Hessing-Lewis, M.; O'Connor, M.I.; Parfrey, L.W. Microeukaryotic communities associated with the seagrass *Zostera marina* are spatially structured. *J. Eukaryot. Microbiol.* **2021**, *68*, e12827. [CrossRef] [PubMed]
90. Jagtap, T.G.; Komarpant, D.S.; Rodrigues, R.S. Status of a seagrass ecosystem: An ecologically sensitive wetland habitat from India. *Wetlands* **2003**, *23*, 161–170. [CrossRef]
91. Novak, R. A Study in Ultra-Ecology: Microorganisms on the seagrass *Posidonia oceanica* (L.) Delile. *Mar. Ecol.* **1984**, *5*, 143–190. [CrossRef]
92. Ettinger, C.L.; Eisen, J.A. Fungi, Bacteria and Oomycota opportunistically isolated from the seagrass, *Zostera marina*. *PLoS ONE* **2020**, *15*, e0236135. [CrossRef]
93. Tasdemir, D.; Scarpato, S.; Utermann-Thüsing, C.; Jensen, T.; Blümel, M.; Wenzel-Storjohann, A.; Welsch, C.; Echelmeyer, V.A. Epiphytic and endophytic microbiome of the seagrass *Zostera marina*: Do they contribute to pathogen reduction in seawater? *Sci. Total Environ.* **2024**, *908*, 168422. [CrossRef]
94. Vohník, M.; Josefiová, J. Novel epiphytic root-fungus symbiosis in the Indo-Pacific Seagrass *Thalassodendron ciliatum* from the Red Sea. *Mycorrhiza* **2024**, 1–15. [CrossRef]
95. Zhang, Y.; Mu, J.; Feng, Y.; Kang, Y.; Zhang, J.; Gu, P.J.; Wang, Y.; Ma, L.F.; Zhu, Y.H. Broad-spectrum antimicrobial epiphytic and endophytic fungi from marine organisms: Isolation, Bioassay and Taxonomy. *Mar. Drugs* **2009**, *7*, 97–112. [CrossRef]
96. Arunpanichlert, J.; Rukachaisirikul, V.; Tadpetch, K.; Phongpaichit, S.; Hutadilok-Towatana, N.; Supaphon, O.; Sakayaroj, J. A Dimeric Chromanone and a Phthalide: Metabolites from the seagrass-derived fungus *Bipolaris* sp. PSU-ES64. *Phytochem. Lett.* **2012**, *5*, 604–608. [CrossRef]
97. Alfattani, A.; Queiroz, E.F.; Marcourt, L.; Leoni, S.; Stien, D.; Hofstetter, V.; Gindro, K.; Perron, K.; Wolfender, J.-L. One-Step bio-guided isolation of secondary metabolites from the endophytic fungus *Penicillium crustosum* using high-resolution semi-preparative HPLC. *Comb. Chem. High Throughput Screen.* **2024**, *27*, 573–583. [CrossRef]
98. Arunpanichlert, J.; Rukachaisirikul, V.; Sukpondma, Y.; Phongpaichit, S.; Supaphon, O.; Sakayaroj, J. A β-resorcylic macrolide from the seagrass-derived fungus *Fusarium* sp. PSU-ES73. *Arch. Pharm. Res.* **2011**, *34*, 1633–1637. [CrossRef] [PubMed]
99. Linn, K.T. Screening of marine endophytic fungi isolated from some seagrasses leaves and their antibacterial activities on *Micrococcus luteus* NITE83297. *J. Myanmar Acad. Arts Sci.* **2018**, *17*, 139–154.
100. Botta, L.; Saladino, R.; Barghini, P.; Fenice, M.; Pasqualetti, M. Production and identification of two antifungal terpenoids from the *Posidonia oceanica* epiphytic Ascomycota *Mariannaea humicola* IG100. *Microb. Cell Fact.* **2020**, *19*, 184. [CrossRef] [PubMed]
101. Qader, M.M.; Hamed, A.A.; Soldatou, S.; Abdelraof, M.; Elawady, M.E.; Hassane, A.S.I.; Belbahri, L.; Ebel, R.; Rateb, M.E. Antimicrobial and antibiofilm activities of the fungal metabolites isolated from the marine endophytes *Epicoccum nigrum* M13 and *Alternaria alternata* 13A. *Mar. Drugs* **2021**, *19*, 232. [CrossRef] [PubMed]
102. Hamed, A.A.; Soldatou, S.; Mallique Qader, M.; Arjunan, S.; Miranda, K.J.; Casolari, F.; Pavesi, C.; Diyaolu, O.A.; Thissera, B.; Eshelli, M.; et al. Screening fungal endophytes derived from under-explored Egyptian marine habitats for antimicrobial and antioxidant properties in factionalised textiles. *Microorganisms* **2020**, *8*, 1617. [CrossRef] [PubMed]
103. Setyati, W.A.; Sedjati, S.; Samudra, A.; Ariyanto, D. Investigation of seagrass-associated fungi as antifouling candidates with anti-bacterial properties. *Jordan J. Biol. Sci.* **2023**, *16*, 323–327. [CrossRef]
104. Notarte, K. Trypanocidal activity, cytotoxicity and histone modifications induced by Malformin A1 isolated from the marine-derived fungus Aspergillus Tubingensis IFM 63452. *Mycosphere* **2017**, *8*, 111–120. [CrossRef]
105. Morell-Rodríguez, G. Potential of Fungal Endophytes from *Thalassia testudinum* Bank Ex KD Koenig as Producers of Bioactive compounds. Ph.D. Thesis, University of Puerto Rico, Mayaguez, Puerto Rico, 2008.
106. Fenical, W.; Jensen, P.R.; Rowley, D.C. Halovir, an Antiviral Marine Natural Product, and Derivatives Thereof. U.S. Patent 983 6,458,766 B1, 1 October 2002.
107. Kaushik, N.; Murali, T.; Sahal, D.; Suryanarayanan, T. A search for antiplasmodial metabolites among fungal endophytes of terrestrial and marine plants of Southern India. *Acta Parasitol.* **2014**, *59*, 745–757. [CrossRef]
108. Notarte, K.I.; Yaguchi, T.; Suganuma, K.; dela Cruz, T.E. Antibacterial, cytotoxic and trypanocidal activities of marine-derived fungi isolated from Philippine macroalgae and seagrasses. *Acta Bot. Croat.* **2018**, *77*, 141–151. [CrossRef]
109. Belofsky, G.N.; Jensen, P.R.; Fenical, W. Sansalvamide: A new cytotoxic cyclic depsipeptide produced by a marine fungus of the genus *Fusarium*. *Tetrahedron Lett.* **1999**, *40*, 2913–2916. [CrossRef]
110. Kasai, Y.; Komatsu, K.; Shigemori, H.; Tsuda, M.; Mikami, Y.; Kobayashi, J. Cladionol A, a polyketide glycoside from marine-derived fungus *Gliocladium* species. *J. Nat. Prod.* **2005**, *68*, 777–779. [CrossRef] [PubMed]
111. Afiyatullov, S.S.; Leshchenko, E.V.; Sobolevskaya, M.P.; Denisenko, V.A.; Kirichuk, N.N.; Khudyakova, Y.V.; Hoai, T.P.; Dmitrenok, P.S.; Menchinskaya, E.S.; Pislyagin, E.A.; et al. New eudesmane sesquiterpenes from the marine-derived fungus *Penicillium thomii*. *Phytochem. Lett.* **2015**, *14*, 209–214. [CrossRef]

112. Afiyatullov, S.S.; Leshchenko, E.V.; Sobolevskaya, M.P.; Antonov, A.S.; Denisenko, V.A.; Popov, R.S.; Khudyakova, Y.V.; Kirichuk, N.N.; Kuz'mich, A.S.; Pislyagin, E.A.; et al. New Thomimarine E from marine isolate of the fungus *Penicillium thomii*. *Chem. Nat. Compd.* **2017**, *53*, 290–294. [CrossRef]
113. Cicatiello, P.; Gravagnuolo, A.M.; Gnavi, G.; Varese, G.C.; Giardina, P. Marine fungi as source of new hydrophobins. *Int. J. Biol. Macromol.* **2016**, *92*, 1229–1233. [CrossRef] [PubMed]
114. Raghukumar, C.; D'souza, T.; Thorn, R.; Reddy, C. Lignin-modifying enzymes of *Flavodon flavus* a Basidiomycete isolated from a coastal marine environment. *Appl. Environ. Microbiol.* **1999**, *65*, 2103–2111. [CrossRef]
115. Mtui, G.; Nakamura, Y. Lignocellulosic enzymes from *Flavodon flavus*, a fungus isolated from Western Indian ocean off the coast of Dar Es Salaam, Tanzania. *Afr. J. Biotechnol.* **2008**, *7*, 3066–3072.
116. Thirunavukkarasu, N.; Jahnes, B.; Broadstock, A.; Govinda Rajulu, M.B.; Murali, T.S.; Gopalan, V.; Suryanarayanan, T.S. Screening marine-derived endophytic fungi for xylan-degrading enzymes. *Curr. Sci.* **2015**, *109*, 112–120.
117. Raghukumar, C.; Mohandass, C.; Kamat, S.; Shailaja, M.S. Simultaneous detoxification and decolorization of molasses spent wash by the immobilized white-rot fungus *Flavodon flavus* isolated from a marine habitat. *Enzyme Microb. Technol.* **2004**, *35*, 197–202. [CrossRef]
118. Arunpanichlert, J.; Rukachaisirikul, V.; Phongpaichit, S.; Supaphon, O.; Sakayaroj, J. Xylariphilone: A new azaphilone derivative from the seagrass-derived fungus *Xylariales* sp. PSU-ES163. *Nat. Prod. Res.* **2016**, *30*, 46–51. [CrossRef] [PubMed]
119. Govers, L.L.; Man In 'T Veld, W.A.; Meffert, J.P.; Bouma, T.J.; van Rijswick, P.C.J.; Heusinkveld, J.H.T.; Orth, R.J.; van Katwijk, M.M.; van der Heide, T. Marine *Phytophthora* species can hamper conservation and restoration of vegetated coastal ecosystems. *Proc. R. Soc. B Biol. Sci.* **2016**, *283*, 20160812. [CrossRef]
120. Hu, Z.; Chen, J.; Liu, Q.; Wu, Q.; Chen, S.; Wang, J.; Li, J.; Liu, L.; Gao, Z. Cyclohexenone derivative and drimane sesquiterpenes from the seagrass-derived fungus *Aspergillus insuetus*. *Chem. Biodivers.* **2023**, *20*, e202300424. [CrossRef] [PubMed]
121. Sánchez-Montesinos, B.; Diánez, F.; Moreno-Gavira, A.; Gea, F.J.; Santos, M. Plant growth promotion and biocontrol of *Pythium ultimum* by saline tolerant *Trichoderma* isolates under salinity stress. *Int. J. Environ. Res. Public Health* **2019**, *16*, 2053. [CrossRef]
122. Sullivan, B.K.; Trevathan-Tackett, S.M.; Neuhauser, S.; Govers, L.L. Review: Host-pathogen dynamics of seagrass diseases under future global change. *Mar. Pollut. Bull.* **2018**, *134*, 75–88. [CrossRef]
123. Hua Tan, M.; Loke, S.; Croft, L.J.; Gleason, F.H.; Lange, L.; Trevathan-Tackett, S.M. First genome of *Labyrinthula*, an opportunistic seagrass pathogen, reveals novel insight into marine protist phylogeny, ecology and CAZyme cell-wall degradation. *Microb. Ecol.* **2021**, *82*, 498–511. [CrossRef]
124. Raghukumar, S. Ecology of the marine Protists, the *Labyrinthulomycetes* (*Thraustochytrids* and *Labyrinthulids*). *Eur. J. Protistol.* **2002**, *38*, 127–145. [CrossRef]
125. Neuhauser, S.; Kirchmair, M.; Gleason, F.H. Ecological roles of the parasitic phytomyxids (plasmodiophorids) in marine ecosystems—A Review. *Mar. Freshw. Res.* **2011**, *62*, 365–371. [CrossRef] [PubMed]
126. Cooke, D.E.L.; Drenth, A.; Duncan, J.M.; Wagels, G.; Brasier, C.M. A molecular phylogeny of *Phytophthora* and related oomycetes. *Fungal Genet. Biol.* **2000**, *30*, 17–32. [CrossRef] [PubMed]
127. Hu, Z.; Zhu, Y.; Chen, J.; Li, C.; Gao, Z.; Li, J.; Liu, L. Sesquiterpenoids with phytotoxic and antifungal activities from a pathogenic fungus *Aspergillus alabamensis*. *J. Agric. Food Chem.* **2022**, *70*, 12065–12073. [CrossRef] [PubMed]
128. Fry, W.E.; Grünwald, N.J. Introduction to Oomycetes. *Plant Health Instr.* **2010**, *10*. [CrossRef]
129. Govers, L.L.; Van Der Zee, E.M.; Meffert, J.P.; Van Rijswick, P.C.J.; Man In T Veld, W.A.; Heusinkveld, J.H.T.; Van Der Heide, T. Copper treatment during storage reduces *Phytophthora* and *Halophytophthora* infection of *Zostera marina* seeds used for restoration. *Sci. Rep.* **2017**, *7*, 43172. [CrossRef]
130. Menning, D.M.; Ward, D.H.; Wyllie-Echeverria, S.; Sage, G.K.; Gravley, M.C.; Gravley, H.A.; Talbot, S.L. Are migratory waterfowl vectors of seagrass pathogens? *Ecol. Evol.* **2020**, *10*, 2062–2073. [CrossRef] [PubMed]
131. Jackson, C.R.; Randolph, K.C.; Osborn, S.L.; Tyler, H.L. Culture dependent and independent analysis of bacterial communities associated with commercial salad leaf vegetables. *BMC Microbiol.* **2013**, *13*, 274. [CrossRef] [PubMed]
132. Turner, T.R.; James, E.K.; Poole, P.S. The Plant Microbiome. *Genome Biol.* **2013**, *14*, 209. [CrossRef]
133. Ettinger, C.L.; Vann, L.E.; Eisen, J.A. Global diversity and biogeography of the *Zostera marina* mycobiome. *Appl. Environ. Microbiol.* **2021**, *87*, e02795-20. [CrossRef]
134. Chen, J.; Zang, Y.; Yang, Z.; Qu, T.; Sun, T.; Liang, S.; Zhu, M.; Wang, Y.; Tang, X. Composition and functional diversity of epiphytic bacterial and fungal communities on marine macrophytes in an intertidal zone. *Front. Microbiol.* **2022**, *13*, 839465. [CrossRef]
135. Banos, S.; Lentendu, G.; Kopf, A.; Wubet, T.; Glöckner, F.O.; Reich, M. A Comprehensive Fungi-specific 18S rRNA gene sequence primer toolkit suited for diverse research issues and sequencing platforms. *BMC Microbiol.* **2018**, *18*, 190. [CrossRef]
136. Ishii, N.; Ishida, S.; Kagami, M. PCR Primers for assessing community structure of aquatic fungi including *Chytridiomycota* and *Cryptomycota*. *Fungal Ecol.* **2015**, *13*, 33–43. [CrossRef]
137. Petersen, L.-E.; Marner, M.; Labes, A.; Tasdemir, D. Rapid metabolome and bioactivity profiling of fungi associated with the leaf and rhizosphere of the baltic seagrass *Zostera marina*. *Mar. Drugs* **2019**, *17*, 419. [CrossRef] [PubMed]
138. Castro González, M.; Gómez-López, D.I. Endophyte mycobiota of *Thalassia testudinum* König in the Colombian Caribbean [Micobiota Endófita de Thalassia Testudinum König En El Caribe Colombiano]. *Bol. Investig. Mar. Y Costeras* **2022**, *52*, 185–192. [CrossRef]

139. Bantoto-Kinamot, V.; Monotilla, A. New report of endophytic Sordariomycetes from the seagrasses of Cebu, Central Philippines. *J. Trop. Life Sci.* **2023**, *13*, 553–562. [CrossRef]
140. Man in 't Veld, W.A.; Rosendahl, K.C.H.M.; Brouwer, H.; de Cock, A.W.A.M. *Phytophthora gemini* sp. Nov., a new species isolated from the halophilic plant *Zostera marina* in the Netherlands. *Fungal Biol.* **2011**, *115*, 724–732. [CrossRef]
141. Poli, A.; Bovio, E.; Ranieri, L.; Varese, G.C.; Prigione, V. News from the Sea: A new genus and seven new species in the pleosporalean families *Roussoellaceae* and *Thyridariaceae*. *Diversity* **2020**, *12*, 144. [CrossRef]
142. Poli, A.; Bovio, E.; Verkley, G.; Prigione, V.; Varese, G.C. *Elbamycella rosea* gen. et sp. nov. (*Juncigenaceae, Torpedosporales*) isolated from the Mediterranean Sea. *MycoKeys* **2019**, *55*, 15–28. [CrossRef] [PubMed]
143. Hu, Z.; Yujiao, Z.; Junjie, C.; Jun, C.; Chunyuan, L.; Zhizeng, G.; Jing, L.; Lan, L. Discovery of novel bactericides from *Aspergillus alabamensis* and their antibacterial activity against fish pathogens. *J. Agric. Food Chem.* **2023**, *71*, 4298–4305. [CrossRef] [PubMed]
144. Camp, E.F.; Suggett, D.J.; Gendron, G.; Jompa, J.; Manfrino, C.; Smith, D.J. Mangrove and seagrass beds provide different biogeochemical services for corals threatened by climate change. *Front. Mar. Sci.* **2016**, *3*, 52. [CrossRef]
145. Viana, I.G.; Artika, S.R.; Moreira-Saporiti, A.; Teichberg, M. Limited trait responses of a tropical seagrass to the combination of increasing pCO2 and Warming. *J. Exp. Bot.* **2023**, *74*, 472–488. [CrossRef] [PubMed]
146. Tisthammer, K.H.; Cobian, G.M.; Amend, A.S. Global biogeography of marine fungi is shaped by the environment. *Fungal Ecol.* **2016**, *19*, 39–46. [CrossRef]
147. Hurtado-McCormick, V.; Kahlke, T.; Petrou, K.; Jeffries, T.; Ralph, P.J.; Seymour, J.R. Regional and microenvironmental scale characterization of the *Zostera muelleri* seagrass microbiome. *Front. Microbiol.* **2019**, *10*, 1011. [CrossRef]
148. Liu, S.; Trevathan-Tackett, S.M.; Jiang, Z.; Cui, L.; Wu, Y.; Zhang, X.; Li, J.; Luo, H.; Huang, X. Nutrient loading decreases blue carbon by mediating fungi activities within seagrass meadows. *Environ. Res.* **2022**, *212*, 113280. [CrossRef]
149. Bongiorni, L. Thraustochytrids, a neglected component of organic matter decomposition and food webs in marine sediments. In *Biology of Marine Fungi*; Raghukumar, C., Ed.; Springer: Berlin/Heidelberg, Germany, 2011; pp. 1–13.
150. Menaa, F.; Wijesinghe, P.A.U.I.; Thiripuranathar, G.; Uzair, B.; Iqbal, H.; Khan, B.A.; Menaa, B. Ecological and industrial implications of dynamic seaweed-associated microbiota interactions. *Mar. Drugs* **2020**, *18*, 641. [CrossRef]
151. Hebert, T.A.; Kuehn, K.A.; Halvorson, H.M. Land use differentially alters microbial interactions and detritivore feeding during leaf decomposition in headwater streams. *Freshw. Biol.* **2023**, *68*, 1386–1399. [CrossRef]
152. Siro, G.; Pipite, A. Mini-Review on the antimicrobial potential of actinobacteria associated with seagrasses. *Explor. Drug Sci.* **2024**, *2*, 117–125. [CrossRef]
153. Boontanom, P.; Chantarasiri, A. Short Communication: Diversity of culturable epiphytic bacteria isolated from seagrass (*Halodule uninervis*) in Thailand and their preliminary antibacterial activity. *Biodiversitas J. Biol. Divers.* **2020**, *21*, 2907–2913. [CrossRef]
154. Srinivasan, R.; Kannappan, A.; Shi, C.; Lin, X. Marine bacterial secondary metabolites: A treasure house for structurally unique and effective antimicrobial compounds. *Mar. Drugs* **2021**, *19*, 530. [CrossRef]
155. Nerva, L.; Ciuffo, M.; Vallino, M.; Margaria, P.; Varese, G.C.; Gnavi, G.; Turina, M. Multiple approaches for the detection and characterization of viral and plasmid symbionts from a collection of marine fungi. *Virus Res.* **2016**, *219*, 22–38. [CrossRef]
156. Gono, C.M.P.; Ahmadi, P.; Hertiani, T.; Septiana, E.; Putra, M.Y.; Chianese, G. A comprehensive update on the bioactive compounds from seagrasses. *Mar. Drugs* **2022**, *20*, 406. [CrossRef]
157. Vasarri, M.; De Biasi, A.M.; Barletta, E.; Pretti, C.; Degl'Innocenti, D. An overview of new insights into the benefits of the seagrass *Posidonia oceanica* for human health. *Mar. Drugs* **2021**, *19*, 476. [CrossRef] [PubMed]
158. Mishra, A.K.; Sahoo, R.; Samantaray, S.S.; Apte, D.; Chowk, S.A.; Singh, S.B. Seagrass ecosystems of India as bioindicators of trace elements. In *Coastal Ecosystems*; Madhav, S., Nazneen, S., Singh, P., Eds.; Springer: Cham, Switzerland, 2020; Volume 38, pp. 45–65.
159. Stipcich, P.; Balmas, V.; Jimenez, C.E.; Oufensou, S.; Ceccherelli, G. Cultivable mycoflora on bleached, decaying and healthy *Posidonia oceanica* leaves in a warm-edge Mediterranean location. *Mar. Environ. Res.* **2023**, *192*, 106188. [CrossRef]
160. The Seagrass Microbiome Project. Available online: https://seagrassmicrobiome.org/ (accessed on 11 July 2024).
161. Han, Q.; Qiu, C.; Zeng, W.; Chen, S.; Zhao, M.; Shi, Y.; Zhang, X. Sediment carbon sequestration and driving factors in seagrass beds from Hainan Island and the Xisha Islands. *Processes* **2023**, *11*, 456. [CrossRef]
162. Fu, C.; Frappi, S.; Havlik, M.N.; Howe, W.; Harris, S.D.; Laiolo, E.; Gallagher, A.J.; Masqué, P.; Duarte, C.M. Substantial blue carbon sequestration in the world's largest seagrass meadow. *Commun. Earth Environ.* **2023**, *4*, 474. [CrossRef]
163. Behera, A.D.; Das, S. Ecological Insights and Potential Application of Marine Filamentous Fungi in Environmental Restoration. *Rev. Environ. Sci. Biotechnol.* **2023**, *22*, 281–318. [CrossRef]
164. Noorjahan, A.; Aiyamperumal, B.; Anantharaman, P. Fungal endophytes from seaweeds and bio-potential applications in agriculture. In *Fungi Bio-Prospects in Sustainable Agriculture, Environment and Nano-Technology*; Elsevier: Amsterdam, The Netherlands, 2021; pp. 83–95.
165. Pang, K.-L.; Jones, E.B.G.; Abdel-Wahab, M.A.; Adams, S.J.; Alves, A.; Azevedo, E.; Bahkali, A.H.; Barata, M.; Burgaud, G.; Caeiro, M.F.; et al. Recent progress in marine mycological research in different countries, and prospects for future developments worldwide. *Bot. Mar.* **2023**, *66*, 239–269. [CrossRef]
166. dela Cruz, T.E.E.; Notarte, K.I.R.; Apurillo, C.C.S.; Tarman, K.; Bungihan, M.E. Biomining fungal endophytes from tropical plants and seaweeds for drug discovery. In *Biodiversity and Biomedicine: Our Future*; Elsevier: Amsterdam, The Netherlands, 2020; pp. 51–62, ISBN 9780128195413.

167. Tarman, K. Marine Fungi as a Source of Natural Products. In *Encyclopedia of Marine Biotechnology*; Wiley: Hoboken, NJ, USA, 2020; pp. 2147–2160, ISBN 9781119143802.
168. Varrella, S.; Barone, G.; Tangherlini, M.; Rastelli, E.; Dell'Anno, A.; Corinaldesi, C. Diversity, ecological role and biotechnological potential of Antarctic marine fungi. *J. Fungi* **2021**, *7*, 391. [CrossRef]
169. Bhatia, S.; Makkar, R.; Behl, T.; Sehgal, A.; Singh, S.; Rachamalla, M.; Mani, V.; Iqbal, M.S.; Bungau, S.G. Biotechnological innovations from ocean: Transpiring role of marine drugs in management of chronic disorders. *Molecules* **2022**, *27*, 1539. [CrossRef] [PubMed]
170. Wang, J.; Qin, Y.; Lin, M.; Song, Y.; Lu, H.; Xu, X.; Liu, Y.; Zhou, X.; Gao, C.; Luo, X. Marine natural products from the Beibu Gulf: Sources, chemistry, and bioactivities. *Mar. Drugs* **2023**, *21*, 63. [CrossRef]
171. Thatoi, H.; Behera, B.C.; Mishra, R.R. Ecological role and biotechnological potential of mangrove fungi: A Review. *Mycology* **2013**, *4*, 54–71. [CrossRef]
172. Chen, H.Y.; Xue, D.S.; Feng, X.Y.; Yao, S.J. Screening and production of ligninolytic enzyme by a marine-derived fungal *Pestalotiopsis* sp. J63. *Appl. Biochem. Biotechnol.* **2011**, *165*, 1754–1769. [CrossRef]
173. Morlighem, J.É.R.L.; Radis-Baptista, G. The place for enzymes and biologically active peptides from marine organisms for application in industrial and pharmaceutical biotechnology. *Curr. Protein Pept. Sci.* **2019**, *20*, 334–355. [CrossRef]
174. Poli, A.; Bovio, E.; Perugini, I.; Varese, G.C.; Prigione, V. *Corollospora mediterranea*: A novel species complex in the Mediterranean Sea. *Appl. Sci.* **2021**, *11*, 5452. [CrossRef]
175. Heckwolf, M.J.; Peterson, A.; Jänes, H.; Horne, P.; Künne, J.; Liversage, K.; Sajeva, M.; Reusch, T.B.H.; Kotta, J. From ecosystems to socio-economic benefits: A systematic review of coastal ecosystem services in the Baltic Sea. *Sci. Total Environ.* **2021**, *755*, 142565. [CrossRef]
176. Setiawan, E.; Chodiantoro, M.R.; Insany, G.F.; Subagio, I.B.; Dewi, N.K.; Muzaki, F.K. Short Communication: Diversity of sponges associated in seagrass meadows at coastal area of Pacitan District, East Java, Indonesia. *Biodiversitas J. Biol. Divers.* **2021**, *22*, 3105–3112. [CrossRef]
177. Hossain, M.S.; Hashim, M. Potential of Earth Observation (EO) technologies for seagrass ecosystem service assessments. *Int. J. Appl. Earth Obs. Geoinf.* **2019**, *77*, 15–29. [CrossRef]
178. Kamal, A.H.M.; Asif, A.-A.; Idris, M.H.; Bhuiyan, M.K.A.; Rahman, A.A. Trends in seagrass research and conservation in Malaysian waters. *J. Trop. Life Sci.* **2023**, *13*, 93–114. [CrossRef]
179. Pham, T.D.; Ha, N.T.; Saintilan, N.; Skidmore, A.; Phan, D.C.; Le, N.N.; Viet, H.L.; Takeuchi, W.; Friess, D.A. Advances in earth observation and machine learning for quantifying blue carbon. *Earth Sci. Rev.* **2023**, *243*, 104501. [CrossRef]
180. Araya-Lopez, R.; de Paula Costa, M.D.; Wartman, M.; Macreadie, P.I. Trends in the application of remote sensing in blue carbon science. *Ecol. Evol.* **2023**, *13*, e10559. [CrossRef]
181. Ivajnšič, D.; Orlando-Bonaca, M.; Donša, D.; Grujić, V.J.; Trkov, D.; Mavrič, B.; Lipej, L. Evaluating seagrass meadow dynamics by integrating field-based and remote sensing techniques. *Plants* **2022**, *11*, 1196. [CrossRef]
182. Zoffoli, M.L.; Gernez, P.; Oiry, S.; Godet, L.; Dalloyau, S.; Davies, B.F.R.; Barillé, L. Remote sensing in seagrass ecology: Coupled dynamics between migratory herbivorous birds and intertidal meadows observed by satellite during four decades. *Remote Sens. Ecol. Conserv.* **2023**, *9*, 420–433. [CrossRef]
183. Picek, L.; Šulc, M.; Matas, J.; Heilmann-Clausen, J.; Jeppesen, T.S.; Lind, E. Automatic fungi recognition: Deep Learning meets Mycology. *Sensors* **2022**, *22*, 633. [CrossRef]
184. Rahman, M.A.; Clinch, M.; Reynolds, J.; Dangott, B.; Meza Villegas, D.M.; Nassar, A.; Hata, D.J.; Akkus, Z. Classification of fungal genera from microscopic images using Artificial Intelligence. *J. Pathol. Inform.* **2023**, *14*, 100314. [CrossRef]
185. Ghosh, S.; Dasgupta, R. *Machine Learning in Biological Sciences*; Springer Nature: Singapore, 2022; ISBN 978-981-16-8880-5.
186. Basheer, M.A.; Abutaleb, K.; Abed, N.N.; Mekawey, A.A.I. Mycosynthesis of silver nanoparticles using marine fungi and their antimicrobial activity against pathogenic microorganisms. *J. Genet. Eng. Biotechnol.* **2023**, *21*, 127. [CrossRef]
187. Rai, M.; Wypij, M.; Trzcińska-Wencel, J.; Yadav, A.; Ingel, I.P.; Avila-Quezada, G.D.; Golińska, P. Myconanotechnology: Opportunities and Challenges. In *Myconanotechnology*, 1st ed.; Rai, M., Golinska, P., Eds.; CRC Press: Boca Raton, FL, USA, 2023; p. 23.
188. Suresh, J.I.; Lydia, N.J.; Atchayadana, U.; Mubina, J. Marine Endophytic Fungi Isolated from Gulf of Mannar—A Source for New Generation of Pharmaceutical Drugs and Biosynthesis of Silver Nanoparticles and Its Antibacterial Efficacy. In *Fungi Bio-Prospects in Sustainable Agriculture, Environment and Nano-Technology*; Elsevier: Amsterdam, The Netherlands, 2021; pp. 175–185.
189. Abdelrahman, S.E.S.A.H.; El Hawary, S.; Mohsen, E.; El Raey, M.A.; Selim, H.M.R.M.; Hamdan, A.M.E.; Ghareeb, M.A.; Hamed, A.A. Bio-fabricated zinc oxide nanoparticles mediated by endophytic fungus *Aspergillus* sp. SA17 with antimicrobial and anticancer activities: In vitro supported by in silico studies. *Front. Microbiol.* **2024**, *15*, 1366614. [CrossRef] [PubMed]
190. Rina, D.Y.; Rozirwan, R.; Hendri, M. Karakteristik Jamur Endofit Pada Lamun *Enhalus acoroides* di Teluk Hurun Lampung Selatan. Ph.D. Thesis, Universitas Sriwijaya, Kota Palembang, Indonesia, 2018.
191. Looney-Patterson, K. (University of Washington, Seattle, WA, USA). Diversity of Fungal Endophytes from *Zostera marina* eelgrass in False Bay Biological Preserve. 2021; (Unpublished Work).
192. Kinamot, V.; Monotilla, A. Colonization and antagonistic activity of endophytic fungi in seagrasses: Understanding endophyte interactions. 2022, *preprint*. [CrossRef]
193. Soperová, B. The Mycobiomes of Vegetative Organs of the Dominant Mediterranean Seagrass *Posidonia oceanica* (*Posidoniaceae, Alismatales*). Master's Thesis, Univerzita Karlova, Staré Město, Czech Republic, 2022.

194. Torta, L.; Burruano, S.; Giambra, S.; Conigliaro, G.; Piazza, G.; Mirabile, G.; Pirrotta, M.; Calvo, R.; Bellissimo, G.; Calvo, S.; et al. Cultivable fungal endophytes in roots, rhizomes and leaves of *Posidonia oceanica* (L.) Delile along the coast of Sicily, Italy. *Plants* **2022**, *11*, 1139. [CrossRef]
195. Vohník, M. Bioerosion and fungal colonization of the invasive foraminiferan *Amphistegina lobifera* in a Mediterranean seagrass meadow. *Biogeosciences* **2021**, *18*, 2777–2790. [CrossRef]
196. Suryanti, S.; Anggoro, S.; Sabdaningsih, A.; Febrianto, S.; Ayuningrum, D. Multi-temporal mapping and recent structures of seagrass community in Panjang Island. *AACL Bioflux* **2022**, *15*, 365–376.
197. Htay, S.A.M.; Linn, K.T. The effects of pH and fermentation media on the antibacterial activity of secondary metabolite producing from marine derived fungi. *J. Myanmar Acad. Arts Sci.* **2021**, *19*, 4B.
198. Newell, S. Fungi and bacteria in or on leaves of Eelgrass (Zostera *marina* L.) from Chesapeake Bay. *Appl. Environ. Microbiol.* **1981**, *41*, 1219–1224. [CrossRef]
199. Hyde, K.D.; Jones, E.B.G. Ecological observations on marine fungi from the Seychelles. *Bot. J. Linn. Soc.* **1989**, *100*, 237–254. [CrossRef]
200. Panno, L.; Voyron, S.; Anastasi, A.; Mussat Sartor, R.; Varese, G.C. Biodiversity of marine fungi associated with the seagrass *Posidonia oceanica. Biol. Mar. Mediterr.* **2011**, *18*, 85–88.
201. Poli, A.; Bovio, E.; Ranieri, L.; Varese, G.C.; Prigione, V. Fungal diversity in the Neptune Forest: Comparison of the mycobiota of *Posidonia oceanica, Flabellia petiolata*, and *Padina pavonica. Front. Microbiol.* **2020**, *11*, 933. [CrossRef]
202. Poli, A.; Vizzini, A.; Prigione, V.; Varese, G.C. Basidiomycota isolated from the Mediterranean Sea-Phylogeny and putative ecological roles. *Fungal Ecol.* **2018**, *36*, 51–62. [CrossRef]
203. Atalla, M.M.; Zeinab, H.K.; Eman, R.H.; Amani, A.Y.; Abeer, A.A.E.A. Screening of some marine-derived fungal isolates for lignin degrading enzymes (LDEs) production. *Agric. Biol. J. N. Am.* **2010**, *1*, 591–599.
204. Khatiwada, S.S. Prevalence and Distribution of Marine *Phytophthora* and *Halophytophthora* in Seagrass Habitats in Victoria, Australia. Master's Thesis, Deakin University, Victoria, Australia, 2019.
205. Man in 't Veld, W.A.; Rosendahl, K.C.H.M.; van Rijswick, P.C.J.; Meffert, J.P.; Boer, E.; Westenberg, M.; van der Heide, T.; Govers, L.L. Multiple *Halophytophthora* spp. and *Phytophthora* spp. Including *P. gemini, P. inundata* and *P. chesapeakensis* sp. nov. isolated from the seagrass *Zostera marina* in the Northern Hemisphere. *Eur. J. Plant Pathol.* **2019**, *153*, 341–357. [CrossRef]
206. Menning, D.M.; Gravley, H.A.; Cady, M.N.; Pepin, D.; Wyllie-Echeverria, S.; Ward, D.H.; Talbot, S.L. Metabarcoding of environmental samples suggest wide distribution of Eelgrass (*Zostera marina*) pathogens in the North Pacific. *Metabarcoding Metagenom* **2021**, *5*, 35–42. [CrossRef]
207. Proctor, T.; Elliott, J. Undescribed oomycete pathogens on *Zostera marina* and *Z. japonica* in the Puget Sound. Summer Research, 353, University of Puget Sound. 2019. Available online: https://www.jstor.org/stable/community.36709228 (accessed on 21 August 2024).
208. Wainwright, B.J.; Zahn, G.L.; Arlyza, I.S.; Amend, A.S. Seagrass-associated fungal communities follow Wallace's Line, but host genotype does not structure fungal community. *J. Biogeogr.* **2018**, *45*, 762–770. [CrossRef]
209. Pan, Y.; Li, G.; Su, L.; Zheng, P.; Wang, Y.; Shen, Z.; Chen, Z.; Han, Q.; Gong, J. Seagrass colonization alters diversity, abundance, taxonomic, and functional community structure of benthic microbial eukaryotes. *Front. Microbiol.* **2022**, *13*, 901741. [CrossRef]

Journal of *Fungi*

MDPI

*Article*

# Xerophilic Aspergillaceae Dominate the Communities of Culturable Fungi in the Mound Nests of the Western Thatching Ant (*Formica obscuripes*)

**Rachelle M. Gross [1,†], Courtney L. Geer [1,†], Jillian D. Perreaux [1], Amin Maharaj [2], Susan Du [2], James A. Scott [3] and Wendy A. Untereiner [1,*]**

[1] Department of Biology, Brandon University, 270-18th Street, Brandon, MB R7A 6A9, Canada; rachelle_gross@outlook.com (R.M.G.); cgeer1010@gmail.com (C.L.G.); jillian_perreaux33@hotmail.com (J.D.P.)

[2] Sporometrics Inc., 219 Dufferin Street, Suite 20-C, Toronto, ON M6K 1Y9, Canada; amaharaj@sporometrics.com (A.M.); sdu@sporometrics.com (S.D.)

[3] Division of Occupational and Environmental Health, Dalla Lana School of Public Health, University of Toronto, 223 College Street, Toronto, ON M5T 1R4, Canada; james.scott@utoronto.ca

* Correspondence: untereiner@brandonu.ca

† Both authors contributed equally to this study. The order of authorship was determined by coin toss.

**Abstract:** The nests of mound-building ants are unexplored reservoirs of fungal diversity. A previous assessment of this diversity in the nests of *Formica ulkei* suggested that water availability may be a determinant of the composition of this mycota. To investigate this question, we recovered 3594 isolates of filamentous Ascomycota from the nests of *Formica obscuripes* and adjacent, non-nest sites, employing Dichloran Rose Bengal agar (DRBA), Dichloran Rose Bengal agar containing glycerol (DRBAG), and malt extract agar containing sucrose (MEA20S). Higher numbers of fungi were isolated from the tops of mounds than from within mounds and non-mound sites. Mound nest soils were dominated by members of the family Aspergillaceae, and up to 50% of the colonies isolated on DRBAG belonged to the genus *Aspergillus*. *Pseudogymnoascus pannorum* and species of *Talaromyces* were also present in higher numbers in mound soils. Species of *Penicillium* were more abundant in non-nest soils, where they accounted for over 66% of isolates on DRBA. All Aspergillaceae assessed for xerotolerance on a medium augmented with glycerol or sucrose were xerophilic. These results, and our observation that the nests of *F. obscuripes* are low-water environments, indicate that water availability influences the structure of the fungal communities in these animal-modified habitats.

**Keywords:** *Aspergillus*; available soil nutrients; low-water-activity habitats; *Penicillium*; *Pseudogymnoascus*; mound-building ants; soil-inhabiting fungi; *Talaromyces*

## 1. Introduction

Mound-building ants (Hymenoptera, Formicidae) build conspicuous above-ground nests composed of excavated soil and organic material in the grasslands and forests of temperate and boreal regions of North America and Eurasia. The organic component of nest mounds, which consists of grass stems, leaf fragments, mosses, conifer needles, and twigs collected by the ants from the surrounding environment, is typically present as a surface thatch that maintains a higher temperature within the nest than in the surrounding soil in the early spring and prevents nests from overheating during the warmer parts of the year [1–7]. In prairie–forest edge ecosystems, this thatch also regulates moisture levels within nests [1,4].

Mound construction dramatically alters the chemical and physical characteristics of soils and in turn influences the composition and size of the microbial communities in nests. Mound nests differ from surrounding soils in pH, porosity, moisture level, organic matter content, and nutrient availability [3,8,9], and harbor communities of bacteria and fungi that differ markedly from those of non-nest soils [6,10–13]. Although the impacts of ant-mediated changes on the soil biota are not well understood, differences in acidity, moisture, and temperature have been hypothesized to have the greatest influence on the composition of microbial communities in mound nests [3,6,9,10].

A previous culture-based assessment of the diversity of fungi in the mound nests of *Formica ulkei* revealed that this mycota was dominated by species of *Aspergillus* (Aspergillaceae) [10], a genus isolated commonly from xeric environments and materials with low-water activities [14,15]. In contrast, non-nest soils contained large populations of *Penicillium* and Ascomycota known to be less tolerant of lower water availability. The dominance of species of *Aspergillus* in the nest mounds of *F. ulkei*, their paucity in non-mound soils, and the observation that the moisture content of the nest mounds of this ant species is lower than adjacent soils [1] suggested that water availability may be a major determinant of the composition of this mycota [10].

The xerophilic mycota of indoor and natural environments have been explored in several recent studies [16–19], but none have examined the potential impact of water availability (i.e., water activity or $a_W$, as inferred by measuring equilibrium relative humidity) on the distribution and abundance of fungi in animal-modified habitats. Here, we report the results of a study undertaken to (i) confirm that the community of culturable, filamentous Ascomycota from the mound nests of the western thatching ant (*Formica obscuripes*) differs from the assemblage of fungi in adjacent, non-nest soils; (ii) test the prediction that soils from nests have a lower water activity than those from non-mound sites; (iii) evaluate if mound nests harbor greater numbers and a higher diversity of xerophilic fungi; and iv) determine the xerotolerances of selected isolates of the most abundant taxa recovered from mound nests (i.e., species of *Aspergillus*, *Penicillium*, *Pseudogymnoascus*, and *Talaromyces*). Although various terms have been employed to describe and distinguish between fungi that prefer, tolerate, or require reduced water activity environments (e.g., extreme/moderate/obligate xerophile, mesophilic, osmophilic, xerotolerant) [20–24], we follow Pitt [25] in defining a xerophilic fungus as a species capable of growth, under at least one set of environmental conditions, at a water activity below 0.85.

## 2. Materials and Methods

Soils from three mound nests of *Formica obscuripes* located in the Brandon Hills Wildlife Management Area in southwestern Manitoba, Canada (UTM coordinates of mound 1 = 14N 434487 5508358; mound 2 = 14N 434504 5508369; mound 3 = 14N 434513 5508352), were sampled on 11 and 19 June 2020. Soils were collected with new (unused) plastic spoons from the uppermost 3 cm of mounds (M) and from within nests to a depth of 15–30 cm below the tops of the mounds (L). Non-mound soils (N) were collected 1 m SW of each mound to a depth of 3 cm using a soil core sampler that was cleaned between sites, as described in Duff et al. [10]. All soils were placed into separate, unused plastic ziplock freezer bags, sealed, transported to the laboratory, and processed within 4 h.

### 2.1. Soil Water Activity, Water Content, pH, and Nutrient Analyses

The water activity ($a_W$) of soil from each site was determined at 25 °C using a WP4C Water Potential Meter (METER Group, Pullman, WA, USA) following the manufacturer's instructions. Soil water content was determined as described by Forster [26]. The portion of each fresh collection used for dilution plating was sieved sequentially through 40 mm, 20 mm, and 2 mm meshes to remove plant debris. The smallest size fraction was retained for preparing serial dilutions. The remaining soils were emptied into clean aluminum pans, air-dried at room temperature (RT, 18–21 °C) for 5 d, and stored in new freezer bags.

The pH of air-dried soils was measured following Forster [27]. Analyses of soil electrical conductivity [28] (1:2 soil/water mixture), particle size distribution [29] (method 3.2.1), and nutrient content, including total inorganic carbon [30], available phosphate and potassium [31], total carbon by combustion [32], and available nitrate and sulfate [33], were performed by ALS Global (ALS Environmental Laboratory, Saskatoon, SK, Canada).

### 2.2. Isolation and Identification of Fungi

Fresh soils were used within 2 h of sieving to prepare 10-fold serial dilutions in sterile 0.05% water agar ranging from $10^{-1}$ to $10^{-7}$. Soil dilutions from $10^{-2}$ to $10^{-7}$ were plated in triplicate on Dichloran Rose Bengal agar (DRBA) [34] containing 25 mg $L^{-1}$ Rose Bengal, 2 mg $L^{-1}$ dichloran, and $KH_2PO_4$, rather than $K_2HPO_4$, Dichloran Rose Bengal agar (DRBAG) containing 18% glycerol, and 2% malt extract agar (MEAS20S) [35] containing 20 g $L^{-1}$ Difco malt extract and 200 g $L^{-1}$ sucrose. All media contained 1.5% agar and were supplemented with 50 mg chlortetracycline hydrochloride $L^{-1}$ and 50 mg streptomycin sulfate $L^{-1}$. The $a_w$ of each isolation medium was determined at 25 °C from three uninoculated 72-h-old plates using a WP4C Water Potential Meter.

Inoculated plates were incubated at RT for 4–5 d, after which all colonies of filamentous fungi were transferred to modified Leonian's agar (MLA) [36] containing 1.5% agar, incubated at RT, and identified to a genus and/or species based on cultural and micromorphological characteristics. Members of the genera *Aspergillus*, *Penicillium*, and *Talaromyces* were identified to species using cultures grown on creatine sucrose agar, Czapek yeast agar, dichloran glycerol agar (DG18), malt extract agar (MEA), and yeast extract agar following Samson et al. [35,37]. The identification of species of *Cephalotrichum*, *Cladosporium*, *Marquandomyces*, *Pseudogymnoascus*, *Purpureocillium*, and *Scopulariopsis* was based on the examination of cultures on DG18, MEA, and MLA incubated at RT for 7 d. All media contained 1.5% agar. Isolates that could be discriminated as separate taxa within genera but not identified to species were assigned numerical designations. Pycnidial and non-sporulating isolates were labeled "pycnidia" and "sterile", respectively. Sporulating isolates that were not identified to the genus level were designated as "undetermined" (see Tables S1–S3). Non-filamentous fungi and Mucoromycota, which were recovered infrequently on all media, were disregarded (see Duff et al. [10]).

Total nucleic acids were extracted from isolates of *Aspergillus*, *Penicillium*, *Pseudogymnoascus*, and *Talaromyces* using a FastDNA Spin Kit for Soil (MP Biomedicals, Irvine, CA, USA) following the manufacturer's instructions. The region spanning the nuclear ribosomal internal transcribed spacer region (nuc rDNA ITS1-5.8S-ITS2 or ITS), 1100–1300 bp of the 3′ end of the nuclear ribosomal 28S rRNA gene (nuc 28S), and portions of the genes encoding for the proteins calmodulin (*CaM*) and ß-tubulin (*TUB*) were amplified using illustra PuReTaq Ready-To-Go PCR Beads (GE Healthcare, Little Chalfont, UK) following the manufacturer's instructions. Amplification primers included (i) ITS4 and ITS5 [38] for the ITS, (ii) LR3F, LR3R, and LR7 [39,40] for the nuc 28S, (iii) Bt2a and Bt2b [41] for the *TUB* gene, and (iv) CMD5 and CMD6 [42] for the *CaM* gene. PCR products were purified using the EZNA Cycle Pure Kit (Omega Biotek, Norcross, Georgia) and quantified using a QuantiFluor dsDNA System (Promega, Madison, WI, USA) following the directions of the manufacturers. Sequencing reactions were performed using a BigDye terminator version 3.1 Cycle Sequencing Kit (Life Technologies, Austin, TX, USA) and the primers listed above.

Sequences were edited in Sequencher 4.7 (Gene Codes, Ann Arbor, Michigan) and compared to ITS sequences in GenBank [43] using BLAST (Basic Local Alignment Search Tool) [44]. The names adopted for taxa (see Table S4) were based on matches to ITS sequences with $\geq$98% sequence similarity. Secondary barcodes (*TUB* and *CaM* sequences) were obtained to verify the identification of isolates of *Aspergillus europaeus*, *A. fructus*, *A. tubingensis*, *Penicillium sanguifluum*, *P. scabrosum*, and *P. yarmokense*.

### 2.3. Assessment of Xerotolerance

Xerotolerances of each sequence isolate of *Aspergillus*, *Penicillium*, *Pseudogymnoascus*, and *Talaromyces* (listed in Table S4) were assessed based on their growth on malt yeast agar (MYA) [45] containing 1.5% agar and increasing molar concentrations of glycerol (MYA + G) or sucrose (MYA + S). We also assessed the xerotolerances of up to two additional representatives of each species.

To prepare MYA + G (1.0 M, 2.0 M, 3.0 M, 4.0 M, 5.0 M, 6.0 M, 7.0 M), the required amount of glycerol ($w/v$) was added to a 1 L volumetric flask. Dry ingredients for MYA, excluding agar, were dissolved in 200 mL of distilled water in a separate beaker, added to the volumetric flask, and mixed with the glycerol. After the final volume was adjusted to 1 L with distilled water, the contents of the flask were mixed and divided equally between two 1 L beakers. The pH of each solution was determined to ensure that they were between 6.8 and 7.0 prior to the addition of agar. MYA + S (0.5 M, 1.0 M, 1.5 M, 2.0 M, 2.5 M, 3.0 M) was prepared in a similar manner, except that the sucrose was dissolved via heating in distilled water and then cooled prior to its addition to a 1 L volumetric flask. Both media were autoclaved, cooled to 50 °C in a water bath, and poured into sterile, 90 mm plastic Petri dishes. Plates of osmoticant-augmented MYA were grouped according to their molar concentrations of glycerol or sucrose, conditioned for 24 h in a laminar flow hood, and inoculated within another 24 h. The $a_w$ of MYA + G and MYA + S at each molar concentration was measured from two uninoculated 72-h-old plates and one uninoculated 7-day-old plate (see Table S5) using a WP4C Water Potential Meter.

Conidial suspensions of isolates of *Aspergillus*, *Penicillium*, and *Talaromyces* were prepared by transferring 2 × 2 mm blocks of mycelium from the edges of sporulating cultures on MLA to 10 mL of a sterile solution containing 0.05% agar, 0.9% NaCl, and 0.05% Tween 20. The conidia of *Pseudogymnoascus pannorum* were suspended in the molten slurry produced by cooling a flame-sterilized transfer tool in the medium near the edges of growing colonies on MLA. Plates of MYA + G and MYA + S were point-inoculated centrally and in triplicate using these suspensions, stacked in Petri dish bags with plates of the same molar concentration of each osmoticant to maintain consistent $a_w$, and incubated for 2 wk at 25 °C. The diameter of the reverse of each colony was measured every second day for 2 wk and used to calculate mean radial growth rates (mm $d^{-1}$) [46]. Standard error was calculated for all means, but only values of 0.1 mm or greater are depicted in graphs. Goodness of fit was assessed using the coefficient of determination ($R^2$).

### 2.4. Statistical Analyses

The car package version 3.0-11 [47] in R version 4.0.3 [48] was used to evaluate simple linear regression models for chemical and physio-chemical soil parameters, including $a_w$, water content, pH, total carbon, total inorganic carbon, available nitrate, available phosphate, available sulfate, available potassium, and electrical conductivity. Significant outcomes were subjected to post hoc tests for multiple pairwise comparisons using the emmeans package version 1.6.2-1 [49]. The same package was used to adjust *p*-values according to the Benjamini–Hochberg (B-H) false discovery rate.

Colony-forming units per gram of soil (CFU $g^{-1}$) and the relative abundance of each species were calculated from the numbers of fungal colonies recovered on DRBA, DRBAG, and MEA20S. Values for relative abundance were utilized to generate stacked bar plots and pie charts using the ggplot2 package version 3.3.5 [50]. Color scales in the viridis package version 0.5.1 [51] were applied to improve the readability of graphics for individuals with common forms of color vision deficiency.

Diversity indices were calculated using the vegan package version 2.5-7 [52]. We employed the car package for multivariate linear models to test the significance of the relationships between the dependent variables (species richness and species diversity), sampling sites (soils from the tops of mounds: M1, M2, M3; soil from within mounds: L1, L2, L3; soils from non-mounds: N1, N2, N3), and isolation media (DRBA, DRBAG, MEA20S). Subsequent post hoc pairwise comparisons were performed for significant outcomes for all combinations of sites using the emmeans package. *p*-values were adjusted according to the B-H false discovery rate employing the same package.

Morisita–Horn similarity indices were calculated employing the vegan package. These data were converted to distance matrices and used to generate dendrograms in R. A simple linear regression model in Microsoft Excel 365 version 16.0 was used to determine if there were significant differences in the number of Aspergillaceae isolated on different media. We then performed post hoc Tukey's honest significant difference tests (HSD) in Excel for all pairwise comparisons of these data to identify significant differences ($p < 0.05$).

A permutational multivariate analysis of variance (PERMANOVA) in R was used to test for statistically significant differences in the composition of fungal communities from the different sampling sites (M, L, N). $p$-values for all pairwise comparisons were adjusted employing the B-H false discovery rate in the emmeans package, and permutational multivariate analyses of dispersion were run to confirm the observed values of significance.

## 3. Results

### 3.1. Soil Water Activity, Water Content, pH, and Nutrient Analyses

The water activity and water content of soils from the tops of nests (M) and within the nests (L) of *Formica obscuripes* were lower than from non-mound (N) sites (Table 1), whereas electrical conductivities and the levels of total carbon, total organic carbon, and available nitrate, phosphorus, potassium, and sulfate were higher (Table 2). Soils from both the tops of mound nests (M) and within mounds (L) differed significantly ($p < 0.05$) from those of non-mound sites (N) with respect to all measured parameters except pH, available nitrate, and electrical conductivity.

**Table 1.** Water activity ($a_w$), water content (WC), and pH of soils (mean ± SD) from the tops of mound nests (M), within nests (L), and non-mound (N) sites.

| Property | M ($n$ = 3) | L ($n$ = 3) | N ($n$ = 3) | Site Effect (d.f. = 2, 6) |
|---|---|---|---|---|
| $a_w$ | $0.75 \pm 0.11^{N}$ | $0.82 \pm 0.03^{N}$ | $1.00 \pm 0.00^{M,L}$ | 11.79 ** |
| WC (%) | $10.93 \pm 0.45^{N}$ | $13.03 \pm 1.46^{N}$ | $19.20 \pm 1.67^{M,L}$ | 32.47 *** |
| pH | $6.33 \pm 0.12^{L,N}$ | $5.89 \pm 0.13^{M,N}$ | $6.86 \pm 0.04^{M,L}$ | 63.85 *** |

Superscript text indicates sites differing significantly in post hoc comparisons of soils; ***, $p < 0.001$; **, $p < 0.01$.

**Table 2.** Chemical properties (mean ± SD) of soils from the tops of mound nests (M), within nests (L), and non-mound (N) sites.

| Property | M ($n$ = 3) | L ($n$ = 3) | N ($n$ = 3) | Site Effect (d.f. = 2, 6) |
|---|---|---|---|---|
| Total carbon (%) | $15.23 \pm 2.35^{N}$ | $11.57 \pm 1.53^{N}$ | $5.71 \pm 0.47^{M,L}$ | 25.77 ** |
| Total organic carbon (%) | $14.20 \pm 2.50^{N}$ | $10.67 \pm 1.26^{N}$ | $3.98 \pm 0.18^{M,L}$ | 30.80 *** |
| Available nitrate (mg kg$^{-1}$) | $88.23 \pm 9.61^{ns}$ | $163.67 \pm 91.08^{N}$ | $4.77 \pm 1.70^{L}$ | 6.78 * |
| Available phosphorus (mg kg$^{-1}$) | $91.33 \pm 30.29^{N}$ | $124.67 \pm 30.37^{N}$ | $3.53 \pm 1.07^{M,L}$ | 19.14 ** |
| Available potassium (mg kg$^{-1}$) | $1173.33 \pm 395.77^{N}$ | $1620.00 \pm 355.95^{N}$ | $339.33 \pm 22.30^{M,L}$ | 13.40 ** |
| Available sulfate (mg kg$^{-1}$) | $27.87 \pm 6.63^{N}$ | $37.00 \pm 8.54^{N}$ | $6.07 \pm 0.49^{M,L}$ | 19.40 ** |
| Electrical conductivity (ds m$^{-1}$) | $0.71 \pm 0.16^{L}$ | $1.74 \pm 0.55^{M,N}$ | $0.32 \pm 0.01^{M}$ | 14.60 ** |

Superscript text indicates sites differing significantly in post hoc comparisons of soils; ***, $p < 0.001$; **, $p < 0.01$; *, $p < 0.05$, $^{ns}$, no significance.

Differences in the sizes of the particles of soils from the tops of mounds (M) and non-mound sites (N) were not significant (Table 3), and there were insufficient quantities of soil from within mounds (L) to perform particle size analysis.

**Table 3.** Particle size comparison of soils from the tops of mound nests (M) and non-mound (N) sites. There were insufficient quantities of soil from within mounds (L) to perform particle size analysis.

| Particle Size | M ($n$ = 3) | N ($n$ = 3) |
|---|---|---|
| % Clay (<4 μm) | 22.6 ± 3.8 | 24.5 ± 2.0 |
| % Silt (4 μm–0.063 mm) | 55.5 ± 2.4 | 49.6 ± 5.4 |
| % Fine sand (0.063–0.2 mm) | 9.2 ± 1.4 | 8.5 ± 0.4 |
| % Coarse sand (0.2–2.0 mm) | 12.6 ± 3.1 | 15.1 ± 2.6 |

### 3.2. Composition, Richness, and Diversity of Fungal Communities

A total of 3594 isolates, including 98 species representing 19 families of Ascomycota, were recovered on DRBA ($a_w$ = 0.998), DRBAG ($a_w$ = 0.944), and MEA20S ($a_w$ = 0.987) from the soils of mound nests (M, L) and non-mound (N) sites (Table 4). The highest number of isolates (1877) was recovered on DRBAG, and most of these (1263) were obtained from soils from the uppermost 3 cm of mounds (M1, M2, M3).

**Table 4.** Numbers of colonies recovered on DRBA, DRBAG, and MEA20S from soils from the tops of mound nests (M), within nests (L), and non-mound (N) sites.

| Medium | M1 | M2 | M3 | L1 | L2 | L3 | N1 | N2 | N3 | Total |
|---|---|---|---|---|---|---|---|---|---|---|
| DRBA ($a_W$ = 0.998) | 139 | 116 | 139 | 168 | 91 | 185 | 22 | 30 | 16 | 906 |
| DRBAG ($a_W$ = 0.944) | 343 | 391 | 529 | 179 | 13 | 124 | 158 | 91 | 49 | 1877 |
| MEA20S ($a_W$ = 0.987) | 147 | 76 | 156 | 89 | 51 | 25 | 111 | 98 | 58 | 811 |
| Total | 629 | 583 | 824 | 436 | 155 | 334 | 291 | 219 | 123 | 3594 |

The structure of the fungal communities between sites differed significantly (PERMANOVA; $p < 0.01$) (Table S6), and species assemblages from mounds (M, L) were more similar to each other than to non-mound (N) sites, with the exception of one community from mound tops (M3) isolated on MEA20S (Figure 1, Figures S1 and S2). Morisita–Horn similarity indices for the fungal communities from the tops of mounds (M), within mounds (L), and non-mound (N) soils differed significantly ($p = 0.0002$), as did the communities isolated on DRBA, DRBAG, and MEA20S (PERMANOVA; $p = 0.022$), but we detected no significant interaction between site and isolation medium.

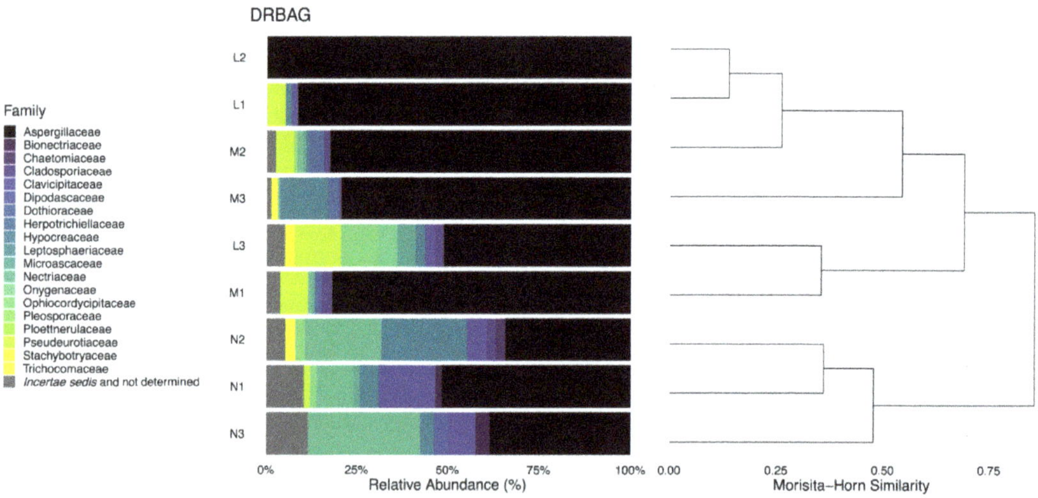


**Figure 1.** Relative abundance of representatives of families of fungi (bar graphs) and hierarchical clustering (dendrogram) of fungal communities from different soils (M = tops of mound nests, L = within nests, N = non-nest mound sites) isolated on DRBAG.

The communities of fungi from mound tops (M) and within mounds (L) on DRBA, DRBAG, and MEA20S were dominated by Aspergillaceae (species of *Aspergillus* and *Penicillium*) (Figure 2, Table 5), a family that accounted for 12.9 (M3 on MEA20S) to 100% (L2 on DRBAG) of the strains recovered from these two sites (Tables S1–S3). The numbers of Aspergillaceae isolated on DRBA, DRBAG, and MEA20S varied among sites ($p = 0.027$), but the post hoc Tukey's HSD test indicated that only DRBAG and MEA20S yielded significantly different numbers of the members of this family. The greatest numbers of Aspergillaceae were isolated on DRBAG, with the highest percentages of the representatives of this family (51.3–100%) found in soils from nests (M, L). This relationship is illustrated in the dendrogram of fungi from different families recovered on DRBAG (Figure 1), which is based on nearly three times the number of Aspergillaceae isolated on DRBA or MEA20S (Figures S1 and S2).

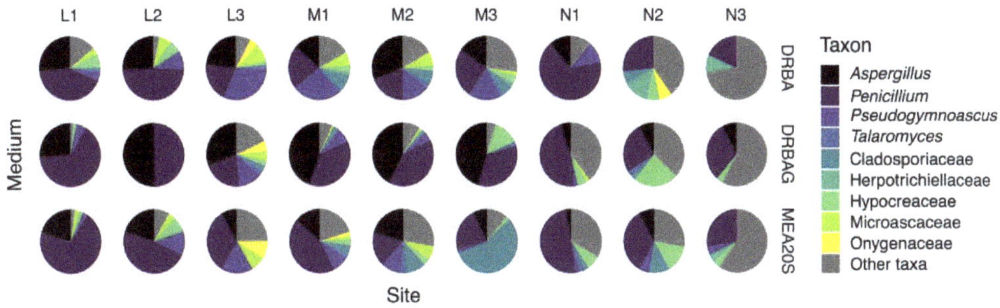


**Figure 2.** Pie charts illustrating the relative abundances of 4 genera and 5 families of fungi from different soils (M = tops of mound nests, L = within nests, N = non-nest mound sites) on three isolation media (DRBA, DRBAG, MEA20S).

**Table 5.** Numbers of isolates belonging to the Aspergillaceae recovered on DRBA, DRBAG, and MEA20S from soils from the tops of mound nests (M), within nests (L), and non-mound sites (N).

| Medium | M1 | M2 | M3 | L1 | L2 | L3 | N1 | N2 | N3 | Total |
|---|---|---|---|---|---|---|---|---|---|---|
| DRBA ($a_w$ = 0.998) | 46 | 69 | 65 | 127 | 71 | 80 | 20 | 5 | 2 | 485 |
| DRBAG ($a_w$ = 0.944) | 287 | 341 | 430 | 167 | 13 | 67 | 91 | 39 | 19 | 1454 |
| MEA20S ($a_w$ = 0.987) | 97 | 36 | 46 | 81 | 34 | 8 | 63 | 50 | 17 | 432 |
| Total | 430 | 446 | 541 | 375 | 118 | 155 | 174 | 94 | 38 | 2371 |

Species of *Penicillium* were the most abundant Aspergillaceae recovered on all isolation media (Tables S1–S3), but a greater number of representatives of this genus were isolated from non-mound (N) soils than from mounds (M, L) on DRBAG and MEA20S. Of the fourteen members of this genus identified to species, the most common were *P. citrinum* (5.1–56.8%) and *P. pasqualense* (1.7–12.9%). *Penicillium citrinum* was recovered from all sites using each isolation medium and was most abundant in soils from mounds (M, L) (i.e., up to 50% on DRBAG and 56.8% on MEA20S).

While less abundant than species of *Penicillium*, members of the genus *Aspergillus* were recovered consistently and in high numbers from the tops of nests (M) and from within mounds (L) on all media. The most abundant species, *A. fructus*, was recovered from all sites on DRBAG and represented 25.6% of isolates from within nests (L) on this medium (Tables S1–S3). *Aspergillus insuetus* and *A. tubingensis* were also abundant in soils from mounds (M, L), with the latter species representing 43.6% of isolates recovered from the tops of mounds (M) on DRBAG. The least abundant member of the genus, *A. europaeus*, was isolated only on DRBA and represented less than 5.6% of isolates recovered on this medium.

Trichocomaceae, represented by *Talaromyces atricola* and *T. neorugulosus*, were less abundant than Aspergillaceae (Tables S1–S3). These species were recovered from soils from mounds (M, L) and non-mound (L) sites (Figures S1 and S2), but they never comprised more than 3.4% of isolates recovered on DRBA, DRBAG, or MEA20S. *Pseudogymnoascus pannorum* (Pseudeurotiaceae) was present in all soils from mounds (M, L) and non-mound (L) sites, but was isolated most frequently from within mounds (L), where it accounted for up to 34.1% of recovered strains on DRBA. Most of the isolates of this species conformed to the descriptions provided by Domsch et al. [53] and Samson et al. [15], but we distinguished a smaller number of strains as "*Ps. pannorum* strain 2" based on their more lightly pigmented colonies on MLA (not illustrated).

Other taxa comprising more than 10% of isolates from any individual sampling site included Chaetothyriales (*Exophiala*), Cladosporiales (*Cladosporium*), Hypocreales (*Acremonium*-like taxa, *Albifimbria, Clonostachys, Cylindrocarpon*, fusarium-like taxa, *Gliocladium, Marquandomyces, Purpureocillium, Tolypocladium, Trichoderma, Volutella*), and Microascales (*Doratomyces, Microascus, Scopulariopsis, Wardomyces*) (Tables S1–S3). Cladosporiales were most abundant in soils from mound tops (M) (up to 56.4%) and non-mound (N) sites (up to 11.1%) on MEA20S, while Hypocreales comprised 11.1–52.6% of strains recovered from non-mounds (N) on all isolation media. The majority of isolates belonging to the latter order (*Marquandomyces marquandii* and *Purpureocillium lilacinum*) were isolated on DRBAG and MEA20S. *Exophiala* represented 9.1–13.3% of isolates in non-mound (N) soils on DRBA, but this genus was less frequently isolated or absent from other sites on DRBAG and MEA20S. Sterile and undetermined fungi were isolated most frequently on DRBA and were less abundant or absent on all media in soils from within mounds (L).

Fungal species richness and diversity were highest in soils from the tops of nests (M) on DRBA and from non-mound (N) sites on DRBAG, respectively, but both measures varied between sites (Table 6). Richness differed significantly between all sites on DRBA ($p < 0.05$), between inner mound (L) and non-mound (N) soils on DRBAG ($p < 0.01$), and between all sites except the tops of mounds (M) and non-mound (N) sites on MEA20S ($p < 0.01$). There were also significant differences in species diversity, as measured by the Shannon index, between mound (M, L) and non-mound (N) soils on DRBAG ($p < 0.05$), and between inner mound (L) and non-mound (N) soils on MEA20S ($p = 0.027$). Simpson diversity and Simpson inverse diversity indices between sites did not differ significantly. Multiple linear regression analyses indicated significant differences in species richness between sites ($p = 0.001$), between fungal communities isolated on different media ($p = 0.045$), and in the interaction between sites and media with respect to species richness ($p = 0.001$) and diversity ($p = 0.014$).

**Table 6.** Species richness and diversity measures (mean ± SE) of communities of fungi isolated from soils from the tops of mound nests (M), within nests (L), and non-mound (N) sites.

| Site | Medium | Species Richness | Shannon Diversity | Simpson Diversity | Simpson Inverse |
|---|---|---|---|---|---|
| M | DRBA | $27.33 \pm 3.48^{L,N}$ | $2.97 \pm 0.10^{ns}$ | $0.93 \pm 0.01^{ns}$ | $13.79 \pm 0.90^{ns}$ |
| | DRBAG | $18.67 \pm 1.45^{ns}$ | $2.05 \pm 0.23^{N}$ | $0.81 \pm 0.05^{ns}$ | $5.96 \pm 1.46^{ns}$ |
| | MEA20S | $22.33 \pm 2.40^{L}$ | $2.65 \pm 0.28^{ns}$ | $0.84 \pm 0.07^{ns}$ | $9.49 \pm 4.02^{ns}$ |
| L | DRBA | $19.67 \pm 2.60^{M,N}$ | $2.43 \pm 0.04^{ns}$ | $0.86 \pm 0.02^{ns}$ | $7.13 \pm 0.87^{ns}$ |
| | DRBAG | $11.67 \pm 4.48^{N}$ | $1.83 \pm 0.43^{N}$ | $0.76 \pm 0.07^{ns}$ | $5.17 \pm 1.77^{ns}$ |
| | MEA20S | $10.67 \pm 0.88^{M,N}$ | $1.96 \pm 0.22^{N}$ | $0.79 \pm 0.08^{ns}$ | $6.10 \pm 1.67^{ns}$ |
| N | DRBA | $10.33 \pm 2.33^{M,L}$ | $2.20 \pm 0.31^{ns}$ | $0.86 \pm 0.06^{ns}$ | $9.36 \pm 2.83^{ns}$ |
| | DRBAG | $25.33 \pm 1.67^{L}$ | $2.99 \pm 0.09^{M,L}$ | $0.93 \pm 0.01^{ns}$ | $15.82 \pm 2.83^{ns}$ |
| | MEA20S | $23.33 \pm 1.76^{L}$ | $2.93 \pm 0.06^{L}$ | $0.93 \pm 0.01^{ns}$ | $14.22 \pm 1.21^{ns}$ |

Superscript text indicates sites differing significantly ($p < 0.05$) in post hoc comparisons of soils; $^{ns}$, no significance.

### 3.3. Assessment of Xerotolerance

Every isolate tested for xerotolerance (listed in Table S4) grew on unaugmented MYA ($a_w = 1.00$) and had mean radial growth rates greater than 1.0 mm $d^{-1}$ on this medium, except *Penicillium charlesii, P. parvulum, Pseudogymnoascus pannorum, Talaromyces atricola*, and *T. neorugulosus* (Figure 3). Mean growth rates of all taxa were higher on MYA containing an osmoticant, and most species grew more rapidly on MYA + 1 M sucrose ($a_w = 0.97$) than on MYA + 1 M glycerol ($a_w = 0.97$) (Table S5).

All members of the Aspergillaceae included in our tests were xerophilic (i.e., capable of growth on media at or below a water activity of 0.85) and most of the species we assessed exhibited faster radial growth on MYA supplemented with sucrose (Figure 3, Table S5). The highest rates of growth were exhibited by *Aspergillus tubingensis* (7.06 mm $d^{-1}$) and *Penicillium thomii* (4.73 mm $d^{-1}$) on MYA + 0.5 M sucrose ($a_w = 0.99$), but species of *Aspergillus* generally grew faster than members of the genus *Penicillium* on MYA amended with the highest tolerated concentration of sucrose (MYA + 2.5 M sucrose, $a_W$ = 0.78). *Talaromyces atricola* and *T. neorugulosus* grew more slowly than Aspergillaceae on glycerol- and sucrose-augmented MYA; both species exhibited limited growth on sucrose-augmented MYA, with a water activity of 0.89, but failed to grow on MYA containing 2.5 M sucrose ($a_W$ = 0.78) or 5 M glycerol ($a_W$ = 0.84).

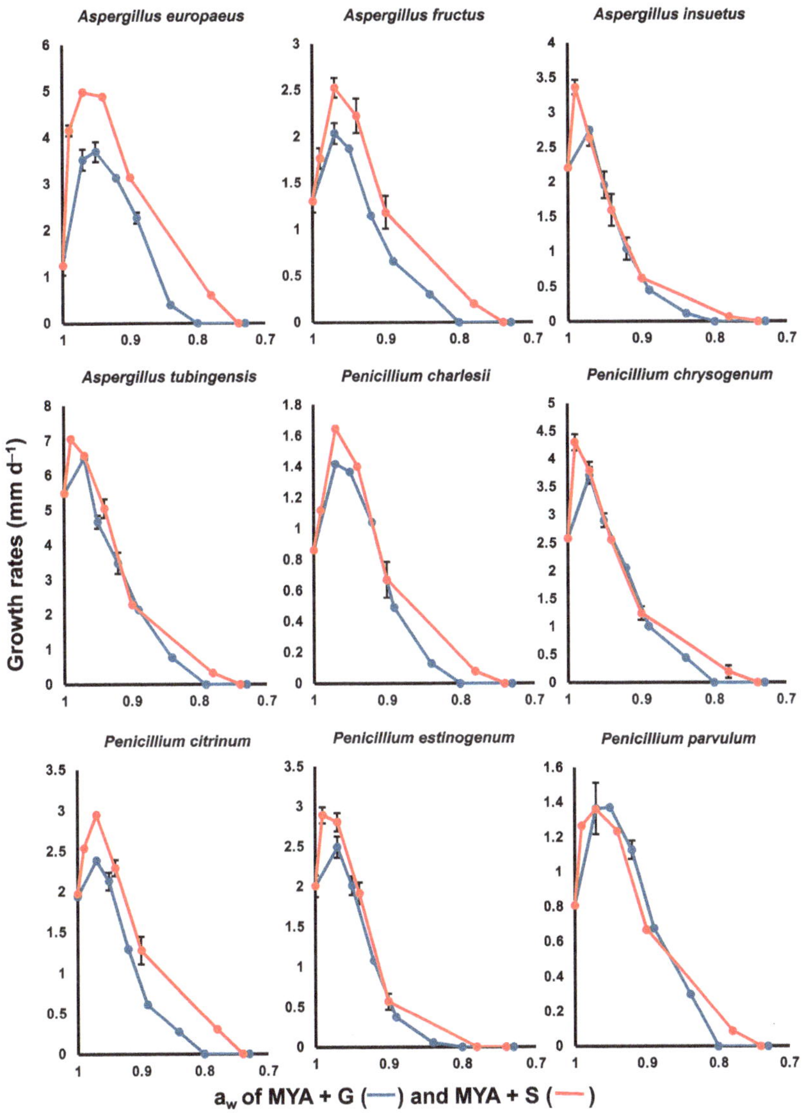


Figure 3. *Cont.*

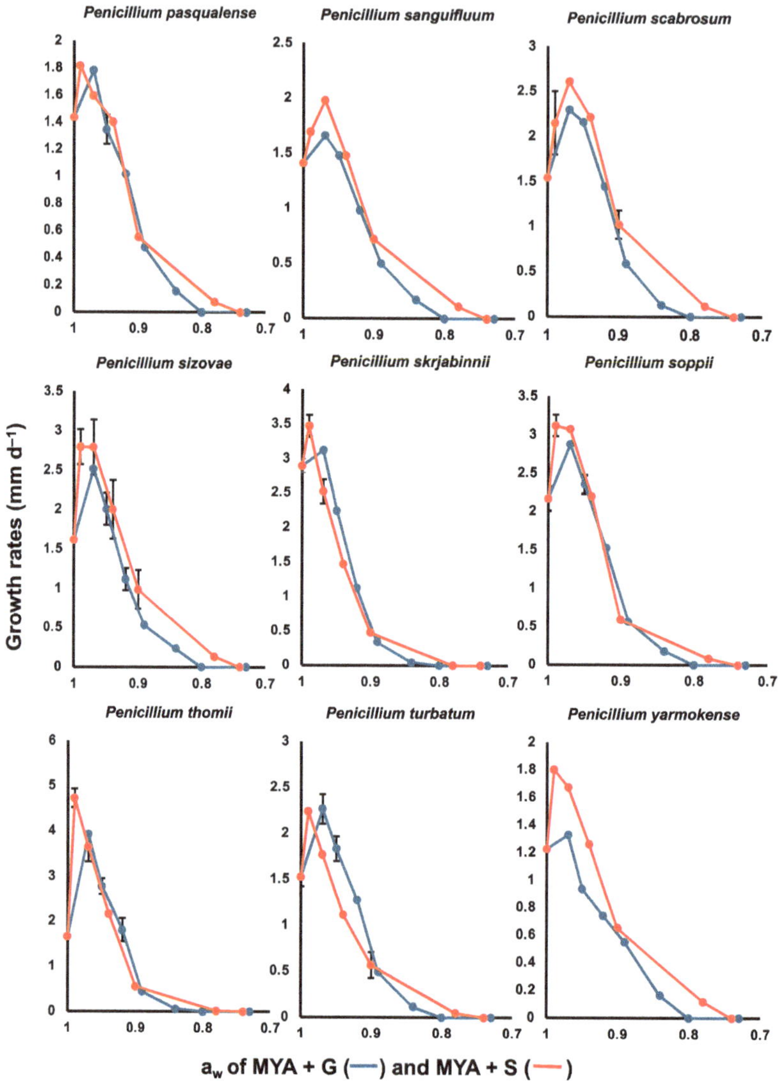


Figure 3. *Cont.*

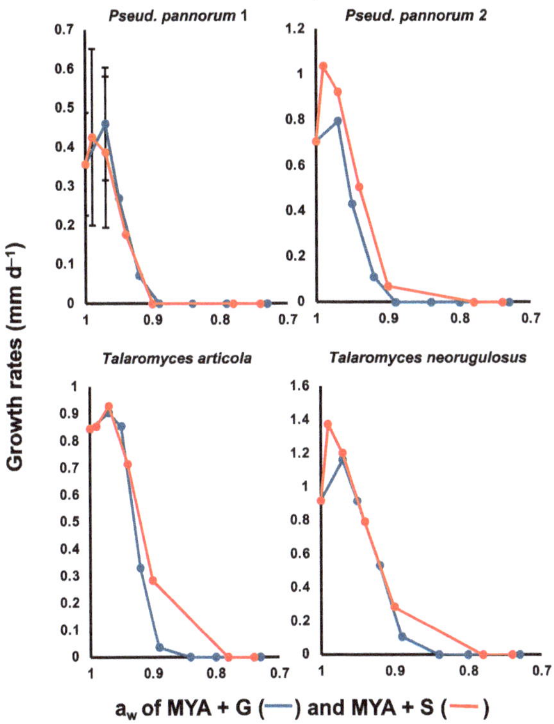


**Figure 3.** Mean growth rates (mm $d^{-1}$ ± SE) of species of *Aspergillus*, *Penicillium*, *Pseudogymnoascus*, and *Talaromyces* on MYA with different water activities ($a_w$) resulting from the addition of glycerol (blue lines) and sucrose (red lines).

*Pseudogymnoascus pannorum* exhibited the slowest rates of growth on MYA and osmoticant-augmented MYA and was the least xerotolerant species assessed. None of the strains of *Ps. pannorum* grew on MYA containing 4.0 M glycerol ($a_w$ = 0.89), but isolates of *Ps. pannorum* strain 2 differed from those identified as *Ps. pannorum* strain 1 in their ability to grow on sucrose-augmented MYA, with a water activity of 0.90. These isolates also exhibited less variable growth responses on MYA and MYA containing up to 1 M of osmoticant ($a_w$ = 1.00 − 0.97) and had faster rates of radial growth (0.71 mm $d^{-1}$ versus 0.36 mm $d^{-1}$ on MYA and 1.04 mm $d^{-1}$ versus 0.43 mm $d^{-1}$ on MYA + 0.5 M sucrose).

## 4. Discussion

### 4.1. The Abundance and Diversity of Fungi in Soils from Mound Nests

The results of this investigation agree with previous studies that recovered higher numbers of fungi from the nests of mound-building species of *Formica* than non-nest sites and reported significant differences in the composition of fungal communities from mound nests and non-nest soils [10,12,54]. We also observed significant differences between the assemblages of fungi recovered on reduced-water-activity media ($a_w$ = 0.944, $a_w$ = 0.987) and that fungal species richness and diversity varied between soils from the tops of mounds (M), soil from within mounds (L), and soils from non-mounds (N).

The present study also documents significant differences in the water characteristics and chemical properties of soils from the mound nests of *F. obscuripes* soils from non-mounds. Soils from the tops of mounds and within mounds had lower water activities ($a^w \leq 0.82$), contained less water as measured by percentage of water content, and were more acidic than non-mound soils. Nest soils also had higher levels of nutrients and electrical conductivities. These results agree with previous reports of ant-mediated changes to soil characteristics, particularly nutrient enrichment, that are attributed largely to the accumulation of organic material by mound-building and soil-dwelling ants [3,4,8,9,55–58].

Our discovery that the mound nests of *F. obscuripes* are reduced-water-activity environments supports the suggestion [10] that water availability is an important factor influencing the structure of fungal communities in the nests of this and possibly other species of *Formica*. This conclusion is reinforced by the recovery of higher numbers of xerophilic Aspergillaceae, a family of Ascomycota predominant in reduced-water-activity and xeric environments [20,22], from mound nests. In our investigation, species of *Aspergillus* and *Penicillium* comprised 29.1–100% of isolates recovered from soils from mound nests (M1, M2, M3, L1, L2, L3) on DRBA ($a_w$ = 0.998) and on media with reduced water activity (DRBAG $a_w$ = 0.944, MEA20S $a_w$ = 0.987).

The genus *Aspergillus* encompasses the most xerotolerant Aspergillaceae, and includes species isolated frequently from alpine and desert soils, the built environment, and a variety of low-water-activity substrates (e.g., dried fruit, stored grain, sugar-preserved foods) [14–16,19,20,22]. Species of *Aspergillus* are also abundant in the nests of soil-inhabiting and mound-building ants [10,11], and in the current study accounted for up to 50% of the fungi recovered from mound nests on DRBAG. All of the members of this genus assessed for xerotolerance in our investigation were capable of growing on MYA with 2.5 M S ($a_w$ = 0.78) and 5.0 M G ($a_w$ = 0.84). *Aspergillus fructus* (*A. versicolor fide* Sklenář et al. [59]), the most abundant member of the genus from the nest mounds of *F. obscuripes*, is a xerophile documented from low-humidity environments, clinical specimens, fruit, house dust, and soil [19,59–62], but it has not been reported previously from animal-modified habitats. *Aspergillus fructus* was present in lower numbers in non-mound soils, while *A. europaeus* and *A. insuetus*, two species found commonly in soils [63,64], were isolated from the nest soils of *F. obscuripes* but never recovered from non-mound sites. These results mirror those of Duff et al. [10], who reported that species of *Aspergillus* accounted for 42.4–80.4% of isolates from the mound nests of *F. ulkei* but were absent or represented only 0.3% of strains from non-nest locations.

Species of *Penicillium* accounted for up to 66.7% of isolates from non-mound sites and were more abundant in soils from non-mound sites than soils from mounds on DRBAG and MEA20S. Only two species, *P. sizovae* and *P. thomii*, were restricted to soils from mounds. Duff et al. [10] also recovered more species of *Penicillium* than *Aspergillus* and reported that isolates of the former genus were more abundant in non-mound soils. Penicillia are common in soils, foods, and the built environment, and many species tolerate low-water activities [15,53]. In the current study, every member of the genus assessed for xerotolerance was capable of growth on MYA with 2.5 M S ($a_w$ = 0.78), and all but two species (*P. estinogenum* and *P. skrjabinii*) grew on MYA of 5.0 M G ($a_w$ = 0.84). Although the xerotolerances of many of the species of *Penicillium* we isolated have not been reported previously, our results support the water-activity limits described for *P. chrysogenum sensu lato* and *P. citrinum* [22,65,66]. The former species has been recovered in high numbers from hypersaline environments [15,67], whereas the latter is reported from cereals, dried foods, decaying plant material, and soil [22,53]. The 12 species of *Penicillium* we recovered from both mound and non-mound soils are not characteristic of reduced-water-activity environments.

*Pseudogymnoascus pannorum* was abundant in both mound and non-mound soils. This slow-growing, psychrotolerant species is common in temperate, arctic, and alpine soils [16,53,68] and exhibits considerable morphological and physiological variation [52,69]. We recovered *Ps. pannorum* from all sites, but the largest numbers of this species were isolated from nest mounds on DRBA. Our finding that most of these isolates resembled the strain we determined to be incapable of growth on media with a water activity at or below $a_w$ = 0.89 (MYA + G) or $a_w$ = 0.90 (MYA + S) (*Ps. pannorum* strain 1) supports the observation that *Ps. pannorum* is not xerophilic [16,65].

*Talaromyces atricola* and *T. neorugulosus* were the least abundant of all of the fungi we assessed for xerotolerance. Both taxa were restricted largely to soils from the mound nests of *F. obscuripes* and grew slowly on osmoticant-augmented media (i.e., $\leq$0.3 mm $d^{-1}$ on MYA + S, $a_w$ = 0.90), supporting the observation that species of *Talaromyces* are generally less xerophilic than members of the Aspergillaceae [22,66,70].

Fungal taxa recovered in large numbers that we did not assess for xerotolerance included members of the orders Cladosporiales and Hypocreales. Xerophily is documented in several species of *Cladosporium* [22,70], and representatives belonging to the complexes *Cladosporium cladosporioides*, *C. herbarum*, and *C. sphaerospermum* accounted for 3.5–50.1% of the isolates recovered from soils from the tops of individual mounds on MEA20S. These and other *Cladosporium* were less abundant or absent in soils from within mounds and non-mound sites. In contrast, members of the Hypocreales were more abundant in non-mound soils, where they represented 11.1–52.6% of strains recovered from individual sites. *Marquandomyces marquandii* and *Purpureocillium lilacinum* were the most frequently isolated members of this order in non-nest soils. Xerotolerance has been reported for both species [71,72], so it is unsurprising that the highest numbers of these taxa were recovered on DRBAG and MEA20S.

### 4.2. Water Activity Is an Important but Poorly Explored Determinant of the Structure of Fungal Communities

Water content, or the mass fraction of water present in materials, has long been recognized as critical in governing the growth of soil-inhabiting microorganisms (for example, see Sewell [73]). Gravimetric water content is determined by the comparison of sample mass before and after drying to constant weight, but this measure is not predictive of the ability of microorganisms to grow when it is used alone and without knowledge of the sorption characteristics of materials and the solute composition of the water contained therein. This is because the solute composition of contained water, particularly the electrochemical characteristics of constituent solutes, determines the osmotic properties of materials and the availability of water that is free to support extracellular microbial processes [74]. The quantity of free water available in a substrate, including adsorbed water held loosely in spaces by capillary action or bound weakly by molecular attraction to other materials, is referred to as water activity [21], and it is directly proportional to equilibrium relative humidity (ERH) measured proximate to the surface [75,76].

The value of water activity for predicting microbial growth was first appreciated by food scientists who understood the importance of water availability in food preservation [72,77] and the close correlation of this measure with certain structural characteristics of food (e.g., crunch and crispness) [78]. Water availability, expressed as $a_w$ or ERH, is also preferred to other measures that describe the biological potential of water in the domains of building science and the analyses of the growth of microorganisms in indoor environments. Building materials vary in their chemical composition, physical complexity, and adsorptive and water-holding capacities [79]. It has also long been recognized that materials with the same water content can possess different water activity values, just as materials with the same water activity values can differ in their water contents [21,70,74,76,80]. Furthermore, because of the hysteretic nature of the sorption curves of many porous substrates, dry materials that take up water will reach water activities that may differ from the values they attain after drying to the same water content following flooding. Finally, because the distribution of water in building materials and other structurally complex substrates is normally not uniform, materials subject to different types of wetting or with different water activities will support the growth of cohorts of organisms with differing requirements for available water [21,23].

Our investigation of the composition of the mycota of the mound nests of *Formica obscuripes* supports the use of both standard or higher-water-activity media and media with reduced water activity to recover fungi from natural environments. Media used commonly in the enumeration of microorganisms from environmental samples and foods have water activities higher than 0.98 [81,82], but the recovery of xerophilic fungi from these substrates is enhanced greatly using media of only incrementally reduced water activity (i.e., $a_w = 0.94 - 0.95$) [17,66,83–85]. This is because the growth optima of both hydrophilic and many xerophilic fungi are above $a_w$ 0.90, even though members of the latter group have lower $a_w$ minima [21]. Given the heterogenous composition of most natural substrates and the expectation that the distribution of water within them is spatially and temporally non-uniform, it is reasonable to assume that xerophilic fungi capable of growing over a range of water activities would be better adapted to reduced water environments than hydrophilic taxa possessing higher $a_w$ minima.

The nests of mound-building ants and other natural, animal-modified habitats are unexplored reservoirs of fungal diversity that warrant further study using high- and reduced-water-activity isolation media. A fuller picture of this diversity would be obtained by identifying all the fungi isolated from these habitats and by using sequence-based approaches to identify micromorphologically similar members of species complexes, non-sporulating and non-culturable taxa, and species that are traditionally under sampled using culture-dependent methods. Finally, a better understanding of the distribution of fungi in natural environments could be achieved by determining the water activities of the substrates from which they are recovered and by employing solute-augmented growth media to assess the xerotolerances of recovered isolates.

**Supplementary Materials:** The following supporting information can be downloaded at: https://www.mdpi.com/article/10.3390/jof10110735/s1, Figure S1: Dendrogram of isolates on DRBA; Figure S2: Dendrogram of isolates on MEA20S; Table S1: Relative abundance and number of recovered fungi on DRBA; Table S2: Relative abundance and number of recovered fungi on DRBAG; Table S3: Relative abundance and number of recovered fungi on MEA20S; Table S4: Accession numbers of isolates; Table S5: Growth rates of isolates; Table S6: PERMANOVA of fungal communities from soils.

**Author Contributions:** Individual contributions to this study are indicated with the authors' initials: Conceptualization, R.M.G., C.L.G. and W.A.U.; methodology, R.M.G., C.L.G., A.M., S.D. and W.A.U.; software, C.L.G. and W.A.U.; validation, R.M.G., C.L.G. and W.A.U.; formal analysis, C.L.G. and W.A.U.; investigation, R.M.G., C.L.G., J.D.P. and W.A.U.; resources, J.A.S. and W.A.U.; writing—original draft preparation, R.M.G. and C.L.G.; writing—review and editing, C.L.G., J.A.S. and W.A.U.; supervision, W.A.U. All authors have read and agreed to the published version of the manuscript.

**Funding:** This study was supported by a Natural Science and Engineering Research Council (NSERC) Discovery Grant to W.A.U. (RGPIN-217253-2013), funding from Canada Student Jobs 2021, and Sporometrics, Inc. We also gratefully acknowledge financial support in the form of an NSERC Undergraduate Summer Research Award to R.M.G. (2020), and funds from the Faculty of Science at BU to support the honors student research projects of R.M.G. and C.L.G. (2020–2021).

**Institutional Review Board Statement:** Not applicable.

**Informed Consent Statement:** Not applicable.

**Data Availability Statement:** All data of this research are reported in the article and Supplementary Materials.

**Acknowledgments:** We thank Gary McNeely (Brandon University) and the reviewers for constructive comments and suggestions that improved this paper. We are also grateful to the staff at the National Identification Service at the Canadian National Collection of Insects, Arachnids and Nematodes in Ottawa, Canada for their assistance in identifying the ants.

**Conflicts of Interest:** Authors Amin Maharaj and Susan Du are employed by Sporometrics Inc. The remaining authors declare that the research was conducted in the absence of any commercial or financial relationships that could be construed as a potential conflict of interest.

## References

1. Sherba, G. Moisture regulation in mound nests of the ant, *Formica ulkei* Emery. *Am. Midl. Nat.* **1959**, *61*, 499–508. [CrossRef]
2. Scherba, G. Mound temperatures of the ant *Formica ulkei* Emery. *Am. Midl. Nat.* **1962**, *67*, 373–385. [CrossRef]

3. Frouz, J.; Jilková, V. The effect of ants on soil properties and processes (Hymenoptera: Formicidae). *Myrmecol. News* **2008**, *11*, 191–199.
4. Jurgensen, M.F.; Finér, L.; Domisch, T.; Kileläinen, J.; Punttila, P.; Ohashi, M.; Niemelä, P.; Sundström, L.; Neuvonen, S.; Risch, A.C. Organic mound-building ants: Their impact on soil properties in temperate and boreal forests. *J. Appl. Entomol.* **2008**, *132*, 266–275. [CrossRef]
5. Kadochová, S.; Frouz, J. Thermoregulation strategies in ants in comparison to other social insects, with a focus on red wood ants (*Formica rufa* group). *F1000 Res.* **2014**, *2*, 280. [CrossRef]
6. Lindström, S.; Timonen, S.; Sundstretöm, L. The bacterial and fungal community composition in time and space in the nest mounds of the ant *Formica exsecta* (Hymenoptera: Formicidae). *MicrobiologyOpen* **2021**, *10*, e1201. [CrossRef]
7. Sorvari, J.; Hartikainen, S. Terpenes and fungal biomass in the nest mounds of *Formica aquilonia* wood ants. *Eur. J. Soil Biol.* **2021**, *105*, 103336. [CrossRef]
8. Frouz, F.; Kalčík, J.; Cudlín, P. Accumulation of phosphorus in nests of red wood ants *Formica s. str. Annal. Zool. Fenn.* **2005**, *42*, 269–275.
9. Jílková, V.; Matejícek, L.; Frouz, J. Changes in the pH and other soil chemical parameters in soil surrounding wood ant (*Formica polyctena*) nests. *Eur. J. Soil Biol.* **2011**, *47*, 72–76. [CrossRef]
10. Duff, L.B.; Urichuk, T.M.; Hodgins, L.N.; Young, J.R.; Untereiner, W.A. Diversity of fungi from the mound nests of *Formica ulkei* and adjacent non-nest soils. *Can. J. Microbiol.* **2016**, *62*, 562–571. [CrossRef]
11. Delgado-Baquerizo, M.; Eldridge, D.J.; Hamonts, K.; Singh, B.K. Ant colonies promote the diversity of soil microbial communities. *ISME J.* **2019**, *13*, 1114–1118. [CrossRef] [PubMed]
12. Lindström, S.; Timonen, S.; Sundstretöm, L.; Johansson, H. Ants reign over a distinct microbiome in forest soil. *Soil Biol. Biochem.* **2019**, *139*, 107529. [CrossRef]
13. Aguilar-Colorado, Á.S.; Rivera-Chávez, J. Ants/nest-associated fungi and their specialized metabolites: Taxonomy, chemistry, and bioactivity. *Rev. Bras. Farmacogn.* **2023**, *33*, 901–923. [CrossRef]
14. Christensen, M.; Tuthill, D.E. *Aspergillus*: An overview. In *Advances in Penicillium and Aspergillus Systematics*; Samson, R.A., Pitt, J.I., Eds.; (NSSA Volume 102); Springer: New York, NY, USA, 1986; pp. 195–209.
15. Samson, R.A.; Houbraken, J.; Thrane, U.; Frisvad, J.C.; Anderson, B. *Food and Indoor Fungi*; CBS-KNAW Fungal Biodiversity Center: Utrecht, The Netherlands, 2010; p. 390.
16. Petrovič, U.; Guande-Cimerman, N.; Zalar, P. Xerotolerant mycobiota from high altitude Annapurna soil, Nepal. *FEMS Microbiol. Lett.* **2000**, *182*, 339–342. [CrossRef] [PubMed]
17. Gunde-Cimerman, N.; Sonjak, S.; Zalar, P.; Frisvad, J.C.; Diderichsen, B.; Plemenitaš, A. Extremophilic fungi in arctic ice: A relationship between adaptation to low temperature and water activity. *Phys. Chem. Earth* **2003**, *28*, 1273–1278. [CrossRef]
18. Micheluz, A.; Manente, S.; Tigini, V.; Prigione, V.; Pinzari, F.; Ravagnon, G.; Varese, G.C. The extreme environments of a library: Xerophilic fungi inhabiting indoor niches. *Internat. Biodeter. Biodegradat.* **2015**, *99*, 1–7. [CrossRef]
19. Visagie, C.M.; Yilmaz, N.; Renaud, J.B.; Sumarah, M.W.; Hubka, V.; Frisvad, J.C.; Chen, A.J.; Meijer, M.; Seifert, K.A. A survey of xerophilic *Aspergillus* from indoor environment, including descriptions of two new section *Aspergillus* species producing eurotium-like sexual states. *MycoKeys* **2017**, *19*, 1–30. [CrossRef]
20. Zak, J.C.; Wildman, H.G. Fungi in stressful environments. In *Biodiversity of Fungi: Inventory and Monitoring*; Mueller, G.M., Bills, G.F., Foster, M.S., Eds.; Elsevier Academic Press: Burlington, VT, USA, 2004; pp. 303–315.
21. Flannigan, B.; Miller, J.D. Microbial growth in indoor environments. In *Microorganisms in Home and Indoor Work Environments: Diversity, Health Impacts, Investigation and Control*, 2nd ed.; Flannigan, B., Samson, R.A., Miller, J.D., Eds.; CRC Press: Boca Raton, FL, USA, 2011; pp. 57–108.
22. Pitt, J.I.; Hocking, A.D. *Fungi and Food Spoilage*, 3rd ed.; Springer: New York, NY, USA, 2009; p. 519.
23. Adams, R.L.; Sylvain, I.; Spilak, M.P.; Taylor, J.W.; Waring, M.S.; Mendell, M.J. Fungal signature of moisture damage in buildings: Identification by targeted approaches with mycobiome data. *Appl. Environ. Microbiol.* **2020**, *86*, e011047-20. [CrossRef]
24. Kujović, A.; Gostinčar, C.; Kavkler, K.; Govedić, N.; Gunde-Cimerman, N.; Zalar, P. Degradation potential of xerophilic and xerotolerant fungi contaminating historic canvas paintings. *J. Fungi* **2024**, *10*, 76. [CrossRef]
25. Pitt, J.I. Xerophilic fungi and the spoilage of foods of plant origin. In *Water Relations of Foods*; Duckworth, R.B., Ed.; Academic Press: London, UK, 1975; pp. 273–307.
26. Forster, J.C. Soil physical analysis. In *Methods in Applied Soil Microbiology and Biochemistry*; Alef, K., Nannipieri, P., Eds.; Academic Press: London, UK, 1995; pp. 105–115.
27. Forster, J.C. Determination of soil pH. In *Methods in Applied Soil Microbiology and Biochemistry*; Alef, K., Nannipieri, P., Eds.; Academic Press: London, UK, 1995; p. 55.
28. Laverty, D.H.; Bollo-Kamara, A. *Recommended Methods of Soil Analyses for Canadian Prairie Agricultural Soils*; Alberta Agriculture, Soils and Animal Nutrition Laboratory: Edmonton, AB, Canada, 1988; p. 38.
29. Soil Survey Staff. *Soil Survey Field and Laboratory Methods Manual*; Soil Survey Investigations Report no. 51, Version 2.0.; Burt, R., Ed.; Department of Agriculture, Natural Resources Conservation Service: Lincoln, NE, USA, 2014; p. 457.
30. Goh, T.B.; Mermut, A.R. Carbonates. In *Soil Sampling and Methods of Analysis*; Carter, M.R., Gregorich, E.G., Eds.; CRC Press: Boca Raton, FL, USA, 2008; pp. 215–223.

31. Quan, P.; Schoenaru, J.J.; Karamanos, R.E. Simultaneous extraction of available phosphorus and potassium with anew soil test: A modification of the Kelowna extraction. *Commun. Soil Sci. Plant Anal.* **2008**, *25*, 627–635. [CrossRef]
32. Skjemstad, J.O.; Baldock, J.A. Total and organic carbon. In *Soil Sampling and Methods of Analysis*; Carter, M.R., Gregorich, E.G., Eds.; CRC Press: Boca Raton, FL, USA, 2008; pp. 225–237.
33. APHA. *Standard Methods for the Examination of Water and Wastewater*, 23rd ed.; Braun-Holland, E., Lipps, W.C., Baxter, T.E., Eds.; American Public Health Association: Washington, DC, USA, 2017.
34. King, A.D., Jr.; Hocking, A.D.; Pitt, J.I. Dichloran-rose bengal medium for enumeration and isolation of molds from foods. *Appl. Environ. Microbiol.* **1979**, *3*, 959–964. [CrossRef] [PubMed]
35. Samson, R.A.; Hoekstra, E.S.; Frisvad, J.C. *Introduction to Food-and Airborne Fungi*, 7th ed.; Centraalbureau voor Schimmelcultures: Utrecht, The Netherlands, 2004; p. 389.
36. Malloch, D. *Moulds: Their Isolation, Cultivation and Identification*; University of Toronto Press: Toronto, ON, Canada, 1981; p. 97.
37. Samson, R.A.; Visagie, C.M.; Houbraken, J.; Hong, S.-B.; Hubka, V.; Klaassen, C.H.W.; Perrone, G.; Seifert, K.A.; Susca, A.; Tanney, J.B.; et al. Phylogeny, identification and nomenclature of the genus *Aspergillus*. *Stud. Mycol.* **2014**, *78*, 141–173. [CrossRef] [PubMed]
38. White, T.J.; Bruns, T.; Lee, S.; Taylor, J.W. Amplification and direct sequencing of fungal ribosomal RNA genes for phylogenetics. In *PCR Protocols: A Guide to Methods and Applications*; Innis, M.A., Gelfand, D.H., Sninsky, J.J., White, T.J., Eds.; Academic Press: San Diego, CA, USA, 1990; pp. 315–322.
39. Vilgalys, R.; Hester, M. Rapid genetic identification and mapping of enzymatically amplified ribosomal DNA from several *Cryptococcus* species. *J. Bacteriol.* **1990**, *172*, 4238–4246. [CrossRef] [PubMed]
40. Vilgalys, R. Conserved Primer Sequences for PCR Amplification of Fungal rDNA. Available online: https://sites.duke.edu/vilgalyslab/rdna_primers_for_fungi/ (accessed on 1 September 2018).
41. Glass, N.L.; Donaldson, G.C. Development of primer sets designed for use with the PCR to amplify conserved genes from filamentous ascomycetes. *Appl. Environ. Microbiol.* **1995**, *61*, 1323–1330. [CrossRef] [PubMed]
42. Hong, S.B.; Cho, H.S.; Shin, H.D.; Frisvad, J.C.; Samson, R.A. Novel *Neosartorya* species isolated from soil in Korea. *Int. J. Syst. Evol. Microbiol.* **2006**, *56*, 477–486. [CrossRef]
43. Clark, K.; Karsch-Mizrachi, I.; Lipman, D.J.; Ostell, J.; Sayers, W.W. GenBank. *Nucleic Acids Res.* **2016**, *44*, D67–D72. [CrossRef]
44. Altschul, S.F.; Thomas, L.; Madden, T.L.; Schaffer, A.A.; Zhang, J.; Zhang, Z.; Miller, W.; Lipman, D.J. Gapped BLAST and PSI-BLAST: A new generation of protein database search programs. *Nucleic Acids Res.* **1997**, *25*, 3389–3402. [CrossRef]
45. Wheeler, K.A.; Hocking, A.; Pitt, J.I. Effects of temperature and water activity on germination and growth of *Wallemia sebi*. *Trans. Br. Mycol. Soc.* **1988**, *90*, 365–368. [CrossRef]
46. Zheng, W.; Lehmann, A.; Ryo, M.; Vályi, K.K.; Rillig, M.C. Growth rate trades off with enzymatic investment in soil filamentous fungi. *Sci. Rep.* **2020**, *10*, 11013. [CrossRef]
47. Fox, J.; Weisberg, S. *An R Companion to Applied Regression*, 3rd ed.; Sage Publications: Thousand Oaks, CA, USA, 2018; p. 608.
48. R Core Team. R: A Language and Environment for Statistical Computing. 2020. Available online: https://www.R-project.org/ (accessed on 1 September 2020).
49. Lenth, R.V. Emmeans: Estimated Marginal Means, Aka Least-Squares Means. In R Package Version 1.6.2-1. 2021. Available online: https://CRAN.Rproject.org//package=emmeans (accessed on 1 September 2020).
50. Wickham, H. ggplot2: Elegant Graphics for Data Analysis. 2016. Available online: https://ggplot2.tidyverse.org (accessed on 1 September 2020).
51. Garnier, S. Viridis: Default Color Maps from 'Matplotlib'. In R Package Version 0.5.1. 2018. Available online: https://CRAN.R-project.org/package=viridis (accessed on 1 September 2020).
52. Oksanen, J.; Blanchet, F.G.; Friendly, M.; Kindt, R.; Legendre, P.; McGlinn, D.; Minchin, P.R.; O'Hara, R.B.; Simpson, G.L.; Solymos, P.; et al. Vegan: Community Ecology Package Version 2.5-7. 2020. Available online: https://CRAN.R-project.org/package=vegan (accessed on 1 September 2020).
53. Domsch, K.H.; Gams, W.; Anderson, T.-H. *Compendium of Soil Fungi*, 2nd ed.; IHW-Verlag: Eching, Germany, 2007; p. 672.
54. Siedlecki, I.; Kochanowski, M.; Pawłowska, J.; Reszotnik, G.; Okrasińska, A.; Wrzosek, M. Ant's nest as a microenvironment: Distinct Mucoromycota (Fungi) community of the red wood ant's (*Formica polyctena*) mounds. *Ecol. Evol.* **2024**, *14*, e70333. [CrossRef]
55. Wali, M.K.; Kannowski, P.B. Prairie ant mound ecology: Interrelationships of microclimate, soils and vegetation. In *Prairie: A Multiple View*; Wali, M.K., Ed.; The University of North Dakota University Press: Grand Forks, ND, USA, 1975; pp. 155–169.
56. Elmes, G.W. Ant colonies and environmental disturbance. *Symp. Zool. Soc. Lond.* **1991**, *63*, 1–13.
57. Cammeraat, E.L.H.; Risch, A.C. The impact of ants on mineral soil properties and processes at different spatial scales. *J. Appl. Entomol.* **2008**, *132*, 285–294. [CrossRef]
58. Farji-Brener, A.G.; Werenkraut, V. The effects of ants on soil fertility and plant performance: A meta-analysis. *J. Anim. Ecol.* **2017**, *86*, 866–877. [CrossRef]
59. Sklenář, F.; Glässnerová, K.; Jurjević, Ž.; Houbraken, J.; Samson, R.A.; Visagie, C.M.; Yilmaz, N.; Gené, J.; Cano, J.; Chen, A.J.; et al. Taxonomy of *Aspergillus* series *Versicolores*: Species reduction and lessons learned about intraspecific variability. *Stud. Mycol.* **2022**, *102*, 53–93. [CrossRef]

60. Jurjevic, Z.; Peterson, S.W.; Horn, B.W. *Aspergillus* section *Versicolores*: Nine new species and multilocus DNA sequence based phylogeny. *IMA Fungus* **2012**, *3*, 59–79. [CrossRef] [PubMed]
61. Sabino, R.; Veríssimo, C.; Parada, H.; Brandão, J.; Viegas, C.; Carolino, E.; Clemons, K.V.; Stevens, D.A. Molecular screening of 246 Portuguese *Aspergillus* isolates among different clinical and environmental sources. *Med. Mycol.* **2014**, *52*, 519–529. [CrossRef]
62. Géry, A.; Rioult, J.P.; Heutte, N.; Séguin, V.; Bonhomme, J.; Garon, D. First characterization and description of *Aspergillus* Series *Versicolores* in French bioaerosols. *J. Fungi* **2021**, *7*, 676. [CrossRef]
63. Houbraken, J.; Due, M.; Varga, J.; Meijer, M.; Frisvad, J.C.; Samson, R.A. Polyphasic taxonomy of *Aspergillus* section *Usti*. *Stud. Mycol.* **2007**, *59*, 107–128. [CrossRef]
64. Hubka, V.; Nováková, A.; Samson, R.A.; Houbraken, J.; Frisvad, J.C.; Sklenár, F.; Varga, J.; Kolařík, M. *Aspergillus europaeus* sp. nov., a widely distributed soil-borne species related to *A. wentii* (section *Cremei*). *Plant Syst. Evol.* **2016**, *302*, 641–650. [CrossRef]
65. Northolt, M.D.; Soentoro, P.S.S. Fungal growth on foodstuffs related to mycotoxin contamination. In *Introduction to Food-Borne Fungi*, 3rd ed.; Samson, R.A., van Reenan Hoekstra, E.S., Eds.; Centraalbureau voor Schimmelcultures: Delft, The Netherlands, 1988; pp. 231–238.
66. Rodríguez-Andrade, E.; Stchigel, A.M.; Terrab, A.; Guarro, J.; Cano-Lira, J.F. Diversity of xerotolerant and xerophilic fungi in honey. *IMA Fungus* **2019**, *10*, 20. [CrossRef]
67. Butinar, L.; Frisvard, J.C.; Gunde-Cimerman, N. Hypersaline waters—A potential source of foodborne toxigenic aspergilli and penicillia. *FEMS Microbiol. Ecol.* **2011**, *7*, 186–199. [CrossRef] [PubMed]
68. Arenz, B.E.; Held, B.W.; Jurgens, J.A.; Farrell, R.L.; Blanchette, R.A. Fungal diversity in soils and historic wood from the Ross Sea Region of Antarctica. *Soil Biol. Biochem.* **2006**, *38*, 3057–3064. [CrossRef]
69. Kochkina, G.A.; Ivanushkina, N.E.; Akimov, V.N.; Gilichinskii, D.A.; Ozerskaya, S.M. Halo- and psychrotolerant *Geomyces* fungi from Arctic cryopegs and marine deposits. *Microbiology* **2007**, *76*, 31–38. [CrossRef]
70. Samson, R.A.; Yilmaz, N.; Houbraken, J.; Spierenburg, H.; Seifert, K.A.; Petersen, S.W.; Varga, J.; Frisvad, J.C. Phylogeny and nomenclature of the genus *Talaromyces* and taxa accommodated in *Penicillium* subgenus *Biverticillium*. *Stud. Mycol.* **2011**, *70*, 159–183. [CrossRef] [PubMed]
71. Yoder, J.A.; Fisher, K.A.; Dobrotka, C.J. A report on *Purpureocillium lilacinum* found naturally infecting the predatory mite, *Balaustium murorum* (Parasitengona: Erythraeidae). *Int. J. Acarol.* **2018**, *44*, 4–5. [CrossRef]
72. Pitt, J.I. An appraisal of identification methods for *Penicillium* species: Novel taxonomic criteria based on temperature and water relations. *Mycologia* **1973**, *65*, 1135–1157. [CrossRef]
73. Sewell, G.W.F. The effect of altered physical conditions of soil on biological control. In *Ecology of Soil-Borne Plant Pathogens: Prelude to Biological Control*; Baker, K.F., Synder, W.C., Eds.; University of California Press: Berkeley, CA, USA, 1963; pp. 479–494.
74. Bell, L.N.; Labuza, T.P. *Moisture Sorption: Practical Aspects of Isotherm Measurement and Use*, 2nd ed.; American Association of Cereal Chemists: St. Paul, MN, USA, 2000; p. 122.
75. Rowan, N.J.; Johnstone, C.M.; McLean, R.C.; Anderson, J.G.; Clarke, J.A. Prediction of toxigenic growth in buildings by using a novel modelling system. *Appl. Environ. Microbiol.* **1999**, *65*, 4814–4821. [CrossRef]
76. Dedesko, S.; Siegel, J.A. Moisture parameters and fungal communities associated with gypsum drywall in buildings. *Microbiome* **2015**, *3*, 71. [CrossRef]
77. Scott, W.J. Water relations of food spoilage microorganisms. *Adv. Food Res.* **1957**, *7*, 87–127.
78. Katz, E.E.; Labuza, T.P. Effect of water activity on sensory crispness and mechanical deformation of snack food products. *J. Food Sci.* **1981**, *46*, 403–409. [CrossRef]
79. Verdier, T.; Coutand, M.; Bertron, A.; Roques, C. A review of indoor microbial growth across building materials and sampling and analysis methods. *Build. Environ.* **2014**, *80*, 136–149. [CrossRef]
80. Adan, O.C.G.; Samson, R.A. (Eds.) Introduction. In *Fundamentals of Mold Growth in Indoor Environments and Strategies for Healthy Living*; Wageningen Academic Publishers: Wageningen, The Netherlands, 2011; pp. 13–38.
81. Hocking, A.D.; Pitt, J.I. Dichloran-glycerol medium for enumeration of xerophilic fungi from low-moisture foods. *Appl. Environ. Microbiol.* **1980**, *39*, 488–492. [CrossRef] [PubMed]
82. Esteban, M.A.; Alcala, M.; Marcos, A.; Fernandez-Salguero, J.; Garcia de Fernando, G.D.; Ordoñez, J.A.; Sanz, B. Water activity of culture media used in food microbiology. *Int. J. Food Sci. Technol.* **1990**, *25*, 464–468. [CrossRef]
83. Segers, F.J.J.; Meijer, M.; Housbraken, J.; Samson, R.A.; Wösten, H.A.B.; Dijksterhuis, J. Xerotolerant *Cladosporium sphaerospermum* are predominant on indoor surfaces compared to other *Cladosporium* species. *PLoS ONE* **2015**, *10*, e0145415. [CrossRef]
84. Hirooka, Y.; Tanney, J.B.; Nguyen, H.D.T.; Seifert, K.A. Xerotolerant fungi in house dust: Taxonomy of *Spiromastix, Pseudospiromastix* and *Sigleria gen. nov.* in Spiromastigaceae (Onygenales, Eurotiomycetes). *Mycologia* **2016**, *108*, 135–156. [CrossRef]
85. Hung, L.L.; Caulfield, S.M.; Miller, D.J. (Eds.) *Recognition, Evaluation, and Control of Indoor Mold*, 2nd ed.; American Industrial Hygiene Association Press: Fairfax, VA, USA, 2020; p. 258.

Journal of *Fungi*

MDPI

*Article*

# Dynamics and Insights into the Unique Ecological Guild of Fungi in Bacteria-Bioaugmented Anaerobic Digesters

**Linda U. Obi [1,2,3], Ashira Roopnarain [1,3], Memory Tekere [3], Jun Zhou [4], Heng Li [5], Yuanpeng Wang [6], Yanlong Zhang [7,8,9] and Rasheed A. Adeleke [1,2,*]**

1 Microbiology and Environmental Biotechnology Research Group, Institute for Soil, Climate and Water, Agricultural Research Council, Arcadia, Pretoria 0083, South Africa; obilindauloma@gmail.com (L.U.O.); roopnaraina@arc.agric.za (A.R.)

2 Unit for Environment Science and Management, North-West University (Potchefstroom Campus), Private Bag X1290, Potchefstroom 2520, South Africa

3 Department of Environmental Sciences, College of Agriculture and Environmental Sciences, University of South Africa (UNISA), P.O. Box 392, Florida 1710, South Africa; tekerm@unisa.ac.za

4 Bioenergy Research Institute, College of Biotechnology and Pharmaceutical Engineering, Nanjing Tech University, Nanjing 211816, China; zhoujun@njtech.edu.cn

5 Key Laboratory of Estuarine Ecological Security and Environmental Health, Tan Kah Kee College, Xiamen University, Zhangzhou 363105, China

6 Key Laboratory for Chemical Biology of Fujian Province, Department of Chemical and Biochemical Engineering, College of Chemistry and Chemical Engineering, Xiamen University, Xiamen 361005, China; wypp@xmu.edu.cn

7 Key Laboratory of the Ministry of Education for Coastal and Wetland Ecosystem, College of the Environment & Ecology, Xiamen University, Xiamen 361102, China; ylzhang@xmu.edu.cn

8 Fujian Key Laboratory of Coastal Pollution Prevention and Control (CPPC), College of Environment & Ecology, Xiamen University, Xiamen 361102, China

9 Fujian Institute for Sustainable Oceans, Xiamen University, Xiamen 361102, China

* Correspondence: rasheed.adeleke@nwu.ac.za; Tel.: +27-18-299-2495

**Abstract:** Anaerobic digesters host a variety of microorganisms, and they work together to produce biogas. While bacterial and archaeal communities have been well explored using molecular techniques, fungal community structures remain relatively understudied. The present study aims to investigate the dynamics and potential ecological functions of the predominant fungi in bacteria-bioaugmented anaerobic digesters. Eight different anaerobic digesters that contained chopped water hyacinth and cow dung as feedstock at 2% total solids were respectively inoculated with eight different bacterial strains and digested anaerobically in controlled conditions. The diversity and dynamics of the fungal community of the digesters before and after digestion were monitored using high-throughput sequencing of the fungal *ITS2* sub-region of the ribosomal gene. The functional potential of the fungal community was predicted using ecological guild analysis. The dominant fungal phyla were (with relative abundance $\geq$1%) Ascomycota and Neocallimastigomycota. Ascomycota exhibited over 90% dominance in all treatments after anaerobic digestion (AD). *Aspergillus* sp. was consistently dominant across treatments during AD, while prominent anaerobic fungal genera *Anaeromyces*, *Cyllamyces*, and *Caeomyces* decreased. Ecological guild analysis at genus level showed that the majority of the identified fungi were saprophytes, and diversity indices indicated decreased richness and diversity after AD, suggesting a negative impact of AD on fungal communities in the anaerobic digesters. The multivariate structure of the fungal communities showed clustering of treatments with similar fungal taxa. The findings from this study provide insights into the fungal ecological guild of different bacteria-bioaugmented anaerobic digesters, highlighting their potentials in bacteria-augmented systems. Identification of an anaerobic fungal group within the phylum Ascomycota, beyond the well-known fungal phylum Neocallimastigomycota, offers a new perspective in optimizing the AD processes in specialized ecosystems.

**Keywords:** fungal microbiome; anaerobic fungi; biogas production; high-throughput sequencing

## 1. Introduction

Anaerobic digestion (AD) is an important process in waste treatment and renewable energy production, providing an environmentally sustainable method for converting organic matter into biogas in an anaerobic digester. In recent times, the extensive application of biogas as a source of renewable energy has risen [1]. The production of biogas in anaerobic digesters through the metabolic process of AD involves the complex synergistic interactions of different microbes including fungi, bacteria, and archaea [2]. Fungi have been observed to be vital microorganisms due to their contribution to the breakdown of organic matter. These microorganisms initiate the degradation process by opening up the cells of complex lignocellulosic substrates for subsequent degradation by bacteria and archaea, thereby accelerating the metabolic process [3,4]. Fungi may not be as prevalent as bacteria and archaea in anaerobic digesters due to environmental conditions, and there are limited studies on fungal communities of anaerobic digesters [2]. However, they demonstrate some metabolic activities, particularly during the early stage of AD, also known as the hydrolytic phase. The hydrolytic phase of AD has been identified as the rate-limiting phase in the AD of lignocellulosic substrates and it involves the metabolic activities of fungal organisms, including the production of extracellular enzymes for the potential breakdown of complex lignocellulosic substrates [5,6]. Inhibition of the hydrolytic step could impede the AD process or result in generating recalcitrant intermediates [7]. To optimize the AD process by increasing microbial activity, the incorporation of active microbial strains into the native microbial community of anaerobic digesters (bioaugmentation) is imperative, to enhance the breakdown of biomass [8]. Bioaugmentation has been implicated in enhancing digester performance and mitigating ammonia and salinity stresses, subsequently improving methane production [9].

Some fungi can initiate cellulose degradation by hydrolyzing the complex crystalline structure of lignocellulose, breaking the core β-1,4-glucan bonds through random depolymerization [10]. These fungi are distributed across different fungal phylogenies and most belong to phyla such as Basidiomycota and Ascomycota, as well as the famous anaerobic fungal phylum Neocallimastigomycota [3]. Bacteria are equally beneficial and predominant in biodigesters in terms of biomass degradation. They are also essential at the initial stages of AD, like hydrolysis, acidogenesis, and acetogenesis, while archaea, which are primarily methanogens, utilize substrates including molecular hydrogen to generate methane in the final process of AD, known as methanogenesis [11,12]. Bacterial and archaeal communities of anaerobic digesters have been widely studied using different molecular techniques that are either focused on the *16S rRNA* gene or focused on the metagenome [2,13–16]. However, only a few studies have reported on the community structure of fungi in bacteria-bioaugmented anaerobic digesters [2,17].

The bioaugmentation of anaerobic digesters with bacteria is ideal due to the resilient and adaptive features of bacteria developed in response to extreme environments. Understanding the metabolic feature of the indigenous and introduced microbial entities of anaerobic digesters is imperative for an efficient AD process. Bioaugmentation with bacteria has been reported to cause a shift in the microbial community of anaerobic digesters, though some studies have reported otherwise [18]. Introducing microbial species with specialized enzymatic capabilities can address key limitations in the anaerobic digestion (AD) of lignocellulosic substrates by enhancing the hydrolysis step, often considered a rate-limiting phase in the process. In this study, the bacterial strains used for bioaugmentation were selected based on their facultative anaerobic characteristics, which allow them to adapt effectively to the digester environment. These strains also demonstrate hydrolytic and cellulolytic abilities when cultivated on sterile carboxymethyl cellulose (CMC) agar, as reported in previous studies [19,20]. *Pseudomonas stutzeri* has shown significant cellulolytic and hydrolytic potential, attributed to the presence of the A1501 cellulase gene [21]. Additionally, *Exiguobacterium* species have exhibited the ability to produce endoglucanase, a form of cellulase, when grown on lignocellulosic substrates such as watermelon peels [22]. *Bacillus cereus* has demonstrated the potential to metabolize lignocellulosic substrates into short-chain fatty acids, a key intermediate in the anaerobic digestion (AD) process [23]. Other cellulolytic bacteria employed in this study include *Lysinibacillus fusiformis*, known for its cellulolytic activity and ability to degrade complex polysaccharides effectively [24]. A well-characterized cellulolytic bacterium, *Serratia marcescens*, can hydrolyze lignocellulose, contributing to the breakdown of plant biomass [25]. *Brevundimonas vesicularis* exhibits cellulase production and plays a role in lignocellulose decomposition [26]. These enzymatic activities accentuate the potential of these bacteria to enhance lignocellulose degradation in AD systems, thereby improving process efficiency and biogas yields.

The intrusive growth and potent fiber-degrading enzymes of fungi including anaerobic fungi (AF) are crucial in the degradation of complex organic compounds. Anaerobic fungi play a critical role in fiber degradation in the gut of herbivores. The incorporation of animal-based substrates has given rise to AF in anaerobic digesters as AF are an essential part of herbivores' manure [18]. However, the survival of AF in such anaerobic environments relies on the operational conditions of the digesters, including but not limited to temperature and retention time. Some studies have established the implications of elevated temperature and extended retention time on the proliferation and survival of cultivable and non-cultivable AF in anaerobic digesters [27,28]. While elevated temperatures can improve the metabolic activities of some anaerobic fungi, extended retention times promote the adaptation and thriving of anaerobic fungi in anaerobic digesters. Combinations of these parameters contribute to the metabolism of organic materials, thus improving digester performance [29]. Since microbial communities, including fungi, play a key role in these metabolic processes, and fungal community shifts have been implicated in the kinetics of AD as well methane yield, understanding their dynamics is essential for improving the efficiency of biogas production [30]. This would pave the way towards possibly optimizing the AD process and enhancing the synergistic interactions between inoculated bacteria and the indigenous fungal community of anaerobic digesters. To understand the dynamics of environmental fungal communities, which include AF, employing a metagenomic approach provides insights into the taxonomic diversity and genetic potential of fungal communities of complex samples like anaerobic digestate [30–32]. Determining fungal community shifts in anaerobic digesters bioaugmented with bacteria is important for understanding the resilience and adaptation of fungi to environmental changes under certain conditions such as bioaugmentation. This study deals with an in-depth evaluation of fungi as a distinct ecological guild within anaerobic digesters bioaugmented with distinct bacterial species. It also evaluates the dynamics of the dominant fungal community structure of bioaugmented anaerobic digesters through DNA metabarcoding, to provide insight into the predicted functional abilities of the fungal microbiome of anaerobic digesters. Anaerobic digestion provides a model for studying these shifts, and insights into fungal dynamics during biogas production could help in the development of strategies for more stable and sustainable biogas systems.

## 2. Materials and Methods

### 2.1. Sampling

Water hyacinth was harvested from Hartbeespoort Dam in the North West province (25°44′51″ S 27°52′1″ E), South Africa, and fresh cow dung was randomly collected from the dairy parlor of the Agricultural Research Council—Animal Production in Gauteng province (25°53′59.6″ S 28°12′51.6″ E), South Africa, using a clean trowel and transferred to anaerobic bags. The sample characteristics were, for water hyacinth, dry matter: 5.97%, volatile solids: 4.46%, pH: 8.11, carbon-to-nitrogen ratio: 14.5; for cow dung, dry matter: 16.8%, volatile solids: 14.16%, pH: 8.34, carbon-to-nitrogen ratio: 23.7. Samples were transported (cow dung in anaerobic bags) to the biogas laboratory of the Agricultural Research Council—Soil Climate and Water for storage. Water hyacinth was stored at −20 °C and cow dung was stored in an air-tight container at 4 °C until usage.

Ten sets of treatments were assembled in 500 mL Schott batch culture bottles fitted with screw caps, each with a working volume of 250 mL. The solid biomass consisted of freshly chopped water hyacinth cut into 2 cm × 2 cm pieces and cow dung at 2% total solids mixed in a 2:1 ratio. The sets differed for the inoculum ($OD_{600}$ 1.5 about $10^9$ CFU/mL) consisting of 5% ($v/v$) of a previously identified and tested single pure bacterial culture [20]. In the first eight sets, a single pure culture ($OD_{600}$ 1.5 about $10^9$ CFU/mL) of the mentioned strains (Table 1) was added, and another set was inoculated with a mix of the previous strains (each of the strains had $OD_{600}$ 1.5 about $10^9$ CFU/mL), i.e., the 'CONS' treatment. To check the influence of living bacterial inocula on the biogas production, the negative control 'CONT' (tenth set), without inoculated bacteria (CONT), was set up (Table 1). The volume of each treatment was bulked to 250 mL with tap water. The treatments were not purged with nitrogen gas to create a conducive environment for the methanogens; however, anaerobiosis was reached by allowing treatments to incubate in sealed screw-capped 500 mL Schott batch culture bottles which prevented oxygen infiltration and promoted consumption of residual oxygen by microbial activity. The experiment was conducted in triplicates. The treatments were subjected to AD at a mesophilic temperature of 30 °C and 120 rpm for a period of 35 days, during which biomethane production was monitored using a Gas Chromatograph (SRI 8610C, CHROMSPEC Canada). The results of the methane/biogas production as well as the dynamics of the bacterial and archaeal (*16S rRNA* gene) communities are outlined by Obi et al. [20].

**Table 1.** Bacterial isolates' ID and bioaugmentation treatments' composition.

| Treatment ID Before AD | Treatment ID After AD | Inoculated Bacteria with GenBank Accession Numbers | Treatment Make-Up |
|---|---|---|---|
| 3H1 | 3H2 | *Pseudomonas stutzeri* (MK104459) | WH + CD + $10^9$ cfu/mL of 3H |
| 3M1 | 3M2 | *Exiguobacterium mexicanum* (MK104464) | WH + CD + $10^9$ cfu/mL of 3M |
| 4F1 | 4F2 | *Bacillus cereus* (MK104469) | WH + CD + $10^9$ cfu/mL of 4F |
| 7B1 | 7B2 | *Lysinibacillus fusiformis* (MK104485) | WH + CD + $10^9$ cfu/mL of 7B |
| 11H1 | 11H2 | *Serratia marcescens* (MK104517) | WH + CD + $10^9$ cfu/mL of 11H |
| 12H1 | 12H2 | *Brevundimonas vesicularis* (MK104523) | WH + CD + $10^9$ cfu/mL of 12H |
| D31B1 | D31B2 | *Acinetobacter iwoffi* (MK104525) | WH + CD + $10^9$ cfu/mL of D31B |
| D31D1 | D31D2 | *Planococcus maritimus* (MK104526) | WH + CD + $10^9$ cfu/mL of D31D |
| CONS1 | CONS2 | *Pseudomonas stutzeri* (MK104459), *Exiguobacterium mexicanum* (MK104464), *Bacillus cereus* (MK104469) *Lysinibacillus fusiformis* (MK104485), *Serratia marcescens* (MK104517), *Brevundimonas vesicularis* (MK104523), *Acinetobacter iwoffi* (MK104525), and *Planococcus maritmus* (MK104526) | WH + CD + $10^9$ cfu/mL of 3H + $10^9$ cfu/mL of 3M + $10^9$ cfu/mL of 4F + $10^9$ cfu/mL of 7B + $10^9$ cfu/mL of 11H + $10^9$ cfu/mL of 12H + $10^9$ cfu/mL of D31B + $10^8$ cfu/mL of D31D |
| CONT1 | CONT2 | No inoculated bacteria | WH + CD |

WH = water hyacinth; CD = cow dung.

### 2.2. Genomic DNA Extraction from Different Bioaugmented Treatments

Samples were collected from each treatment, including the control, before and after AD to explore the fungal community structure of the anaerobic digesters. Following the method outlined by Mukhuba et al. [33] with modifications, a 2 mL volume of pooled digester samples (composite samples) was centrifuged at 10,000× *g* for 3 min to collect the cells, and the pellets that settled at the bottom of the tube were used for genomic DNA extraction. Genomic DNA was extracted from the collected samples using the DNeasy PowerSoil kit (Qiagen, Carlsbad, CA, USA) as per the manufacturer's instructions and the isolated DNA was standardized to a concentration of 5 ng/µL using a Qubit 2.0 fluorometer (Invitrogen, Carlsbad, CA, USA).

#### 2.2.1. Preparation of Internal Transcribed Spacer (*ITS*) Gene Library and Analysis

To perform taxonomic profiling of the fungal *ITS2* region, a set of primers, *ITS3* (5′-CAHCGATGAAGAACGYRG-3′) and *ITS4* (5′-TTCCTSCGCTTATTGATATGC-3′), was used [34–36]. The 5′-end of each primer was attached with Illumina overhang adapters (TCGTCGGCAGCGTCAGATGTGTATAAGAGACAG and GTCTCGTGGGCTCGGAGAT-GTGTATAAGAGACAG) (Illumina, Inc., San Diego, CA, USA). Amplification via PCR was performed using 12.5 ng of DNA template, 12.5 µL of One Taq 2X Master Mix with Standard Buffer (MO482S, New England Biolabs. Inc. Ipswich, MA, USA), 0.2 µM (1 µL) of each of the primers, and PCR-grade water in a 25 µL reaction. The PCR was conducted in a Bio Rad thermal cycler (Bio-Rad Laboratories, Hercules, CA, USA) with the following conditions: initial denaturation at 94 °C for 30 s followed by 35 cycles of 94 °C for 30 s, 65 °C for 30 s, and 68 °C for 1 min, with a final extension at 68 °C for 5 min. The amplicons were confirmed on 1% agarose gel, and preparation of amplicon libraries was carried out according to Illumina's protocol (Illumina, Inc., San Diego, CA, USA) for 2 × 300 bp paired-end reads. Sequencing was performed using the Illumina MiSeq sequencer (Illumina, Inc., San Diego, CA, USA) at the Agricultural Research Council—Biotechnology Platform (ARC—BTP), Pretoria, South Africa. Subsequent to adapter and primer sequence trimmings, raw fastq ITS sequences were obtained and subjected to bioinformatics analyses.

#### 2.2.2. Bioinformatics Analyses for Fungal Community in Anaerobic Digesters

In analyzing the fungal *ITS* sequences, an automated PIPITS pipeline was utilized [37]. A tab-delimited text file was generated for all paired-end reads from the raw sequence directory using PIPITS_GETREADPAIRSLIST. PIPITS-PREP activated its action by joining the paired-end sequences on the overlapping regions with PEAR [38], and the FAST QUALITY FILTER FASTX-Toolkit was employed to quality-filter the sequences [39]. A total of 1,667,405 reads were produced from all the treatments (20 samples) for the PIPITS_prep study. Of these, 1,320,289 reads were joined, and 1,305,701 reads passed the quality filter. To eliminate redundancy, representative sequences were obtained and the *ITS2* sub-region was extracted from 1,198,541 sequences with the aid of ITSx [40]. Reads that did not contain an *ITS* region were discarded. Short sequences (<100 bp) were also removed during processing, and the resulting sequences were clustered into 1654 operational taxonomic units (OTUs) and 454 phylotypes at a 97% sequence similarity threshold. De novo VSEARCH was used to detect and remove chimeras from the representative sequences [41]. Taxonomic assignment to representative sequences was performed using the Ribosomal Database Project (RDP) Classifier (version 2.1) [42] against the UNITE UCHIME fungal ITS reference dataset (version 7.1) [43–45]. Prior to analysis, the generated fungal taxonomic data were normalized to 63056 sequences per sample using the median sequencing depth. The fungal ecological guild at the genus taxonomic rank was envisaged using FUNGuild, an open annotation tool (http://www.stbates.org/guilds/app.php [46] (accessed on 1 March 2020)).

Correlograms were generated using the Hmisc (v. 4.4-2), Corr (v.0.4.3), and Corrplot (v. 0.84) packages of R software. Correlation analysis was based on the Spearman rank correlation coefficient. The correlation matrix was reordered according to the correlation coefficient using the "hclust" method. Correlation network analysis was generated using the phyloseq (v.1.38.0), dyplr (v.1.1.2), and ggplot2 (v.3.4.4) packages of R software. The threshold was set at 0.3 to include only significant correlations in the network.

#### 2.2.3. Data Availability

The raw sequences generated from this study have been submitted to the Sequence Read Archive (SRA) at NCBI as part of the BioProject PRJNA704473 (https://www.ncbi.nlm.nih.gov/sra/PRJNA704473, accessed on 1 March 2020), as an experiment entitled "Dynamics of fungal community structure during biogas production". These sequences can be accessed through the biosample accession numbers from SAMN18042783 to SAMN18042802, as well as the SRA accession numbers from SRR13779324 to SRR13779343.

### 2.3. *Statistical Analysis*

The fungal community structure was examined using non-metric dimensional scaling (NMDS) in multivariate spaces. Differences among treatment groups in these spaces were evaluated based on the Bray–Curtis dissimilarity. All statistical analyses were performed using R software (v.4.03) and Excel 2013 unless stated otherwise. Construction of NMDS was executed with the vegan (v. 2.5.7), ggplot2 (v. 3.4.4), and dyplr (v. 1.1.2) packages of R software (https://cran.r-project.org/, accessed on 1 March 2020).

## 3. Results

### 3.1. *Batch Culture Analysis for Methane Production*

The results of the methane/biogas production as well as the dynamics of the prokaryotic (bacteria and archaea) communities are detailed in Figure 1 [20]. The study investigated the effects of bioaugmentation on the diversity of the prokaryotic community structure in anaerobic digesters. It was found that the treatment inoculated with 11H (*Serratia marcescens*) produced the highest cumulative methane yield of 0.68 L, representing a 45.6% increase compared to the consortium treatment, which produced the lowest methane yield (Figure 1).

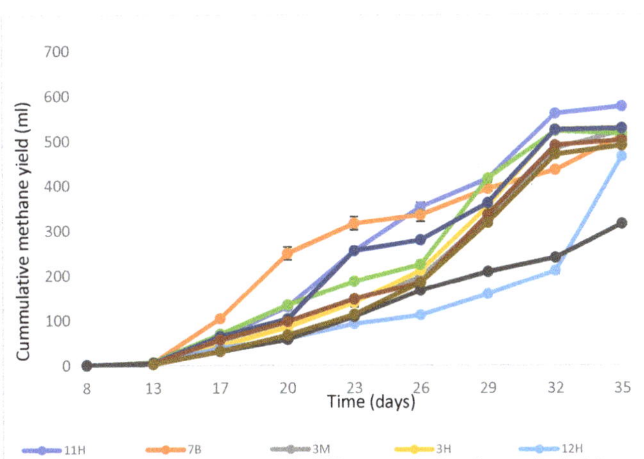


**Figure 1.** Cumulative methane yield of AD of water hyacinth and cow dung inoculated with pure cultures (adapted from [20].

However, bioaugmentation had a negligible effect on the overall prokaryotic community structure. The bacterial community, initially dominated by the genus *Pseudomonas*, shifted to *Bacteroides* after AD. No archaeal community was detected before AD, but after AD, *Methanosarcina* dominated the archaeal communities across all treatments.

### 3.2. High-Throughput Metabarcoding of ITS Genes from Different Bioaugmented Treatments

Variations in taxonomic classification at the phylum, class, and genus levels were observed within the fungal community of anaerobic digesters subjected to bacterial bioaugmentation. The relatively abundant fungal phyla, including Ascomycota, Basidiomycota, and Neocallimastigomycota, were identified across all treatments. Notably, Ascomycota accounted for 60–99% of the relatively abundant sequences within the fungal phyla, with increased dominance observed in most treatments after AD (Figure 2). Ascomycota exhibited a remarkable dominance, constituting approximately 99% of the fungal composition in about 60% of the treatments after AD. This prevalence was particularly evident in treatments bioaugmented with specific bacterial strains such as *Pseudomonas stutzeri* (3H2), *Exiguobacterium mexicanum* (3M2), *Lysinibacillus fusiformis* (7B2), *Serratia marcescens* (11H2), *Acinetobacter iwoffii* (D31B2), and *Planococcus maritimus* (D31D1). In these treatments, the relative abundance of Ascomycota ranged from 11% (before AD) to 25% (after AD).

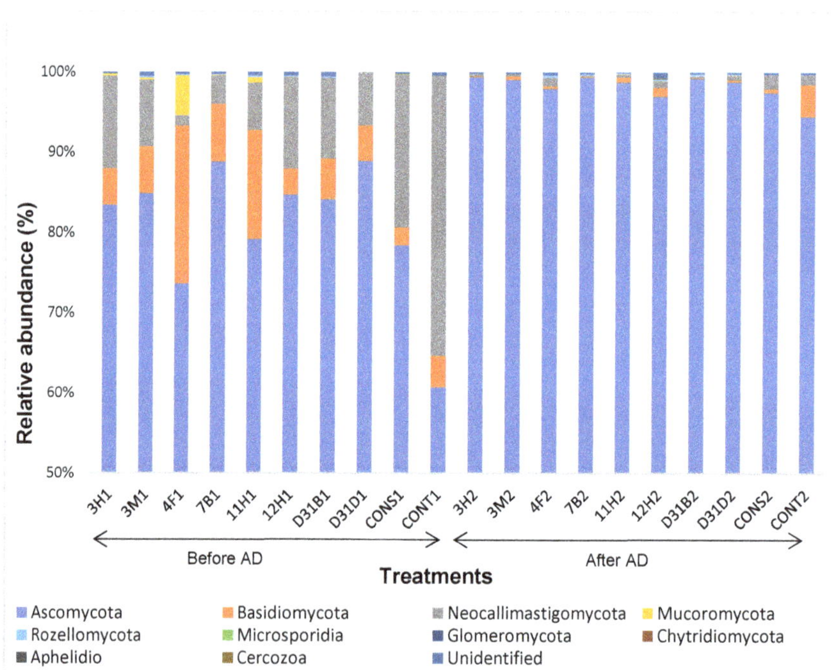


**Figure 2.** Relative abundance (≥1%) of fungal phyla in treatments before AD and after AD. Phylotypes with an average relative abundance of less than 1% and those unclassified at the phylum level were excluded from the plot.

Contrastingly, the control treatment (CONT2), which was not bioaugmented, recorded a 54% difference (increase) in the relative abundance of Ascomycota after AD. This underscores the impact of bacterial bioaugmentation on shaping the taxonomic dynamics of fungal communities in anaerobic digesters. This observation highlights the importance of bacterial bioaugmentation as a tool for influencing the microbial community structure of anaerobic digesters. The shift suggests that bacterial bioaugmentation probably suppressed some fungal populations by reducing their reliance on fungal-associated hydrolysis, thus leading to bacterial-dominated hydrolysis. In addition to the control treatment CONT2, the remarkable presence of Basidiomycota was noted only in treatment 12H2 (post-AD). The presence of Neocallimastigomycota and Basidiomycota markedly decreased after AD, except in the case of the control treatment (CONT2). In CONT2, the abundance of Basidiomycota remained remarkably consistent after AD, whereas Neocallimastigomycota decreased by 97%. The substantial reduction in the presence of Neocallimastigomycota, a notable anaerobic fungus, in the same treatment after AD might be attributed to the absence of bioaugmented bacteria.

Examining the class level, the prevalent fungal classes across all treatments exhibit a similarity in composition to the taxonomic ranking at the phylum level. The dominant classes include Eurotiomycetes, Sordariomycetes, and Dothideomycetes, along with Tremellomycetes, with Eurotiomycetes (Appendix A, Figure A1) standing out as the most abundant class in the bioaugmentation process of anaerobic digesters across all treatments, both before and after AD. Neocallimastigomycetes, abundant in treatments before AD, showed a decrease in abundance after AD, a trend also observed in Tremellomycetes. Conversely, the appearance of Leotiomycetes after AD was observed in all treatments except 3M2 and 12H. Saccharomycetes were exclusively observed in treatment 3H2 after AD, while Agaricomycetes were present only in the control treatment CONT1 (before AD), and exhibited an increase in CONT2 (after AD).

Further examination of fungal genera suggests that AD did not impact the abundance of obligate aerobic fungi such as *Thermomyces, Acremonium, Aspergillus, Chaetomium*, and *Microascus*. However, there was a decrease in the prevalence of anaerobic fungal genera such as *Anaeromyces, Cyllamyces*, and *Caecomyces* following AD, except in treatment D31D1. This is evident from the dominance of *Aspergillus* observed in all treatments throughout the AD period, as depicted in Figure 3 and Appendix A. The presence and absence of the aforementioned fungal genera align with the observed patterns at the phylum and class taxonomic levels, particularly concerning Necallimastigomycota and Neocallimastigomyces. Sordariomycetes and Dothideomycetes were relatively dominant across the treatments, further highlighting the importance of these fungal classes in AD (Appendix B).

The majority of fungi that belong to the ecological guild are mainly saprotrophs and endophytes, as depicted in Figure 4, and were abundant before and after AD. Fungal parasites and plant pathogens, which were abundant before AD, decreased after AD in all treatments except for treatment D31D1. Treatment D31D1 had the lowest abundance of fungi, and the guild of fungi classified as animal pathogens showed similar levels of abundance across all treatments as no major change was observed before and after AD.

The color intensity and the size of the circles (Figure 5A) are proportional to the correlation coefficients. A positive correlation is denoted by +1 or values closer to +1, while −1 shows strong negative correlations. *Chaetomium* displayed strong positive correlations with *Microascus* and positive correlations with *Anaeromyces, Caecomyces, Cyllamyces*, and *Thermomyces*. In the same vein, *Thermomyces* showed strong positive correlations with *Aspergillus* and correlated positively with a few dominant fungi (*Chaetomium* and *Microascus*) (Figure 5A). *Anaeromyces* correlated strongly with *Caecomyces* and *Cyllamyces*. Positive correlations indicate relative abundance. Correlations with a $p$-value < 0.01 were considered statistically significant and are presented as blue circles, while correlation coefficient values that are insignificant are left blank (Figure 5B). In the correlation network (Figure 5C), only correlations with a correlation coefficient of 0.3 or higher are considered significant and are included in the network. The yellow-colored *Anaeromyces* in treatment 4F1 is more central in the network with thick edges (Figure 5C), indicating multiple stronger connections with other fungal genera, especially *Acremonium* in 11H. *Acremonium*, on the other hand, showed fewer connections, which indicates a limited association with other genera. The thick edges (lines) show short ecological distances, while longer distances (thin lines) reflect greater dissimilarity, i.e., weaker correlations.

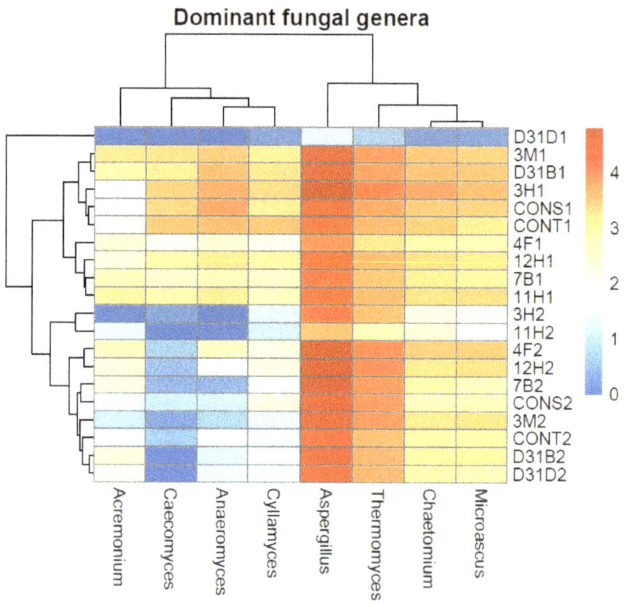


**Figure 3.** Relative abundance of dominant (≥1%) fungal phylotypes at the genus taxonomic rank before AD and after AD. Phylotypes with an average relative abundance of less than 1% and those unclassified at the genus level were excluded from the plot. The plot was generated using the average relative abundance for each group.

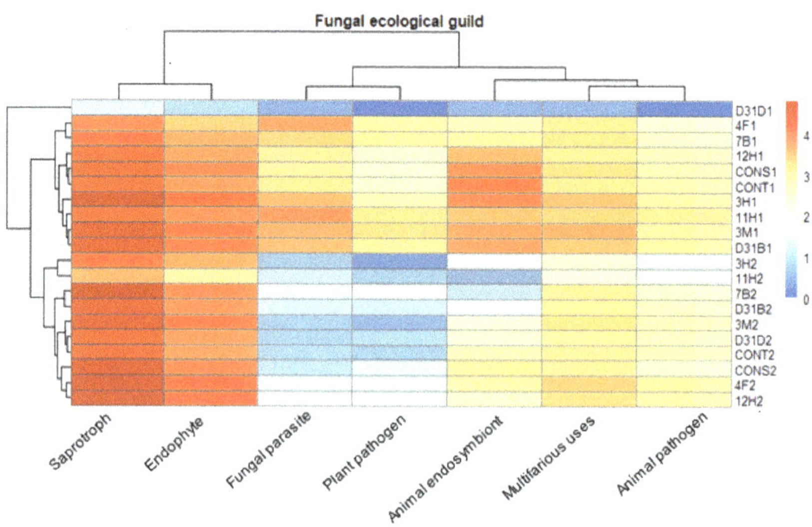


**Figure 4.** Ecological guild of fungi. Ecological guild was generated with FUNGuild database. Guild allotted as 'multifarious uses' comprises fungal sequences that were allotted as animal pathogens, plant pathogens, endophytes, and saprotrophs.

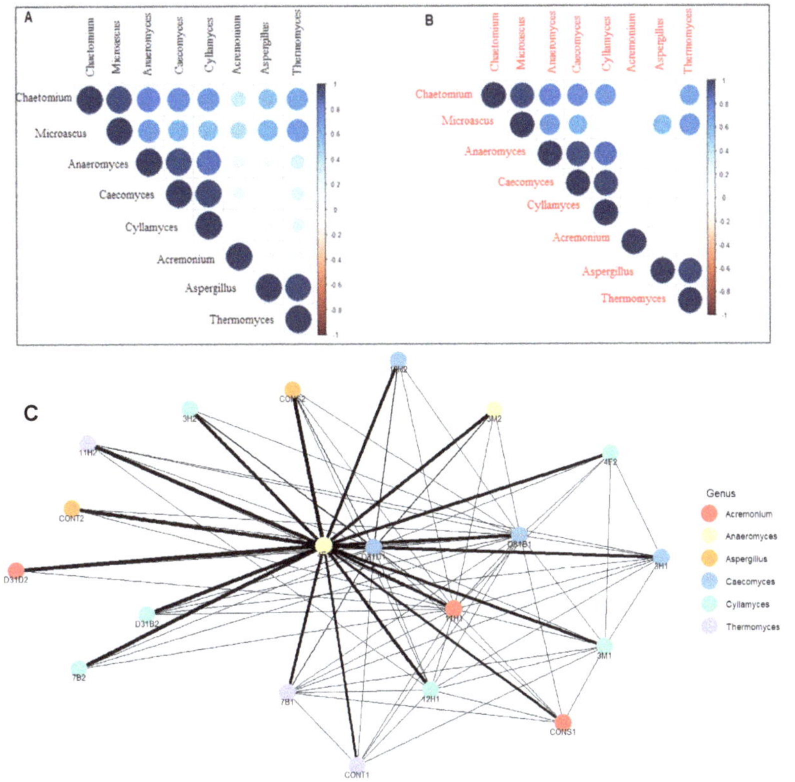


**Figure 5.** Correlation: (**A**) Correlogram showing the correlations among the dominant fungal genera. (**B**) Correlogram showing significance test for correlated fungi. (**C**) Correlation network analysis of fungal communities visualized at genus level.

The description of fungal richness and diversity showed that the richness of the fungi decreased after AD. However, some dominant genera such as *Cyllamyces*, *Caecomyces*, *Anaeromyces*, and *Acremonium* decreased after AD, corresponding to the information in Figure 2 and Appendix A. A decrease in the fungal community diversity was observed after AD (Figure 6) and the Shannon–Weiner index (H′) was used to measure the diversity within the fungal community, incorporating both species richness and evenness. It indicated moderate diversity, suggesting a more even distribution of fungi before AD (Figure 6). The after-AD treatments showed low diversity in composition as just a few fungi dominated the fungal community after AD, with the majority of fungi belonging to particular phyla, classes, or genera.

Communities with similar fungi were clustered together. Multivariate ordination methods explored the fungal community structure. Treatments that are closer together have comparable fungal structures (Figure 7), indicating the ecological distances between them. A separation of clusters between the treatments before AD and after AD was observed. However, treatment D31D1 clustered with the after-AD treatments, indicating similarity in their fungal diversity (Figure 7). The stress value of the non-metric multidimensional scaling (NMDS) plot, depicted in Figure 7, is 0.03. The stress plot, which provides additional information, can be referred to in Appendix C.

Insights into the fungal genera and AD dynamics over time suggest a change in the composition of the fungal community over time (days). In Figure 8, Aspergillus and

Caecomyces are positive associated with PC2 ($y$-axis), while Cyllamyces is negatively linked to PC2 (negative scores). Chaetomium and Acremonium are located closer to the origin, indicating weaker contributions compared to other genera. The top contributors to PC1 include day 23, day 26, and day 29, while the top contributors to PC2 are day 9 and day 32.

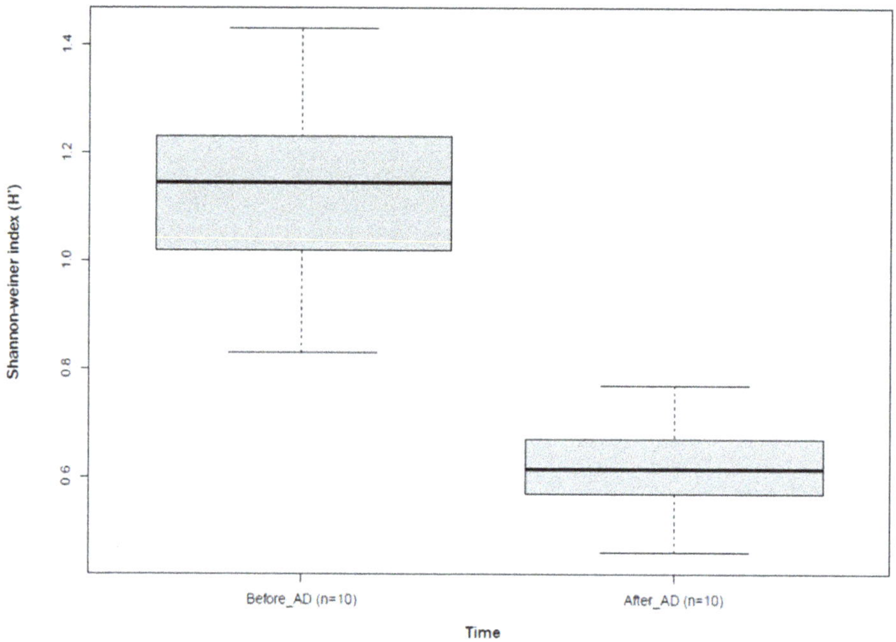


**Figure 6.** Description of fungal community diversity with Shannon–Weiner index.

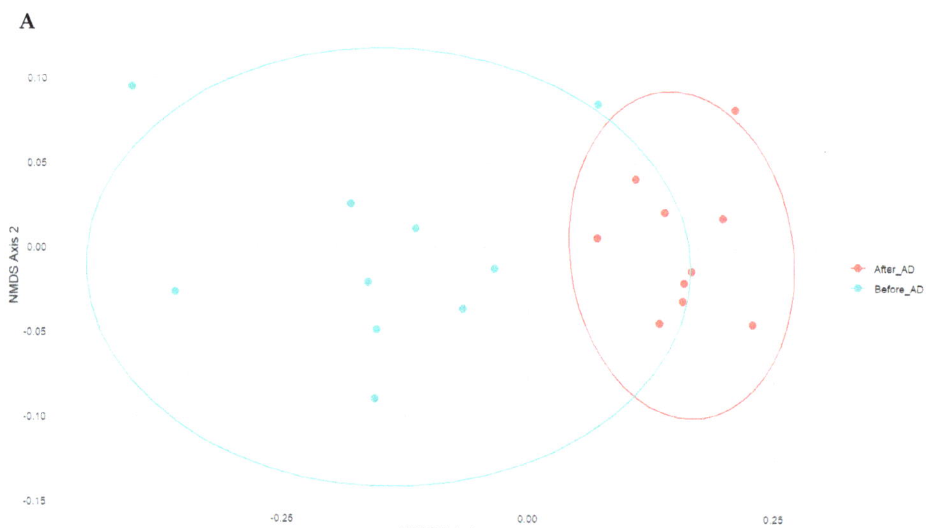


**Figure 7.** *Cont.*

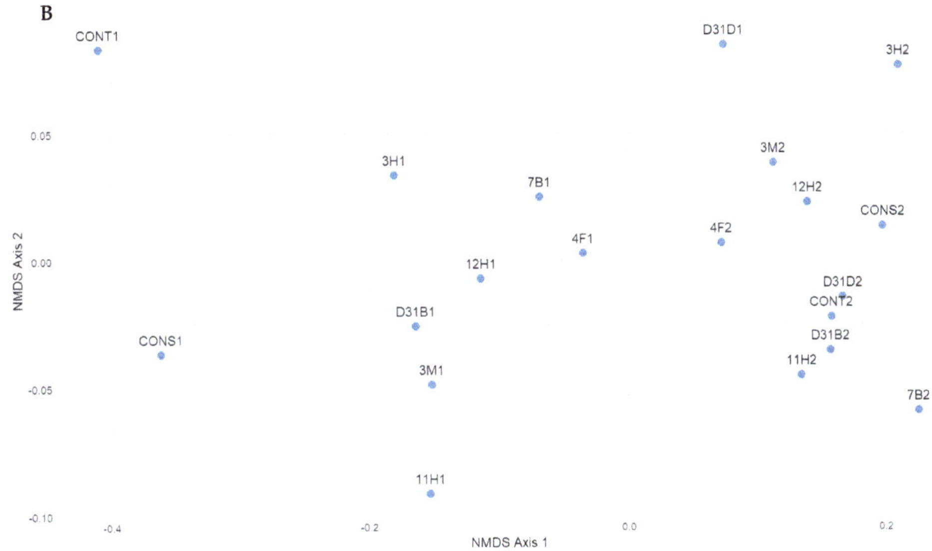


**Figure 7.** Bray–Curtis dissimilarity between fungal communities at the genus taxonomic rank. (**A**) NMDS to envisage the multivariate structure of the fungal communities at the genus taxonomic rank before and after AD. (**B**) NMDS showing the treatments the fungi belong to. The stress value of the NMDS plot is 0.03 (Appendix C).

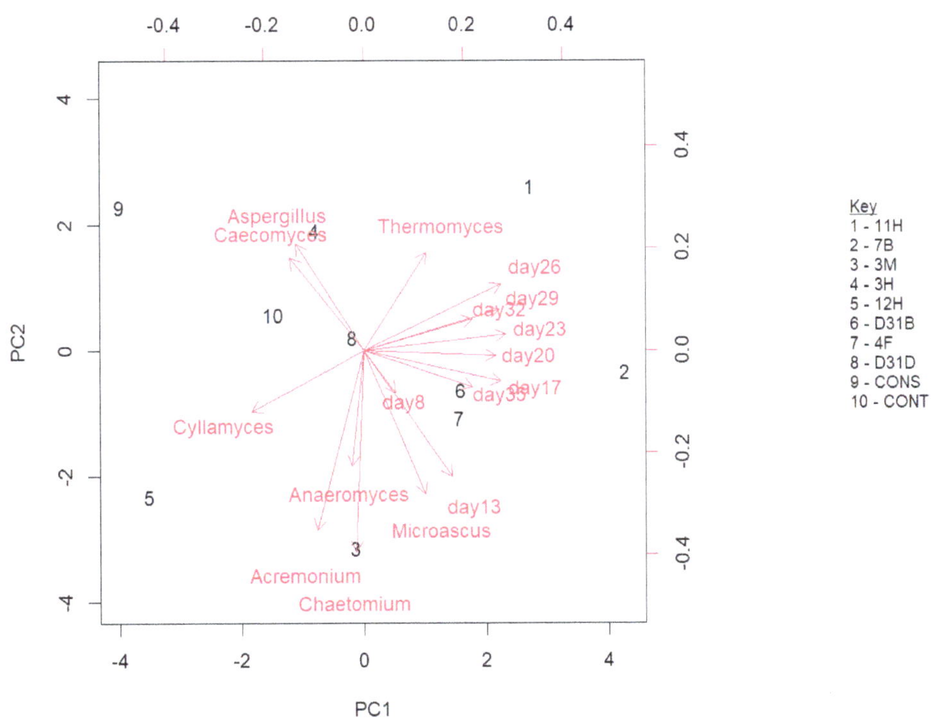


**Figure 8.** Principal component analysis (PCA) to envisage relationship between anaerobic digestion dynamics and fungi at genus taxonomic rank over time.

## 4. Discussion

There are limited studies on the core fungal community structures in bacterial-bioaugmented anaerobic digesters, despite fungi being an essential part of the microbiota involved in the anaerobic digestion of lignocellulosic substrates. In this study, the structure and composition of the basic fungal community of different bacterial-bioaugmented anaerobic digesters were investigated by sequencing of the *ITS* region. Presently, the *ITS* region has the greatest number of reference sequences in the GenBank and the use of *ITS2* as a molecular marker stems from its conservation within fungi. However, the existing reference sequences in the public database have a low rate of sequence identification for basal fungal lineages, making some taxa unrepresentative in the reference database [12,47].

Findings from the present study highlight the specific fungal taxa that thrive in the anaerobic digestion environment and suggest their potential significance in this process. Ascomycota was the dominant fungal phylum across all treatments. This observation aligns with the findings of Sun et al. [48], who also reported the predominance of Ascomycota-associated fungi in anaerobic digesters. Neocallimastigomycota was relatively abundant before AD but decreased after AD, and a similar trend was also observed for Basidiomycota. The decrease in the abundance of Neocallimastigomycota, a key anaerobic fungus, in the same treatment after AD could be attributed to a shift in the native anaerobic fungal community, which could be a response to changes in environmental or operational conditions and metabolic by-products as well as substrate availability [27,28]. Several studies have shown Neocallimastigomycota to be the predominant anaerobic fungal phylum [12,49,50]. The fungi in this group are known as obligate anaerobes, and their high abundance before AD is likely because their natural habitat is in the digestive system of herbivores such as cows [12,51]. This corresponds with the findings of Zhang et al. [52], who reported a decreased abundance of Neocallimastigomycota in the ruminal microbiota of ruminants upon the inclusion of lignocellulosic materials in their diet. However, it contradicts the findings of Langer et al. [2], as the phylum Neocallimastigomycota was absent in the analyzed anaerobic digesters with cow dung as part of the substrates. Neocallimastigomycota has been reported to degrade lignocellulose while co-existing with bacteria and methanogens during anaerobic digestion for the production of biogas [53]. The fungal phylum Basidiomycota was dominant before AD but reduced after AD in most treatments; however, the prevalence of Basidiomycota in the control treatments, both before and after AD, might be due to the lack of inoculated bacteria, thus affecting the fungal community dynamics in bioaugmented anaerobic digesters. Eurotiomycetes emerged as the dominant fungal class both before and after AD, indicating their resilience and adaptability within the system. Sordariomycetes and Dothideomycetes were relatively dominant across the treatments, further highlighting the importance of these fungal classes in AD. The dominant genera (which belonged solely to Ascomycota and Neocallimastigomycota) across all treatments included *Aspergillus*, *Thermomyces*, *Chaetomium*, and *Microascus*. However, this contrasts with the findings of Dollhofer et al. [49], who reported different genera, *Neocallimastix* and *Piromyces*, to be the most abundant genera in anaerobic digesters that had substrates such as sugar beets, silage grass, etc., and the digestion temperature ranged between 38 and 53 °C. Although there exists a similarity to this study in terms of the inclusion of substrates such as cow dung, the present study showed the relative abundance of *Neocallimastix* and *Piromyces* to be less than 1%. The reduced abundance of anaerobic fungal genera (*Anaeromyces, Cyllamyces, Caeomyces*) after AD may suggest limited substrate availability or a shift in microbial interactions within the AD environment. This shift in fungal composition suggests a dynamic response to the AD process.

The majority of fungi, as indicated by the ecological guild classification (Figure 4), were identified as saprophytes (saprotrophs), and their presence in anaerobic digesters has been previously studied [54]. The abundance of saprophytic fungi in the anaerobic digesters signifies the abundance of a diverse range of enzymes and subsequently enhanced organic matter decomposition. This highlights the functional alignment of fungal communities (saprophytes) for the degradation of organic matter. Saprophytes contribute to the stability of the AD process by promoting a balanced microbial community as they work in synergistic interaction with prokaryotic communities in anaerobic digesters [55]. *Aspergillus* sp., being saprophytes, demonstrated dominance across all treatments during AD. The abundance of saprophytes in the digester may not directly correlate with their metabolic efficiency since saprophytes typically thrive in aerobic environments. This is evident in the cumulative methane yield from the anaerobic co-digestion of water hyacinth and cow dung inoculated with pure bacteria isolates, as reported by Obi et al. [20]. The abundance of *Aspergillus* across all treatments suggests their potential to utilize nitrate as an oxidant in an ATP-generating process during AD. This allows for their proliferation but not optimal growth under oxygen-limited conditions [56,57]. Liu et al. [58] outlined the ability of different species of *Aspergillus* to produce different types of hydrolases that catalyze the rate-limiting phase (hydrolysis) of the AD of the lignocellulosic substrates. This further confirms their presence and potential activities in anaerobic digesters. However, the drastic reduction in their community diversity (Figure 7) after AD could be related to the limited oxygen concentration of the anaerobic digesters. Another guild that exhibited relatively high abundance after AD was the endophytes, which are non-pathogenic microorganisms. Their presence in the potential digestate indicates its suitability for use as a soil ameliorant. The reduced abundance of fungal parasites and pathogens in the treatments following AD suggests the potential benefits of AD in effectively mitigating or suppressing the presence of these pathogenic organisms, making the resulting digestate a safer option for utilization as a possible soil ameliorant. However, the resilience of the animal pathogen guild across treatments is a concern in managing zoonotic risks relating to waste management and agriculture. Only a few recent studies have explored fungi in anaerobic reactors [3,4,27,49]. To the best of our knowledge, this is one of the first studies to characterize the ecological guilds of fungal communities in bacteria-bioaugmented anaerobic digesters focusing on their potential suitability for agricultural and environmental applications.

The correlograms presented in Figure 5 offer a comprehensive visualization of the correlation patterns within the dataset. The predominance of positive correlations between *Thermomyces* and other dominant fungi is a key finding that suggests potential ecological and functional relationships. This analysis contributes to our understanding of the relationship between microorganisms in this specialized ecosystem and highlights areas for further exploration and research. The strong connection (thick lines in the correlation network analysis, Figure 5C) of fungal genus *Anaeromyces* in treatment 4F1 indicates ecological distances from other genera, specifically *Acremonium* in 11H. A short ecological distance and strong positive correlations (as indicated by the thick lines) portray the ability of the organisms involved to co-occur and respond to similar environmental conditions. The genera connected by thick lines are the key players within the network, likely exerting greater influence on the community dynamics than others due to their strong interactions. The diversity indices used to assess fungal richness and diversity revealed a decrease in both measures following AD, indicating a negative impact of AD on fungal communities. This observation suggests that the AD process may favor certain fungal taxa. This could affect the overall functional potential of the fungal microbiome. The treatment bioaugmented with 4F1 (Bacillus cereus) exhibited the most connections to other treatments. Although genera such as *Caecomyces* and *Cyllamyces* are connected to multiple treatments, their thin edges indicate a weak association with other linked genera. Treatments CONT2 and D31D1 showed the presence of different fungal communities due to their multiple connections, although these connections were weak. Less connected genera like *Thermomyces* and *Acremonium* show unique fungal taxa that thrive under special ecological conditions. The multivariate ordination methods applied in this study suggest a comprehensive view of the fungal community structure and its dynamics before and after AD. The observed clustering patterns and separation of treatments before and after AD highlight the ecological shifts that occur during this process. The clustering of treatment D31D1 (a 'before-AD' treatment) with the 'after-AD' treatments shows comparable fungi and emphasizes the ecological distances, highlighting the complexity of fungal interactions and community dynamics in specialized ecosystems like anaerobic digesters. These insights contribute to our understanding of the stability and adaptability of the fungal microbiome of anaerobic digesters and how fungal communities respond to and influence the AD process, paving the way for more targeted research into optimizing and managing these systems for various applications, including biogas production and waste management. The shift in the fungal community could be due to microbial interactions, as earlier suggested, or environmental changes in the digesters. The dominance of *Thermomyces* and *Caecomyces* during the later stages (day 26 and day 29) of AD indicated their possible metabolic roles towards the later stages of AD. The location of *Chaetomium* and *Acremonium* closer to the origin (Figure 8) shows their weaker contributions to the fungal dynamics compared to other genera. The contribution of day 23, day 26, and day 29 (Figure 1) to PC1 reflects the important stages of AD where a significant shift was observed in the digester performance or fungal community. The mid-to-late stages of AD (day 23–day 29) are fundamental to the microbial community dynamics that drive the variability in PC1 (Appendix E, Figure A5). This indicates periods of optimal changes in the digester. Day 9 and day 32 contributed strongly to PC2, indicating their link to distinct fungal community shifts, while low contributors to PC2 include days 29, 23, and 26 (Appendix F, Figure A6).

## 5. Conclusions

The findings of this study describe the distribution of fungi as well as the ecological guild, thereby contributing to our understanding of the community structure of the fungal ecosystem of anaerobic digesters. They also reflect the resilience and adaptability of Ascomycota and their ability to thrive in an unfavorable environment despite their ecological nature. The relative increase in Ascomycota after AD in non-bioaugmented digesters shows the potential of bacteria bioaugmentation in modeling the fungal community dynamics of anaerobic digesters. However, further research could employ absolute quantification methods such as qPCR to track changes in fungal population sizes over time. The study also revealed additional fungal phyla, beyond the well-known anaerobic Neocallimastigomycota, within anaerobic digesters. These findings expand our knowledge on the use of high-throughput metabarcoding approaches to explore the microbial ecology in specialized ecosystems, with implications for optimizing AD processes and harnessing the full potential of these systems for biogas production and waste management. However, a limitation of this study was that fungal community analysis was performed on pooled samples from three replicated digesters rather than each replicate separately. While the *ITS* region is widely used for fungal identification and phylogenetic studies, it is a possible limitation in this study as targeting other regions such as 28S rRNA or 18S rRNA could provide more information in the fungal community analysis. Exploring synergistic cross-kingdom interactions between bacteria and fungi is a major perspective for fungal research in AD field. This includes further investigation of fungal-driven pretreatment of lignocellosic substrates to enhance its AD for optimal biogas production. Incorporation of transcriptomic or proteomics approaches will give precision to improved strategies for enhancing the efficiency and stability of AD systems while promoting sustainable bioenergy production.

Further investigation could focus on the effect of a different potential bacterial/fungal inoculum on different substrates.

**Author Contributions:** L.U.O.: conceptualization, methodology, validation, investigation, writing—original draft, visualization, data curation, writing—review and editing; M.T.: writing—review and editing, supervision; A.R.: conceptualization, validation, resources, writing—review and editing, supervision, funding acquisition; H.L.: validation, methodology; J.Z.: validation, methodology; Y.W.: validation, methodology; Y.Z.: validation, methodology; R.A.A.: conceptualization, validation, resources, writing—review and editing, supervision, project administration, funding acquisition. All authors have read and agreed to the published version of the manuscript.

**Funding:** This work was funded by the Water Research Commission (WRC, grant no. K5/2543) of South Africa. The funding source had no role in the collection, design, analysis, data interpretation, writing, or the decision to submit the manuscript for publication.

**Data Availability Statement:** The raw sequences are available at https://www.ncbi.nlm.nih.gov/sra/PRJNA704473 (accessed on 1 March 2020).

**Acknowledgments:** We are grateful to the University of South Africa (UNISA) and the Agricultural Research Council (ARC), South Africa.

**Conflicts of Interest:** The authors declare no conflicts of interest.

## Appendix A

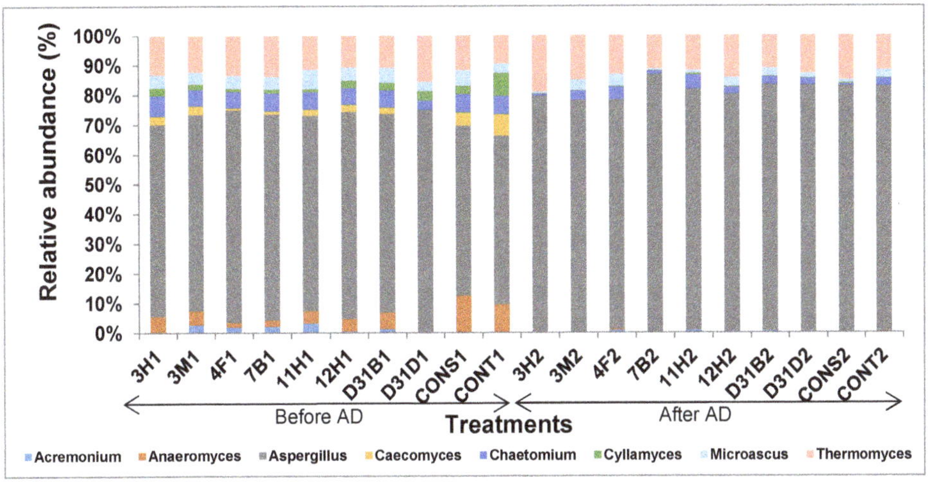


**Figure A1.** Relative abundance of dominant fungal class in treatments before and after batch anaerobic digestion. Unclassified phylotypes at the class level were omitted.

## Appendix B

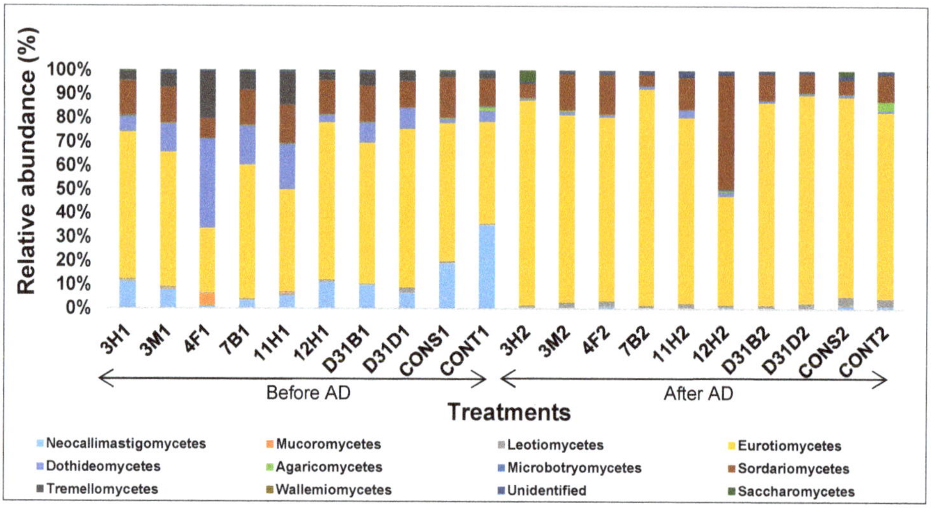


**Figure A2.** Relative abundance of dominant fungal genera in treatments before and after batch anaerobic digestion. Unclassified phylotypes at the genus level were omitted.

## Appendix C

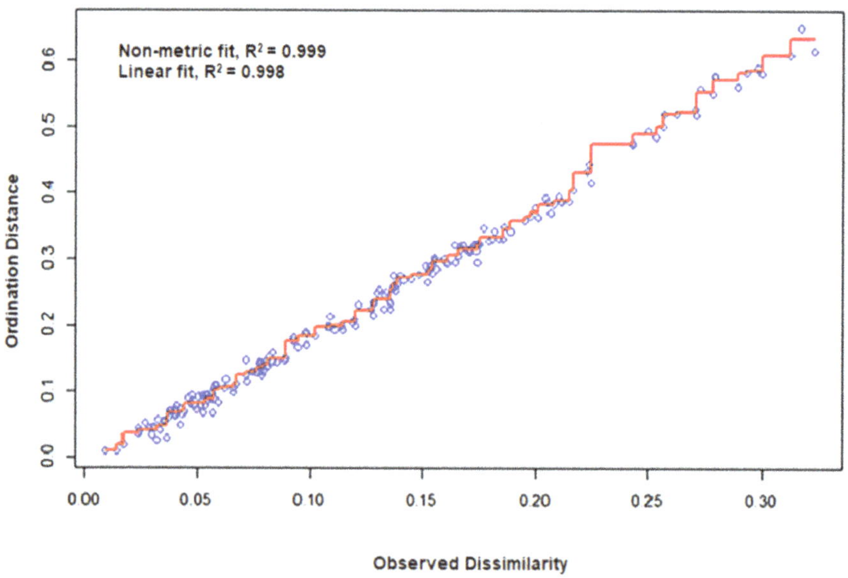


**Figure A3.** Stress plot for the NMDS plot in Figure 5. Plot was generated to ascertain the integrity of the ordination fit. The 'stressplot' function of the vegan package of the R software was used to create the stress plot.

## Appendix D

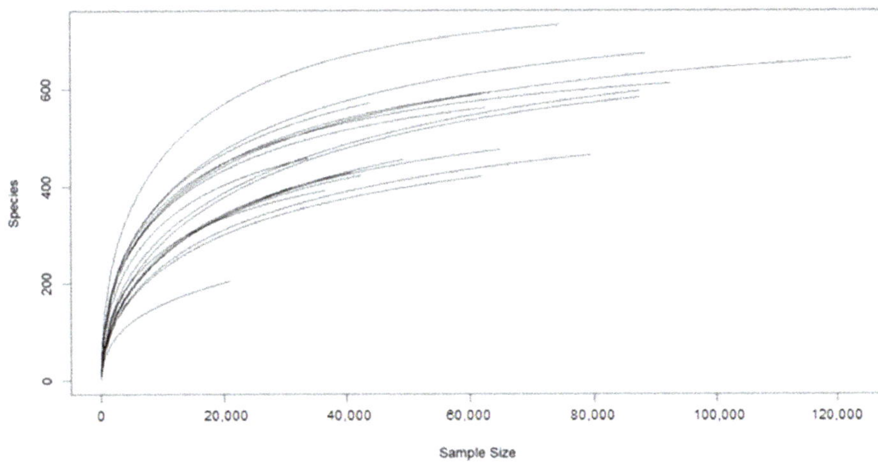


**Figure A4.** Rarefaction curve of fungi.

## Appendix E

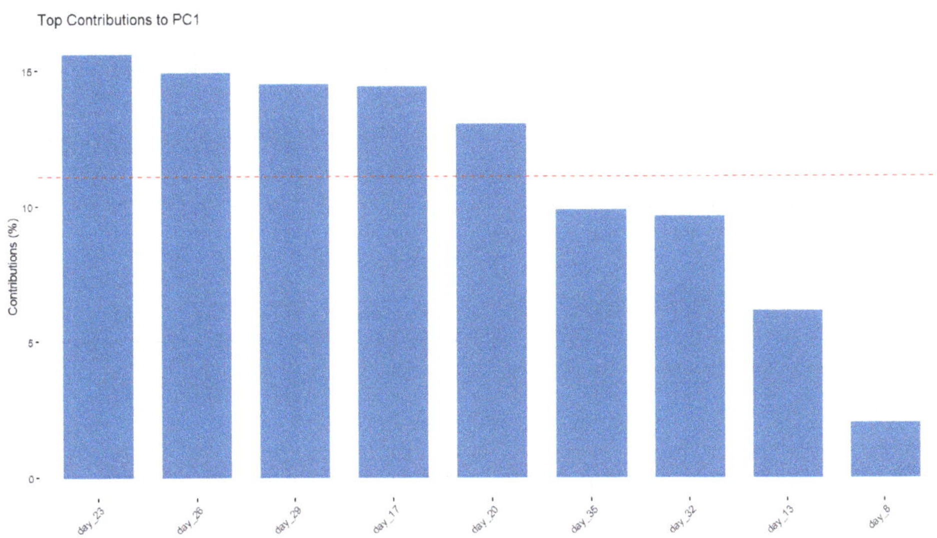


**Figure A5.** Top contributions of samples (days) to the second principal component (PC1) from the principal component analysis (PCA).

## Appendix F

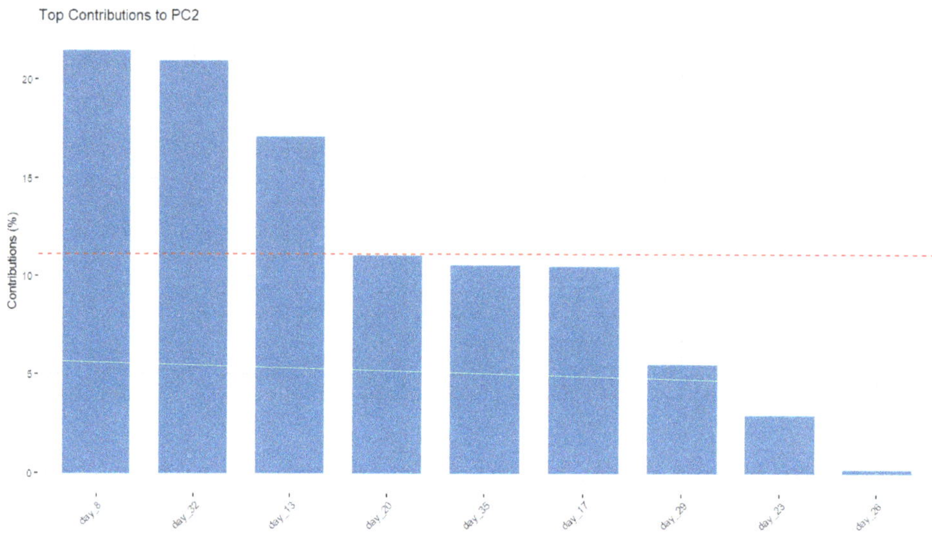


**Figure A6.** Top contributions of samples (days) to the second principal component (PC2) from the principal component analysis (PCA).

## References


1. Suman, A. Role of renewable energy technologies in climate change adaptation and mitigation: A brief review from Nepal. *Renew. Sustain. Energy Rev.* **2021**, *151*, 111524. [CrossRef]
2. Langer, S.G.; Gabris, C.; Einfalt, D.; Wemheuer, B.; Kazda, M.; Bengelsdorf, F.R. Different response of bacteria, archaea and fungi to process parameters in nine full-scale anaerobic digesters. *Microb. Biotechnol.* **2019**, *12*, 1210–1225. [CrossRef] [PubMed]
3. Young, D.; Dollhofer, V.; Callaghan, T.M.; Reitberger, S.; Lebuhn, M.; Benz, J.P. Isolation, identification and characterization of lignocellulolytic aerobic and anaerobic fungi in one- and two-phase biogas plants. *Bioresour. Technol.* **2018**, *268*, 470–479. [CrossRef] [PubMed]
4. Yang, X.; Zhang, Z.; Li, S.; He, Q.; Peng, X.; Du, X.; Feng, K.; Wang, S.; Deng, Y. Fungal dynamics and potential functions during anaerobic digestion of food waste. *Environ. Res.* **2022**, *212*, 113298. [CrossRef]
5. Seppälä, S.; Wilken, S.E.; Knop, D.; Solomon, K.V.; O'malley, M.A. The importance of sourcing enzymes from non-conventional fungi for metabolic engineering and biomass breakdown. *Metab. Eng.* **2017**, *44*, 45–59. [CrossRef]
6. Caruso, M.C.; Braghieri, A.; Capece, A.; Napolitano, F.; Romano, P.; Galgano, F.; Altieri, G.; Genovese, F. Recent updates on the use of agro-food waste for biogas production. *Appl. Sci.* **2019**, *9*, 1217. [CrossRef]
7. Ren, Y.; Yu, M.; Wu, C.; Wang, Q.; Gao, M.; Huang, Q.; Liu, Y. A comprehensive review on food waste anaerobic digestion: Research updates and tendencies. *Bioresour. Technol.* **2018**, *247*, 1069–1076. [CrossRef]
8. Tian, H.; Yan, M.; Treu, L.; Angelidaki, I.; Fotidis, I.A. Hydrogenotrophic methanogens are the key for a successful bioaugmentation to alleviate ammonia inhibition in thermophilic anaerobic digesters. *Bioresour. Technol.* **2019**, *293*, 122070. [CrossRef]
9. Gállego-Bravo, A.K.; García-Mena, J.; Piña-Escobedo, A.; López-Jiménez, G.; Gutiérrez-Castillo, M.E.; Tovar-Gálvez, L.R. Monitoring of a microbial community during bioaugmentation with hydrogenotrophic methanogens to improve methane yield of an anaerobic digestion process. *Biotechnol. Lett.* **2023**, *45*, 1339–1353. [CrossRef] [PubMed]
10. Bernardi, A.V.; de Gouvêa, P.F.; Gerolamo, L.E.; Yonamine, D.K.; Balico, L.d.L.d.L.; Uyemura, S.A.; Dinamarco, T.M. Functional characterization of GH7 endo-1,4-β-glucanase from Aspergillus fumigatus and its potential industrial application. *Protein Expr. Purif.* **2018**, *150*, 1–11. [CrossRef] [PubMed]
11. Greening, C.; Geier, R.; Wang, C.; Woods, L.C.; E Morales, S.; McDonald, M.J.; Rushton-Green, R.; Morgan, X.C.; Koike, S.; Leahy, S.C.; et al. Diverse hydrogen production and consumption pathways influence methane production in ruminants. *ISME J.* **2019**, *13*, 2617–2632. [CrossRef]
12. Vinzelj, J.; Joshi, A.; Insam, H.; Podmirseg, S.M. Employing anaerobic fungi in biogas production: Challenges & opportunities. *Bioresour. Technol.* **2020**, *300*, 122687. [CrossRef]

13. Sundberg, C.; Al-Soud, W.A.; Larsson, M.; Alm, E.; Yekta, S.S.; Svensson, B.H.; Sørensen, S.J.; Karlsson, A. 454 pyrosequencing analyses of bacterial and archaeal richness in 21 full-scale biogas digesters. *FEMS Microbiol. Ecol.* **2013**, *85*, 612–626. [CrossRef] [PubMed]
14. Abendroth, C.; Vilanova, C.; Günther, T.; Luschnig, O.; Porcar, M. Eubacteria and archaea communities in seven mesophile anaerobic digester plants in Germany. *Biotechnol. Biofuels* **2015**, *8*, 87. [CrossRef] [PubMed]
15. Langer, S.G.; Ahmed, S.; Einfalt, D.; Bengelsdorf, F.R.; Kazda, M. Functionally redundant but dissimilar microbial com-munities within biogas reactors treating maize silage in co-fermentation with sugar beet silage. *Microb. Biotechnol.* **2015**, *8*, 828–836. [CrossRef] [PubMed]
16. Westerholm, M.; Isaksson, S.; Karlsson Lindsjö, O.K.; Schnürer, A. Microbial community adaptability to altered temperature conditions determines the potential for process optimisation in biogas production. *Appl. Energy* **2018**, *226*, 838–848. [CrossRef]
17. Yang, S.; Li, L.; Peng, X.; Zhang, R.; Song, L. Eukaryotic community composition and dynamics during solid waste decomposition. *Appl. Microbiol. Biotechnol.* **2022**, *106*, 3307–3317. [CrossRef]
18. Tsapekos, P.; Kougias, P.; Vasileiou, S.; Treu, L.; Campanaro, S.; Lyberatos, G.; Angelidaki, I. Bioaugmentation with hydrolytic microbes to improve the anaerobic biodegradability of lignocellulosic agricultural residues. *Bioresour. Technol.* **2017**, *234*, 350–359. [CrossRef]
19. Obi, L.U.; Tekere, M.; Roopnarain, A.; Sanko, T.; Maguvu, T.E.; Bezuidenhout, C.C.; Adeleke, R.A. Whole genome sequence of Serratia marcescens 39_H1, a potential hydrolytic and acidogenic strain. *Biotechnol. Rep.* **2020**, *28*, e00542. [CrossRef] [PubMed]
20. Obi, L.U.; Tekere, M.; Roopnarain, A.; Adeleke, R.A. Bioaugmentation Strategies to Enhance Methane Production From Lignocellulosic Substrates: Dynamics Of The Prokaryotic Community Structure. In Proceedings of the 30th European Biomass Conference and Exhibition, Online, 9–12 May 2022.
21. Al Makishah, N.H.; Elfarash, A.E. Molecular characterization of cellulase genes in *Pseudomonas stutzeri*. *Electron. J. Biotechnol.* **2022**, *59*, 55–61. [CrossRef]
22. Baltaci, M.O.; Omeroglu, M.A.; Albayrak, S.; Adiguzel, G.; Adiguzel, A. Production of Endoglucanase by Exiguobacterium mexicanum OB24 Using Waste Melon Peels as Substrate. *An. Acad. Bras. Cienc.* **2022**, *94*, e20220151. [CrossRef]
23. Liao, Y.; Wu, S.; Zhou, G.; Mei, S.; Yang, Z.; Li, S.; Jin, Z.; Deng, Y.; Wen, M.; Yang, Y. Cellulolytic *Bacillus cereus* produces a variety of short-chain fatty acids and has potential as a probiotic. *Microbiol. Spectr.* **2024**, *12*, e0326723. [CrossRef]
24. Biswas, S.; Paul, D.; Bhattacharjee, A. Cellulolytic Potential of Lysinibacillus fusiformis Strain WGI4 Isolated From White Grub Beetle Phyllophaga sp. (Coleoptera: Scarabaeidae) Larvae Gut. *Proc. Zool. Soc.* **2024**, *77*, 116–125. [CrossRef]
25. Kumar, H.N.; Mohana, N.C.; Rakshith, D.; Abhilash, M.; Satish, S. Multicomponent assessment and optimization of the cellulase activity by Serratia marcescens inhabiting decomposed leaf litter soil. *Sustain. Chem. Pharm.* **2023**, *31*, 100951. [CrossRef]
26. Hu, X.; Yu, J.; Wang, C.; Chen, H. Cellulolytic bacteria associated with the gut of *Dendroctonus armandi* larvae (Coleoptera: Curculionidae: Scolytinae). *Forests* **2014**, *5*, 455–465. [CrossRef]
27. Dollhofer, V.; Dandikas, V.; Dorn-In, S.; Bauer, C.; Lebuhn, M.; Bauer, J. Accelerated biogas production from lignocellulosic biomass after pre-treatment with *Neocallimastix frontalis*. *Bioresour. Technol.* **2018**, *264*, 219–227. [CrossRef] [PubMed]
28. Stoyancheva, G.; Kabaivanova, L.; Hubenov, V.; Chorukova, E. Metagenomic Analysis of Bacterial, Archaeal and Fungal Diversity in Two-Stage Anaerobic Biodegradation for Production of Hydrogen and Methane from Corn Steep Liquor. *Microorganisms* **2023**, *11*, 1263. [CrossRef]
29. Mutungwazi, A.; Ijoma, G.N.; Matambo, T.S. The significance of microbial community functions and symbiosis in enhancing methane production during anaerobic digestion: A review. *Symbiosis* **2021**, *83*, 1–24. [CrossRef]
30. Miranda, F.M.; Azevedo, V.C.; Ramos, R.J.; Renard, B.Y.; Piro, V.C. Hitac: A hierarchical taxonomic classifier for fungal ITS sequences compatible with QIIME2. *BMC Bioinform.* **2024**, *25*, 228. [CrossRef]
31. Sneha, M.J.X.; Thangavel, M.; Mani, I.; Rajapriya, P.; Ponnuraj, N.; Pandi, M. Endophytic Fungal Diversity in *Hardwickia binata*: Bridging the Gap between Traditional and Modern Techniques. *Microbiol. Res.* **2024**, *15*, 823–840. [CrossRef]
32. Viotti, C.; Chalot, M.; Kennedy, P.G.; Maillard, F.; Santoni, S.; Blaudez, D.; Bertheau, C. Primer pairs, PCR conditions, and peptide nucleic acid clamps affect fungal diversity assessment from plant root tissues. *Mycology* **2024**, *15*, 255–271. [CrossRef] [PubMed]
33. Mukhuba, M.; Roopnarain, A.; Adeleke, R.; Moeletsi, M.; Makofane, R. Comparative assessment of bio-fertiliser quality of cow dung and anaerobic digestion effluent. *Cogent Food Agric.* **2018**, *4*, 1435019. [CrossRef]
34. White, T.J.; Bruns, T.; Lee, S.; Taylor, J. Amplification and direct sequencing of fungal ribosomal RNA genes for phylogenetics. In *PCR Protocols: A Guide to Methods and Applications*; Innis, M.A., Gelfand, D.H., Sninsky, J.J., White, T.J., Eds.; Academic Press: New York, NY, USA, 1990; Volume 18, pp. 315–322.
35. Tedersoo, L.; Bahram, M.; Põlme, S.; Kõljalg, U.; Yorou, N.S.; Wijesundera, R.; Ruiz, L.V.; Vasco-Palacios, A.M.; Thu, P.Q.; Suija, A.; et al. Global diversity and geography of soil fungi. *Science* **2014**, *346*, 1256688. [CrossRef]

36. Ezeokoli, O.T.; Mashigo, S.K.; Paterson, D.G.; Bezuidenhout, C.C.; Adeleke, R.A. Microbial community structure and relationship with physicochemical properties of soil stockpiles in selected South African opencast coal mines. *Soil Sci. Plant Nutr.* **2019**, *65*, 332–341. [CrossRef]
37. Gweon, H.S.; Oliver, A.; Taylor, J.; Booth, T.; Gibbs, M.; Read, D.S.; Griffiths, R.I.; Schonrogge, K. PIPITS: An automated pipeline for analyses of fungal internal transcribed spacer sequences from the Illumina sequencing platform. *Methods Ecol. Evol.* **2015**, *6*, 973–980. [CrossRef]
38. Zhang, M.; Sun, H.; Fei, Z.; Zhan, F.; Gong, X.; Gao, S. Fastq_clean: An optimized pipeline to clean the Illumina sequencing data with quality control. In Proceedings of the 2014 IEEE International Conference on Bioinformatics and Biomedicine (BIBM), Belfast, UK, 2–5 November 2014; pp. 44–48. [CrossRef]
39. Gordon, A.; Hannon, G. Fastx-Toolkit, FASTQ/A Short-Reads Preprocessing Tools. FASTX-Toolkit. [Computer software]. Available online: https://github.com/agordon/fastx_toolkit (accessed on 1 March 2020).
40. Bengtsson-Palme, J.; Ryberg, M.; Hartmann, M.; Branco, S.; Wang, Z.; Godhe, A.; De Wit, P.; Sánchez-García, M.; Ebersberger, I.; de Sousa, F.; et al. Improved software detection and extraction of ITS1 and ITS2 from ribosomal ITS sequences of fungi and other eukaryotes for analysis of environmental sequencing data. *Methods Ecol. Evol.* **2013**, *4*, 914–919. [CrossRef]
41. Rognes, T.; Flouri, T.; Nichols, B.; Quince, C.; Mahé, F. VSEARCH: A versatile open source tool for metagenomics. *PeerJ* **2016**, *2016*, e2584. [CrossRef]
42. Wang, Q.; Garrity, G.M.; Tiedje, J.M.; Cole, J.R. Naïve Bayesian Classifier for Rapid Assignment of rRNA Sequences into the New Bacterial Taxonomy. *Appl. Environ. Microbiol.* **2007**, *73*, 5261–5267. [CrossRef] [PubMed]
43. Kõljalg, U.; Nilsson, R.H.; Abarenkov, K.; Tedersoo, L.; Taylor, A.F.S.; Bahram, M.; Bates, S.T.; Bruns, T.D.; Bengtsson-Palme, J.; Callaghan, T.M.; et al. Towards a unified paradigm for sequence-based identification of fungi. *Mol. Ecol.* **2013**, *22*, 5271–5277. [CrossRef] [PubMed]
44. Nilsson, R.H.; Tedersoo, L.; Ryberg, M.; Kristiansson, E.; Hartmann, M.; Unterseher, M.; Porter, T.M.; Bengtsson-Palme, J.; Walker, D.M.; de Sousa, F.; et al. A comprehensive, automatically updated fungal ITS sequence dataset for reference-based chimera control in environmental sequencing efforts. *Microbes Environ.* **2015**, *30*, 145–150. [CrossRef]
45. Sha, S.P.; Suryavanshi, M.V.; Tamang, J.P. Mycobiome diversity in traditionally prepared starters for alcoholic beverages in India by high-throughput sequencing method. *Front. Microbiol.* **2019**, *10*, 348. [CrossRef] [PubMed]
46. Nguyen, N.H.; Song, Z.; Bates, S.T.; Branco, S.; Tedersoo, L.; Menke, J.; Schilling, J.S.; Kennedy, P.G. FUNGuild: An open annotation tool for parsing fungal community datasets by ecological guild. *Fungal Ecol.* **2016**, *20*, 241–248. [CrossRef]
47. Henske, J.K.; Gilmore, S.P.; Knop, D.; Cunningham, F.J.; Sexton, J.A.; Smallwood, C.R.; Shutthanandan, V.; Evans, J.E.; Theodorou, M.K.; O'malley, M.A. Transcriptomic characterization of Caecomyces churrovis: A novel, non-rhizoid-forming lignocellulolytic anaerobic fungus. *Biotechnol. Biofuels* **2017**, *10*, 305. [CrossRef]
48. Sun, W.; Yu, G.; Louie, T.; Liu, T.; Zhu, C.; Xue, G.; Gao, P. From mesophilic to thermophilic digestion: The transitions of anaerobic bacterial, archaeal, and fungal community structures in sludge and manure samples. *Appl. Microbiol. Biotechnol.* **2015**, *99*, 10271–10282. [CrossRef] [PubMed]
49. Dollhofer, V.; Callaghan, T.M.; Griffith, G.W.; Lebuhn, M.; Bauer, J. Presence and transcriptional activity of anaerobic fungi in agricultural biogas plants. *Bioresour. Technol.* **2017**, *235*, 131–139. [CrossRef]
50. Wilken, S.E.; Saxena, M.; Petzold, L.R.; O'malley, M.A. In silico identification of microbial partners to form consortia with anaerobic fungi. *Processes* **2018**, *6*, 7. [CrossRef]
51. Drake, H.; Ivarsson, M. The role of anaerobic fungi in fundamental biogeochemical cycles in the deep biosphere. *Fungal Biol. Rev.* **2018**, *32*, 20–25. [CrossRef]
52. Zhang, J.; Shi, H.; Wang, Y.; Li, S.; Cao, Z.; Ji, S.; He, Y.; Zhang, H. Effect of dietary forage to concentrate ratios on dynamic profile changes and interactions of ruminal microbiota and metabolites in holstein heifers. *Front. Microbiol.* **2017**, *8*, 2206. [CrossRef] [PubMed]
53. Thongbunrod, N.; Chaiprasert, P. Potential of enriched and stabilized anaerobic lignocellulolytic fungi coexisting with bacteria and methanogens for enhanced methane production from rice straw. *Biomass-Convers. Biorefinery* **2024**, *14*, 8229–8250. [CrossRef]
54. Alanbagi, R.A.; Alshuwaili, F.E.; Stephenson, S.L. Fungi associated with forest floor litter in northwest Arkansas. *Curr. Res. Environ. Appl. Mycol.* **2019**, *9*, 25–35. [CrossRef]
55. Srikanth, M.; Sandeep, T.S.R.S.; Sucharitha, K.; Godi, S. Biodegradation of plastic polymers by fungi: A brief review. *Bioresour. Bioprocess.* **2022**, *9*, 1–10. [CrossRef] [PubMed]
56. Taubitz, A.; Bauer, B.; Heesemann, J.; Ebel, F. Role of respiration in the germination process of the pathogenic mold Aspergillus fumigatus. *Curr. Microbiol.* **2007**, *54*, 354–360. [CrossRef]

57. Kumar, M.; Kumar, H.; Topno, R.K.; Kumar, J. Analysis of impact of anaerobic condition on the aflatoxin production in Aspergillus parasiticus Speare. *Agric. Sci. Dig. A Res. J.* **2019**, *39*, 75–78. [CrossRef]
58. Liu, X.; Jiang, Z.; Ma, S.; Yan, Q.; Chen, Z.; Liu, H. High-level production and characterization of a novel β-1,3-1,4-glucanase from Aspergillus awamori and its potential application in the brewing industry. *Process. Biochem.* **2020**, *92*, 252–260. [CrossRef]

Journal of
*Fungi*

MDPI

*Article*

# Study on the Biological Characteristics of Dark Septate Endophytes under Drought and Cadmium Stress and Their Effects on Regulating the Stress Resistance of *Astragalus membranaceus*

**Duo Wang [1], Yali Xie [1], Wanyi Zhang [1], Li Yao [1], Chao He [2,*] and Xueli He [1,*]**

[1] School of Life Sciences, Hebei University, Baoding 071002, China; wd18740746653@163.com (D.W.); xieyali998@163.com (Y.X.); zwaner041117@126.com (W.Z.); yaoli202202@163.com (L.Y.)

[2] Institute of Medicinal Plant Development, Chinese Academy of Medical Sciences & Peking Union Medical College, Beijing 100193, China

* Correspondence: hc891215@126.com (C.H.); xlh3615@126.com (X.H.)

**Abstract:** *Astragalus membranaceus* is a famous traditional medicinal plant. However, drought and cadmium (Cd) pollution are the main abiotic stress factors that affect plant growth and yield and the ability to improve the host's stress resistance through the use of beneficial endophytic fungi. To evaluate the tolerance of dark septate endophytes (DSE) to various abiotic stresses, 10 DSE strains [*Microsphaeropsis cytisi* (*Mc*), *Alternaria alstroemeriae* (*Aa*), *Stagonosporopsis lupini* (*Sl*), *Neocamarosporium phragmitis* (*Np*), *Paraphoma chlamydocopiosa* (*Pc*), *Macrophomina phaseolina* (*Mp'*), *Papulaspora equi* (*Pe*), *Alternaria tellustris* (*At*), *Macrophomina pseudophaseolina* (*Mp*), and *Paraphoma radicina* (*Pr*)] were investigated under different drought and Cd stressors in vitro by using solid-plate cultures and liquid-shaker cultures in the current study. The experiments involved using varying concentrations of PEG (0, 9, 18, and 27%) and $Cd^{2+}$ (0, 25, 50, and 100 mg/L) to simulate different stress conditions on DSE. Additionally, the effect of DSE (*Np* and *At*) on the growth of *A. membranaceus* at different field water capacities (70% and 40%) and at different $CdCl_2$ concentrations (0, 5, 10, and 15 mg Cd/kg) in soil was studied. The results demonstrated that the colony growth rates of *Aa*, *Np*, *Pc*, *Mp'*, and *Mp* were the first to reach the maximum diameter at a PEG concentration of 18%. *Aa*, *Np*, and *At* remained growth-active at 100 mg Cd/L. In addition, *Aa*, *Np*, and *At* were selected for drought and Cd stress tests. The results of the drought-combined-with-Cd-stress solid culture indicated that the growth rate of *Np* was significantly superior to that of the other strains. In the liquid culture condition, the biomasses of *Np* and *Aa* were the highest, with biomasses of 1.39 g and 1.23 g under the concentration of 18% + 25 mg Cd/L, and *At* had the highest biomass of 1.71 g at 18% + 50 mg Cd/L concentration, respectively. The CAT and POD activities of *Np* reached their peak levels at concentrations of 27% + 50 mg Cd/L and 27% + 25 mg Cd/L, respectively. Compared to the control, these levels indicated increases of 416.97% and 573.12%, respectively. *Aa*, *Np*, and *At* positively influenced SOD activity. The glutathione (GSH) contents of *Aa*, *Np*, and *At* were increased under different combined stressors of drought and Cd. The structural-equation-modeling (SEM) analysis revealed that *Aa* positively influenced biomass and negatively affected Cd content, while *Np* and *At* positively influenced Cd content. Under the stress of 40% field-water capacity and the synergistic stress of 40% field-water capacity and 5 mg Cd/kg soil, *Np* and *At* significantly increased root weight of *A. membranaceus*. This study provides guidance for the establishment of agricultural planting systems and has good development and utilization value.

**Keywords:** dark septate endophyte; performance; tolerance stress; *Astragalus membranaceus*; drought stress; cadmium pollution

## 1. Introduction

Drought and heavy-metal pollution are the most prevalent and detrimental abiotic stressors that significantly impede plant growth and survival [1–3]. As a non-essential element for plants, cadmium (Cd) is recognized as one of the most toxic heavy-metal pollutants, with the potential to degrade agricultural product quality, diminish yields, and contaminate the food chain, thereby posing a significant threat to human health. Soil Cd levels can vary widely, particularly in China, ranging from 0.003 to 9.57 mg/kg [3]. The compounding effects of drought and Cd stress, which are frequently encountered in many regions, further amplify the challenges for plant life [4,5]. Such dual stressors can lead to nutritional imbalances, a reduction in chlorophyll synthesis, and the disruption of enzymatic activities and metabolic functions, all of which severely impact plant growth and agricultural productivity. It is evident that the study of single stress factors is insufficient to address the demands of sustainable agricultural development, necessitating a comprehensive understanding of the synergistic effects of multiple abiotic stressors.

*Astragalus membranaceus*, a renowned traditional medicinal plant, is celebrated for its dried roots, which are rich in bioactive compounds, including flavonoids, saponins, and polysaccharides. These constituents are known to mitigate the effects of hyperglycemia, exhibit anti-aging properties, and possess anti-tumor activities [6]. Beyond its medicinal significance, *A. membranaceus* demonstrates resilience to harsh environmental conditions such as drought and cold, contributing to ecological benefits through its windbreak and sand-fixation capabilities, thereby enhancing the ecological environment. However, the prevalence of concurrent drought and Cd contamination in various regions poses a significant challenge, necessitating research into the plant's adaptability and potential remediation strategies in the face of these dual stressors. Research reports have underscored a concerning issue regarding the quality of Chinese medicinal materials, revealing that eight types, including *A. membranaceus*, have been identified as having Cd content exceeding permissible levels in 27 provinces and regions across China, with an alarming over-standard rate of 17.73%—1.4 times the established Cd limit standard. This issue is particularly pronounced in Northwest China, characterized by low precipitation and high evaporation rates, where as much as 20.00% of *A. membranaceus* samples from Shaanxi province have been found to exceed the Cd safety threshold [7]. Addressing the enhancement of plant tolerance to the combined stressors of drought and Cd contamination is thus an urgent challenge in the field. Studies have suggested that the introduction of beneficial microbes [8], such as dark septate endophytes (DSE), may offer a promising avenue for improving plant stress resilience and overall growth, warranting further investigation for sustainable agricultural practices and medicinal plant cultivation.

DSEs constitute a diverse group of ascomycetes fungi known for their ability to colonize the roots of living plants without inducing any apparent detrimental effects. Characterized by the presence of microsclerotia and pigmented hyphae, these endophytic fungi are ubiquitous in nearly all natural ecosystems, particularly thriving under challenging conditions such as those found in saline, polluted, and arid environments [9]. Empirical research has demonstrated that DSEs can significantly enhance the host plants' water absorption capacity, promote growth, and bolster their stress tolerance [10,11]. Under drought conditions, DSEs can form a complex and continuous network within the plant's root system to improve underground and above-ground water transportation and enhance the drought resistance of plants [12]. Under heavy-metal pollution conditions, DSEs hinder the migration of heavy-metal ions in plants by their adsorption abilities and improve plant tolerance to heavy-metal stress [13,14].

DSEs have currently been reported to play a role in modulating the physiological and biochemical stress responses in plants, but the existing literature primarily focuses on their effects under single-stress conditions. For example, under drought stress, DSEs have been shown to augment the biomass and active compound levels in *Isatis indigotica*, as well as to positively influence the morphology, biomass, physiological traits, and bioactive content of licorice plants. The inoculation of DSEs in wheat under drought conditions not only promotes growth but also diminishes the water consumption of seedlings, thereby facilitating water-efficient agricultural practices [15–17]. In the context of heavy-metal stress, DSEs have been demonstrated to significantly increase the biomass and height of maize plants exposed to Cd, with higher Cd sequestration in the root cell walls of DSE-inoculated plants, indicating an enhanced tolerance to Cd. The underlying mechanism involves alterations in root morphology and the facilitation of Cd binding to cell walls and phosphates. Furthermore, DSEs have been reported to enhance Cd tolerance in rice, reducing its accumulation in the root system and translocation to the stem. This process is associated with the upregulation of the SNARE Syntaxin 1 gene, which is implicated in the regulation and alleviation of Cd accumulation by DSEs [18–20].

However, plants are often subjected to the combined stresses of drought and Cd pollution. The screening of DSE strains with combined resistance to drought and Cd is not only crucial for application in phytoremediation but also for the regulation of plant stress resistance. At present, there have been few studies on DSE-resistant strains; however, the screening and application of DSE-resistant strains will be the main research direction and goal in the future [8,21]. To elucidate the mechanisms underlying the tolerance of DSEs to synergistic environmental stresses, this study designed in vitro assays incorporating both single-factor and compound-factor conditions, utilizing varying concentrations of PEG-6000 to simulate osmotic stress and a range of Cd concentrations to simulate heavy-metal stress. These assays were conducted to assess the influence of combined stressors on the growth performance of ten distinct DSE strains. Concurrently, field experiments were established to examine the impact of DSE under different soil water capacities and Cd levels on the growth of *A. membranaceus*. The objectives of this study encompass the exploration of the following topics: (1) How does DSE tolerate drought stress in vitro? (2) How does DSE tolerate Cd stress in vitro? (3) How does DSE tolerate combined drought and Cd stress in vitro? (4) What is the effect of DSE on the growth of *A. membranaceus* under drought and Cd single and synergistic stresses?

## 2. Materials and Methods

### 2.1. Fungal Materials of DSE

In this study, 10 DSE strains were isolated from the roots of different plants (Table 1) and isolated from the fine roots of *Glycyrrhiza uralensis*, *Isatis indigotica*, *Astragalus membranaceus*, *Lycium ruthenicum*, *Dendranthema morifolium*, and *Salvia miltiorrhiza* [22–27].

**Table 1.** List of DSE fungal strains used in this study.

| Strain | Host Plant | Geographic Location | Acc. nr |
|---|---|---|---|
| *Microsphaeropsis cytisi* | *Glycyrrhiza uralensis* | Chifeng in Inner Mongolia Autonomous Region | ON413886 |
| *Stagonosporopsis lupini* | *Glycyrrhiza uralensis* | Chifeng in Inner Mongolia Autonomous Region | OP363602 |
| *Neocamarosporium phragmitis* | *Glycyrrhiza uralensis* | Jingtai, Gansu Province | ON413888 |
| *Alternaria alstroemeriae* | *Isatis indigotica* | Anguo, Hebei Province | MZ505449 |
| *Macrophomina pseudophaseolina* | *Astragalus membranaceus* | Anguo, Hebei Province | MZ506881 |
| *Paraphoma radicina* | *Astragalus membranaceus* | Anguo, Hebei Province | MT723853 |
| *Papulaspora equi* | *Lycium ruthenicum* | Minqin, Gansu Province | MW548086 |
| *Alternaria tellustris* | *Lycium ruthenicum* | Anxi, Gansu Province | OM936046 |
| *Paraphoma chlamydocopiosa* | *Dendranthema morifolium* | Anguo, Hebei Province | MT723851 |
| *Macrophomina phaseolina* | *Salvia miltiorrhiza* | Yaozhou, Shaanxi Province | OR434038 |

Root segments of these plants were selected for surface disinfection. They were rinsed several times with sterile water, sterilized with 75% ethanol for 5 min, sterilized with sodium hypochlorite for 5 min, rinsed several times with sterile water, and dried on sterile filter paper. The root segments were cultured in potato-dextrose-agar medium (PDA medium, selected from Beijing Aoboxing Bio-Technology Co., Ltd., Beijing, China) with antibiotic supplements (ampicillin and streptomycin sulfate) in Petri dishes. Sterilized root samples were incubated in the dark at 27 °C and observed daily. When black mycelia grew around the surface of the root segment, the growth condition was good. Fresh mycelia were selected from a super-clean worktable, placed in a new PDA medium for purification, and then cultured in the dark at 27 °C. Fresh mycelium (approximately 50 mg) was scraped from each colony, and colony DNA was extracted with a genomic DNA isolation kit (Beijing Suolaibao Technology Co., Ltd., SolarBio, Beijing, China). The primers ITS4 (5′-TCCTCCGCTTATTGATATGC-3′) and ITS5 (5′-GGAAGTAAAAGTCGTAACAAGG-3′) were applied to amplify the colony DNA, and the amplified products were sequenced. The BLAST tool in NCBI (https://www.ncbi.nlm.nih.gov/, 2 May 2024) was used for comparison, and a phylogenetic tree was constructed using the MEGA version 6.0 software [28]. The ITS1-5.8S-ITS2 ribosomal DNA sequences of these DSE strains were uploaded to GenBank with the following specific accession numbers, and their names were *Microsphaeropsis cytisi* (*Mc*), *Alternaria alstroemeriae* (*Aa*), *Stagonosporopsis lupini* (*Sl*), *Neocamarosporium phragmitis* (*Np*), *Paraphoma chlamydocopiosa* (*Pc*), *Macrophomina phaseolina* (*Mp′*), *Papulaspora equi* (*Pe*), *Alternaria tellustris* (*At*), *Macrophomina pseudophaseolina* (*Mp*), and *Paraphoma radicina* (*Pr*), respectively. The strains were stored at 4 °C in the Mycorrhizal Biology Laboratory of Hebei University.

### 2.2. Growth Conditions

The seeds of *A. membranaceus* were collected from the Gansu Province of China and stored at 4 °C. Uniform and full *A. membranaceus* seeds were selected and rinsed three times with distilled water. The seeds were soaked in distilled water for 12 h, placed in a seedling tray with a small amount of distilled water, then placed in an incubator at 25 °C under shade and set aside. The seeds germinated in incubators for one week before cultivation. The growth substrate used was a mixture of 1:2 (*W*:*W*) sand (less than 2 mm) and soil. A completely randomized three-factor (three DSE inoculation treatments × two drought stress treatments × four Cd concentrations) block group experimental design was used. Factor 1 was the DSE strain, where the inoculation treatments included two distinct strains, designated as *Np* and *At*, alongside a blank control group that received no DSE inoculation. Factor 2 was the soil moisture treatment. One simulating normal moisture levels was set at 70% field water capacity, and the other induced drought stress at 40% field water capacity. These treatments were designed to evaluate the plants' responses to varying water availabilities. Additionally, Factor 3 introduced a soil Cd stress treatment, where we prepared solutions of cadmium chloride ($CdCl_2$) at concentrations of 67.5 mg/L, 135 mg/L, and 202.5 mg/L and added 100 mL of each to the soil substrate to create a gradient of Cd solution, and the blank control group was added with an equal amount of distilled water. Four levels of Cd stress were set at 0, 5, 10, and 15 mg Cd/kg, respectively. Four replicates were set up for each treatment group, with a total of 96 pots. Different single and synergistic stress-treatment groups were established (Table 2).

**Table 2.** Experimental treatment scheme.

| Groups | Treatment Group | Methods |
|---|---|---|
| Control group | CK | 70% field water capacity |
| | Cd1 | 5 mg Cd/kg soil |
| Cd stress group | Cd2 | 10 mg Cd/kg soil |
| | Cd3 | 15 mg Cd/kg soil |
| Drought stress group | D | 40% field water capacity |
| | DCd1 | 40% field water capacity and 5 mg Cd/kg soil |
| Drought-Cd interaction stress group | DCd2 | 40% field water capacity and 10 mg Cd/kg soil |
| | DCd3 | 40% field water capacity and 15 mg Cd/kg soil |
| Inoculation group | *Np* | Inoculation with *N. phragmitis* |
| | *At* | Inoculation with *A. tellustris* |
| Non-inoculation group | *C* | No DSE inoculation |

A total of 800 g of mixed substrate was weighed and placed in plastic pots (mouth diameter of 13 m, bottom diameter of 10 m, height of 12 cm); a 7 mm diameter hole punch was used to intercept the fungus cake from the PDA medium, and four pieces were placed on the top layer of the substrate in each pot and finally covered with 550 g of mixed substrate. Four pieces of blank PDA medium were taken in the same way as the DSE blank control group [29]. Four well-grown and uniform *A. membranaceus* seedlings were selected for planting in each pot, and after planting, all potted plants were placed in an artificial climate incubator with a photoperiod of 14/10 h, 24/22 °C (day/night) and an average relative humidity of 60%. The experiment started on 25 February 2024 and concluded with the harvest on 25 June 2024 for a period of 4 months. After 45 d, *A. membranaceus* was treated with various stress treatments. During the experimental period, water loss was regularly replenished with distilled water, and soil moisture was maintained by regular weighing. The position of the seedling pots was randomly changed weekly to ensure that they were not affected by positional effects. The height of each plant before harvest was recorded. Above-ground plants and underground roots were harvested separately. The roots were rinsed in tap water and then three times in distilled water, drained, and weighed.

### 2.3. Drought Tolerance of DSE Strains

A 120 mL PDA medium was prepared and amended with polyethylene glycol (PEG-6000) and phytagel to imitate a controlled osmotic-stress environment. The ratio of PEG-6000 to phytagel was maintained at 17:1, with incremental additions of 0, 11.87, 26.34, and 44.38 g of PEG-6000 to establish a range of osmotic pressures, corresponding to PEG mass fractions of 0%, 9%, 18%, and 27% [30]. The medium was heated and agitated to ensure the uniform distribution of PEG-6000. Following a 14-day cultivation period, a 5 mm diameter inoculum of DSE colonies was extracted from the peripheral region and placed at the center of the PDA medium with varying drought-stress gradients. Subsequently, the inoculated plates were incubated at 27 °C for 14 d in a dark, inverted incubator. Each treatment was replicated three times to ensure experimental rigor.

### 2.4. Cd Tolerance of DSE Strains

The $CdCl_2 \cdot 2.5H_2O$ was prepared with a Cd ion concentration of 10 mg/mL. Each 250 mL conical bottle was filled with 100 mL PDA medium, and 10 mg/mL of Cd ion master batch was added at 0, 0.25, 0.5, and 1 mL successively. The concentration gradients of Cd ion were set as 0, 25, 50, and 100 mg/L [31]. After the DSE colonies were cultivated for 14 d, a 5 mm diameter fungal inoculum was taken from their periphery. The medium, adjusted to reflect a spectrum of Cd concentration gradients, was centrally positioned for uniformity. Subsequent to inoculation, the cultures were maintained under a 14-day incubation regimen at 27 °C within a dark, inverted incubator, ensuring consistent and controlled environmental conditions. Each experimental condition was replicated three times.

### 2.5. Drought and Cd Tolerance of DSE Strains

#### 2.5.1. Solid Culture of DSE

Drought-and-Cd-combined stress was simulated by adding PEG-6000 and Cd ion master batch at a concentration of 10 mg/mL to the PDA medium at different ratios, and the combined stress concentration gradients were set to 0, 9% + 25 mg Cd/L, 9% + 50 mg Cd/L, 9% + 100 mg Cd/L, 18% + 25 mg Cd/L, 18% + 50 mg Cd/L, 18% + 100 mg Cd/L, 27% + 25 mg Cd/L, 27% + 50 mg Cd/L, and 27% + 100 mg Cd/L [30]. Colonies of DSE, after 14 d of cultivation, were sampled for a fungal inoculum of 5 mm in diameter from their periphery. These were placed at the center of the PDA medium, which was prepared with varying gradients of drought-and-Cd-combined stress concentrations. The inoculated medium was then incubated at 27 °C for a period of 14 d in a dark, inverted incubator. A control group (CK), representing conventional culture conditions without any additional treatment, was included alongside ten experimental treatments, each with three replicates.

#### 2.5.2. Liquid Culture of DSEs

Each 250 mL conical flask contained 120 mL of Potato Dextrose Water liquid medium (PD liquid medium, sourced from Qingdao Hope Bio-Technology Co., Ltd., Qingdao, China). To this medium, a PEG solid and a 10 mg/mL master batch of Cd ions were incorporated in varying proportions to achieve a consistent level of combined stress as previously described. After the DSE colonies were cultivated for 14 d, three fungal inoculations (5 mm in diameter) were removed from their periphery, inoculated into a PD liquid medium, and incubated on a shaker at a constant temperature (27 °C, 150 r/min) for 14 d. A control group (CK), representing conventional culture conditions without any additional treatment, was included alongside ten experimental treatments, each with three replicates. Following incubation in liquid media, the mycelium and culture liquid were separated using a SHD-III type circulating water multi-purpose vacuum pump. Randomly, two sections of the fresh mycelia were separated. One portion was utilized to measure the levels of melanin, soluble protein, superoxide dismutase (SOD), glutathione (GSH), malondialdehyde (MDA), peroxidase (POD) activity, and catalase (CAT). To determine the Cd content and soluble sugar, when the other portion reached a consistent weight, it was weighed and dried at 80 °C.

### *2.6. DSE Growth and Cd Content*

The growth of DSE colonies under various conditions of drought and Cd stress, both individually and in combination, was monitored by employing the cross-measurement technique to determine the colony diameter. This assessment was conducted every 24 h over a 14-day period [32]. The tolerance index (TI) of the DSE strain was calculated by measuring the colony growth after treatment (cm) divided by the control colony growth (cm) [33]. The concentration causing a 50% growth inhibition (IC50) of DSE strains was calculated by the method of Medina-Armijo et al. [31]. On the 8th day of culture, photos were taken to record the colony morphology. The total biomass of DSE strains was obtained by converting the water content ratio of part of the mycelium, and the total biomass = total fresh weight × dry weight/partial fresh weight.

The Cd content of the dry mycelium was determined using an inductively coupled plasma optical emission spectrometer (ICP-OES, Beijing Jianling Technology Co., Ltd., Beijing, China). A grey or white residue was formed by heating the dried hyphae to 550 °C in a crucible that had been weighed. HClO4 (1 mL) was added. The acid treatment process was repeated after heating the hyphae on a heat plate. Concentrated HNO3 in the amount of 1 mL was used to dissolve the remainder, diluted and fixed to 25 mL (grade 1 water) [14].

### *2.7. Determination of DSE Antioxidant Enzyme Activity*

Mycelial SOD activity was determined by a nitro-tetrazolium chloride blue-light reduction method [34]. Firstly, 0.2 g of fresh mycelium was weighed using an electronic balance and placed in a 5 mL centrifuge tube. Subsequently, 4 mL of 50 mM phosphate buffer (pH 7.8) was added in an ice bath. The sample was then ground with a high-throughput tissue grinder and centrifuged at 10,000× *g* and 4 °C for 10 min. Then, 0.3 mL of the supernatant was aspirated, and 3.8 mL of a 50 mM phosphate buffer (pH 7.8), 0.3 mL of methionine, 0.3 mL of nitro-tetrazolium blue chloride, and 0.3 mL of riboflavin were added in turn. The light and dark response groups served as controls and were placed in a light incubator under 4000 Lx fluorescent illumination for a duration of 20 min. Subsequently, the absorbance was measured at 560 nm using a spectrophotometer (model 752 N, Shanghai INESA Instrument Analytical Instruments Co., Ltd., Shanghai, China).

Mycelial POD activity was determined by the guaiacol colorimetric method [35], using a change in A470 of 0.01 per minute as one unit (U) of peroxidase activity. Firstly, 0.1 g of fresh mycelium was weighed in a 5 mL centrifuge tube with an electronic balance, and a 5-fold amount of 50 mM phosphate buffer (pH 7.8) was added in an ice bath; then, the sample was ground with a high-throughput tissue grinder, followed by centrifugation at

15,000× *g* and 4 °C for 10 min. Then, 1 mL of 0.3% hydrogen peroxide ($H_2O_2$), 0.95 mL of 0.2% guaiacol, and 1 mL of 50 mM phosphate buffer (pH 7.8) were sequentially added to a 3 mL reaction system. The reaction was initiated by adding 0.05 mL of enzyme solution, followed by shaking well, and the reaction was immediately timed. A blank tube without $H_2O_2$ was used to adjust to zero to measure the absorbance value, and the absorbance value was measured at 470 nm by spectrophotometer.

Mycelial CAT activity was quantified spectrophotometrically by monitoring ultraviolet (UV) absorbance [36]. A decrease of 0.01 in absorbance at 240 nm per minute was defined as the activity unit (U) for catalase. Firstly, 0.1 g of fresh mycelium was weighed in a 5 mL centrifuge tube with an electronic balance, followed by the addition of a 5-fold amount of 50 mM phosphate buffer (pH 7.8) in an ice bath; the sample was ground with a high-throughput tissue grinder, followed by centrifugation at 15,000× *g* and 4 °C for 10 min. The reaction was initiated by the sequential introduction of 1 mL of 0.3% $H_2O_2$, followed by the addition of 1.95 mL water (grade 1) and 0.05 mL of an enzyme solution into a 3 mL reaction system. Upon mixing, the reaction was promptly timed, and the contents were thoroughly agitated to ensure homogeneity. The absorbance value was measured by zero adjustment of a blank tube without $H_2O_2$, and the absorbance value was measured at 240 nm by a spectrophotometer.

### 2.8. Determination of DSE Resistance Parameters

The NaOH extraction method was used to determine the mycelial melanin concentration [37]. Firstly, 0.05 g of fresh mycelium was weighed in 1 mol $L^{-1}$ sodium hydroxide in an electronic balance and heated at 100 °C for 5 h. Then, the cooled mycelium was filtered, and 7 M hydrochloric acid (pH 2.0) was added. Following the precipitate's extraction, it was washed, 1 mol $L^{-1}$ sodium hydroxide was used to dissolve it, and it was centrifuged at 10,000× *g* for 15 min to determine the amount of melanin extracted [38]. A standard curve was established by spectrophotometer at 459 nm, and melanin content was calculated.

The mycelial-soluble-protein concentration was determined by using the colorimetric method of Thomas Brilliant Blue [39]. Firstly, 0.2 g of fresh mycelium was weighed in a 5 mL centrifuge tube with an electronic balance, then 4 mL of 50 mM phosphate buffer (pH 7.8) was added, and the sample was ground with a high-throughput tissue grinder, followed by centrifugation at 10,000× *g* for 10 min. The supernatant was aspirated for 0.1 mL, 0.9 mL of distilled water was added, 5 mL of Caumas Brilliant Blue was added, and the distilled water was used as a control. Then, the absorbance value at 595 nm was measured with a spectrophotometer.

The concentration of mycelial soluble sugars was measured using the anthrone colorimetric method [40]. Firstly, 0.05 g dry mycelium was weighed in a 5 mL centrifuge tube with an electronic balance, and after grinding, 4 mL of 80% alcohol was added, and the supernatant was transferred to a 10 mL centrifugal tube with a water bath at 80 °C for 30 min, with the above steps repeated. Subsequently, a small amount of activated charcoal was added, and a water bath at 80 °C was performed for 30 min. The sample was diluted with distilled water to a final volume of 10 mL. A 250 μL aliquot was then filtered and transferred into a glass test tube. Subsequently, 5 mL of anthrone reagent was added, and the mixture was subjected to a boiling-water bath for 10 min. Distilled water served as a control in this assay. The absorbance was measured at 625 nm using a spectrophotometer.

The 5,5-dithiobis-(2-nitrobenzoic acid) (DTNB) method was used to measure the concentration of GSH in mycelial [41]. Firstly, 0.2 g of fresh mycelium was weighed in a 5 mL centrifuge tube with an electronic balance, 4 mL of 10% trichloroacetic acid was added, and the sample was ground with a high-throughput tissue grinder, followed by centrifugation at 10,000× *g* for 10 min. Then, 0.25 mL of the supernatant was aspirated in a glass test tube, and 2.6 mL of NaH2PO4 and 0.15 mL of dithiobis-(2-nitrobenzoic acid) were added in order, and then the glass test tube was shaken well. Thereafter, the reaction was kept at 30 °C for 10 min, and the absorbance value was determined using a spectrophotometer at 412 nm.

Mycelial MDA concentration was determined using the thiobarbituric-acid method [42]. Firstly, 0.2 g of fresh mycelium was weighed in a 5 mL centrifuge tube with an electronic balance, 4 mL of 10% trichloroacetic acid was added, and the sample was ground with a high-throughput tissue grinder, followed by centrifugation at 10,000× g for 10 min. Then, 2 mL of the supernatant was aspirated in a glass test tube, and 2 mL of 0.5% thiobarbituric acid (TBA) was added to a boiling-water bath for 20 min, followed by cooling quickly, and spectrophotometers were used to measure the absorbance at 450, 532, and 600 nm. MDA content (μmol/gFW) = [6.452 × (OD532-OD600) − 0.559 × OD450] × VT/(V1 × FW), in which VT—the total volume of extracted enzyme solution, mL; V1—the total volume of extracted solution used for the determination, mL; FW—Fresh weight of the sample, g.

### 2.9. Statistical Analysis

In this study, a two-way analysis of variance (ANOVA) was conducted to evaluate the impact of combined drought and Cd stress on various performance and stress tolerance metrics of DSE strains, including biomass, Cd content, melanin, osmoregulatory substances, and antioxidant-enzyme activity. The statistical analysis was performed using SPSS 25.0 software, with significant differences identified through Duncan's test ($p < 0.05$). Data processing was conducted using Excel 2021, while box scatter plots and correlation heat maps were generated with Origin Version 2021 software. The concentrations of PEG and Cd that corresponded to a 50% growth inhibition ($IC_{50}$) for the DSE strains were determined utilizing GraphPad Prism version 8.0 software. Variance partitioning analysis (VPA) was implemented to dissect the effects of drought and Cd as individual factors on the performance and tolerance indices of DSE strains, employing the "vegan" package within RStudio version 1.3.1073 software. Furthermore, the influence of DSE strains on antioxidant, osmotic-regulatory, and drought-resistance indices was elucidated through a Structural Equation Model (SEM) using AMOS 21.0 software.

## 3. Results

### 3.1. Effect of Drought Stress on Morphology and Growth of DSE Colonies

The colony morphology of 10 DSE strains under drought stress after 8 d of PDA medium cultivation is shown in Figure 1A–J. The colony growth rates of *At* decreased with the increase of drought gradient. *Aa*, *Np*, *Pc*, *Mp'*, and *Mp* increased and then decreased in colony growth rate with the increase of drought stress and were the first to reach the maximum diameter at a PEG concentration of 18% (Figure 2). *Pr* showed a faster growth rate under drought stress, and the mycelium cultured under 9%, 18%, and 27% PEG concentrations preferentially attained its maximum diameter on the fourth day. With the increase of PEG-6000 concentration, except *Pe*, all DSE strains showed different degrees of drought resistance. The tolerance index (TI) of DSE strains to varying PEG concentrations was assessed, along with the estimation of the 50% inhibitory concentration ($IC_{50}$) after an 8-day incubation period. An incubation time of 8 d was used as an intermediate reference (Table 3). *Pc*, *Mp'*, *Mp*, and *Pr* strains had the highest tolerance to PEG concentrations. After 8 d of incubation in a PDA medium, *Aa* was the next most tolerant to PEG, with a TI of 0.85 at 27% PEG concentration. In the corresponding model, the IC50 estimate for *At* was 32.84% PEG concentration, which exceeds the highest concentration of PEG tested in the study (Figure 3(C1)).

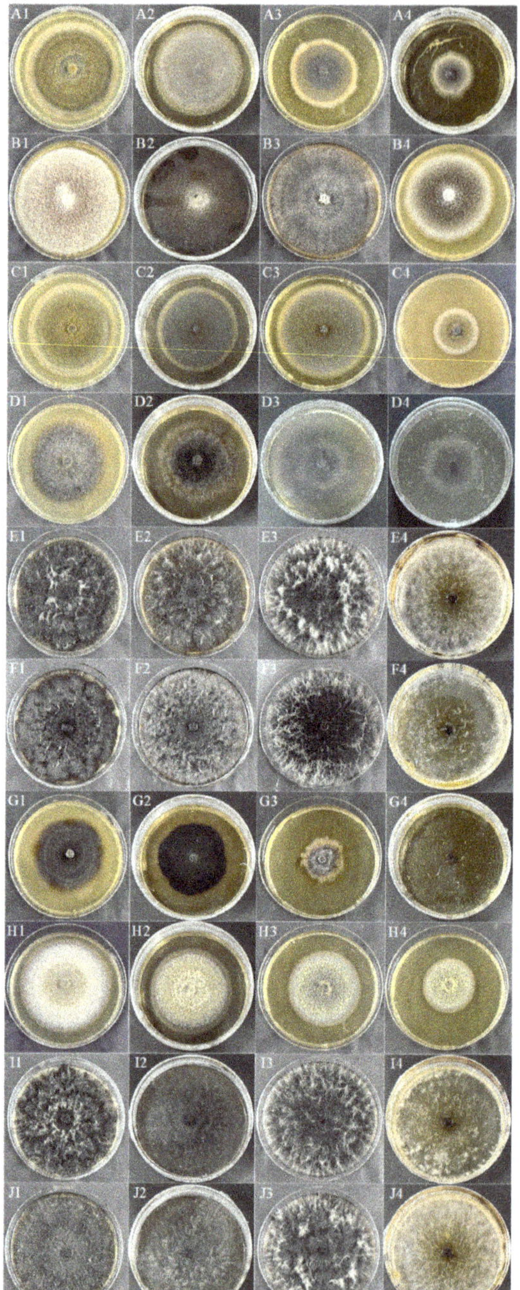


**Figure 1.** DSE colony morphology under different PEG concentrations. (**A–J**) stands for *M. cytisi* (*Mc*), *A. alstroemeriae* (*Aa*), *S. lupini* (*Sl*), *N. phragmitis* (*Np*), *P. chlamydocopiosa* (*Pc*), *M. phaseolina* (*Mp'*), *P. equi* (*Pe*), *A. tellustris* (*At*), *M. pseudophaseolina* (*Mp*), and *P. radicina* (*Pr*). **1–4** indicates PEG gradients of 0, 9, 18, and 27%, respectively.

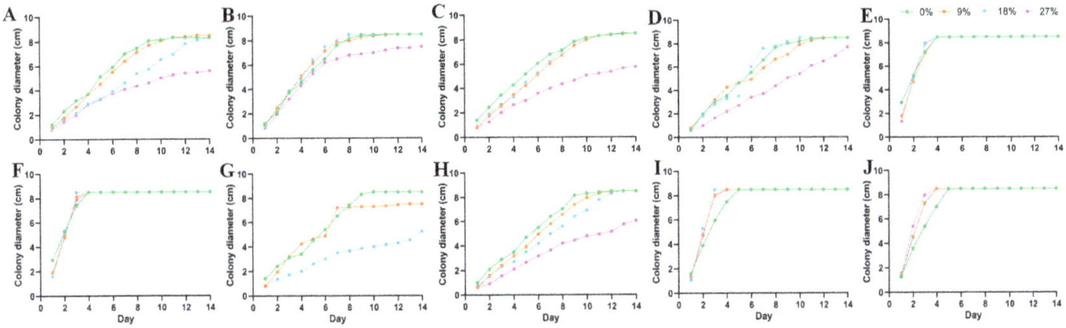


**Figure 2.** Colony growth diameters of 10 DSE strains under different PEG concentrations. (**A–J**) stands for *M. cytisi* (*Mc*), *A. alstroemeriae* (*Aa*), *S. lupini* (*Sl*), *N. phragmitis* (*Np*), *P. chlamydocopiosa* (*Pc*), *M. phaseolina* (*Mp'*), *P. equi* (*Pe*), *A. tellustris* (*At*), *M. pseudophaseolina* (*Mp*), and *P. radicina* (*Pr*).

**Table 3.** Fungal TI and $IC_{50}$ to PEG and $Cd^{2+}$ after the incubation of PDA medium exposed to the different tested concentrations for 8 days. IC50 was estimated by fitting the TI data to a logistic model ($r^2 > 0.9$). [a] $0.80 < r^2 < 0.9$; [b] $r^2 \leq 0.80$. Tolerance Index categories: Very low tolerance (0.00–0.39); Low tolerance (0.40–0.59); Moderate tolerance (0.60–0.79); High tolerance (0.80–0.99); Very high tolerance (≥1) [33].

| Strain | TI to PEG (%) | | | $IC_{50}$ to PEG | TI to $Cd^{2+}$ (mg Cd/L) | | | $IC_{50}$ to $Cd^{2+}$ |
|---|---|---|---|---|---|---|---|---|
| | 9 | 18 | 27 | (%) | 25 | 50 | 100 | (mg Cd/L) |
| *Microsphaeropsis cytisi* | 0.95 | 0.72 | 0.58 | 30.84 (27.16–38.12) | 0.22 | 0.00 | 0.00 | ~22.53 |
| *Alternaria alstroemeriae* | 0.99 | 1.06 | 0.85 | ~28.51 [b] | 0.74 | 0.40 | 0.27 | 44.71 (39.98–49.92) |
| *Stagonosporopsis lupini* | 0.94 | 0.98 | 0.61 | 28.36 (~ 30.65) | 0.52 | 0.21 | 0.00 | 26.23 (24.19–28.14) |
| *Neocamarosporium phragmitis* | 0.87 | 1.02 | 0.57 | ~27.24 [a] | 0.54 | 0.39 | 0.33 | 29.58 (20.95–36.66) |
| *Paraphoma chlamydocopiosa* | 1.00 | 1.00 | 1.00 | - | 0.48 | 0.27 | 0.13 | 23.51 (22.30–24.66) |
| *Macrophomina phaseolina* | 1.00 | 1.00 | 1.00 | - | 0.43 | 0.20 | 0.14 | 19.25 (15.98–21.97) |
| *Papulaspora equi* | 0.97 | 0.49 | 0.00 | 17.97 | 0.43 | 0.22 | 0.00 | 22.07 (17.19–25.80) |
| *Alternaria tellustris* | 0.93 | 0.80 | 0.60 | 32.84 (30.34–36.63) | 1.01 | 0.87 | 0.56 | 109.30 (99.61–125.00) |
| *Macrophomina pseudophaseolina* | 1.00 | 1.00 | 1.00 | - | 0.43 | 0.20 | 0.00 | 22.08 (19.36–24.35) |
| *Paraphoma radicina* | 1.00 | 1.00 | 1.00 | - | 0.47 | 0.28 | 0.11 | 23.28 (21.77–24.69) |

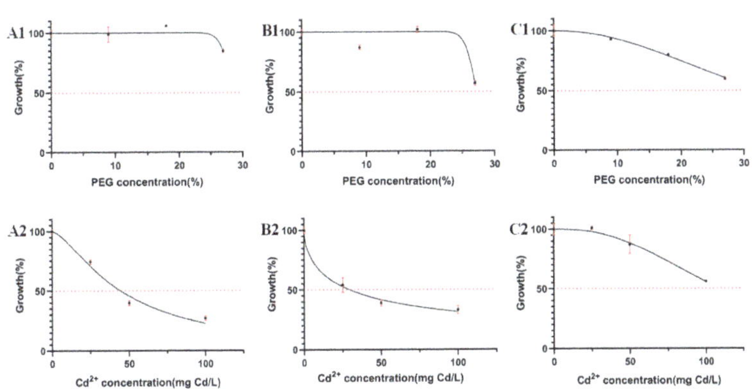


**Figure 3.** Effect of PEG concentration (**A1–C1**) and Cd concentration (**A2–C2**) on the mean and standard deviation (solid bars, $n = 3$) of the tolerance index (TI) of selected stress-tolerant strains measured in radial growth in PDA medium after 8 d of incubation. The solid line corresponds to the fitted logistic model used to determine the $IC_{50}$ values. (**A–C**) stands for *A. alstroemeriae* (*Aa*), *N. phragmitis* (*Np*), and *A. tellustris* (*At*).

### 3.2. Effect of Cd Stress on Morphology and Growth of DSE Colonies

The colony morphology of 10 DSE strains under different Cd concentrations after 8d of PDA medium cultivation is presented in Figure 4A–J. With the increase of Cd stress, the adaptability changes of colony morphology varied among different strains, and the color of the *Mc*, *Aa*, *Sl*, *Np*, and *Pe* colonies gradually became lighter. The mycelia of *Pc*, *Mp'*, *At*, *Mp*, and *Pr* gradually changed from loose to dense and darkened in color. The colony diameter of the strain decreased with increasing Cd concentration. As the concentration of Cd rose, *Mc* and *Mp* stopped growing when the concentration reached 100 mg Cd/L. *Sl*, *Pc*, *Mp'*, *Pe*, and *Pr* grew slowly at 100 mg Cd/L. *Aa*, *Np*, and *At* remained growth-active at 100 mg Cd/L (Figure 5). The tolerance index (TI) of DSE strains to a range of Cd concentrations was determined, alongside the calculation of the 50% inhibitory concentration ($IC_{50}$) following an 8-day period. An incubation time of 8 days was used as an intermediate reference (Table 3). The IC50 estimates for *Aa*, *Np*, and *At* were 48 mg Cd/L, 30 mg Cd/L, and 109 mg Cd/L, respectively (Figure 3(A2)–(C2)). From the one-factor Cd stress results, it was concluded that *Aa*, *Np*, and *At* had better Cd tolerance.

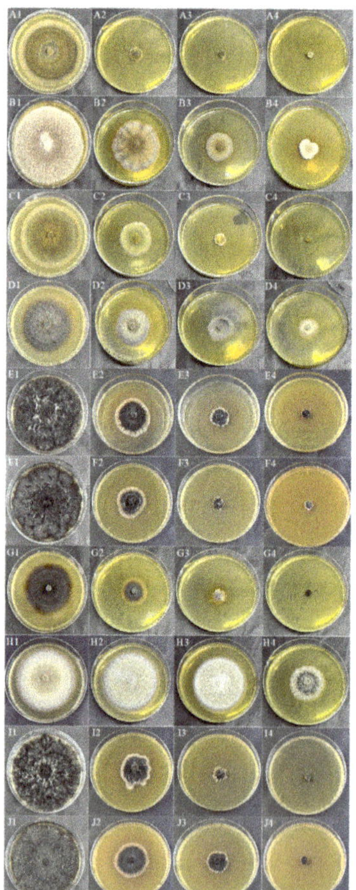

**Figure 4.** DSE colony morphology under different Cd stresses. (**A–J**) stands for *M. cytisi* (*Mc*), *A. alstroemeriae* (*Aa*), *S. lupini* (*Sl*), *N. phragmitis* (*Np*), *P. chlamydocopiosa* (*Pc*), *M. phaseolina* (*Mp'*), *P. equi* (*Pe*), *A. tellustris* (*At*), *M. pseudophaseolina* (*Mp*), and *P. radicina* (*Pr*). **1–4** indicates Cd stress gradients of 0, 25, 50, and 100 mg/L, respectively.

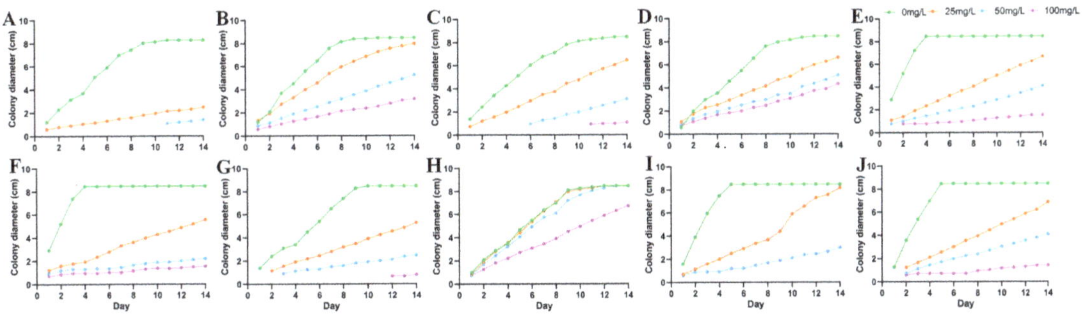


**Figure 5.** Growth diameters of 10 DSE strains under different Cd stresses. (**A–J**) stands for *M. cytisi* (*Mc*), *A. alstroemeriae* (*Aa*), *S. lupini* (*Sl*), *N. phragmitis* (*Np*), *P. chlamydocopiosa* (*Pc*), *M. phaseolina* (*Mp'*), *P. equi* (*Pe*), *A. tellustris* (*At*), *M. pseudophaseolina* (*Mp*), and *P. radicina* (*Pr*).

### 3.3. *Effects of Combined Stress of Drought and Cd on Morphology and Physiological Indices of DSE Colonies*

#### 3.3.1. Morphological Parameters and Biomass of DSE Colonies

Combined with the results of single-factor drought and Cd stress, three DSE strains *Aa*, *Np*, and *At*, were selected for a drought-and-Cd-combined stress test. After 8 d of PDA medium cultivation, the colony morphology of three DSE strains under different combined stresses of drought and Cd was measured (Figure 6). The colony color of *Aa* was grayish-white and dense when there was no stress. The colony color changed from black to grayish-white with the increase of stress degree (Figure 6A). The *Np* strain had a black colony color with aerial mycelium on the surface in the absence of stress, and with increasing stress, the mycelium on the edge of the colony was sparse, the edge was covered with white downy hairs, and the aerial mycelium increased. *Np* grew fastest after 8 days at a 9% + 25 mg Cd/L concentration (Figure 6B). In the absence of stress, the color of the colonies of the *At* strain was whitish with dense colonies. At a concentration of 9% + 25 mg Cd/L, colonies darkened in color, and with increasing stress, the color of the colonies changed from black to white, with white tomentum in the middle (Figure 6C). The colony diameter of the three DSE strains is shown in Figure 7. As combined stress increased, the colony diameter of *Aa* declined in comparison to the control; *Np* preferentially culminated in a maximum diameter after 10 d at 9% + 50 mg Cd/L and 18% + 25 mg Cd/L combined stress. The colony diameter of *At* strains revealed a tendency of first growing and then declining with the increase of combined stress, and the diameter preferentially culminated in a maximum at 9% + 25 mg Cd/L combined stress. The results showed that the *Np* strain had strong drought-and-Cd-combined stress tolerance.

The biomass of *Aa* was higher under all combined stress than under the control treatment, but with the aggravation of combined stress, the biomass of *Aa* increased first and then decreased and reached the highest value of 1.23 g at the combined concentration of 18% + 25 mg Cd/L, a significant increase of 61.84% compared with the control (Figure 8A). The biomass of *At* and *Np* revealed a growing and subsequently declining trend with the aggravation of combined stress. The *At* reached the highest biomass of 1.71 g at the combined concentration of 18% + 50 mg Cd/L, which was a significant increase of 64.42% compared with the control. The *Np* reached the highest biomass of 1.39 g at the combined concentration of 18% + 25 mg Cd/L, which was a significant increase of 117.19% compared with the control. The biomass of *Np* increased at 9% + 25 mg Cd/L, 18% + 25 mg Cd/L, and 18% + 50 mg Cd/L compared with the control, but there was no significant difference between other treatments and the control.

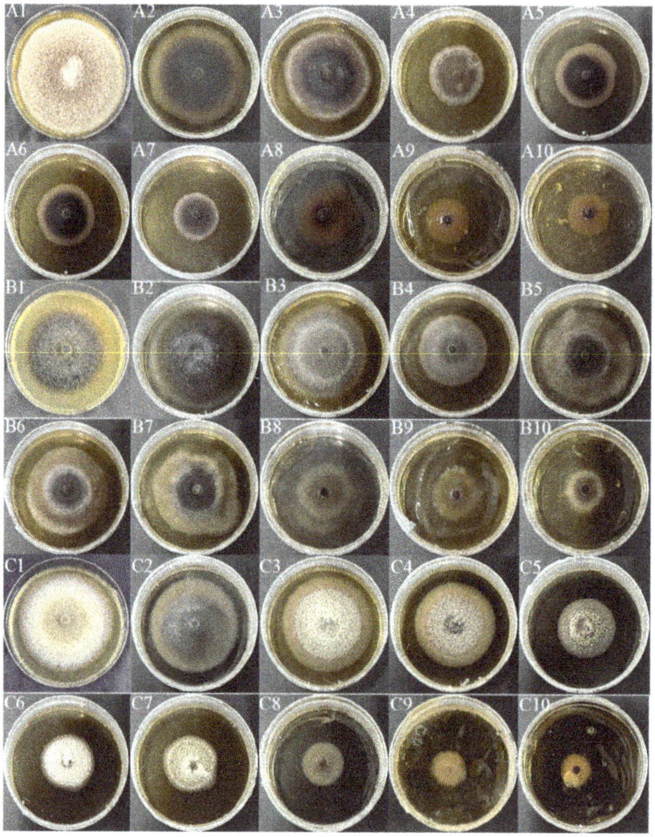


**Figure 6.** DSE colony morphology under different combined stress of drought and Cd. (**A–C**) stands for *A. alstroemeriae* (*Aa*), *N. phragmitis* (*Np*), and *A. tellustris* (*At*). **1–10** indicate that the combined stress gradients of drought and Cd are 0, 9% + 25 mg Cd/L, 9% + 50 mg Cd/L, 9% + 100 mg Cd/L, 18% + 25 mg Cd/L, 18% + 50 mg Cd/L, 18% + 100 mg Cd/L, 27% + 25 mg Cd/L, 27% + 50 mg Cd/L, and 27% + 100 mg Cd/L, respectively.

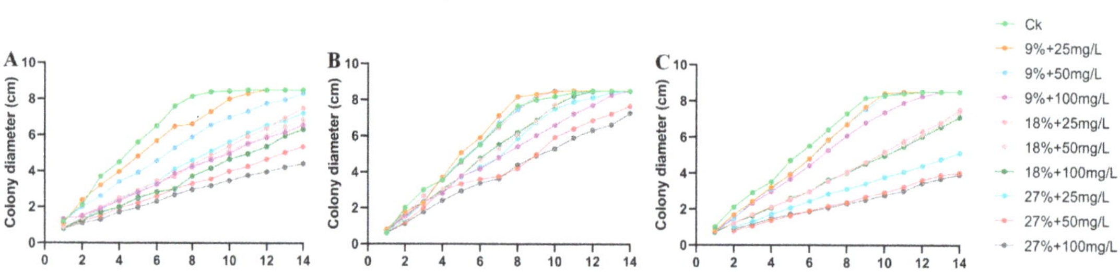


**Figure 7.** Colony diameter of three DSE strains under different combined stress of drought and Cd. (**A–C**) stands for *A. alstroemeriae* (*Aa*), *N. phragmitis* (*Np*), and *A. tellustris* (*At*). 1–10 indicate that the combined stress gradients of drought and Cd are 0, 9% + 25 mg Cd/L, 9% + 50 mg Cd/L, 9% + 100 mg Cd/L, 18% + 25 mg Cd/L, 18% + 50 mg Cd/L, 18% + 100 mg Cd/L, 27% + 25 mg Cd/L, 27% + 50 mg Cd/L, and 27% + 100 mg Cd/L, respectively.

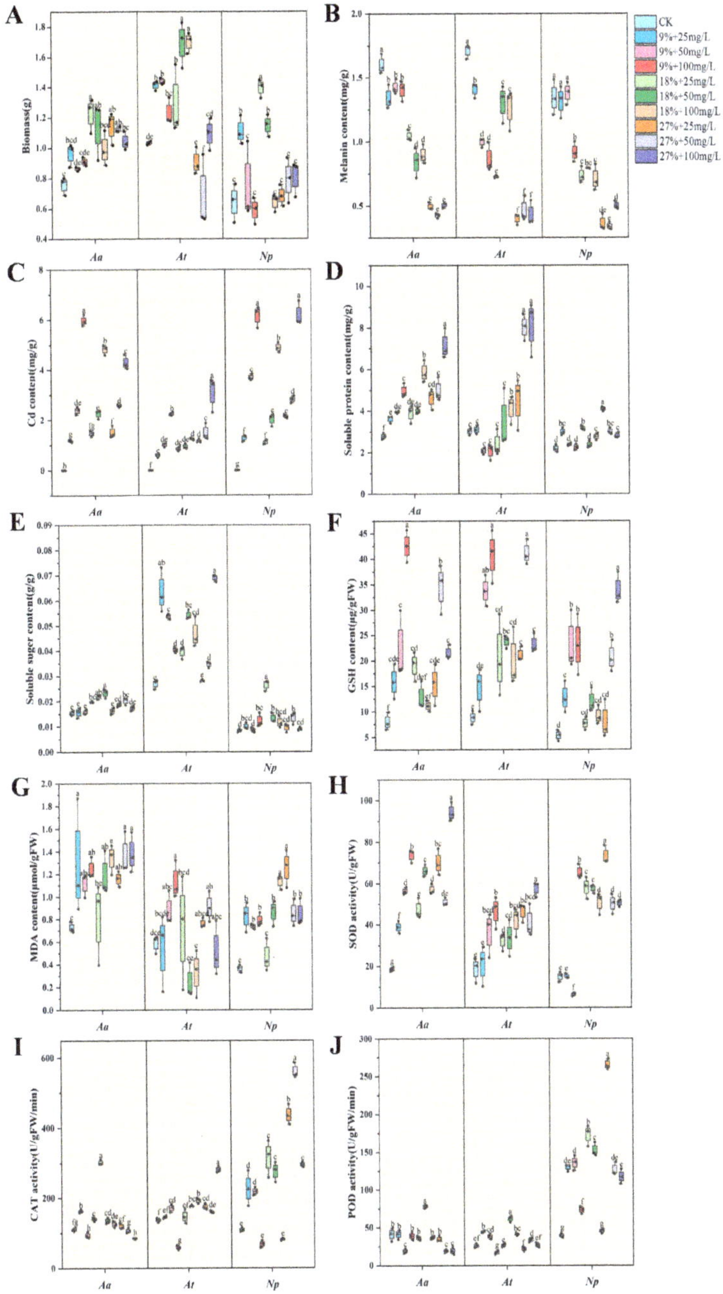


**Figure 8.** Physiological and growth indexes of DSE strains in liquid cultures. The abbreviations in the figure are *A. alstroemeriae* (*Aa*), *A. tellustris* (*At*), and *N. phragmitis* (*Np*). Biomass (**A**), melanin content (**B**), Cd content (**C**), soluble protein content (**D**), soluble sugar content (**E**), GSH content (**F**), MDA content (**G**), SOD activity (**H**), CAT activity (**I**), and POD activity (**J**) of DSE under drought and Cd stress. Means followed by the different letter(s) within each column are significantly different at $p < 0.05$.

### 3.3.2. Melanin and Cd Content of DSE Strains

With a rise in stress level, the melanin content of *Np* first increased before declining and attained the maximum value of 1.38 mg/g at the combined concentration of 9% + 50 mg Cd/L, which increased by 2.84% as compared to the control (Figure 8B). However, the melanin content of *Aa* and *At* decreased with the enhancement of combined stress, and the difference was significant compared with the control. With an increase in Cd concentration, the Cd content of DSE strains gradually increased (Figure 8C). The Cd content of *Aa* was 1.18, 2.38, 5.97, 1.47, 2.25, 4.82, 1.51, 2.57, and 4.28 mg/g, the Cd content of *At* was 0.59, 1.02, 2.20, 0.88, 0.94, 1.27, 1.23, 1.78, and 2.97 mg/g, and the Cd content of *Np* was 1.28, 3.69, 6.15, 1.09, 2.01, 4.87, 2.14, 2.81, and 6.18 mg/g, respectively.

### 3.3.3. Soluble Substance Content of DSE Strains

The soluble protein content of *Aa* and *Np* strains increased with the increase of combined stress (Figure 8D). Compared with the control, the soluble protein content of *Aa* under all combined stress treatments increased and reached the maximum value in 27% + 100 mg Cd/L treatment, which was 2.55 times that of the control and increased by 155.36% compared with the control. At the combined concentration of 18% + 100 mg Cd/L, 27% + 25 mg Cd/L, 27% + 50 mg Cd/L, and 27% + 100 mg Cd/L, the soluble protein content of *At* was increased compared to the control and reached the maximum value at the combined concentration of 27% + 100 mg Cd/L, which was increased by 181.94% in comparison to the control. No significant differences were observed between the other treatments and the control group. However, the soluble protein content in the *Np* peaked under the 27% + 25 mg/L Cd treatment, exhibiting a substantial increase of 87.16% relative to the control.

The soluble sugar content of *Aa* and *Np* strains showed an increasing and then decreasing trend with the increase of combined stress (Figure 8E). The combined concentration of *Aa* was maximum at 18% + 50 mg Cd/L, reaching a maximum of 0.023 g/g. The *Np* was higher than the control at the combined concentration of 18% + 25 mg Cd/L, reaching a maximum of 0.026 g/g. The soluble sugar content of *At* was higher in the other treatments (except 27% + 25 mg Cd/L) when compared with the control. The maximum value was 0.07 g/g when the combined concentration was 27% + 100 mg Cd/L, which was 2.33 times that of the control.

### 3.3.4. GSH and MDA Content of DSE Strains

The GSH content of DSE strains was increased under all the combined stress treatments compared to the control (Figure 8F). The *Aa* reached the maximum value of 42.56 μg/g FW at the combined concentration of 9% + 100 mg Cd/L. At a combined concentration of 27% + 50 mg Cd/L, the *At* reached a maximum value of 37.88 μg/g. The combined concentration of *Np* was 27% + 100 mg Cd/L, which was 6.41 times higher than the control.

The MDA content of *Aa* and *Np* increased with each treatment compared with the control except for the combined concentration of 18% + 25 mg Cd/L (Figure 8G). The MDA content of *Aa* reached the maximum value in the 27% + 100 mg Cd/L treatment, which was 1.89 times that of the control. The MDA content of the *Np* reached a maximum of 1.25 μmol/g FW in the 27% + 25 mg Cd/L treatment, which was 3.47 times higher than the control. The MDA content of *At* reached a maximum value of 1.09 μmol/g FW at 9% + 100 mg Cd/L combined concentration, which increased by 94.64% compared with the control.

### 3.3.5. Antioxidant Enzyme Activity of DSE Strains

With the intensification of combined stress, the SOD activity of *Aa* and *At* increased compared with the control (Figure 8H). At a combined concentration of 27% + 100 mg Cd/L, the SOD activity of *Aa* and *At* was more than that of the control, and the SOD activity showed the maximum value, which increased by 407.22% and 241.12%, respectively. The SOD activity of *Np* showed a decreasing and then increasing trend with the increase in the degree of stress and reached the maximum value at 27% + 25 mg Cd/L, which increased 390.87% compared with the control. Compared with the control, SOD activity decreased by 56.78% when the combined concentration was 9% + 50 mg Cd/L, and its activity was greatly reduced.

The CAT activity of *Aa* showed an increasing and then decreasing trend with the increase of combined stress and reached the maximum value at 18% + 25 mg Cd/L concentration, which increased by 174.25% compared with the control (Figure 8I). In addition to 9% + 100 mg Cd/L treatment, CAT activity of *At* increased compared with the control and reached the maximum value at 27% + 100 mg Cd/L combined concentration, which was 2.07 times that of the control. Except for the combined concentrations of 9% + 100 mg Cd/L and 18% + 100 mg Cd/L, the *Np* showed an increase in CAT activity in other treatments compared to the control and reached a maximum of 560.55 U/g FW/min at 27% + 50 mg Cd/L, which was an increase of 416.97% compared to the control.

The POD activity of *Aa* reached the maximum value of 78.45 U/g FW/min under moderate stress (18% + 50 mg Cd/L), which increased by 89.04% compared with the control (Figure 8J). Under 50 mg Cd/L stress, POD activity of *At* under 9%, 18%, and 27% drought stress was increased compared with the control, which were 46.19%, 134.16%, and 29.17% of the control, respectively. The POD activity of *Np* was higher than that of the control under all combined stress treatments. POD activity reached the maximum value of 265.41 U/g FW/min under the 27% + 25 mg Cd/L treatment, which was increased by 573.12% compared with the control. The POD activity of *Np* under 25 mg Cd/L stress and 9%, 18%, and 27% drought stress was increased by 227.97%, 338.22%, and 573.12%, respectively, compared with the control.

### 3.4. Relationship between DSE Growth and Physiological Indicators

The relationship between biomass, Cd content, GSH and melanin, SP, SS, MDA, SOD, CAT, and POD was further analyzed for each DSE (Figure 9), and the correlations between the individual metrics varied depending on the DSE.

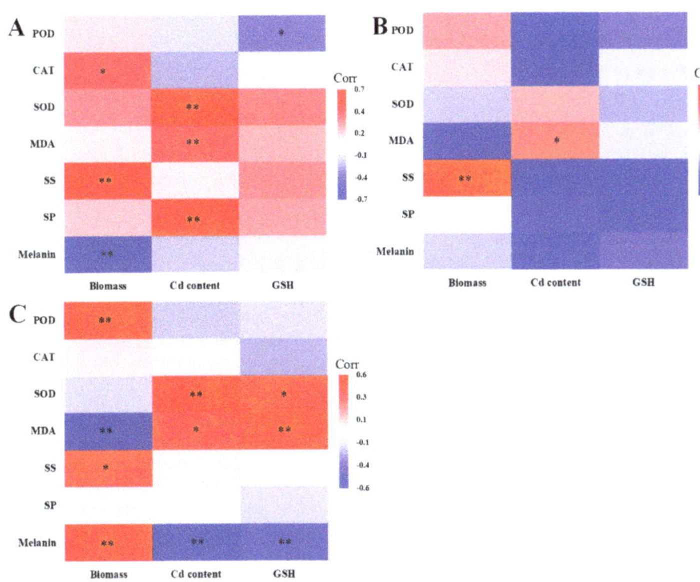


**Figure 9.** Relationship between DSE growth and physiological indicators. (**A–C**) stands for *A. alstroemeriae* (*Aa*), *N. phragmitis* (*Np*), and *A. tellustris* (*At*). SS represents soluble sugar, and SP represents soluble protein. The various symbols above the bars indicate significant differences between DSE growth and physiological indices (* $p < 0.05$, ** $p < 0.01$).

The biomass of *Aa* was positively correlated with CAT activity and soluble sugar content and negatively correlated with melanin content. The Cd content of *Aa* was positively correlated with SOD activity, MDA, and soluble protein content. GSH content of *Aa* was negatively correlated with POD activity (Figure 9A). The biomass of *Np* was positively correlated with the soluble sugar content. The Cd content of *Np* was positively correlated with MDA content (Figure 9B). The biomass of *At* was positively correlated with POD activity, soluble sugar, and melanin content but negatively correlated with MDA content. Cd and GSH content of *At* were positively correlated with SOD activity and MDA content and negatively correlated with melanin content (Figure 9C).

### 3.5. Variation Partitioning Analyses

The effects of drought-and-Cd stress on the physiological and growth indicators of DSE were estimated by variance partitioning analysis, and the contribution rates of various factors to the differences of DSE strains were quantitatively assessed (Figure 10). The combined explanation of drought-and-Cd stress on the biomass of *Aa* was 69.6%. The individual explanations were 18.9% and 8.2%, respectively. The interaction between drought and Cd accounted for 42.5% (Figure 10(A1)). The combined explanation of drought-and-Cd stress on the biomass of *Np* was 28.4%, and the individual explanations were 28.9% and 24.7%, respectively (Figure 10(B1)). The combined explanation of drought-and-Cd stress on the biomass of *At* was 65.3%. The individual explanations were 63.8% and 67.9%, respectively (Figure 10(C1)).

The variance in melanin production by strain *Aa*, attributed to combined drought-and-Cd stress, was explained by 95.5% of the model, with individual contributions of 0.5% for drought and 0.9% for Cd stress alone and an interaction effect accounting for 94% (Figure 10(A2)). For strain *Np*, the combined stress factors explained 85% of the variance in melanin levels, with individual contributions of 0.6% for drought and 0.1% for Cd stress, and the interaction effect was responsible for 84.3% (Figure 10(B2)). In the case of strain *At*, the combined stress factors explained 76.8% of the variance in melanin production, with Cd stress alone accounting for 1.5% and the interaction between drought and Cd stress contributing to 76% of the explained variance (Figure 10(C2)).

The impact of combined drought-and-Cd stress on the antioxidant-enzyme activity in the *Aa* strain was found to be 33.9%, with individual contributions of 28.2% attributed to drought and 27.4% to Cd stress alone (Figure 10(A3)). For the *Np* strain, the combined stress factors explained 61% of the variance in antioxidant-enzyme activity, with individual contributions of 0.8% for drought and 5.2% for Cd stress, and the interaction between the two stressors accounted for 55% (Figure 10(B3)). In the case of the *At* strain, the combined stress factors explained 32% of the variance in antioxidant-enzyme activities, with Cd stress alone contributing 1.3% and the interaction between drought and Cd stress accounting for 30.9% (Figure 10(C3)).

The effects of combined drought and Cd stress on the physiological indices of strains *Aa*, *Np*, and *At* were 8.7%, 17.2%, and 21.1%, respectively. The interactions between the two stressors were 13.8%, 19.4%, and 21%, respectively, with Cd stress alone explaining 1.4% of the physiological indices in the *At* strain (Figure 10(A4)–(C4)).

In terms of Cd content, the combined stress factors explained 27.8% of the variance in the *Aa* strain, with individual contributions of 13.1% for drought and 7.2% for Cd stress, and the interaction between the two factors was 7.5% (Figure 10(A5)). For the *Np* strain, the combined stress factors explained 24.2% of the variance in Cd content, with drought stress contributing 3.1% and the interaction between drought and Cd stress accounting for 21.5% (Figure 10(B5)). For the *At* strain, the combined explanation of drought and Cd stress on Cd content was 43.2%, with the interaction between the two stressors contributing a significant 45.6% (Figure 10(C5)).

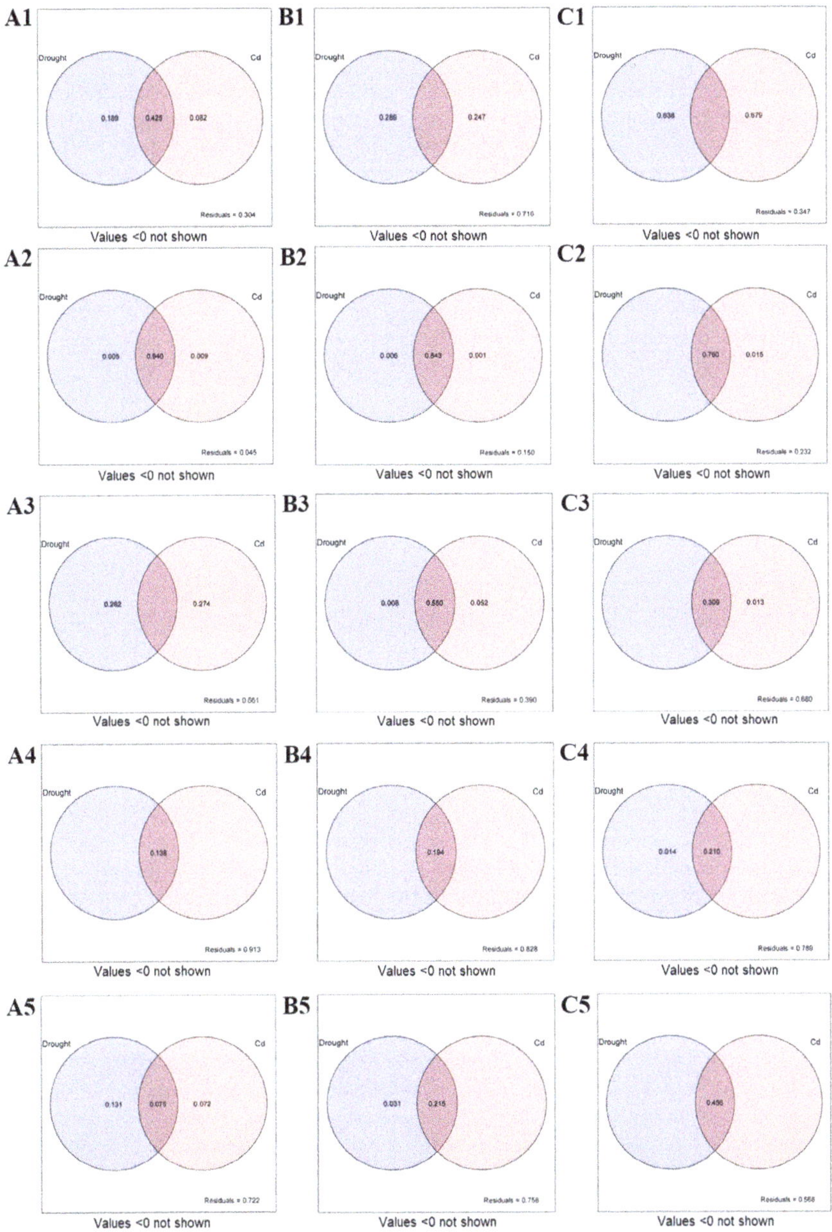


**Figure 10.** The variance-partitioning analysis of drought and Cd stress on growth and physiological indices of DSE. (**A–C**) represents the variance-partitioning analysis plots of *A. alstroemeriae* (*Aa*), *N. phragmitis* (*Np*), and *A. tellustris* (*At*) under combined stress, respectively. **1** represents the effect of drought and Cd stress on the biomass. **2** represents the effect of drought and Cd stress on the melanin content. **3** represents the effect of drought and Cd combined on the antioxidant-enzyme activity. **4** represents the effect of drought and Cd stress on the SP, SS, MDA, and GSH. **5** represents the effect of drought and Cd stress on the Cd content.

### 3.6. Correlation Analyses

Based on the correlation according to the coefficients of correlation (R-values), the association between DSE and all examined parameters was evaluated using a SEM model. *Aa* positively influenced SOD, soluble sugar, and biomass and negatively affected Cd content. The SOD activity of *Aa* had a positive impact on soluble protein. The CAT activity of *Aa* had a positive impact on soluble sugar and biomass. The POD activity of *Aa* had positive effects on soluble sugar and adverse impacts on soluble protein and GSH. The soluble protein of *Aa* had a positive effect on Cd content (Figure 11A).

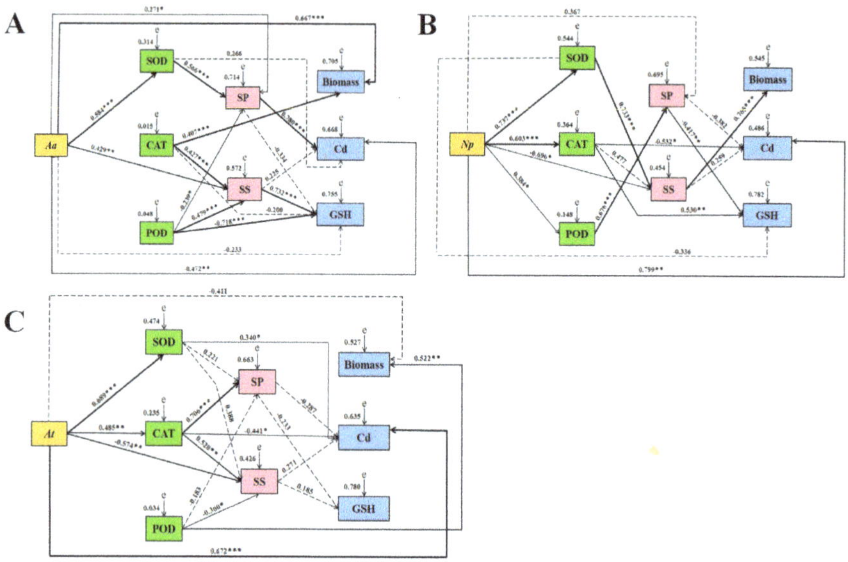

**Figure 11.** Structural-equation modeling of causal relationships between DSE and enzyme activity, soluble matter, biomass, Cd, and GSH content. (**A–C**) represents structural-equation models of *A. alstroemeriae* (*Aa*), *N. phragmitis* (*Np*), and *A. tellustris* (*At*), respectively. SS represents soluble sugar, and SP represents soluble protein. Results of fitness of influencing factors of *Aa*: (**A**) $\chi^2$ = 8.500, *df* = 6, *p* = 0.004, RMSEA (root mean square error of approximation) = 0.120, GFI (goodness-of-fit index) = 0.942, IFI (incremental fit index) = 0.987, CFI (comparative fit index) = 0.984; Results of fitness of influencing factors of *Np*: (**B**) $\chi^2$ = 28.978, *df* = 6, *p* = 0.001, RMSEA = 0.363, GFI = 0.870, IFI = 0.883, CFI = 0.862; Results of fitness of influencing factors of *At*: (**C**) $\chi^2$ = 21.358, *df* = 6, *p* = 0.002, RMSEA = 0.297, GFI = 0.877, IFI = 0.913, CFI = 0.895. Whether the path of action between various factors was significant or not is indicated by the solid and dotted lines, respectively. The numbers near the arrows represent the normalized path coefficients, and the width of the solid line represents the strength of the effect between the various factors (* $p < 0.05$, ** $p < 0.01$, and *** $p < 0.001$).

The *Np* had a positive effect on SOD, CAT, POD, and Cd content. *Np* negatively affected soluble sugars. The SOD of *Np* positively influenced soluble sugars. The CAT of *Np* negatively affected Cd content. The POD of *Np* positively influenced soluble protein. The soluble protein of *Np* had a negative effect on GSH. The soluble sugars of *Np* positively influenced biomass (Figure 11B).

The *At* positively influenced SOD, CAT, and Cd content. *At* negatively affected soluble sugars. The SOD activity of *At* positively influenced Cd content. The CAT activity of *At* positively influenced soluble protein and soluble sugar and negatively affected Cd content. The POD activity of *At* positively influenced biomass and negatively affected soluble sugar (Figure 11C).

### 3.7. Growth Parameter

DSE had significant effects on the growth of *A. membranaceus* seedlings under drought and Cd stress (Figure 12A). Specifically, the effect of *Np* and *At* inoculation on shoot height was not significant (Figure 12B). Under Cd2, D, and DCd1 treatments, inoculation with *At* significantly increased root weight by 86.88, 124.36, and 56.78%, respectively, compared with non-inoculated counterparts. Under D, DCd1 and DCd3 treatments, inoculation with *Np* significantly increased root weight by 100.57, 32.54 and 97.04%, respectively, compared with non-inoculated counterparts (Figure 12C).

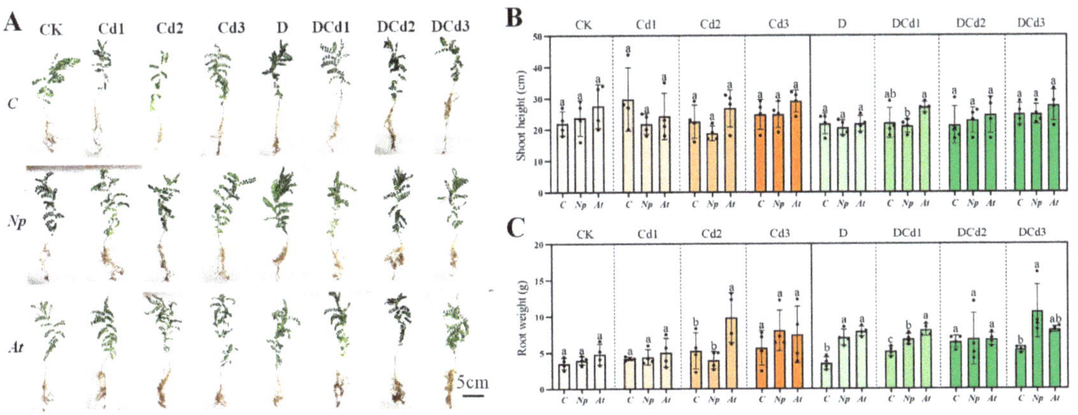

**Figure 12.** Effect of different DSEs on growth-morphological parameters of *Astragalus membranaceus* seedlings' growth under synergistic stress of drought and Cd. Growth picture (**A**), shoot height (**B**), and root weight (**C**) of *A. membranaceus* under drought and Cd stress. The abbreviations in the figure are non-inoculated plants(*C*), *N. phragmitis* (*Np*), and *A. tellustris* (*At*). Means followed by the different letter(s) within each column are significantly different at $p < 0.05$.

## 4. Discussion

### 4.1. Effects of Drought Stress on the Performance of DSE

All nine strains of DSE, except *Pe*, grew well under a drought-stress environment in this study. The growth rates of *Aa*, *Np*, *Pc*, *Mp'*, and *Mp* surpassed those of the control under an 18% drought stress. Corresponding research has indicated that a PEG-6000 concentration ranging from 10% to 20% is conducive to the biomass accumulation of ectomycorrhizal fungi [43]. Under mild or moderate drought conditions, most DSE strains exhibited darkened colonies, whereas under severe drought stress, most DSE strains exhibited lightened colonies. This phenotypic plasticity is likely an adaptive response of DSE strains to stressful environments, potentially enhancing the efficiency of nutrient and water uptake [44]. Resistant cultivars can drive divergence in the ecological roles of the cultivated fungi *Mortierella alpina* and *Epicoccum nigrum*, and successful colonization of root surfaces by *M. alpina* enhances the resilience of wheat drought stress through the activation of drought-responsive genes [45]. The combination of the arbuscular mycorrhizal fungi (AMF) (*Glomus* spp.) with energy grasses also improves the adaptation of *Saccharum arundinaceum* to marginal lands with drought-affected soils [46]. It has been shown that the symbiosis of AMF enhanced the tobacco plants' secondary metabolism and overall growth pattern under severe drought stress [47]. It has been shown that two Cd-tolerant and plant-growth-promoting actinomycete strains, *Streptomyces* sp. and *Nocardiopsis* sp., were isolated from metal-contaminated soils. Both actinomycete strains can be used as effective agents for phytoremediation of soil contaminated with Cd under drought conditions [48].

### 4.2. *Effects of Cd Stress on the Performance of DSE*

With the increasing concentration of Cd, the alterations in colony morphology among strains exhibited strain-specific responses. Notably, strains *Aa*, *Np*, and *At* demonstrated sustained activity even at 100 mg Cd/L, indicative of their superior Cd tolerance. This resilience is likely attributed to the inherent capacity of certain DSEs to strongly adsorb heavy metals, thereby impeding the mobility of heavy-metal ions. Among these, *At* displayed the most rapid growth rate under varying Cd stress levels. The observed differences in tolerance and growth kinetics are postulated to stem from the distinct biochemical attributes and physiological adaptations of DSEs in response to Cd-induced stress [49]. The IC50 estimates for *Aa*, *Np*, and *At* were 48 mg Cd/L, 30 mg Cd/L, and 109 mg Cd/L, respectively. Studies have reported that endophytic fungi isolated from barley roots, such as *Alternaria* sp., exhibit exceptional tolerance to Cd, a finding that corroborates the results presented here [50]. The study showed that *Np* was isolated from saline areas [51]. In this study, *Neocamarosporium* sp. was also screened for Cd tolerance, which enriched the strain resource library. Medina-Armijo et al. [31] screened *Exophiala crusticola* for Cr tolerance. DSE species were isolated and characterized from the roots of *Medicago sativa* and *Ammopiptanthus mongolicus*, and the results showed enhanced growth and tolerance to Cd in the host plants [52]. Furthermore, by promoting the conversion of Cd into chemical forms that have low activity, DSE inoculation has been demonstrated to increase plants' ability to withstand Cd stress and may mitigate the detrimental effects of Cd toxicity on plant growth [19]. Lin et al. [53] characterized the bioaccumulation of Zn and Cd by *Streptomyces zinciresistens*. Xue and Wang [54] found that inoculating soil with Cd-resistant *Actinomycetes* flora can reduce Cd accumulation in rice plants.

### 4.3. *Effects of Combined Drought and Cd Stress on the Performance of DSE*

In the current study, solid-plate-culture experiments showed that *Np* and *At* promoted fungal growth under low combined stress with increasing combined concentration, while high combined stress showed inhibition for the growth of DSE strains. The reason may be that DSEs are adaptive to the combined stress environment, which showed low promotion and high inhibition, and *Np* had more drought-and-Cd-combined tolerance. Liquid-shaker-culture experiments showed that the biomass of *Aa*, *At*, and *Np* strains achieved maximum values at 18% + 25 mg Cd/L, 18% + 50 mg Cd/L, and 18% + 25 mg Cd/L, respectively. Therefore, the moderately stressed might be more suitable for the growth of DSE strains [55].

Melanin, a crucial constituent of the cell wall in DSEs, serves to decelerate the rate of cellular water loss, thereby augmenting the fungus's survival and competitive edge in harsh environmental conditions [56–58]. However, in this study, with the increase of combined stress, the melanin content of *Aa*, *At*, and *Np* strains decreased, which indicated that the melanin content of different DSE strains had different effects on improving the stress resistance of fungi. The results showed that melanin accumulation was not an important character of heavy-metal tolerance in DSEs [59].

Glutathione (GSH) is an important antioxidant in living organisms, which not only scavenges free radicals and suboxide ion and reduces oxidative stress but also facilitates the maintenance of cellular homeostasis and contributes to the detoxification of harmful substances such as Cd [60,61]. In this study, the GSH content in strains *Aa*, *At*, and *Np* was observed to increase under all combined stress treatments when compared to the control group. However, as the intensity of the combined stress escalated, the GSH content in strains *Aa* and *At* initially rose and then declined. This biphasic response may be attributed to the role of GSH in cellular defense mechanisms, where it participates in the neutralization of reactive oxygen species (ROS). The combined stress likely elevated intracellular GSH levels, which, given its sulfhydryl structure, can react with ROS. Consequently, as GSH becomes engaged in these detoxification processes, its cellular concentration diminishes. A related study showed that two plants of the *Atriplex atacamensis* and *A. halimus* further reduced growth parameters but not GSH and proline contents under the combined stress of Cu, NaCl, and PEG, showing positive tolerance responses [62]. Furthermore, DSE has been shown to enhance antioxidant activity in plants, leading to increased levels of GSH and putrescine under heat-stress conditions [63]. Overall, this research indicates that DSEs play a significant role in enhancing plant tolerance to environmental stresses through mechanisms involving GSH and other antioxidants.

MDA, which membrane lipids create in reaction to reactive-oxygen species, is a trustworthy indicator of the degree of damage to the plasma membrane [64,65]. In this study, the MDA content of *Aa* and *Np* increased under all combined stress. The MDA content of *At* decreased compared with the control under moderate combined stress, indicating that *At* had a certain ability to weaken the accumulation of membrane-lipid-peroxidation products to resist the adverse effects of combined stress, and DSE cells were damaged when it exceeded a certain range. As shown by the correlation heat map, there was a positive connection between the MDA content of *Aa*, *Np*, *At* and the Cd concentration. It has been shown that there is a positive correlation between leaf MDA content and Cd concentration in the substrate [13], which is similar to the results of this study.

Organisms can adapt to a stressful environment by adding the content of soluble proteins and soluble sugars, thus raising the concentration of cell fluid [66–68]. The study findings demonstrate that the soluble protein content of DSE strains varied with increasing levels of combined stress. The soluble protein content in strains *At* and *Np* exhibited a pattern of decrease followed by an increase under the combined stress conditions, in contrast to the control group. The soluble protein content in strain *Aa* consistently increased across all stress treatments. The soluble sugar content of DSE strains was higher than that of the control treatment under all combined stress. Both *Aa* and *Np* reached their maximum values under moderate compound stress, whereas the soluble sugar content of *At* reached its maximum value under heavy combined stress. Organisms often adapt to stress-induced osmotic imbalances by actively augmenting the levels of osmoregulatory substances, which serves to ameliorate the detrimental impacts of heavy-metal exposure.

CAT, SOD, and POD are the most important antioxidant enzymes that can remove ROS free radicals in bacteria to protect cell membranes from various stress damage [69,70]. It has been proven that under extreme stress, DSE isolated from the root system of desert plants increased SOD activity [71]. In the current study, under 27% + 100 mg Cd/L treatment, the SOD activity of the *Aa* and *At* strains was greatly increased, suggesting that the strains can stimulate SOD activity to reduce toxic accumulation under combined stress. In contrast to the control, the SOD activity of *Np* demonstrated a pattern of decreasing and then increasing with the strengthening of the combined stress, and the CAT and POD of *Np* showed higher activities compared to both *Aa* and *At*. The *Aa* and *At* showed lower CAT and POD activities compared to the control at a combined concentration of 27% + 100 mg Cd/L and 9% + 100 mg Cd/L, respectively. Possibly due to the accumulation of $H_2O_2$ from DSE in the organisms caused by the combined stress, the oxidative system in the organisms was exacerbated. The CAT activity of the plants inoculated with DSE increased at all Cd concentrations, implying that this enzyme plays a crucial role in DSE-induced protection against oxidative stress in plants [13]. Tyagi et al. [72] discovered that when wheat plants were under drought stress, AM fungi increased the SOD activity. Additionally, the study by Muhammad et al. [73,74] indicated that melatonin significantly increased the activity of POD, CAT, and SOD in maize seedlings under drought stress, leading to improved drought tolerance. Overall, these studies emphasize the significance of antioxidant enzymes, including POD, CAT, and SOD, in increasing the resistance of plants to drought and other environmental stresses. The SEM model analysis revealed that *Np* and *At* could positively affect SOD and CAT activities and synergistically resist the combined stress.

### 4.4. Effects of DSE on Growth Performance of *Astragalus membranaceus* under Drought and Cd Stress

DSE is widespread in natural ecosystems and can help plants to resist a variety of stressful environments, especially under harsh conditions (e.g., saline, contaminated habitats, and arid ecosystems) [9]. Inoculation with DSE is beneficial for plant growth and resistance, but different DSE strains vary in their ability to promote host plants [75]. It was shown that under moderate- and high-salt stress conditions, *P. macrospinosa* and *Cadophora* sp. improved the shoot and root growth of tomato plants after 6 weeks of inoculation [76]. Under single and synergistic stresses of drought and salt, inoculation with *F. mosseae* significantly increased the fresh weight of hemp plants. However, inoculation with *F. Mosseae* had no significant effect on hemp plant height [77]. The study showed that inoculation with *N. phragmitis* under drought stress significantly increased the plant height of *Lycium ruthenicum* [78]. Compared with non-inoculated treatments, there was no significant difference in inoculation with *N. phragmitis* on the shoot height of *A. membranaceus* in all treatments of this experiment. It may be that the same DSE shows different effects on different plants. However, inoculation with *N. phragmitis* under D, DCd1, and DCd3 treatments significantly increased the root weight of *A. membranaceus*. Compared with non-inoculated treatments, root weight was significantly increased inoculation of *A. tellustris* under Cd2, D, and DCd1 treatments. The effect of inoculation with DSE on the growth of *A. membranaceus* under drought and Cd stress varied depending on the type of DSE but was generally beneficial.

## 5. Conclusions

In this study, we found that the colony growth rates of *A. alstroemeriae, N. phragmitis, P. chlamydocopiosa, M. phaseolina,* and *M. pseudophaseolina* were the first to reach the maximum diameter at a PEG concentration of 18%. *A. alstroemeriae, N. phragmitis,* and *A. tellustris* still possessed growth activity as the concentration of Cd rose. In addition, we investigated the performance and tolerance parameters of *A. alstroemeriae, N. phragmitis,* and *A. tellustris* strains under combined drought and Cd stress. The results of solid cultures under combined stress showed that the growth rate of *N. phragmitis* was significantly better than that of other strains. In the liquid culture condition, *A. alstroemeriae* positively influenced biomass and negatively affected Cd content, while *N. phragmitis* and *A. tellustris* positively influenced Cd content. *A. alstroemeriae, N. phragmitis,* and *A. tellustris* positively influenced SOD activity, and MDA content was positively correlated with Cd content. The GSH contents of *A. alstroemeriae, N. phragmitis,* and *A. tellustris* were increased under the combined stress of drought and Cd. Inoculation of *N. phragmitis* and *A. tellustris* promotes plant growth under synergistic drought and Cd stress. The utilization of DSEs to enhance plant growth under different combined stresses of drought and Cd has significant potential in the future.

## 6. Patents

A patent entitled "A DSE strain with drought-and-Cd-combined tolerance and its application in improving plant stress resistance" has been published and is currently in the application stage.

**Author Contributions:** Conceptualization, D.W. and X.H.; methodology, D.W., Y.X., L.Y. and W.Z.; formal analysis, D.W. and C.H.; data curation, D.W. and L.Y.; investigation, D.W.; writing—original draft preparation, D.W.; writing—review and editing, X.H. and C.H.; visualization, D.W., Y.X. and W.Z.; project administration, C.H. and X.H.; funding acquisition, C.H. and X.H. All authors have read and agreed to the published version of the manuscript.

**Funding:** This research was supported by the National Key R&D Program of China (No. 2022YFC3501501), the Natural Science Foundation of Hebei Province (No. H2022201056), and Central Guidance for Local Scientific and Technological Development Funding Projects (No. 236Z2904G).

**Institutional Review Board Statement:** Not applicable.

**Informed Consent Statement:** Not applicable.

**Data Availability Statement:** The original contributions presented in the study are included in the article, further inquiries can be directed to the corresponding authors.

**Conflicts of Interest:** The authors declare no conflict of interest.

## References


1. Sintaha, M.; Man, C.K.; Yung, W.S.; Duan, S.; Li, M.W.; Lam, H.M. Drought stress priming improved the drought tolerance of soybean. *Plants* **2022**, *11*, 2954. [CrossRef] [PubMed]
2. Caser, M.; D'Angiolillo, F.; Chitarra, W.; Lovisolo, C.; Ruffoni, B.; Pistelli, L.; Pistelli, L.; Scariot, V. Ecophysiological and phytochemical responses of *Salvia sinaloensis* Fern. to drought stress. *Plant Growth Regul.* **2018**, *84*, 383–394. [CrossRef]
3. Gao, Y.; Li, H.P.; Song, Y.; Zhang, F.L.; Yang, Z.G.; Yang, Y.; Grohmann, T. Response of glutathione pools to cadmium stress and the strategy to translocate cadmium from roots to leaves (*Daucus carota* L.). *Sci. Total Environ.* **2022**, *823*, 153575. [CrossRef] [PubMed]
4. Liu, X.R.; Chen, L.H.; Zhang, J.; He, S.H. Effects of drought and cadmium pollution on the physiology and cadmium enrichment of *Pennisetum sinese*. *Acta Bot. Boreal.—Occid. Sin.* **2019**, *39*, 277–284.
5. Cheng, Y.X.; Qiu, L.Z.; Shen, P.K.; Wang, Y.Q.; Li, J.L.; Dai, Z.Y.; Qi, M.F.; Zhou, Y.; Zou, Z.K. Transcriptome studies on cadmium tolerance and biochar mitigating cadmium stress in muskmelon. *Plant Physiol. Biochem.* **2023**, *197*, 107661. [CrossRef] [PubMed]
6. Ji, B.Y.; Xuan, L.S.; Zhang, Y.X.; Zhang, G.Q.; Meng, J.; Mu, W.R.; Liu, J.J.; Paek, K.Y.; Park, S.Y.; Wang, J.; et al. Advances in biotechnological production and metabolic regulation of *Astragalus membranaceus*. *Plants* **2023**, *12*, 1858. [CrossRef] [PubMed]
7. Zhao, R. *Systematic Evaluation of Heavy Metal Pollution of Eight Chinese Herbal Medicines in China*; Beijing University of Chinese Medicine: Beijing, China, 2016.
8. Islam, M.; Sandhi, A. Heavy metal and drought stress in plants: The role of microbes-A review. *Gesunde Pflanz.* **2023**, *75*, 695–708. [CrossRef]
9. Xie, L.L.; He, X.L.; Wang, K.; Hou, L.F.; Sun, Q. Spatial dynamics of dark septate endophytes in the roots and rhizospheres of *Hedysarum scoparium* in northwest China and the influence of edaphic variables. *Fungal Ecol.* **2017**, *26*, 135–143. [CrossRef]
10. Akhtar, N.; Wani, A.K.; Dhanjal, D.S.; Mukherjee, S. Insights into the beneficial roles of dark septate endophytes in plants under challenging environment: Resilience to biotic and abiotic stresses. *World J. Microbiol. Biotechnol.* **2022**, *38*, 79. [CrossRef]
11. Newsham, K.K. A meta-analysis of plant responses to dark septate root endophytes. *New Phytol.* **2011**, *190*, 783–793. [CrossRef]
12. Li, X.; He, X.L.; Hou, L.F.; Ren, Y.; Wang, S.J.; Su, F. Dark septate endophytes isolated from a xerophyte plant promote the growth of *Ammopiptanthus mongolicus* under drought condition. *Sci. Rep.* **2018**, *8*, 7896. [CrossRef] [PubMed]
13. Wang, J.L.; Li, T.; Liu, G.Y.; Smith, J.M.; Zhao, Z.W. Unraveling the role of dark septate endophyte (DSE) colonizing maize (*Zea mays*) under cadmium stress: Physiological, cytological and genic aspects. *Sci. Rep.* **2016**, *6*, 22028. [CrossRef]
14. Xu, M.H.; Li, X.; Ye, Q.N.; Gong, F.; He, X.L. Occurrence of dark septate endophytes in *Phragmites australis* in Baiyang Lake and their resistance to Cd stress. *Pedosphere* **2023**, *34*, 484–496. [CrossRef]
15. Li, M.; Hou, L.F.; Liu, J.Q.; Yang, J.Y.; Zuo, Y.L.; Zhao, L.L.; He, X.L. Growth-promoting effects of dark septate endophytes on the non-mycorrhizal plant *Isatis indigotica* under different water conditions. *Symbiosis* **2021**, *85*, 291–303. [CrossRef]
16. He, C.; Wang, W.; Hou, J. Plant Growth and soil microbial impacts of enhancing licorice with inoculating dark septate endophytes under drought stress. *Front. Microbiol.* **2019**, *10*, 2277. [CrossRef]
17. Li, X.; Liu, Y.X.; Ye, Q.N.; Xu, M.H.; He, X.L. Application of desert DSEs to nonhost plants: Potential to promote growth and alleviate drought stress of wheat seedlings. *Agriculture* **2022**, *12*, 1539. [CrossRef]
18. Xiao, Y.; Dai, M.X.; Zhang, G.Q.; Yang, Z.X.; He, Y.M.; Zhan, F.D. Effects of the dark septate endophyte (DSE) *Exophiala pisciphila* on the growth of root cell wall polysaccharides and the cadmium content of *Zea mays* L. under cadmium stress. *J. Fungi* **2021**, *7*, 1035. [CrossRef]
19. Chen, S.; Zhang, G.; Liang, X.; Wang, L.; Li, Z.; He, Y.; Li, B.; Zhan, F. A Dark septate endophyte improves cadmium tolerance of maize by modifying root morphology and promoting cadmium binding to the cell wall and phosphate. *J. Fungi* **2023**, *9*, 531. [CrossRef]
20. Su, Z.Z.; Dai, M.D.; Zhu, J.N.; Liu, X.H.; Li, L.; Zhu, X.M.; Wang, J.Y.; Yuan, Z.L.; Lin, F.C. Dark septate endophyte *Falciphora oryzae*-assisted alleviation of cadmium in rice. *J. Hazard. Mater.* **2021**, *419*, 126435. [CrossRef]
21. Bi, Y.L.; Xie, L.L. Functions of arbuscular mycorrhizal fungi and dark septate endophytes in ecological restoration. *Acta Microbiol. Sin.* **2021**, *61*, 58–67.
22. Yang, X.R. Species Diversity and Drought Tolerance of Dark Septate Endophytes from Different Germplasm of *Glycyrrhiza uralensis* Fisch. Master's Thesis, Hebei University, Baoding, China, 2023.
23. Yao, J.J. Effects of DSE Fungi Combined with Organic Residues on Growth and Nutrient Absorption of *Isatis indigotica* under Drought Stress. Master's Thesis, Hebei University, Baoding, China, 2022.
24. Ren, Y.F. Study on Growth Promoting and Drought Resistance Potential of *Astragalus membranaceus* by Double Inoculation of DSE Fungi and *Trichoderma* spp. Master's Thesis, Hebei University, Baoding, China, 2022.
25. Shi, J.X. Spatiotemporal Distribution and Drought Promotion of Dark Septate Endophytes Fungi in *Lycium ruthenicum* in Northwest China. Master's Thesis, Hebei University, Baoding, China, 2021.
26. Wei, M. Resources Distribution and Salt Tolerance of Dark Septate Endophytes in *Lycium ruthenicum* in Northwest Desert Region. Master's Thesis, Hebei University, Baoding, China, 2022.
27. Han, L. Species Diversity and Cd Tolerance of Dark Septate Endophytes in Medicinal Plant in Anguo. Hebei Province. Doctoral Thesis, Hebei University, Baoding, China, 2022.

28. Li, M.; He, C.; Wei, M.; Long, J.M.; Wang, J.R.; Yang, X.R.; Wang, K.H.; He, X.L. Temporal and spatial dynamics and functional metabolism of dark septate endophytes of *Gymnocarpos przewalskii* Maxim. in Northwest Desert, China. *Appl. Soil. Ecol.* **2024**, *194*, 105194. [CrossRef]
29. Ban, Y.H.; Xu, Z.Y.; Yang, Y.R.; Zhang, H.H.; Chen, H.; Tang, M. Effect of dark septate endophytic fungus *Gaeumannomyces cylindrosporus* on plant growth, photosynthesis and Pb tolerance of maize (*Zea mays* L.). *Pedosphere* **2017**, *27*, 283–292. [CrossRef]
30. Bai, J.H.; Jia, X.M.; Wu, Y.Q.; Wang, Y.K.; Song, W.Y.; Liu, Y.N. Ability of DSE against abiotic stresses and improving drought resistance of *Solanum tuberosu*. *Crops* **2023**, *39*, 150–159.
31. Medina-Armijo, C.; Isola, D.; Illa, J.; Puerta, A.; Viñas, M.; Prenafeta-Boldú, F.X. The metallotolerance and biosorption of As (V) and Cr (VI) by black fungi. *J. Fungi* **2024**, *10*, 47. [CrossRef] [PubMed]
32. Tremarin, A.; Longhi, D.A.; Salomão, B.; Aragão, G.M.F. Modeling the growth of *Byssochlamys fulva* and *Neosartorya fischeri* on solidified apple juice by measuring colony diameter and ergosterol content. *Int. J. Food Microbiol.* **2015**, *193*, 23–28. [CrossRef] [PubMed]
33. Melati, I.; Rahayu, G.; Surono; Effendi, H.; Henny, C.; Susanti, E. Chromium (VI) bioremediation potential of dark septate endophytic (DSE) fungi. *IOP Conf. Ser. Earth Environ. Sci.* **2023**, *1201*, 012077. [CrossRef]
34. Ginnopol, C.N.; Ries, S.K. Superoxide dismutase: II. purification and quantitative relationship with water-soluble protein in seedling. *Plant Physiol.* **1997**, *59*, 315–318.
35. Gao, J.F. *Experimental Techniques in Plant Physiology*; World Book Publishing House: Xi'an, China, 2000; pp. 137–202.
36. Cai, Q.S. *Experiments in Plant Physiology*; China Agricultural University Press: Beijing, China, 2013.
37. Gaber, D.A.; Berthelot, C.; Camehl, I.; Kovács, G.M.; Blaudez, D.; Franken, P. Salt stress tolerance of dark septate endophytes is independent of melanin accumulation. *Front. Microbiol.* **2020**, *11*, 562931. [CrossRef]
38. Ellis, D.H.; Griffiths, D.A. The location and analysis of melanins in the cell walls of some soil fungi. *Can. J. Microbiol.* **1974**, *20*, 1379–1386. [CrossRef]
39. Zhang, C.; Liu, F.; Kong, W.W.; He, Y. Application of visible and near-infrared hyperspectral imaging to determine soluble protein content in oilseed rape leaves. *Sensors* **2015**, *15*, 16576–16588. [CrossRef]
40. Zhang, Z.L.; Qu, W.J. *Laboratory Instruction of Plant Physiology*; Higher Education Press: Beijing, China, 2003.
41. Li, Z.L.; Min, D.D.; Fu, X.D.; Zhao, X.M.; Wang, J.H.; Zhang, X.H.; Li, F.J.; Li, X.A. The roles of *SlMYC2* in regulating ascorbate-glutathione cycle mediated by methyl jasmonate in postharvest tomato fruits under cold stress. *Sci. Hortic.* **2021**, *288*, 110406. [CrossRef]
42. Jiao, W.X.; Liu, X.; Chen, Q.M.; Du, Y.M.; Li, Y.Y.; Yue, F.L.; Dong, X.Q.; Fu, M.R. Epsilon-poly-L-lysine (ε-PL) exhibits antifungal activity *in vivo* and *in vitro* against Botrytis cinerea and mechanism involved. *Postharvest Biol. Technol.* **2020**, *168*, 111270. [CrossRef]
43. Lu, Z.K.; Li, M.; Ding, G.J. Growth characteristics of three ectomycorrhizal fungi under PEG-6000 stress. *J. Northwest Forest. Univ.* **2020**, *35*, 151–158.
44. Yu, J.; He, X.L.; Zhao, L.L.; Su, F. Colonization and community composition of dark endophytic fungi in Fengfeng mining area of Hebei. *J. Fungal Res.* **2018**, *16*, 228–238.
45. Yue, H.; Sun, X.M.; Wang, T.T.; Zhang, A.L.; Han, D.J.; Wei, G.H.; Song, W.N.; Shu, D.T. Host genotype-specific rhizosphere fungus enhances drought resistance in wheat. *Microbiome* **2024**, *12*, 44.
46. Mirshad, P.P.; Puthur, J.T. Arbuscular mycorrhizal association enhances drought tolerance potential of promising bioenergy grass (*Saccharum arundinaceum* Retz.). *Environ. Monit. Assess.* **2016**, *188*, 425. [CrossRef]
47. Begum, N.; Akhtar, K.; Ahanger, M.A.; Iqbal, M.; Wang, P.; Mustafa, N.S.; Zhang, L. Arbuscular mycorrhizal fungi improve growth, essential oil, secondary metabolism, and yield of tobacco (*Nicotiana tabacum* L.) under drought stress conditions. *Environ. Sci. Pollut. Res.* **2021**, *28*, 45276–45295. [CrossRef] [PubMed]
48. Silambarasan, S.; Logeswari, P.; Vangnai, A.S.; Kamaraj, B.; Cornejo, P. Plant growth-promoting actinobacterial inoculant assisted phytoremediation increases cadmium uptake in *Sorghum bicolor* under drought and heat stresses. *Environ. Pollut.* **2022**, *307*, 119489. [CrossRef]
49. Zhu, L.L.; Li, T.; Wang, C.J.; Zhang, X.R.; Xu, L.J.; Xu, R.B.; Zhao, Z.W. The effects of dark septate endophyte (DSE) inoculation on tomato seedlings under Zn and Cd stress. *Environ. Sci. Pollut. Res.* **2018**, *25*, 35232–35241. [CrossRef]
50. Shadmani, L.; Jamali, S.; Fatemi, A. Isolation, identification, and characterization of cadmium-tolerant endophytic fungi isolated from barley (*Hordeum vulgare* L.) roots and their role in enhancing phytoremediation. *Braz. J. Microbiol.* **2021**, *52*, 1097–1106. [CrossRef]
51. Gonçalves, M.F.M.; Aleixo, A.; Vicente, T.F.L.; Esteves, A.C.; Alves, A. Three new species of *Neocamarosporium* isolated from saline environments: *N. aestuarinum* sp. nov., *N. endophyticum* sp. nov. and *N. halimiones* sp. nov. *Mycosphere* **2019**, *10*, 608–621.
52. Hou, L.F.; Yu, J.; Zhao, L.L.; He, X.L. Dark septate endophytes improve the growth and the tolerance of *Medicago sativa* and *Ammopiptanthus mongolicus* under cadmium stress. *Front. Microbiol.* **2020**, *10*, 3061. [CrossRef] [PubMed]
53. Lin, Y.B.; Wang, X.Y.; Wang, B.P.; Mohamad, O.; Wei, G.H. Bioaccumulation characterization of zinc and cadmium by *Streptomyces zinciresistens*, a novel actinomycete. *Ecotoxicol. Environ. Saf.* **2012**, *77*, 7–17. [CrossRef] [PubMed]
54. Xue, S.P.; Wang, X.H. Inoculation of soil with cadmium-resistant *Actinomycetes* flora reduces cadmium accumulation in rice (*Oryza Sativa* L.). *Int. J. Environ. Pollut. Remed.* **2021**, *9*, 1–11.

55. Li, X.; He, C.; He, X.L.; Su, F.; Hou, L.F.; Ren, Y.; Hou, Y.T. Dark septate endophytes improve the growth of host and non-host plants under drought stress through altered root development. *Plant Soil* **2019**, *439*, 259–272. [CrossRef]
56. Bell, A.A.; Wheeler, M.H. Biosynthesis and functions of fungal melanins. *Annu. Rev. Phytopathol.* **1986**, *24*, 411–451. [CrossRef]
57. Fernandez, C.W.; Koide, R.T. The function of melanin in the ectomycorrhizal fungus *Cenococcum geophilum* under water stress. *Fungal Ecol.* **2013**, *6*, 479–486. [CrossRef]
58. Eisenman, H.C.; Greer, E.M.; McGrail, C.W. The role of melanins in melanotic fungi for pathogenesis and environmental survival. *Appl. Microbiol. Biotechnol.* **2020**, *104*, 4247–4257. [CrossRef]
59. Berthelot, C.; Zegeye, A.; Gaber, D.A.; Chalot, M.; Franken, P.; Kovács, G.M.; Leyval, C.; Blaudez, D. Unravelling the role of melanin in Cd and Zn tolerance and accumulation of three dark septate endophytic species. *Microorganisms* **2020**, *8*, 537. [CrossRef]
60. Zechmann, B. Compartment-specific importance of glutathione during abiotic and biotic stress. *Front. Plant Sci.* **2014**, *5*, 566. [CrossRef]
61. Zhan, F.D.; Li, B.; Jiang, M.; Qin, L.; Wang, J.X.; He, Y.M.; Li, Y. Effects of a root-colonized dark septate endophyte on the glutathione metabolism in maize plants under cadmium stress. *J. Plant Interact.* **2017**, *12*, 421–428. [CrossRef]
62. Orrego, F.; Ortiz-Calderón, C.; Lutts, S.t.; Ginocchio, R. Growth and physiological effects of single and combined Cu, NaCl, and water stresses on *Atriplex atacamensis* and *A. halimus. Environ. Exp. Bot.* **2020**, *169*, 103919. [CrossRef]
63. Bi, Y.L.; Xue, Z.K. Dark septate endophyte inoculation enhances antioxidant activity in *Astragalus membranaceus* var *mongholicus* under heat stress. *Physiol. Plant.* **2023**, *175*, e14054. [PubMed]
64. Zhang, Y.N.; Luan, Q.F.; Jiang, J.M.; Li, Y.J. Prediction and utilization of malondialdehyde in exotic pine under drought stress using near-infrared spectroscopy. *Front. Plant Sci.* **2021**, *12*, 735275. [CrossRef] [PubMed]
65. Espanany, A.; Fallah, S.; Tadayyon, A. Seed priming improves seed germination and reduces oxidative stress in black cumin (*Nigella sativa*) in presence of cadmium. *Ind. Crops Prod.* **2016**, *79*, 195–204. [CrossRef]
66. Fu, R.; Zhang, H.Y.; Liang, X.Y.; Li, M.; Li, J.L.; Song, Y.J.; Li, R.X.; Wang, X.Y. Physiological response of *Mesembryanthemum cordifolium* L. F. to salt and drought stresses. *Shandong Agr. Sci.* **2019**, *53*, 17–21.
67. Chen, C.S.; Xie, Z.X.; Liu, X.J. Effects of drought and salt interactions on the growth of winter wheat seedlings and their physiological characteristics of stress tolerance. *Chin. J. Appl. Ecol.* **2009**, *20*, 811–816.
68. Zhao, Q.Q.; Xu, H. Explore the changes of thiol compounds and antioxidant enzymes in *Oudemansiella* radicata mycelium under cadmium and lead stresses. *J. Sichuan Univ. (Nat. Sci. Ed.)* **2014**, *51*, 1051–1055.
69. Dastogeer, K.M.G. Influence of fungal endophytes on plant physiology is more pronounced under stress than well-watered conditions: A meta-analysis. *Planta* **2018**, *248*, 1403–1416. [CrossRef]
70. Xie, Y.J.; Wang, B.; Liang, X.H.; Han, Z.D. Effects of drought stress on reactive oxygen metabolism and protective enzyme activities in *Glycyrrhiza* seedlings. *J. Agr. Sci.* **2008**, *29*, 19–22.
71. Zuo, Y.L.; Hu, Q.N.; Liu, J.Q.; He, X.L. Relationship of root dark septate endophytes and soil factors to plant species and seasonal variation in extremely arid desert in Northwest China. *Appl. Soil Ecol.* **2022**, *175*, 104454. [CrossRef]
72. Tyagi, J.; Varma, A.; Pudake, R.N. Evaluation of comparative effects of arbuscular mycorrhiza (*Rhizophagus intraradices*) and endophyte (*Piriformospora indica*) association with finger millet (*Eleusine coracana*) under drought stress. *Eur. J. Soil Biol.* **2017**, *81*, 1–10. [CrossRef]
73. Muhammad, I.; Yang, L.; Ahmad, S.; Mosaad, I.S.M.; Al-Ghamdi, A.A.; Abbasi, A.M.; Zhou, X.B. Melatonin application alleviates stress-induced photosynthetic inhibition and oxidative damage by regulating antioxidant defense system of maize: A meta-analysis. *Antioxidants* **2022**, *11*, 512. [CrossRef] [PubMed]
74. Muhammad, I.; Yang, L.; Ahmad, S.; Farooq, S.; Khan, A.; Muhammad, N.; Ullah, S.; Adnan, M.; Ali, S.; Liang, Q.P.; et al. Melatonin-priming enhances maize seedling drought tolerance by regulating the antioxidant defense system. *Plant Physiol.* **2023**, *191*, 2301–2315. [CrossRef] [PubMed]
75. Mayerhofer, M.S.; Kernaghan, G.; Harper, K.A. The effects of fungal root endophytes on plant growth: A meta-analysis. *Mycorrhiza* **2013**, *23*, 119–128. [CrossRef] [PubMed]
76. Gaber, D.A.; Berthelot, C.; Blaudez, D.; Kovács, G.M.; Franken, P. Impact of dark septate endophytes on salt stress alleviation of tomato plants. *Front. Microbiol.* **2023**, *14*, 1124879. [CrossRef] [PubMed]
77. Yuan, H.P.; Si, H.; Ye, Y.S.; Ji, Q.Y.; Wang, H.Y.; Zhang, Y.H. Arbuscular mycorrhizal fungi-mediated modulation of physiological, biochemical, and secondary metabolite responses in hemp (*Cannabis sativa* L.) under salt and drought stress. *J. Fungi* **2024**, *10*, 283. [CrossRef]
78. He, C.; Han, T.T.; Tan, L.; Li, X.E. Effects of dark septate endophytes on the performance and soil microbia of *Lycium ruthenicum* under drought stress. *Front. Plant Sci.* **2022**, *13*, 898378. [CrossRef]

Journal of *Fungi*

MDPI

*Article*

# The Metallotolerance and Biosorption of As(V) and Cr(VI) by Black Fungi

**Cristy Medina-Armijo [1,2,*], Daniela Isola [3], Josep Illa [4], Anna Puerta [1], Marc Viñas [1] and Francesc X. Prenafeta-Boldú [1,*]**

1 Program of Sustainability in Biosystems, Institute of Agrifood Research and Technology (IRTA), 08140 Caldes de Montbui, Spain
2 Faculty of Pharmacy and Food Sciences, University of Barcelona, 08028 Barcelona, Spain
3 Department of Economics, Engineering, Society and Business Organization (DEIM), University of Tuscia, 01100 Viterbo, Italy
4 Department of Computing and Industrial Engineering, University of Lleida, 25001 Lleida, Spain
* Correspondence: cristy.medina@irta.cat (C.M.-A.); francesc.prenafeta@irta.cat (F.X.P.-B.)

**Abstract:** A collection of 34 melanized fungi isolated previously from anthropogenic contaminated sites were assessed for their tolerance to toxic concentrations of As(V) and Cr(VI) anions. Three strains of the species *Cyphellophora olivacea*, *Rhinocladiella similis*, and *Exophiala mesophila* (Chaetothyriales) were identified as hyper-metallotolerant, with estimated $IC_{50}$ values that ranged from 11.2 to 16.9 g $L^{-1}$ for As(V) and from 2.0 to 3.4 g $L^{-1}$ for Cr(VI). *E. mesophila* and *R. similis* were selected for subsequent assays on their biosorption capacity and kinetics under different pH values (4.0 and 6.5) and types of biomass (active and dead cells and melanin extracts). The fungal biosorption of As(V) was relatively ineffective, but significant removal of Cr(VI) was observed from liquid cultures. The Langmuir model with second-order kinetics showed maximum sorption capacities of 39.81 mg $Cr^{6+}$ $g^{-1}$ for *R. similis* and 95.26 mg $Cr^{6+}$ $g^{-1}$ for *E. mesophila* on a dry matter basis, respectively, while the kinetic constant for these two fungi was $1.32 \times 10^{-6}$ and $1.39 \times 10^{-7}$ g (mg $Cr^{6+}$ min)$^{-1}$. Similar experiments with melanin extracts of *E. mesophila* showed maximum sorption capacities of 544.84 mg $Cr^{6+}$ $g^{-1}$ and a kinetic constant of $1.67 \times 10^{-6}$ g (mg $Cr^{6+}$ min)$^{-1}$. These results were compared to bibliographic data, suggesting that metallotolerance in black fungi might be the result of an outer cell-wall barrier to reduce the diffusion of toxic metals into the cytoplasm, as well as the inner cell wall biosorption of leaked metals by melanin.

**Keywords:** Chaetothyriales; *Exophiala mesophila*; fungal melanin; heavy metal bioremediation; hexavalent chromium tolerance; Langmuir biosorption isotherm; pentavalent arsenic tolerance; second-order kinetic model

## 1. Introduction

Heavy metals and metalloids (HMMs) are found naturally in the Earth's crust at relatively diffused amounts, but they might become concentrated as a result of anthropogenic activities such as mining, agriculture, and husbandry, as well as several industrial manufacturing processes. Because of their inherent high toxicity, pollution with HMMs has become a pressing public health issue in many parts of the world [1]. Metal ions persist in the environment due to the bioaccumulation tendency of living organisms and their limited capacity to metabolize them into less toxic forms [2]. Once absorbed into the cell, HMMs bind to vital components, such as structural proteins, enzymes, and nucleic acids, impairing several fundamental metabolic functions [3].

Groundwater contaminated with arsenic is still poisoning millions of people, primarily from developing countries, and has been listed by the WHO as one of the ten chemicals of major public health concern [4,5]. Arsenic exposure to humans can be attributed to various sources, including natural deposits, agricultural pesticides, and industrial effluents [5]. Natural geochemical processes can also lead to toxic and subtoxic concentrations of arsenic in groundwater [6]. Arsenic-contaminated water typically contains both arsenate and arsenite species, As(V) and As(III). Water pollution with chromium is gaining more consideration because it is globally widespread and also represents an important public health concern [7,8]. The majority of chromium released into the environment originates from industrial activities, particularly from the processing and manufacturing of chemicals, minerals, steel, leather tanning, textile dyeing, metallurgical operations, and various other industrial processes [7]. The most stable oxidation states of chromium in water correspond to the hexavalent Cr(VI) and trivalent Cr(III) species, but concerns are primarily related to Cr(VI) owing to its high biotoxicity.

A number of technical approaches for the remediation of HMM pollution and water purification have been proposed [9]. These solutions should rely on technically and economically viable methods, such as the biosorption of HMMs into readily available biomass sources. Many reports in the literature describe the capacity of pure cultures of bacteria [10,11], algae [12], aquatic plants [13], and fungi [14,15] to remove HMM ions from aqueous solutions. Such bioremediation alternatives are an interesting option for the decontamination of water and soil because these processes require few reagents and energy, they generate relatively low amounts of toxic waste products, and because biosorption is highly effective in reducing HMMs at relatively low concentrations [1,13]. However, the biotechnological removal of HMMs still faces some scalability problems associated with the difficulty to find suitable microorganisms that are able to cope with a highly oligotrophic and toxic environment while having a high HMM biosorption potential.

In this context, one particular group of ascomycetes known as black fungi (BF), because of their common polymorphism as yeast-like, hyphal, and meristematic growth, have been identified recently for their capacity to bind heavy metals to their cell wall [16,17]. These fungi are characterized by a high morphological and metabolic plasticity and poly-extremotolerant traits, which allow them to colonize a diverse range of often divergent and uncommon habitats [18]. The main defining character of BF is a strongly melanized cell wall, which confers them protection against extreme environmental conditions, such as exposure to UV and ionizing radiation, desiccation, cold/hot temperatures, high salinity, and oligotrophic environments [19,20]. This latter trait has also been linked to the ability to use toxic volatile hydrocarbons as the only source of carbon and energy [21,22]. As for their identities, BF primarily fall into two main phylogenetic groups: the Eurotiomycetes (Chaetothyriales) and the Dothideomycetes (Dothideales, Cladosporiales, and a few other related orders in the Mycophaerellales) [18,23].

The structure of fungal melanin is somewhat similar to soil humic acids with respect to volatile compounds released upon pyrolysis and amino acid hydrolysis [24]. In particular, the melanin of BF is composed of short-distance non-hydrolysable strong carbon–carbon bonds based on 1,8-dihydroxynaphthalene (DHN), modified with different functional groups, such as carboxyl, phenolic, hydroxyl, and amino [25,26], which provide many potential binding or biosorption sites for metal ions [27]. Special attention has been given to two main functions attributed to melanin in relation to metal ions: as a reservoir for the temporary storage and release of certain nutrients, and as chelating agents of HMMs for protecting the cell against metal toxicity. However, the biodiversity of metallotolerant BF and their HMM biosorption potential have seldom been investigated.

In this study, we screened a collection of BF species isolated previously from diverse anthropogenic-polluted sites for their tolerance to As(V) and Cr(VI), selected, respectively, as a model heavy metal and metalloid. The biosorption potential was determined on selected metallotolerant strains, using living fungal cultures, dead biomass, and melanin extracts. The obtained results could give new insights into the biology of BF and might contribute to the development of biotechnological applications.

## 2. Materials and Methods

### 2.1. Biological Material

A culture collection of 34 BF strains confidently identified at the species level were used (Table 1). Most of these strains were isolated during previous studies [21,28,29] from car fuel dispensers and tanks, stone buildings exposed to pollution and toxic biocides, and washing machines. These strains are currently maintained at the Culture Collection of Fungi from Extreme Environments (CCFEE) at the Tuscia University in Viterbo, the Westerdijk Fungal Biodiversity Center (formerly Centralbureau voor Schimmelcultures—CBS), and the Institute of Agrifood Science and Technology (IRTA).

**Table 1.** List of melanized fungal strains used in this study.

| Species | Phylogenetic Group | Isolation Source | Strain nr [a] | Accession nr [b] |
|---|---|---|---|---|
| *Aulographina pinorum* | Asterinales | Diesel pump | CCFEE 6222 | OR660094 |
| *A. pinorum* | Asterinales | Diesel pump | CCFEE 6230 | MZ573423 |
| *Aureobasidium melanogenum* | Dothideales | Diesel car tank | CCFEE 6213 | OR660095 |
| *Au. melanogenum* | Dothideales | Diesel car tank | CCFEE 6234 | OR660096 |
| *Au. pullulans* | Dothideales | Gasoline car tank | CCFEE 5876 | JX681059 |
| *Au. pullulans* | Dothideales | Gasoline pump | CCFEE 6244 | OR660097 |
| *Cladosporium herbarum* [c] | Cladosporiales | Diesel pump | CCFEE 6193 | MZ573426 |
| *Cl. herbarum* | Cladosporiales | Diesel pump | CCFEE 6192 | OR660098 |
| *Coniosporium uncinatum* | Dothideomycetes i.s. [d] | Gasoline car tank | CCFEE 5820 | JX681057 |
| *Co. uncinatum* | Dothideomycetes i.s. | Gasoline pump | CCFEE 6149 | MZ573424 |
| *Cyphellophora olivacea* | Chaetothyriales | Biocide-treated monument | CCFEE 6619 | MT472271 |
| *Exophiala angulospora* | Chaetothyriales | Biocide-treated monument | CCFEE 6620 | MT472272 |
| *E. crusticola* | Chaetothyriales | Gasoline pump | CCFEE 6188 | OR660099 |
| *E. equina* | Chaetothyriales | Washing machine soap dispenser | CCFEE 5883 | JX681045 |
| *E. heteromorpha* | Chaetothyriales | Gasoline pump | CCFEE 6240 | MZ573439 |
| *E. heteromorpha* | Chaetothyriales | Diesel car tank | CCFEE 6150 | OR660100 |
| *E. lecanii-corni* | Chaetothyriales | Washing machine soap dispenser | CCFEE 5688 | OR660101 |
| *E. mesophila* | Chaetothyriales | Washing machine soap dispenser | CCFEE 5690 | JX681043 |
| *E. mesophila* | Chaetothyriales | Glued ceramics | IRTA M2-F10 | OR660102 |
| *E. oligosperma* | Chaetothyriales | Human patient | CBS 725.88 | AY163551 |
| *E. oligosperma* | Chaetothyriales | Diesel car tank | CCFEE 6139 | MZ573441 |
| *E. phaeomuriformis* | Chaetothyriales | Diesel car tank | CCFEE 6242 | MZ573445 |
| *E. xenobiotica* | Chaetothyriales | Gasoline car tank | CCFEE 5784 | OR660103 |
| *E. xenobiotica* | Chaetothyriales | Bathroom wet cell | CCFEE 5985 | JX681024 |
| *E. xenobiotica* | Chaetothyriales | Diesel car tank | CCFEE 6143 | OR660104 |
| *E. xenobiotica* | Chaetothyriales | Gasoline pump | CCFEE 6142 | OR660105 |
| *Knufia epidermis* | Chaetothyriales | Diesel car tank | CCFEE 6138 | MZ573455 |
| *K. epidermis* | Chaetothyriales | Gasoline car tank | CCFEE 5813 | JX681055 |
| *K. epidermis* | Chaetothyriales | Diesel car tank | CCFEE 6198 | OR660106 |
| *K. epidermis* | Chaetothyriales | Diesel car tank | CCFEE 6366 | OR660107 |
| *Neohortaea acidophila* | Mycospharellales | Lignite | CBS 113389 | OL739260 |
| *Rhizosphaera kalkholffii* | Dothideales | Diesel car tank | CCFEE 6144 | OR660108 |
| *Rhinocladiella similis* | Chaetothyriales | Diesel car tank | CCFEE 6361 | MZ573467 |
| *Scolecobasidium* cft *globale* | Venturiales | Diesel car tank | CCFEE 6363 | MZ573464 |

[a] CCFEE: Culture Collection of Fungi from Extreme Environments, Tuscia University; CBS: fungal collection of the Westerdijk Fungal Biodiversity Institute; IRTA: microbial collection of the Laboratory of Environmental Microbiology, Institute of Agrifood Research and Technology. [b] ITS1-5.8S-ITS2 ribosomal DNA sequences deposited in GenBank. [c] the strain reported as *Cladosporium herbarum* should be read as belonging to the corresponding species complex. [d] *Incertae sedis* (uncertain taxonomic placement).

### 2.2. Metallotolerance Assays

The capacity of the collected fungal strains to grow in the presence of increasing concentrations of As(V) and Cr(VI) was evaluated on solid cultures. Into Petri dishes (12 cm in diameter), we poured potato dextrose agar (PDA; Condalab, Torrejón de Ardoz, Spain) supplemented with 2.5, 5.0, 7.5, 10.0, or 12.5 g of $As^{5+}$ $L^{-1}$ from sodium arsenate ($Na_2HAsO_4{\cdot}7H_2O$; Thermo Fisher Scientific, Kandel, Germany) and 0.1, 0.5, 1.0, 1.5, and

2.5 g of $Cr^{6+}$ $L^{-1}$ from potassium dichromate ($K_2Cr_2O_7$; Scharlab ExpertQ®, Sentmenat, Spain). Each BF strain was inoculated six times with an inoculation loop on each agar plate containing a defined concentration of As(V) or Cr(VI). The radial growth of each fungal colony was measured by averaging orthogonal diameters, and this value was then averaged again for all six colonies from every single plate. These measurements were repeated after 7, 15, 30, 45, and 60 days of incubation, which was performed at 25 °C under dark conditions. This prolonged timeframe was established to account for the potential long-term adaptation of slow-growing BF strains. Unamended control plates were also included. The tolerance index (TI) was calculated for every strain by dividing the measured growth when exposed to the metal in relation to the control plates.

### 2.3. Production of Fungal Biomass

Fungal biomass for biosorption experiments on selected BF strains was produced in 0.5 L batch liquid cultures incubated at room temperature under shaking conditions (80 rpm) for 10 days. Yeast extract (4 g $L^{-1}$) was supplied as the carbon and energy source, and macronutrients were added in the form of 4.5 g of $KH_2PO_4$, 0.5 g of $K_2HPO_4$, 2.0 g of $NH_4Cl$, and 0.1 mg of $MgSO_4{\cdot}7H_2O$ per liter. Mineral micronutrients were added in the form of 2 mL of a stock solution that contained 120 mg of $FeCl_3$, 50 mg of $H_3BO_3$, 10 mg of $CuSO_4{\cdot}5H_2O$, 10 mg of KI, 45 mg of $MnSO_4{\cdot}H_2O$, 20 mg of $Na_2MoO_4{\cdot}H_2O$, 75 mg of $ZnSO_4{\cdot}H_2O$, 50 mg of $CoCl_2{\cdot}6H_2O$, 20 mg of $AlK(SO_4)_2{\cdot}12H_2O$, 13.25 g of $CaCl_2{\cdot}H_2O$, and 10 g of NaCl per liter. Spore suspensions of pure cultures (0.5 mL, $>10^6$ CFU $mL^{-1}$) were used as inoculant. After 10 days of incubation, the fungal biomass was harvested by centrifugation (20 min at 4000 rpm) and washed with milliQ sterilized water. This process was repeated three times before using fungal biomass in batch biosorption assays.

### 2.4. Extraction and Purification of Melanin

Melanin was extracted from pre-grown BF cultures by adapting an acid hydrolysis method used previously with other fungi [30,31]. Briefly, harvested fungal biomass was homogenized in 100 mL of 1 M NaOH (120 rpm for 10 min) and treated with hot alkali (1 M NaOH at 121 °C for 20 min). The resulting suspension was centrifuged at 10,000 rpm for 10 min to remove fungal biomass, and the brownish liquid fraction was acidified to pH 2.5 with HCl 6 N and incubated for 12 h at 100 °C. The resulting black precipitate was centrifuged (4000 rpm for 20 min) and washed with deionized water three times. The precipitate was then lyophilized at a pressure of 0.7 mBar at −50 °C for 24 h, and the obtained melanin powder was kept at −20 °C until use.

### 2.5. Biosorption Assays

The capacity of the fungal biomass to accumulate As(V) and Cr(VI) ions was tested on a oligotrophic liquid medium to minimize growth. Consequently, the number of sorption sites remained consistent. This medium contained a buffer of 35 mM $K_2HPO_4/NaH_2PO_4{\cdot}2H_2O$ (pH 6.5), glucose (0.3%), and yeast extract (0.01%), along with 20 to 200 mg $L^{-1}$ of As(V) or Cr(VI), added as $Na_2HAsO_4.7H_2O$ or $K_2Cr_2O_7$, respectively. Biomass from living and heat-inactivated fungal cultures (on a dry matter basis, approx. 50 mg DM) was resuspended into the liquid mineral medium and incubated in serum flasks (30 mL) under sterile conditions on a shaker (80 rpm at 25 °C). Experiments were carried out in triplicate, and liquid samples were taken after 2, 5, 7, 14, 21, and 30 days of incubation to measure the concentration of As(V) and Cr(VI). Biosorption experiments were repeated at pH 4.0 by adding 0.1 M HCl.

The assays for the biosorption of Cr(VI) onto melanin extracts were performed with 0.3 g of melanin powder resuspended in 50 mL of a solution (pH 6.5; 150 rpm; 25 °C). The initial solution's Cr(VI) concentration was 30 mg $Cr^{6+}$ $L^{-1}$ and incubations lasted up to 72 h. Incubations were carried out in triplicate, and liquid samples were taken regularly to measure the time-course evolution of the concentration of Cr(VI).

### 2.6. Analytical Methods

The content of total arsenic in liquid culture supernatant was determined using Flame Atomic Absorption Spectroscopy (Model SpectrAA-110, Varian, Mulgrave, Australia). Chromium (VI) in the liquid fraction was determined with a colorimetric method based on the reaction with the complexing agent 1,5-diphenylcabazide (Sigma-Aldrich, St. Louis, MI, USA) that forms a purple-violet-colored complex, which was quantified by measuring the absorbance at a wavelength of 540 nm using a spectrophotometer (model EMC-11S UV brand, Duisburg, Germany).

### 2.7. Numerical Methods and Statistical Analysis

The software GraphPad Prism version 8.0 (GraphPad Software, San Diego, CA, USA) was used for performing one-way ANOVA (multiple comparisons Tukey test) and for the calculation of the metal concentration causing a 50% inhibition of fungal growth ($IC_{50}$). After a given time of incubation, $IC_{50}$ was calculated by fitting the tolerance index ($TI$) measurements at different HMM concentrations ($C_t$) to a two-parameter Hill logistic model ($IC_{50}$ and slope factor $h$; Equation (1)).

$$TI = \frac{1}{\left(1 + \frac{IC_{50}}{C_t}\right)^h} \tag{1}$$

As for the biosorption assays, the specific amount of the HMMs that is taken up at any given time ($q_t$) by a defined amount of fungal biomass $M_a$, when incubated at an initial concentration $C_0$ in batch incubations of liquid volume $V$, is a function of the remaining HMM concentration ($C_t$) as the experimentally measured variable, as described by Equation (2).

$$q_t = \frac{V \cdot (C_0 - C_t)}{M_a} \tag{2}$$

The sorption rate is usually described either as first- or second-order kinetics (Equation (3)), where $k_i$ is the kinetic constant at the considered $i$th order, and $q_e$ is the adsorbed metal fraction when reaching equilibrium with the concentration $C_e$ of metal in solution. The integration of Equation (3), under the hypothesis of constant $C_e$, by considering second-order kinetics gives Equation (4), which relates $q_t$ with time. Equation (4) is used to produce a linear plot ($t \cdot {q_t}^{-1}$) versus $t$ with experimental data that allows the calculation of the two parameters $k_2$ and $q_e$ from the intercept and slope.

$$\frac{dq_t}{dt} = k_i \cdot (q_e - q_t)^i \tag{3}$$

$$\frac{t}{q_t} = \frac{1}{k_2 \cdot q_e^2} + \frac{t}{q_e} \tag{4}$$

When the concentration of metal in solution varies with time, the adsorbed fraction at equilibrium $q_e$ is usually modelled as a function of liquid equilibrium concentration $C_e$ by the Langmuir or Freundlich isotherm (Equations (5) and (6)). Both functions have two unknown parameters, $q_{\max}$ and $K_L$ in the Langmuir equation and $K_F$ and $n$ in the Freundlich equation.

$$q_e = q_{max} \frac{K_L \cdot C_e}{1 + K_L \cdot C_e} \tag{5}$$

$$q_e = K_L \cdot C_e^{1/n} \tag{6}$$

The substitution in Equation (3) of $q_e$ by Equations (5) or (6), the Langmuir or Freundlich isotherms named here as $q_e(C_t)$, and $q_t$ by Equation (2) results in the differential Equation (7). This new expression describes the temporal evolution of the concentration in the liquid phase depending on the assumed sorption isotherm model and kinetic order.

$$\frac{dC_t}{dt} = -k_i \cdot (\frac{M_a}{V} \cdot q_e(C_t) \cdot (C_0 - C_t))^i \quad (7)$$

For the given set of 3 parameters, Equation (7) was numerically integrated along time and the goodness of the fit to the measured $C_t$ values evaluated as the sum of square errors. A Matlab (version R2017a) routine was used for this purpose to find the parameter values that produced the best fit.

## 3. Results and Discussion

### 3.1. The Metallotolerance of Black Fungi

Of the 34 studied BF (Table 1), 28 strains were able to grow on some of the tested concentrations of As(V) and/or Cr(VI) (Table 2). Seventeen strains (50%) displayed measurable growth on agar cultures exposed to As(V) at the maximum tested concentration of 12.5 g of $As^{5+}$ $L^{-1}$, while eight strains (23.5%) showed no growth under the minimum tested concentration of 2.5 g of $As^{5+}$ $L^{-1}$ (Table 2). The first group of As(V)-tolerant strains primarily belonged to the Chaetothyriales (*Cyphellophora olivacea, Exophiala crusticola, E. equina, E. lecanii-corni, E. mesophila, E. oligosperma, E. phaeomuriformis, E. xenobiotica,* and *Rhinocladiella similis*), though some fungi from the Dothidiomycetes were found to be metallotolerant as well (*Aulographina pinorum, Aureobasidium melanogenum, Coniosporium uncinatum,* and *Neohortaea acidophila*). A few chaetotyrialean species were also found to be slightly As(V) tolerant (*E. angulospora, E. heteromorpha,* and *Knufia epidermis*), but representatives of the order Dothidiomycetes were predominant in this group (*Cladosporium herbarum, Scolecobasidium globalis,* and *Rhizosphaera kalkhoffii*). As for Cr(VI), only 5 chaetothyrialean strains (*Rhinocladiella similis, Cyphellophora olivacea, Exophiala mesophila, E. crusticola,* and *E. lecanii-corni*) were able to grow at the highest tested concentration of 2.5 g of $Cr^{6+}$ $L^{-1}$, while 12 strains showed measurable growth at the lowest tested concentrations of 0.1 g of $Cr^{6+}$ $L^{-1}$ (*Au. pullulans, C. uncinatum,* and *Cl. herbarum* in the Dothidiomyces; *K. epidermis, E. equina,* and *E. phaeomuriformis* in the Chaetothyriales).

The tolerance index (TI) of fungi to the tested concentrations of As(V) and Cr(VI) and the estimated 50% inhibitory concentration of these compounds ($IC_{50}$) after 30 days of incubation are summarized in Table 2. The 30-day incubated period was taken as an intermediate reference between short-term and long-term toxicity. When considering all incubation times, most fungi showed a negative correlation between TI values and exposure time to high HMM concentrations (Supplementary Table S1), which indicates that HMM toxicity tends to manifest after prolonged incubations. A few chaetothyrialean strains stood out from the rest because of their remarkable level of metallotolerance, such as *Exophiala crusticola* CCFEE 6188. This particular strain had the highest tolerance to As(V), with average TI values that after 30 days ranged from 1.00 at 2.5 g of $As^{5+}$ $L^{-1}$ to 0.50 at 12.5 g of $As^{5+}$ $L^{-1}$. The corresponding modelled $IC_{50}$ estimate was 10.0 g of $As^{5+}$ $L^{-1}$. The number of available strains of *E. crusticola* is relatively small, so little information is available on the ecophysiology of the species. The type strain was isolated from a biological soil crust sample of the Colorado Plateau and Great Basin desert [32]. More recently, this species has also been reported in the Atacama Desert, near Calama in Chile [33], which is among the driest sites in the world and lies close to a large open-pit copper mine. This environment illustrates the polyextremophilic nature of chaetothyrialean fungi to withstand drought, UV radiation, and exposure to toxic metals.

**Table 2.** Fungal TI and $IC_{50}$ to As(V) and Cr(VI) after the incubation of agar cultures exposed to the highest and lowest tested concentrations during 30 days. $IC_{50}$ was estimated by fitting the TI data to a logistic model ($r^2 > 0.9$ unless stated otherwise). The strains from Table 1 that are not shown did not display any growth on the tested As(V) and Cr(VI) concentrations.

| Species | Strain Nr | TI to As(V) | | $IC_{50}$ to As(V) | TI to Cr(VI) | | $IC_{50}$ to Cr(VI) |
|---|---|---|---|---|---|---|---|
| | | (2.5 g $L^{-1}$) | (12.5 g $L^{-1}$) | (g $As^{5+}$ $L^{-1}$) | (0.1 g $L^{-1}$) | (2.5 g $L^{-1}$) | (g $Cr^{6+}$ $L^{-1}$) |
| A. pinorum | CCFEE 6222 | 0.61 | – [a] | 3.87 (3.06–4.58) | – | – | – |
| | CCFEE 6230 | 0.44 | – | 1.70 (0.67–2.56) | – | – | – |
| Au. melanogenum | CCFEE 6213 | 0.49 | – | 2.42 (2.15–2.68) | – | – | – |
| Au. melanogenum | CCFEE 6234 | 0.66 | 0.31 | 4.14 (3.39–4.83) | – | – | – |
| Au. pullulans | CCFEE 6244 | 0.28 | – | 0.96 (0.69–1.21) | 0.44 | – | 0.07 (0.05–0.09) |
| Cl. herbarum | CCFEE 6193 | 0.56 | – | 2.61 (2.27–2.90) | – | – | – |
| Cl. herbarum | CCFEE 6192 | – | – | – | 0.24 | – | 0.00 (0.0002–0.008) |
| C. uncinatum | CCFEE 5820 | 0.36 | 0.22 | 0.75 (0.28–1.26) | 0.37 | – | 0.04 (0.02–0.06) |
| C. olivacea | CCFEE 6619 | 0.86 | 0.48 | 12.94 (11.53–15.09) | 0.47 | 0.19 | 0.08 (0.05–0.12) |
| E. crusticola | CCFEE 6188 | 1.00 | 0.50 | 10.04 (8.94–11.61) [b] | 0.73 | 0.32 | 0.56 (0.36–0.84) [b] |
| E. equina | CCFEE 5883 | 0.15 | – | 0.03 (0.0003–0.15) | 0.43 | – | 0.04 [c] |
| E. lecanii-corni | CCFEE 5688 | 0.32 | – | 0.20 (0.001–0.71) | 0.65 | 0.46 | 1.01 (0.82–1.30) [c] |
| E. mesophila | CCFEE 5690 | 0.42 | – | 1.50 (0.96–1.98) | – | – | – |
| | IRTA M2-F10 | 0.81 | 0.34 | 6.03 (5.19–6.94) [b] | 0.44 | 0.30 | 0.05 (0.02–0.09) |
| E. oligosperma | CBS 725.88 | 0.49 | – | 2.49 (1.62–3.20) | – | – | – |
| | CCFEE 6139 | 0.19 | – | 0.16 (0.02–0.44) | – | – | – |
| E. phaeomuriformis | CCFEE 6242 | 0.31 | – | 0.70 (0.32–1.09) | 0.32 | – | 0.01 (0.003–0.03) |
| E. xenobiotica | CCFEE 5784 | 0.32 | – | 0.67 (0.15–1.28) | – | – | – |
| | CCFEE 5985 | 0.50 | 0.24 | 2.92 (1.59–3.96) [b] | – | – | – |
| | CCFEE 6142 | 0.33 | – | 0.39 (0.04–0.93) | – | – | – |
| | CCFEE 6143 | 0.38 | – | 0.67 (0.16–1.25) | – | – | – |
| K. epidermis | CCFEE 5813 | 0.50 | 0.24 | 2.92 (1.59–3.96) [b] | – | – | – |
| | CCFEE 6138 | 0.14 | – | 0.01 (~–0.15) | – | – | – |
| | CCFEE 6198 | 0.42 | – | 1.93 (1.54–2.27) | – | – | – |
| | CCFEE 6366 | 0.23 | – | 0.25 (0.06–0.53) | – | – | – |
| N. acidophila | CBS 113389 | 0.41 | 0.52 | ~ | – | – | – |
| R. similis | CCFEE 6361 | 0.65 | 0.56 | 48.26 (16.14–~) [b] | 0.60 | 0.40 | 0.25 (0.18–0.33) |
| Scolecobasidium | CCFEE 6363 | 0.26 | 0.15 | ~ | – | – | – |

[a]: no significant growth; [b] $0.80 < r^2 < 0.90$; [c] $r^2 \leq 0.80$.

Conversely, the tolerance of *E. crusticola* CCFEE 6188 to Cr(VI) was lower when compared to other fungi, as it had a TI of 0.73 at 0.1 g of $Cr^{6+}$ $L^{-1}$ after 30 days of exposure but it grew scarcely at the higher tested concentrations (its $IC_{50}$ was 0.56 g of $Cr^{6+}$ $L^{-1}$). There were three other strains that displayed a remarkable tolerance to both As(V) and Cr(VI), showing limited inhibition with respect to the control (Figure 1): *R. similis* CCFEE 6361, *C. olivacea* CCFEE 6619, and *E. mesophila* IRTA M2-F10. Their TI values were above 0.4 at the tested concentrations of As(V) and Cr(VI), and the $IC_{50}$ was in the range of 6–48 g of $As^{5+}$ $L^{-1}$, though the latest estimate must be taken with caution as it is well above the highest tested concentration, 0.05–0.25 g of $Cr^{6+}$ $L^{-1}$ after 30 days of exposure (Table 2). These strains were generally characterized by TI values at the highest As(V) and Cr(VI) tested concentrations that tended to increase with longer incubation times, up to 60 days (Supplementary Table S1). This phenomenon indicates a progressive longer-term adaptation to HMMs of these particular strains, in contrast to most of the tested BF collection.

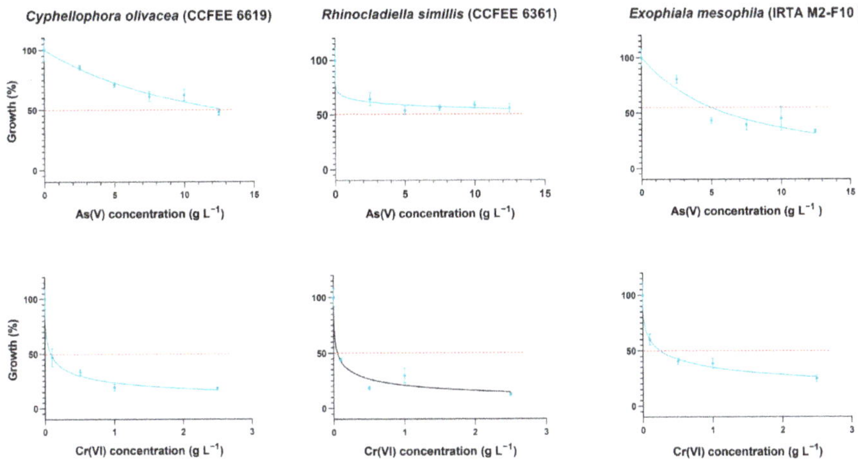


**Figure 1.** Effect of the concentration of As(V) (**top**) and Cr(VI) (**bottom**) on the average and standard deviation (solid bars, $n$ = 6) of the tolerance index (TI) of selected metallotolerant fungal strains, measured from the radial growth in agar cultures after 30 days of incubation. Solid lines correspond to a fitted logistic model for the determination of the $IC_{50}$ value.

Literature data on the fungal toxicity of HMMs are scarce, particularly with melanized fungi despite their polyextremophilic nature and association to toxic chemicals [21]. There are some reports on the fungal toxicity of arsenic oxyanions, but those have often been performed at a rather low milligram per liter range (between 10 and 500 mg $L^{-1}$) [34–38]. A few surveys were carried out at the gram per liter range, however. A screening of fungi isolated from arsenic-polluted soil for As(V) tolerance resulted in the selection of five strains, dubbed as "hyper-tolerant", that had TI values at 10 g of $As^{5+}$ $L^{-1}$ that ranged from 0.19 to 0.31 for *Aspergillus* sp., *Neocosmospora* sp., *Rhizopus* sp., and *Penicillium* sp., and up to 0.96 for an unidentified sterile fungus [15]. Singh et al. [39] reported nine fungal strains that were tolerant to As(V) up to 10 g of $As^{5+}$ $L^{-1}$, which belonged to the genera *Trichoderma*, *Aspergillus*, *Rhizopus*, *Microdochium*, *Chaetomium*, *Myrothecium*, *Stachybotrys*, *Rhizomucor*, and *Fusarium*. However, no quantitative tolerance parameters were derived from this study.

Chromium has a complex valence layer that produces different oxidation states that interact with specific nutrients, accentuating the toxicity of this metal [3]. Several previous reports have corroborated the severity of Cr(VI) toxicity to fungi, when compared to As(V). In general, Cr(VI) is more toxic than As(V) to fungi because of its higher reactivity and capacity to generate oxygen reactive species (ROS), which can disrupt several metabolic functions [3,8]. For example, out of 14 isolates from tannery effluents contaminated with Cr(VI), only 1 strain of *Trichoderma viride* (fam. Hypocreaceae) was able to show some growth at 1 g of $Cr^{6+}$ $L^{-1}$ when cultured under laboratory conditions [40]. Other species in this genus have been evaluated for Cr(VI) tolerance, such as a *T. harzianum* strain isolated from an HMM-polluted mine [41], which displayed a TI at 1 g of $Cr^{6+}$ $L^{-1}$ as low as 0.024. Other *Aspergillus* spp. were also tested in that study (*A. sclerotiorum*, *A. aculeatus*, and *A. niger*) and yielded higher TI values, between 0.12 and 0.67, depending on the strain and the tested Cr(VI) concentration. Another strain of *T. viride* isolated from tannery wastewaters displayed TI values of 1.15, 0.13, and 0.08, as determined from the biomass production from liquid cultures, after 21 days of incubation with 50, 500, and 1000 mg of $Cr^{6+}$ $L^{-1}$ [40]. A second Cr(VI)-tolerant strain identified as *Penicillium citrinum* showed a similar profile. Interestingly, those fungi seemed to be biostimulated at low concentrations of Cr(VI), both in terms of biomass production and secreted laccase enzymes.

None of the previously mentioned taxa are BF and, in fact, there are few reports quantifying the metallotolerance within this particular group of fungi. The chaetothyrialean *Exophiala pisciphila* has been isolated repeatedly from the roots of plants growing on soils that are polluted with heavy metals, and in vitro analyses have shown that this fungus tolerates concentrations of Pb(II), Cd(II), and Zn(II) at an $IC_{50}$ of 0.8, 0.3, and 1.5 g $L^{-1}$, respectively [42]. Concerning the tolerance to As(V), one study with liquid cultures of different strains of *E. sideris* isolated from HMM-polluted environments reported $IC_{50}$ values between 2.0 and 3.7 g of $As^{5+}$ $L^{-1}$, depending on the isolate [43]. These latter results are in the As(V) $IC_{50}$ range found in our study for the strains that belong to the *Exophiala* genus (Table 2). As for specific accounts on the tolerance of BF to Cr(VI), a minimum inhibitory Cr(VI) concentration of 300 mg of $Cr^{6+}$ $L^{-1}$ was determined for the growth of a strain identified as *Cladosporium perangustum* (fam. Cladosporiaceae) [44].

Some of the BF included in this study encompassed multiple strains from the same species (i.e., two strains of *Au. pullulans*, *C. uncinatum*, *A. pinorum*, *E. heteromorpha*, *E. mesophila*, and *E. oligosperma*, and four of *E. xenobiotica* and *K. epidermis*). Comparing growth inhibition patterns among these strains revealed a wide intraspecific variability in their tolerance to HMMs (Table 2). Examples of the most disparate cases include *Au. pullulans* and *C. uncinatum*, with one of the two tested strains (CCFEE 6244 and CCFEE 5820) of each species able to grow in the presence of As(V) and Cr(VI), while the other two (CCFEE 5876 and CCFEE 6149) did not show any growth at all. The "hyper-tolerant" *E. mesophila* strain IRTA M2-F10 also differed significantly from the conspecific CCFEE 5690.

This observation deserves further investigation to verify whether the recorded variability in metal tolerance is intrinsic to the species considered or if it is due to the lack of knowledge in the identification of related species. On the one hand, a detailed molecular analysis could be useful in better determining the position occupied by the two strains within the large group of *Au. pullulans*. On the other hand, it could be valuable in defining the close relatives of *C. uncinatum*, which are currently unknown.

On the contrary, metallotolerance appears to be relatively conserved in the four tested strains of *E. xenobiotica*, which were all able to grow at 12.5 g of $As^{5+}$ $L^{-1}$ and displayed $IC_{50}$ values of 0.7–3 g of $As^{5+}$ $L^{-1}$, but none of them grew in the presence of Cr(VI). The two available strains of *E. oligosperma* displayed a somewhat similar growth pattern in the presence of As(V), with $IC_{50}$ values of 0.2–2.5 g of $As^{5+}$ $L^{-1}$, but neither grew on any of the tested concentrations of Cr(VI). These strains are very similar to each other when comparing their ITS sequences.

At the intragenus level, significant variability in HMM tolerance is also manifested in *Exophiala*, which, along the observed diverse degrees of metallotolerance, also includes species that, like *E. heteromorpha*, were consistently unable to grow at any tested concentration of As(V) and Cr(VI). The observed intra- and interspecific differences in HMM tolerance might be attributed to diverse degrees of adaptation to HMM-polluted environments, due to specific genetic changes, epigenetic regulation, and phenotypic adaptations to stressful conditions [30]. Melanization is often cited as a feature that confers tolerance to HMMs because of its capacity to absorb toxic metals [28,45]. However, all tested strains were conspicuously melanized and were isolated from similar environments exposed to toxic chemicals. Hence, besides melanin production and adaptation to toxic environments, other factors must contribute to the tolerance of HMMs in BF. Earlier studies have proposed an array of multiple mechanisms that enable fungi to cope with HMMs [3,46,47], such as reducing the basal energy for metabolism, activating protein protection and DNA repair against oxidative stress, enhancing iron and sulfur acquisition, transforming metal species to less toxic or volatile metabolites, detoxifying free radicals, and through homeostasis.

Exposure to As(V) and Cr(VI) caused macroscopic morphological changes in fungal growth that were visible on agar colonies, as shown in Figure 2 for the metallotolerant *R. similis* CCFEE 6361, *C. olivacea* CCFEE 6619, and *E. mesophila* IRTA M2-F10. Fungal biomass tended to display a stronger dark pigmentation upon metal exposure, which might be attributable to an increased level of melanin biosynthesis as a defensive mechanism. A few strains like those of *E. mesophila* also formed colonies that were irregular in shape and displayed coarser edges under stressful conditions, a phenomenon known as meristematic growth. This morphological plasticity has also been observed with *E. oligosperma* CCFEE 6327 when grown at 35 °C under laboratory conditions, a temperature close to its upper temperature growth limit [48]. A similar morphology is also manifested in the case of opportunistic mammal infections, defined then as muriform cells [49]. The isodiametric growth and the aggregated, compact shape of fungal microcolonies ensure the optimal surface/volume ratio, minimizing the direct exposure to external stressors [50]. This type of growth is also expressed as an adaptation in the closely related lithobiontic black fungi (known as rock-inhabiting fungi, RIF) [51].

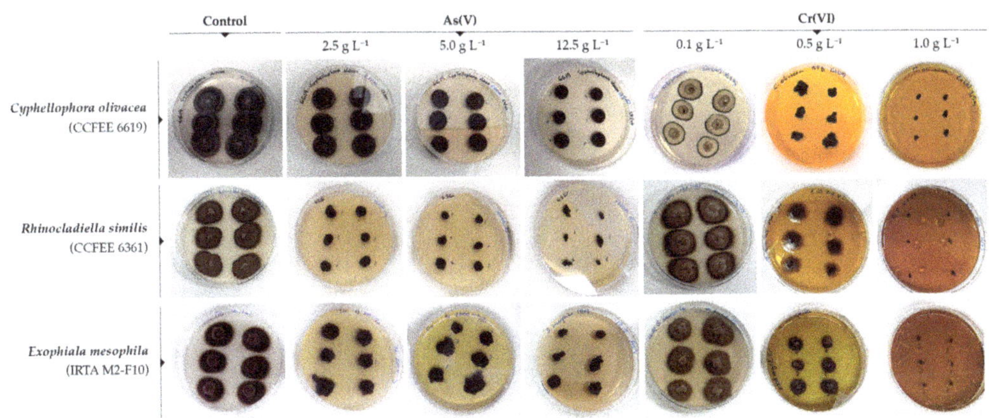


**Figure 2.** Macromorphological effects of the exposure to increasing concentrations of As(V) and Cr(VI) on fungal agar cultures of the metallotolerant black fungi.

### 3.2. Biosorption Assays

Two of the strains screened previously for metallotolerance, *R. similis* CCFEE 6361 and *E. mesophila* IRTA M2-F10, were selected for subsequent biosorption experiments because of their intrinsic tolerance to both As(V) and Cr(VI) and for their easy cultivation in liquid media for producing biomass. After 30 days of incubation of pre-grown fungal liquid cultures with As(V) and Cr(VI) (approx. 50 mg of DM $L^{-1}$; an initial HMM concentration of 20 mg $L^{-1}$), the HMM content remaining in the supernatant was measured (Table 3). From these results, it was evident that BF biomass had a comparatively low absorption capacity for As(V) when compared to Cr(VI), as removal efficiencies for the first did not exceed 10%, while for the second, they were higher than 80% under similar test conditions.

The specific As(V) removal capacity of living cultures of *E. mesophila* and *R. similis* incubated for 30 days at pH 6.5 was 1.07 and 1.34 mg of $As^{5+}$ g $DM^{-1}$. The difference between the two fungi was not statistically significant ($p > 0.05$). However, the biosorption of As(V) by heat-inactivated biomass of *E. mesophila* was significantly lower than that by living cultures of the same fungus. Despite the apparently modest As(V) biosorption results with the tested BF, previous similar studies with this metalloid have yielded even lower numbers. Different fungal species in the genera *Neocosmospora*, *Sordaria*, *Rhizopus*, and *Penicillium* displayed biosorption capacities that ranged from 0.009 to 0.016 mg $As^{5+}$ g $DM^{-1}$ [15]. Other authors claimed that cultures of fungi belonging to *Aspergillus*, *Fusarium*, *Rhizomucor*, and *Emericella* were able to absorb between 0.023 and 0.259 mg $As^{5+}$ g $DM^{-1}$ depending on the strain [39]. Those experiments were performed within the pH range of 4–7 used in our study, and all used strains correspond to fungi that are not conspicuously melanized. Hence, fungal melanization might indeed improve As(V) biosorption by cultures of BF.

**Table 3.** Fungal HMM biosorption capacity and removal efficiency. Incubations lasted 30 days, and results are expressed as the average and standard deviation of three independent experiments. One-way ANOVA comparisons were performed on the specific removal capacity for every metal, and non-significant differences are indexed ($n = 3$; $p < 0.05$). Significance letters for the Tukey test were added to the specific removal capacity data.

| Fungus (Strain) | Type of Biomass | pH | Parameter (Units) | As(V) | Cr(VI) |
|---|---|---|---|---|---|
| *R. similis* | Living culture | 6.5 | Biomass (mg DM) | 48.07 ± 6.05 | 43.33 ± 2.95 |
| | | | Final concentration in solution (mg $L^{-1}$) | 18.30 ± 0.31 | 3.48 ± 1.33 |
| | | | Specific removal capacity (mg g $DM^{-1}$) | 1.07 ± 0.32 $^{AC}$ | 11.50 ± 1.56 $^{A}$ |
| | | | Removal efficiency (%) | 8.5 | 82.6 |
| | | 4.0 | Biomass (mg DM) | 45.28 ± 5.08 | 49.67 ± 0.01 |
| | | | Final concentration in solution (mg $L^{-1}$) | 19.53 ± 0.31 | <0.01 |
| | | | Specific removal capacity (mg g $DM^{-1}$) | 0.33 ± 0.24 $^{B}$ | ≥12.79 |
| | | | Removal efficiency (%) | 2.3 | 100 |
| *E. mesophila* | Living culture | 6.5 | Biomass (mg DM) | 44.33 ± 1.91 | 46.90 ± 1.23 |
| | | | Final concentration in solution (mg $L^{-1}$) | 18.04 ± 0.57 | 1.16 ± 0.25 |
| | | | Specific removal capacity (mg g $DM^{-1}$) | 1.34 ± 0.42 $^{A}$ | 12.57 ± 3.00 $^{A}$ |
| | | | Removal efficiency (%) | 9.8 | 94.2 |
| | | 4.0 | Biomass (mg DM) | 46.57 ± 6.12 | 53.57±1.55 |
| | | | Final concentration in solution (mg $L^{-1}$) | 19.67 ± 0.21 | <0.01 |
| | | | Specific removal capacity (mg g $DM^{-1}$) | 0.21 ± 0.12 $^{B}$ | 11.20 |
| | | | Removal efficiency (%) | 1.67 | 100 |
| | Dead cells | 6.5 | Biomass (mg DM) (mg $L^{-1}$) | 42.03 ± 0.63 | 46.40 ± 0.46 |
| | | | Final concentration in solution | 19.41 ± 0.25 | 5.16 ± 0.66 |
| | | | Specific removal capacity (mg g $DM^{-1}$) | 0.41 ± 0.17 $^{BC}$ | 9.59 ± 0.47 $^{A}$ |
| | | | Removal efficiency (%) | 2.92 | 74.2 |

Fungal biosorption patterns changed completely when Cr(VI) was used. Equivalent incubations with this metal showed that the specific removal capacity after 30 days of incubation was slightly higher for *E. mesophila* (12.57 mg $Cr^{6+}$ g $DM^{-1}$) than for *R. similis* (11.50 g $Cr^{6+}$ mg $DM^{-1}$). As with As(V), the heat inactivation of cultures of *E. mesophila* caused a reduction in the biosorption capacity (9.59 g $Cr^{6+}$ mg $DM^{-1}$). However, differences in the specific Cr(VI) biosorption capacity in all tested fungi and incubation conditions were statistically not significant ($p > 0.05$). Previous reports on the specific Cr(VI) biosorption capacity of fungi correspond primarily to modified biomass, and relatively few records with living fungal cultures are available. Some fungi isolated from samples of sludge and industrial effluents contaminated with heavy metals (*Trichoderma viride*, *T. longibrachiatum*, *Aspergillus niger*, and *Phanerochaete chrysosporium*) displayed biosorption capacities that ranged from 0.03 to 0.55 mg $Cr^{6+}$ g $DM^{-1}$ when incubated (150 rpm at 28 °C) with potato dextrose broth containing 50 mg $Cr^{6+}$ $L^{-1}$ for 4 days [52]. Lotlikar et al. [53] isolated three strains from Arabian Sea sediments, identified as *Purpureocillium lilacinum*, *Aspergillus sydowii*, and *A. terreus*, that were able to grow with 300 mg $Cr^{6+}$ $L^{-1}$ (pH 5, shaken conditions, and room temperature) and biosorbed 8, 10, and 13 mg $Cr^{6+}$ g $DM^{-1}$, respectively, after 20 days of incubation. Reports on Cr(VI) biosorption by BF are very limited; liquid cultures of *Aureobasidium pullulans* growing on the acid hydrolysate of peat containing HMMs (200 rpm; 26 °C; pH 6.0) were able to absorb 0.77 mg $Cr^{6+}$ g $DM^{-1}$ after 160 h of incubation [54].

The sorption of As(V) and Cr(VI) onto organic materials has been described as a pH-dependent phenomenon [55,56]. At acidic conditions, As(V) exists primarily in the form of dihydrogen arsenate ($H_2AsO_4^-$), while Cr(VI) is present as chromate ($CrO_4^{2-}$). As the pH increases, $H_2AsO_4^-$ transforms into hydrogen arsenate ($HAsO_4^{2-}$) or arsenate ($AsO_4^{3-}$) and $CrO_4^{2-}$ transforms into dichromate ($Cr_2O_7^{2-}$), resulting in an increase in the net negative charge. The pH also influences the surface charge of the sorbent, so the point of zero charge (pzc)—the pH at which the net charge of the total absorbent's surface is equal to zero—for fungal biomass is in the range of pH 4.0–4.5 [57,58]. Above these pH values, the surface becomes negatively charged, exacerbating its electrostatic repulsion towards As(V) and Cr(VI) anions. Interestingly, fungal incubations at pH 4.0 significantly ($p < 0.05$) enhanced the biosorption of Cr(VI), but not that of As(V), compared to analogous experiments at pH 6.5 (Table 3).

The HMM biosorption capacity of heat-inactivated biomass of *E. mesophila* was lower than with equivalent viable cultures (Table 3). Several investigations on the passive adsorption process on the cell surface have been performed using dead fungal biomass [59–61]. It has been claimed that the biosorption of dried biomass increases the contact surface between HMMs and the metal binding sites of the fungal cell wall [62]. However, heat-inactivated biomass may still keep the structures of the fungal cell relatively intact. A comparison between living cultures and dried pulverized biomass of a group of fungi (*Aspergillus foetidus, A. niger, A. terricola, Acremonium strictum, Paecilomyces variotii, Phanerochaete chrysosporium, Aureobasidium pullulans*, and *Cladosporium resinae*) claimed that while the Cr(VI) biosorption capacity in the first ranged from 0.1 to 3.0 mg $Cr^{6+}$ $g^{-1}$-DM, in the second, it increased up to 11.2 mg $Cr^{6+}$ $g^{-1}$-DM (pH 3; 100 rpm; 28.1 °C) [63]. Literature accounts on the use of inactivated fungal biomass as HMM sorbents have often included the addition of specific coadjuvants that mitigate the electrostatic repulsion between the sorbent and sorbate. A *Penicillium chrysogenum* was modified with three different surfactants (amines) to increase the removal capacities of modified biomass from 37.85 to 56.07 mg $As^{5+}$ $g^{-1}$ at pH 3 [64].

### 3.3. Biosorption Isotherms and Kinetics

The differential Equation (7) describing the first- and second-order kinetics, integrating the Langmuir and Freundlich isotherms, was fitted to experimental data on the time-course evolution of metal concentration in the supernatant. Changes with As(V) were too subtle to fit any sorption model with enough confidence, but the experimental data on the time-course depletion of Cr(VI) showed that the best fits were obtained with second-order sorption kinetics and the Langmuir sorption isotherm (Figure 3). Previous similar studies that compared different models to describe the fungal biosorption of metals showed that the Langmuir isotherm and second-order sorption kinetics displayed the best coefficients of determination [65–67]. The Langmuir isotherm is a theoretical approximation that considers a finite monolayer of available adsorption sites onto a homogeneous surface. This assumption implies that the sorbent has a limited surface-dependent maximum theoretical sorption capacity ($q_{\max}$), and that there is neither interaction nor transmigration of metal ions after monolayer adsorption.

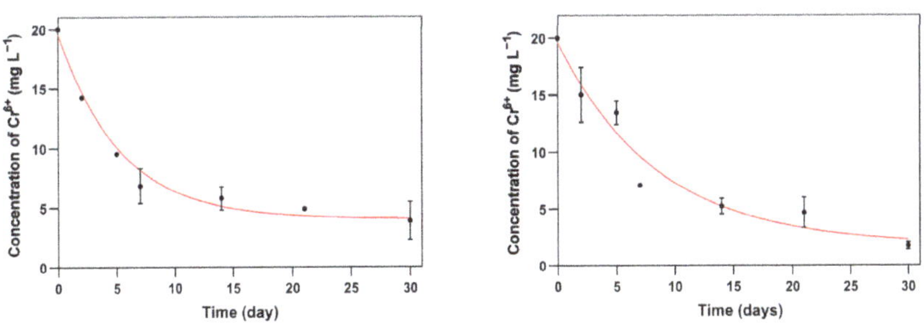


**Figure 3.** Fits of experimental data of Cr(VI) biosorption on living cultures (pH 6.5) of *Rhinocladiella similis* (**left**) and *Exophiala mesophila* (**right**) to the second-order kinetics and the Langmuir isotherm model (differential Equation (7)). Experimental data correspond to the average and standard deviation of three replicates.

The fitted parameters of Langmuir isotherms and second-order kinetics for the biosorption of Cr(VI) with the tested BF are summarized and compared with bibliographic data in Table 4. The "theoretical" maximum sorption capacity $q_{max}$ of *R. similis* and *E. mesophila* (39.81 and 95.26 mg $Cr^{+6}$ g $DM^{-1}$) is above those seen previously with several other non-melanized fungi in analogous experiments. Most of those assays were performed with inactivated/modified fungal biomass and at a low pH, conditions that might favor biosorption, but under which process scaling-up into practical applications is challenging. In what concerns the half-saturation concentration (the metal concentration in equilibrium at which the biosorbed metal equals half of $q_{max}$), which is the reverse of the affinity constant $K_L$, both tested BF displayed rather similar values of 8.90 and 5.09 mg $Cr^{6+}$ $L^{-1}$, respectively, for *R. similis* and *E. mesophila*. The affinity to Cr(VI) of biomass from hyaline species tested previously is quite low in general when compared to BF (Table 4).

The second-order kinetic constant $k_2$, which corresponds to the specific biosorption rate, is one order of magnitude lower for *E. mesophila* when compared to *R. similis*, $1.39 \times 10^{-7}$ versus $1.32 \times 10^{-6}$ g DM (mg $Cr^{6+}$ min)$^{-1}$. The higher $q_{max}$ and lower $k_2$ of *E. mesophila* when compared to *R. similis* could be related to the higher Cr(VI) tolerance of the former over the latter (Table 2). Several previous accounts on the second-order kinetics of Cr(VI) biosorption by non-melanogenic fungi have reported $k_2$ values above $10^{-2}$ g DM (mg $Cr^{6+}$ min)$^{-1}$ (Table 4). Interestingly, previous studies with melanized fungal structures, like lyophilized cells of the BF *Cladosporium cladosporioides* and spores of *Aspergillus niger*, also yielded comparatively low biosorption rates and high substrate affinity.

In order to gain a deeper insight into the role of fungal melanin when exposed to Cr(VI), biosorption experiments were repeated with melanin extracts from *E. mesophila* (Figure 4). The obtained fitted parameter values are between 1 and 2 orders of magnitude higher than those measured with living cultures of the same fungus ($q_{max}$ = 544.84 mg $Cr^{6+}$ g $DM^{-1}$; $K_L$ = 0.0075 L $mg^{-1}$; and $k_2 = 1.67 \times 10^{-6}$ g DM (mg $Cr^{6+}$ min)$^{-1}$). Considering that the mass of the extracted melanin corresponded to 12.5% of the fungal biomass, on a dry matter basis (125 mg g $DM^{-1}$), the potential contribution of melanin to the biosorption capacity of whole cells could be as high as 71.5%. These results confirm that fungal melanin plays a vital role in the biosorption of HMMs. Furthermore, the $q_{max}$ of melanin is well above that of several tested Cr(VI) organic absorbents, from raw and modified lignocellulosic materials [68] to advanced carbon nanomaterials [69].

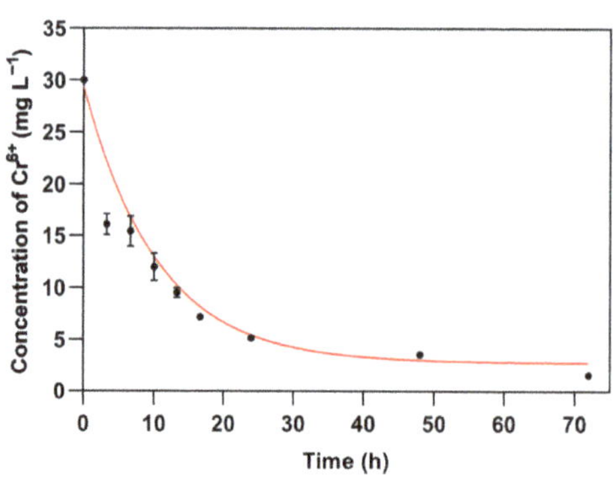


**Figure 4.** Fit of experimental data of Cr(VI) biosorption on melanin extracts of *Exophiala mesophila* to the second-order kinetics and the Langmuir isotherm model (differential Equation (7)). Experimental data correspond to the average and standard deviation of three replicates.

**Table 4.** Fitted parameters for modelling the biosorption of Cr(VI) using the second-order kinetic model and/or the Langmuir isotherm by different fungal species.

| Fungus [a] | pH | C (mg $L^{-1}$) | Fitted Equation | Model Parameters | | | | Source |
|---|---|---|---|---|---|---|---|---|
| | | | | $k_2$ (g (mg min)$^{-1}$) | $q_{max}$ (mg $g^{-1}$) | $K_L$ (L $mg^{-1}$) | $r^2$ | |
| *Rhinocladiella similis* (CS) | 6.5 | 25 | Equation (7) | $1.32 \times 10^{-6}$ | 39.81 | 0.1124 | 0.977 | This study |
| *Exophiala mesophila* (CS) | 6.5 | 25 | Equation (7) | $1.39 \times 10^{-7}$ | 95.26 | 0.1964 | 0.924 | This study |
| *Exophiala mesophila* (MEs) | 6.5 | 25 | Equation (7) | $1.67 \times 10^{-6}$ | 544.84 | 0.0075 | 0.969 | This study |
| *Cladosporium cladosporioides* (LB) | 2.0 | 25 | Equation (4) | $7.50 \times 10^{-5}$ | 28.90 | – | 0.991 | [70] |
| *Aspergillus niger* (SS) | 2.0 | 100 | Equation (4)<br>Equation (5) | $5.76 \times 10^{-4}$<br>– | 56.15<br>47.33 | –<br>0.5416 | 0.994<br>0.999 | [71] |
| *Aspergillus niger* (DB) | 2.0 | 27 | Equation (4)<br>Equation (5) | 3.38<br>– | 6.45<br>71.9 | –<br>0.031 | 0.998<br>0.999 | [67] |
| *Lentinus sajor-caju* (CS) | 2.0 | 30 | Equation (4)<br>Equation (5) | $3.39 \times 10^{-2}$<br>– | 20.80<br>23.32 | –<br>0.0133 | 0.994<br>0.993 | [72] |
| *Ustilago maydis* (DB) | 5.5–6.5 | 25 | Equation (4)<br>Equation (5) | $1.37 \times 10^{-2}$<br>– | 1.95<br>17.16 | –<br>0.0090 | ns [b]<br>0.965 | [73] |
| *Mucor hiemalis* (DB) | 2.0 | 100 | Equation (4)<br>Equation (5) | $5.5 \times 10^{-1}$<br>– | 30.5<br>47.4 | –<br>0.0307 | 0.993<br>0.999 | [74] |
| *Ganoderma applanatum* (DB) | 2.0 | 25 | Equation (4)<br>Equation (5) | $7.4 \times 10^{-1}$<br>– | 16.13<br>200 | –<br>0.002 | 0.999<br>0.999 | [75] |
| *Rhizopus* sp. (DB) | 2.0 | 25 | Equation (4)<br>Equation (5) | $1.13 \times 10^{-2}$<br>– | 5.4509<br>8.0589 | –<br>0.7730 | 0.986<br>0.841 | [65] |

[a] fungal species and pre-treatment of the biomass: living cell suspension (CS); melanin extracts (MEs); lyophilized biomass (LB); spore suspension (SS); dead biomass (DB). [b] not determined/not shown.

The results from this study are consistent with the hypothesis that fungal melanin plays an active role in the biosorption of toxic HMM, such as As(V) and Cr(VI). Melanin is an amorphous polymeric structure that concentrates in the fungal cell wall and offers a great number of heterogeneous binding sites to the sorption of metal ions [28]. Melanin is also a conductive material that mediates electron transport between fungal biomass and the solute, which might contribute to chemisorption by sharing or exchanging electrons between the sorbent and sorbate [76]. A previous study on the binding of copper by melanin extracts, intact cells, and albino mutants of the BF *Cladosporium resinae* and *Aureobasidium pullulans* demonstrated that the metal uptake capacity was higher in melanin extracts, followed by intact cells and, finally, in albino cells [77].

However, other studies have disputed the role of melanin as a metal biosorbent. An early study found no significant differences in the binding of copper by cultures of the melanized plant pathogen *Gaeumannomyces graminis*, either grown at low concentrations of this metal or additionally supplemented with tricyclazole, an inhibitor of DHN-melanin [78]. More recently, it has been proposed that melanin may have a role in binding metals and protecting fungi from toxic metals, but the main mechanisms might not necessarily be related to sorption processes [79,80]. Potisek [80] used different strains of dark septate endophytes of *Cadophora* spp., with different melanin contents, to investigate their tolerance to cadmium. The melanin content was positively correlated with a higher cadmium tolerance, but the accumulation of this metal was not. More contentious studies on

*Exophiala pisciphila* even claimed that the inhibition of melanization in this fungus did not cause any remarkable effect on the tolerance of metal ions [43].

Indeed, the high inter- and intraspecific variability in metallotolerance observed among the studied black fungi (Table 2) suggest that there must be other physiologic mechanisms to cope with the toxic effects of toxic metals and metalloids, rather than relying solely on melanization. In this regard, the comparatively high biosorption capacity of BF to HMMs, in combination with low sorption rates, could be the result of an evolved strategy to, firstly, minimize the penetration of toxic metals into the cell and, secondly, sequester those that would eventually leak into the cytoplasm.

## 4. Conclusions

Several published studies have revealed that fungal biomass has good performance as a biosorbent of heavy metals and metalloids, in comparison to commercial materials such as ion exchange resins, activated carbon, and metal oxides. Fungal biosorption depends on parameters such as the used fungi and metal species but also the pH, temperature, biomass pre-treatments, and the presence of various ligands in solution. The fungal cell wall fraction seems to play an important role in the sorption of heavy metals, though the biosorption mechanisms are understood only to a limited extent. The ability of fungi to tolerate and biosorb specific heavy metals and metalloids has previously been evaluated and reviewed, as is the case for As(V) and Cr(VI), but the literature on the metallotolerance of extremophilic black fungi is still comparatively scarce.

To our knowledge, the present work provides the first in-depth survey of the metallotolerance of a strain collection of BF isolated previously from polluted sites. The obtained results indicate that there is broad inter- and intraspecific variability in metallotolerance, using As(V) and Cr(VI) as model HMMs. Living cultures of two hyper-metallotolerant strains of *Rhinocladiella similis* and *Exophiala mesophila* displayed a significant biosorption capacity but, conversely, sorption rates were comparatively slow in relation to other non-black fungi. This phenomenon could be explained by the interplay of two protective processes: an outer cell wall barrier and the inner cell wall biosorption of leaked metals and metalloids by melanin. These findings might contribute to the development of future strategies for the bioremediation of HMM pollution and need to be further investigated. Black fungi could also be valuable bioindicators of HMMs in natural and anthropogenic environments.

**Supplementary Materials:** The following supporting information can be downloaded at: https://www.mdpi.com/article/10.3390/jof10010047/s1, Table S1_Metallotoletance.

**Author Contributions:** Individual contributions to this study are indicated with the authors' initials: Conceptualization, C.M.-A. and F.X.P.-B.; methodology, C.M.-A. and A.P.; software, C.M.-A. and J.I.; validation, J.I. and F.X.P.-B.; formal analysis, C.M.-A. and F.X.P.-B.; investigation, C.M.-A. and F.X.P.-B.; resources, D.I. and M.V.; writing—original draft preparation, C.M.-A.; writing—review and editing, F.X.P.-B., J.I. and D.I.; supervision, F.X.P.-B. All authors have read and agreed to the published version of the manuscript.

**Funding:** This research has been partly funded by the project AGAUR (Generalitat de Catalunya) through the Consolidated Research Group SOSBIO (ref. 2021 SGR 01568). Cristy Medina-Armijo was a recipient of the grant ANID Chile.

**Institutional Review Board Statement:** Not applicable.

**Informed Consent Statement:** Not applicable.

**Data Availability Statement:** All data of this research are reported in the main text and Supplementary material.

**Acknowledgments:** We acknowledge Alexia Douillard, from the University of Angers, and Marta Manyà, from the Autonomous University of Barcelona, for laboratory assistance during her undergraduate internship. This research was supported by the CERCA Programme/Generalitat de Catalunya.

**Conflicts of Interest:** The authors declare no conflicts of interest.

## References


1. Zhang, Q.; Wang, C. Natural and human factors affect the distribution of soil heavy metal pollution: A Review. Water. *Air. Soil. Pollut.* **2020**, *231*, 1–13. [CrossRef]
2. Siddiquee, S.; Saallah, S.; Rovina, K.; Al Azad, S.; Naher, L.; Suryani, S.; Chaikaew, P. Heavy metal contaminants removal from wastewater using the potential filamentous fungi biomass: A review. Artic. *J. Microb. Biochem. Technol.* **2015**, *7*, 384–393. [CrossRef]
3. Viti, C.; Marchi, E.; Decorosi, F.; Giovannetti, L. Molecular mechanisms of Cr (VI) resistance in bacteria and fungi. *FEMS Microbiol. Rev.* **2014**, *38*, 633–659. [CrossRef]
4. Podgorski, J.; Berg, M. Global threat of arsenic in groundwater. *Science* **2020**, *368*, 845–850. [CrossRef]
5. *WHO/SDE/WSH/03.04/75*; Arsenic in Drinking-Water—Background Document for Development of WHO Guidelines for Drinking-Water Quality. World Health Organization: Geneva, Switzerland, 2019.
6. Sappa, G.; Ergul, S.; Ferranti, F. Geochemical modeling and multivariate statistical evaluation of trace elements in arsenic contaminated groundwater systems of Viterbo area, (Central Italy). *SpringerPlus* **2014**, *3*, 237. [CrossRef]
7. Tumolo, M.; Ancona, V.; De Paola, D.; Losacco, D.; Campanale, C.; Massarelli, C.; Uricchio, V.F. Chromium pollution in European water, sources, health risk, and remediation strategies: An overview. *Int. J. Environ. Res. Public. Health* **2020**, *17*, 5438. [CrossRef]
8. Sun, Q.; Li, Y.; Shi, L.; Hussain, R.; Mehmood, K.; Tang, Z.; Zhang, H. Heavy metals induced mitochondrial dysfunction in animals: Molecular mechanism of toxicity. *Toxicology* **2022**, *469*, 153136. [CrossRef]
9. Fu, F.; Wang, Q. Removal of heavy metal ions from wastewaters: A review. *J. Environ. Manag.* **2011**, *92*, 407–418. [CrossRef]
10. Costa, F.; Tavares, T. Bioremoval of Ni and Cd in the presence of diethylketone by fungi and by bacteria—A comparative study. *Int. Biodeterior. Biodegrad.* **2017**, *120*, 115–123. [CrossRef]
11. Kang, C.H.; Kwon, Y.J.; So, J.S. Bioremediation of heavy metals by using bacterial mixtures. *Ecol. Eng.* **2016**, *89*, 64–69. [CrossRef]
12. Zeraatkar, A.K.; Ahmadzadeh, H.; Talebi, A.F.; Moheimani, N.R.; McHenry, M.P. Potential use of algae for heavy metal bioremediation, a critical review. *J. Environ. Manag.* **2016**, *181*, 817–831. [CrossRef]
13. Ali, S.; Abbas, Z.; Rizwan, M.; Zaheer, I.E.; Yavas, I.; Ünay, A.; Abdel-Daim, M.M.; Bin-Jumah, M.; Hasanuzzaman, M.; Kalderis, D. Application of floating aquatic plants in phytoremediation of heavy metals polluted water: A review. *Sustainability* **2020**, *12*, 1927. [CrossRef]
14. Acosta-Rodríguez, I.; Cardenás-González, J.F.; Pérez, A.S.R.; Oviedo, J.T.; Martínez-Juárez, V.M. Bioremoval of different heavy metals by the resistant fungal strain *Aspergillus niger*. *Bioinorg. Chem. Appl.* **2018**, *2018*, 3457196. [CrossRef]
15. Srivastava, P.K.; Vaish, A.; Dwivedi, S.; Chakrabarty, D.; Singh, N.; Tripathi, R.D. Biological removal of arsenic pollution by soil fungi. *Sci. Total Environ.* **2011**, *409*, 2430–2442. [CrossRef]
16. da Silva, N.M.; Reis, G.F.; Costa, F.D.F.; Grisolia, M.E.; Geraldo, M.R.; Lustosa, B.P.R.; Lima, B.J.F.D.S.; Weiss, V.A.; de Souza, E.M.; Li, R.; et al. Genome sequencing of *Cladophialophora exuberans*, a novel candidate for bioremediation of hydrocarbon and heavy metal polluted habitats. *Fungal Biol.* **2023**, *127*, 1032–1042. [CrossRef]
17. Oh, J.J.; Kim, J.Y.; Kim, Y.J.; Kim, S.; Kim, G.H. Utilization of extracellular fungal melanin as an eco-friendly biosorbent for treatment of metal-contaminated effluents. *Chemosphere* **2021**, *272*, 129884. [CrossRef]
18. Teixeira, M.M.; Moreno, L.F.; Stielow, B.J.; Muszewska, A.; Hainaut, M.; Gonzaga, L.; Abouelleil, A.; Patané, J.S.L.; Priest, M.; Souza, R.; et al. Exploring the genomic diversity of black yeasts and relatives (Chaetothyriales, Ascomycota). *Stud. Mycol.* **2017**, *86*, 1–28. [CrossRef]
19. Selbmann, L.; Egidi, E.; Isola, D.; Onofri, S.; Zucconi, L.; de Hoog, G.S.; Chinaglia, S.; Testa, L.; Tosi, S.; Balestrazzi, A.; et al. Biodiversity, evolution and adaptation of fungi in extreme environments. *Plant Biosyst.* **2013**, *147*, 237–246. [CrossRef]
20. Selbmann, L.; Isola, D.; Fenice, M.; Zucconi, L.; Sterflinger, K.; Onofri, S. Potential extinction of antarctic endemic fungal species as a consequence of global warming. *Sci. Total Environ.* **2012**, *438*, 127–134. [CrossRef]
21. Isola, D.; Selbmann, L.; de Hoog, G.S.; Fenice, M.; Onofri, S.; Prenafeta-Boldú, F.X.; Zucconi, L. Isolation and screening of black fungi as degraders of volatile aromatic hydrocarbons. *Mycopathologia* **2013**, *175*, 369–379. [CrossRef]
22. Prenafeta-Boldú, F.X.; de Hoog, G.S.; Summerbell, R.C. Fungal communities in hydrocarbon degradation. In *Microbial Communities Utilizing Hydrocarbons and Lipids: Members, Metagenomics and Ecophysiology*; Handbook of Hydrocarbon and Lipid Microbiology; McGenity, T., Ed.; Springer: Berlin/Heidelberg, Germany, 2019; pp. 1–36.
23. Abdollahzadeh, J.; Groenewald, J.Z.; Coetzee, M.P.A.; Wingfield, M.J.; Crous, P.W. Evolution of lifestyles in Capnodiales. *Stud. Mycol.* **2020**, *95*, 381. [CrossRef]
24. Butler, M.J.; Gardiner, R.B.; Day, A.W. Fungal melanin detection by the use of copper sulfide-silver. *Mycologia* **2005**, *97*, 312–319. [CrossRef]
25. Prota, G. *Melanins and Melanogenesis*; Academic Press: San Diego, CA, USA, 1992.
26. Kumar, C.G.; Mongolla, P.; Pombala, S.; Kamle, A.; Joseph, J. Physicochemical characterization and antioxidant activity of melanin from a novel strain of *Aspergillus bridgeri* ICTF-201. *Lett. Appl. Microbiol.* **2011**, *53*, 350–358. [CrossRef]
27. Fogarty, R.V.; Tobin, J.M. Fungal melanins and their interactions with metals. *Enzym. Microb. Technol.* **1996**, *19*, 311–317. [CrossRef]

28. Isola, D.; Scano, A.; Orrù, G.; Prenafeta-Boldú, F.X.; Zucconi, L. Hydrocarbon-Contaminated sites: Is there something more than *Exophiala xenobiotica*? New insights into black fungal diversity using the long cold incubation method. *J. Fungi* **2021**, *7*, 817. [CrossRef]
29. Isola, D.; Zucconi, L.; Cecchini, A.; Caneva, G. Dark-pigmented biodeteriogenic fungi in etruscan hypogeal tombs: New data on their culture-dependent diversity, favouring conditions, and resistance to biocidal treatments. *Fungal Biol.* **2021**, *125*, 609–620. [CrossRef]
30. Selvakumar, P.; Rajasekar, S.; Periasamy, K.; Raaman, N. Isolation and characterization of melanin pigment from *Pleurotus cystidiosus* (telomorph of antromycopsis macrocarpa). *World J. Microbiol. Biotechnol.* **2008**, *24*, 2125–2131. [CrossRef]
31. Suwannarach, N.; Kumla, J.; Watanabe, B.; Matsui, K.; Lumyong, S. Characterization of melanin and optimal conditions for pigment production by an endophytic fungus, *Spissiomyces endophytica* SDBR-CMU319. *PLoS ONE* **2019**, *14*, e0222187. [CrossRef]
32. Bates, S.T.; Reddy3, G.S.N.; Garcia-Pichel, F. *Exophiala crusticola* Anam. Nov. (affinity Herpotrichiellaceae), a novel black yeast from biological soil crusts in the western United States. *Int. J. Syst.* **2006**, *56*, 2697–2702. [CrossRef]
33. Madrid, H.; Gené, J.; Quijada, L.; Cantillo, T.; Gacitúa, R.; Valdés, J.; Sánchez, C.; Prenafeta-Boldú, F.; Wijayawardene, N.; Silva, V.; et al. *Exophiala atacamensis* sp. nov. and *E. crusticola* from the Atacama desert, northern Chile. *Sydowia* **2023**, *75*, 181–192.
34. Vala, A.K. Tolerance and removal of arsenic by a facultative marine fungus *Aspergillus candidus*. *Bioresour. Technol.* **2010**, *101*, 2565–2567. [CrossRef]
35. Mukherjee, A.; Das, D.; Kumar Mondal, S.; Biswas, R.; Kumar Das, T.; Boujedaini, N.; Khuda-Bukhsh, A.R. Tolerance of arsenate-induced stress in *Aspergillus niger*, a possible candidate for bioremediation. *Ecotoxicol. Environ. Saf.* **2010**, *73*, 172–182. [CrossRef]
36. Nam, I.H.; Murugesan, K.; Ryu, J.; Kim, J.H. Arsenic (As) removal using *Talaromyces* sp. KM-31 isolated from As-contaminated mine soil. *Minerals* **2019**, *9*, 568. [CrossRef]
37. Ceci, A.; Spinelli, V.; Massimi, L.; Canepari, S.; Persiani, A.M. Fungi and arsenic: Tolerance and bioaccumulation by soil saprotrophic species. *Appl. Sci.* **2020**, *10*, 3218. [CrossRef]
38. Oladipo, O.G.; Awotoye, O.O.; Olayinka, A.; Bezuidenhout, C.C.; Maboeta, M.S. Heavy metal tolerance traits of filamentous fungi isolated from gold and gemstone mining sites. *Braz. J. Microbiol.* **2018**, *49*, 29–37. [CrossRef]
39. Singh, M.; Srivastava, P.K.; Verma, P.C.; Kharwar, R.N.; Singh, N.; Tripathi, R.D. Soil fungi for mycoremediation of arsenic pollution in agriculture soils. *J. Appl. Microbiol.* **2015**, *119*, 1278–1290. [CrossRef]
40. Zapana-Huarache, S.V.; Romero-Sánchez, C.K.; Gonza, A.P.D.; Torres-Huaco, F.D.; Rivera, A.M.L. Chromium (VI) Bioremediation potential of filamentous fungi isolated from Peruvian Tannery industry effluents. *Braz. J. Microbiol.* **2020**, *51*, 271–278. [CrossRef]
41. Liaquat, F.; Munis, M.F.H.; Haroon, U.; Arif, S.; Saqib, S.; Zaman, W.; Khan, A.R.; Shi, J.; Che, S.; Liu, Q. Evaluation of metal tolerance of fungal strains isolated from contaminated mining soil of Nanjing, China. *Biology* **2020**, *9*, 469. [CrossRef]
42. Zhang, Y.; Zhang, Y.; Liu, M.; Shi, X.; Zhao, Z. Dark septate endophyte (DSE) fungi isolated from metal polluted soils: Their taxonomic position, tolerance, and accumulation of heavy metals in vitro. *J. Microbiol.* **2008**, *46*, 624–632. [CrossRef]
43. Seyedmousavi, S.; Badali, H.; Chlebicki, A.; Zhao, J.; Prenafeta-boldú, F.X.; De Hoog, G.S. *Exophiala sideris*, a novel black yeast isolated from environments polluted with toxic alkyl benzenes and arsenic. *Fungal Biol.* **2011**, *115*, 1030–1037. [CrossRef]
44. Sharma, S.; Malaviya, P. Bioremediation of Tannery wastewater by chromium resistant novel fungal consortium. *Ecol. Eng.* **2016**, *91*, 419–425. [CrossRef]
45. Fomina, M.; Gadd, G.M. Metal sorption by biomass of melanin-producing fungi grown in clay-containing medium. *J. Chem. Technol. Biotechnol.* **2003**, *78*, 23–34. [CrossRef]
46. Liu, S.; Chen, M.; Cao, X.; Li, G.; Zhang, D.; Li, M.; Meng, N.; Yin, J.; Yan, B. Chromium (VI) removal from water using cetylpyridinium chloride (CPC)-modified montmorillonite. *Sep. Purif. Technol.* **2020**, *241*, 116732. [CrossRef]
47. Acevedo-Aguilar, F.J.; Espino-Saldaña, A.E.; Leon-Rodriguez, I.L.; Rivera-Cano, M.E.; Avila-Rodriguez, M.; Wrobel, K.; Wrobel, K.; Lappe, P.; Ulloa, M.; Gutiérrez-Corona, J.F. Hexavalent chromium removal in vitro and from industrial wastes, using chromate-resistant strains of filamentous fungi indigenous to contaminated wastes. *Can. J. Microbiol.* **2006**, *52*, 809–815. [CrossRef]
48. Isola, D.; Bartoli, F.; Meloni, P.; Caneva, G.; Zucconi, L. Black fungi and stone heritage conservation: Ecological and metabolic assays for evaluating colonization potential and responses to traditional biocides. *Appl. Sci.* **2022**, *12*, 2038. [CrossRef]
49. Seyedmousavi, S.; Netea, M.G.; Mouton, J.W.; Melchers, W.J.G.; Verweij, P.E.; de Hoog, G.S. Black yeasts and their filamentous relatives: Principles of pathogenesis and host defense. *Clin. Microbiol. Rev.* **2014**, *27*, 527–542. [CrossRef]
50. Harutyunyan, S.; Muggia, L.; Grube, M. Black fungi in lichens from seasonally arid habitats. *Stud. Mycol.* **2008**, *61*, 83. [CrossRef]
51. Sterflinger, K.; De Hoog, G.S.; Haase, G. Phylogeny and ecology of meristematic Ascomycetes. *Stud. Mycol.* **1999**, *1999*, 5–22.
52. Joshi, P.K.; Swarup, A.; Maheshwari, S.; Kumar, R.; Singh, N. Bioremediation of heavy metals in liquid media through fungi isolated from contaminated sources. *Indian J. Microbiol.* **2011**, *4*, 482–487. [CrossRef]
53. Lotlikar, N.P.; Damare, S.R.; Meena, R.M.; Linsy, P.; Mascarenhas, B. Potential of marine-derived fungi to remove hexavalent chromium pollutant from culture broth. *Indian J. Microbiol.* **2018**, *2*, 182–192. [CrossRef]
54. Radulović, M.D.; Cvetković, O.G.; Nikolić, S.D.; Dordević, D.S.; Jakovljević, D.M.; Vrvić, M.M. Simultaneous production of pullulan and biosorption of metals by *Aureobasidium pullulans* strain CH-1 on peat hydrolysate. *Bioresour. Technol.* **2008**, *14*, 6673–6677. [CrossRef] [PubMed]
55. Unceta, N.; Séby, F.; Malherbe, J.; Donard, O.F.X. Chromium speciation in solid matrices and regulation: A review. *Anal. Bioanal. Chem.* **2010**, *397*, 1097–1111. [CrossRef] [PubMed]

56. Chiban, M.; Zerbet, M.; Carja, G.; Sinan, F. Application of low-cost adsorbents for arsenic removal: A Review. *J. Environ. Chem. Ecotoxicol.* **2012**, *4*, 91–102. [CrossRef]
57. Mohamed, L.A.; Aniagor, C.O.; Taha, G.M.; Abou-Okeil, A.; Hashem, A. Mechanistic investigation of the mass transfer stages involved during the adsorption of aqueous lead onto *Scopulariopsis brevicompactum* fungal biomass. *Environ. Chall.* **2021**, *5*, 100373. [CrossRef]
58. Long, D.-D.; Wang, Q.; Han, J.-R. Biosorption of copper (II) from aqueous solutions by sclerotiogenic *Aspergillus oryzae* G15. *Water Environ. Res.* **2017**, *89*, 703–713. [CrossRef]
59. Benila Smily, J.R.M.; Sumithra, P.A. Optimization of chromium biosorption by fungal adsorbent, *Trichoderma* sp. BSCR02 and its desorption studies. *Hayati J. Biosci.* **2017**, *24*, 65–71. [CrossRef]
60. Velmurugan, P.; Shim, J.; You, Y.; Choi, S.; Kamala-Kannan, S.; Lee, K.J.; Kim, H.J.; Oh, B.T. Removal of zinc by live, dead, and dried biomass of *Fusarium* spp. Isolated from the abandoned-metal mine in South Korea and its perspective of producing nanocrystals. *J. Hazard. Mater.* **2010**, *182*, 317–324. [CrossRef]
61. Shroff, K.A.; Vaidya, V.K. Kinetics and equilibrium studies on biosorption of nickel from aqueous solution by dead fungal biomass of *Mucor hiemalis*. *Chem. Eng. J.* **2011**, *171*, 1234–1245. [CrossRef]
62. Ayele, A.; Haile, S.; Alemu, D.; Kamaraj, M. Comparative utilization of dead and live fungal biomass for the removal of heavy metal: A concise review. *Sci. World J.* **2021**, *2021*, 5588111. [CrossRef]
63. Ahluwalia, S.S.; Goyal, D. Removal of Cr(VI) from aqueous solution by fungal biomass. *Eng. Life Sci.* **2010**, *10*, 480–485. [CrossRef]
64. Loukidou, M.X.; Matis, K.A.; Zouboulis, A.I.; Liakopoulou-Kyriakidou, M. Removal of As(V) from wastewaters by chemically modified fungal biomass. *Water Res.* **2003**, *37*, 4544–4552. [CrossRef] [PubMed]
65. Espinoza-Sánchez, M.A.; Arévalo-Niño, K.; Quintero-Zapata, I.; Castro-González, I.; Almaguer-Cantú, V. Cr(VI) Adsorption from aqueous solution by fungal bioremediation based using *Rhizopus* sp. *J. Environ. Manag.* **2019**, *251*, 109595. [CrossRef] [PubMed]
66. Subbaiah, M.V.; Kalyani, S.; Reddy, G.S.; Boddu, V.M.; Krishnaiah, A. Biosorption of Cr(VI) from aqueous solutions using *Trametes versicolor polyporus* Fungi. *J. Chem.* **2008**, *5*, 499–510. [CrossRef]
67. Samuel, S.M.; Abigail, M.E.A.; Chidambaram, R. Isotherm modelling, kinetic study and optimization of batch parameters using response surface methodology for effective removal of Cr(VI) using fungal biomass. *PLoS ONE* **2015**, *10*, e0116884. [CrossRef] [PubMed]
68. Miretzky, P.; Cirelli, A.F. Cr(VI) and Cr(III) Removal from aqueous solution by raw and modified lignocellulosic materials: A review. *J. Hazard. Mater.* **2010**, *180*, 1–19. [CrossRef]
69. Aigbe, U.O.; Osibote, O.A. A review of hexavalent chromium removal from aqueous solutions by sorption technique using nanomaterials. *J. Environ. Chem. Eng.* **2020**, *8*, 104503. [CrossRef]
70. Garza-González, M.T.; Ramírez-Vázquez, J.E.; García-Hernández, M.D.L.Á.; Cantú-Cárdenas, M.E.; Liñan-Montes, A.; Villarreal-Chiu, J.F. Reduction of chromium (VI) from aqueous solution by biomass of *Cladosporium cladosporioides*. *Water Sci. Technol.* **2017**, *76*, 2494–2502. [CrossRef]
71. Ren, B.; Zhang, Q.; Zhang, X.; Zhao, L.; Li, H. Biosorption of Cr(VI) from aqueous solution using dormant spores of *Aspergillus niger*. *RSC Adv.* **2018**, *8*, 38157–38165. [CrossRef]
72. Arica, M.Y.; Bayramoğlu, G. Cr(VI) Biosorption from aqueous solutions using free and immobilized biomass of *Lentinus sajor-caju*: Preparation and kinetic characterization. *Colloids Surfaces A Physicochem. Eng. Asp.* **2005**, *253*, 203–211. [CrossRef]
73. Serrano-Gómez, J.; Olguín, M.T. Separation of Cr(VI) from aqueous solutions by adsorption on the microfungus *Ustilago maydis*. *Int. J. Environ. Sci. Technol.* **2015**, *12*, 2559–2566. [CrossRef]
74. Tewari, N.; Vasudevan, P.; Guha, B.K. Study on biosorption of Cr(VI) by *Mucor hiemalis*. *Biochem. Eng. J.* **2005**, *23*, 185–192. [CrossRef]
75. Pourkarim, S.; Ostovar, F.; Mahdavianpour, M.; Moslemzadeh, M. Adsorption of chromium(VI) from aqueous solution by artist's bracket fungi. *Sep. Sci. Technol.* **2017**, *52*, 1733–1741. [CrossRef]
76. Thaira, H.; Raval, K.; Manirethan, V.; Balakrishnan, R.M. Melanin nano-pigments for heavy metal remediation from water. *Sep. Sci. Technol.* **2019**, *54*, 265–274. [CrossRef]
77. Gadd, G.M.; de Rome, L. Biosorption of copper by fungal melanin. *Appl. Microbiol. Biotechnol.* **1988**, *29*, 610–617. [CrossRef]
78. Caesar-Tonthat, T.C.; Van Ommen, F.; Geesey, G.G.; Henson, J.M. Melanin production by a filamentous soil fungus in response to copper and localization of copper sulfide by sulfide-silver staining. *Appl. Environ. Microbiol.* **1995**, *61*, 1968–1975. [CrossRef]
79. Zhan, F.; He, Y.; Yang, Y.; Li, Y.; Li, T.; Zhao, Z. Effects of tricyclazole on cadmium tolerance and accumulation characteristics of a dark septate endophyte (DSE), *Exophiala pisciphila*. *Bull. Environ. Contam. Toxicol.* **2016**, *96*, 235–241. [CrossRef]
80. Potisek, M.; Likar, M.; Vogel-Mikuš, K.; Arčon, I.; Grdadolnik, J.; Regvar, M. 1,8-Dihydroxy naphthalene (DHN)-melanin confers tolerance to cadmium in isolates of melanised dark septate endophytes. *Ecotoxicol. Environ. Saf.* **2021**, *222*, 112493. [CrossRef]

Journal of
*Fungi*

MDPI

*Article*

# Genotypic and Phenotypic Diversity of *Kluyveromyces marxianus* Isolates Obtained from the Elaboration Process of Two Traditional Mexican Alcoholic Beverages Derived from Agave: Pulque and Henequen (*Agave fourcroydes*) Mezcal

**Patricia Lappe-Oliveras [1,†], Morena Avitia [2,†], Sara Darinka Sánchez-Robledo [3], Ana Karina Castillo-Plata [3], Lorena Pedraza [4], Guillermo Baquerizo [5] and Sylvie Le Borgne [6,*]**

1 Laboratorio de Micología, Instituto de Biología, Universidad Nacional Autónoma de México, Ciudad Universitaria, Ciudad de México 04510, Mexico; lappe@ib.unam.mx
2 Laboratorio Nacional de Ciencias de la Sostenibilidad (LANCIS), Instituto de Ecología, Universidad Nacional Autónoma de México, Ciudad Universitaria, Ciudad de México 04510, Mexico; morena.avitia@iecologia.unam.mx
3 Posgrado en Ciencias Naturales e Ingeniería, Universidad Autónoma Metropolitana-Unidad Cuajimalpa, Avenida Vasco de Quiroga 4871, Santa Fe Cuajimalpa, Ciudad de México 05348, Mexico; saradsr11@gmail.com (S.D.S.-R.); anncastilloplata@gmail.com (A.K.C.-P.)
4 Departamento de Ingeniería Química, Industrial y de Alimentos, Universidad Iberoamericana CDMX, Prolongación Paseo de la Reforma 880, Lomas de Santa Fe, Ciudad de México 01219, Mexico; lorena.pedraza@ibero.mx
5 Instituto de Investigaciones en Medio Ambiente Xabier Gorostiaga S.J., Universidad Iberoamericana Puebla, Boulevard del Niño Poblano 2901, Reserva Territorial Atlixcáyotl, San Andrés Cholula 72810, Puebla, Mexico; g.baquerizo@outlook.com
6 Departamento de Procesos y Tecnología, Universidad Autónoma Metropolitana-Unidad Cuajimalpa, Avenida Vasco de Quiroga 4871, Santa Fe Cuajimalpa, Ciudad de México 05348, Mexico
* Correspondence: sylvielb@cua.uam.mx
† These authors contributed equally to this work.

**Abstract:** Seven *Kluyveromyces marxianus* isolates from the elaboration process of pulque and henequen mezcal were characterized. The isolates were identified based on the sequences of the D1/D2 domain of the 26S rRNA gene and the internal transcribed spacer (ITS-5.8S) region. Genetic differences were found between pulque and henequen mezcal isolates and within henequen mezcal isolates, as shown by different branching patterns in the ITS-5.8S phylogenetic tree and $(GTG)_5$ microsatellite profiles, suggesting that the substrate and process selective conditions may give rise to different *K. marxianus* populations. All the isolates fermented and assimilated inulin and lactose and some henequen isolates could also assimilate xylose and cellobiose. Henequen isolates were more thermotolerant than pulque ones, which, in contrast, presented more tolerance to the cell wall-disturbing agent calcofluor white (CFW), suggesting that they had different cell wall structures. Additionally, depending on their origin, the isolates presented different maximum specific growth rate ($\mu_{max}$) patterns at different temperatures. Concerning tolerance to stress factors relevant for lignocellulosic hydrolysates fermentation, their tolerance limits were lower at 42 than 30 °C, except for glucose and furfural. Pulque isolates were less tolerant to ethanol, NaCl, and Cd. Finally, all the isolates could produce ethanol by simultaneous saccharification and fermentation (SSF) of a corncob hydrolysate under laboratory conditions at 42 °C.

**Keywords:** *Kluyveromyces marxianus*; agave; genetic diversity; cell wall; carbon sources; stress tolerance; lignocellulosic hydrolysates; ethanol

## 1. Introduction

The nonconventional yeast *Kluyveromyces marxianus* has attractive characteristics for industrial applications, such as a high growth rate; thermotolerance; a broad spectrum of carbon sources utilization, including xylose, lactose, and inulin; and the ability to produce ethanol, aroma compounds, enzymes, vaccines, bioactive molecules, fructo-oligosaccharides, fatty acids, and single-cell proteins (SCP) [1–5]. SCP, which are the proteins produced by various microorganisms such as bacteria, yeast, algae, and fungi, are a source of easily digestible protein with a balanced amino acid composition. In particular, yeast SCP have attracted attention as valuable nutrients that can be produced from inexpensive waste substrates, including agricultural and industrial wastes [6]. *K. marxianus* has also been considered for use as a probiotic [7] and in wine production [8]. This yeast is a promising cell factory for biorefinery applications using lignocellulosic biomass hydrolysates and dairy industry lactose-rich effluents as feedstocks [9–11]. Due to its thermotolerance, *K. marxianus* has been considered of interest for the simultaneous saccharification and fermentation (SSF) of lignocellulosic biomass hydrolysates in which the fermentation must be carried out at high temperatures (>40 °C) close to the optimal temperature for enzymatic saccharification (50 °C) [12–14].

*K. marxianus* has most frequently been isolated from dairy environments, in raw milk [15], Pecorino di Farindola cheese [16], and traditional fermented milk products from different parts of the world such as kefir [17]. This association with dairy products is due its capability to metabolize lactose. This species has also been isolated from plants such as agave [18]; fruits such as overripened mango pulp [19]; vegetables such as rotting onions [20]; and agro-industrial residues in sugar mills [21], sugarcane bagasse hydrolysates [22], blue agave bagasse [23], coffee wet processing wastewater [24], and distillery effluents and molasses [25]. *K. marxianus* is also part of the microbiota involved in the fermentation of cereal-based African fermented beverages [26], French cider [27], Georgian wine [28], Brazilian cachaça [29], and Mexican agave-based spirits tequila and mezcal [30–32].

As emphasized by several authors, *K. marxianus* exhibits a substantial genetic and physiological diversity, as illustrated by the variety of habitats from which it can be isolated [1,33,34]. This diversity has mainly been studied in strains isolated from dairy products [16,35–37]. However, it has been recently proposed that *K. marxianus* strains isolated from agave or associated with agave-based fermentations may represent a divergent clade compared to strains from dairy environments and other habitats [38].

The objectives of the present study were to genotypically and phenotypically characterize *K. marxianus* isolates obtained from the elaboration processes of pulque and henequen mezcal, two agave-based alcoholic beverages, as well as to evaluate their tolerance to different stress conditions and their ability to produce ethanol from a corncob hydrolysate by using a SSF procedure.

Pulque is an ancient Mesoamerican non-distilled beverage (4–7% of ethanol) made from several agave species cultivated in the Central Mexican plateau [39]. The plant is not cooked; instead, the fresh plant sap (*aguamiel*) is extracted directly from the plant by scrapping the cavity made in the center of the agave stem. The obtained sap is fermented for 24 h to several days by adding a portion of previously fermented pulque (called *semilla*) as the inoculum and directly consumed after the fermentation process has finished [40]. Mezcal is a spirit distillate made from the stems or cores (called *piñas*) of a variety of agave species, which are slowly cooked in large pit ovens, then crushed in stone mills to obtain a sweet juice that is fermented and finally distilled [41]. Henequen (*Agave fourcoydes*) is an agave species native to the Yucatan Peninsula [42].

## 2. Materials and Methods

### 2.1. Yeasts Strains and Growth Conditions

Seven isolates of *K. marxianus* recovered from different stages of the production process of pulque and henequen mezcal were obtained from the Microbial Culture Collection (yeasts and molds) at the Laboratorio de Micromicetos (C006) of the Instituto de Biología of the Universidad Nacional Autónoma de México (Table 1). The artisanal pulque was produced in Santa Mónica (Hidalgo State, Mexico: 20°18′31.542″ N, 98°13′8.076″ W) [43] and henequen mezcal at the GeMBio Laboratory, Centro de Investigación Científica de Yucatán A.C. in Merida, Yucatan State, Mexico [44], from samples collected in Tixpéhual (Yucatán State, Mexico: 20°57′39.321″ N, 89°26′21.671″ W). Axenic yeast cultures were preserved by two methods: (a) in aqueous suspensions at 4 °C and (b) frozen at −80 °C in YPD broth (20 g/L bacteriological peptone, 10 g/L yeast extract, and 20 g/L of glucose) supplemented with 25% (*v*/*v*) glycerol. Unless otherwise indicated, cultures were performed in YPD broth or in YPD agar and incubated at 30 °C.

**Table 1.** Origin of the *K. marxianus* isolates. Circles at the left of each isolate key are colored according to their origin.

| Isolate | Origin |
|---|---|
| •Kmx11 | Pulque seed or inoculum |
| •Kmx14 | Base of a freshly cut henequen leaf |
| •Kmx15 | Fermented (168 h) pulque |
| •Kmx16 | Non-fermented cooked henequen juice |
| •Kmx21 | Non-fermented cooked henequen juice |
| •Kmx22 | Fermented (48 h) cooked henequen juice |
| •Kmx24 | Cooked henequen stem |

### 2.2. DNA Extraction, Molecular Identification, and Phylogenetic Analysis

Genomic DNA was extracted using the ZR Fungal/bacterial DNA Miniprep kit (Zymo Research Corp, Irvine, CA, USA) according to the manufacturer's instructions. The concentration and purity of the DNA were determined using the NanoDrop 2000 spectrophotometer (Thermo Scientific, Waltham, MA, USA) with absorbance readings at 230, 260, and 280 nm. DNA integrity was evaluated by electrophoresis in 0.8% (*w*/*v*) agarose gels stained with GelRed (Biotium, Fremont, CA, USA). The D1/D2 domain of the 26S rRNA gene was amplified using the forward NL1 (5′-GCA TAT CAA TAA GCG GAG GAA AAG-3′) and reverse NL4 (5′-GGT CCG TGT TTC AAG ACG G-3′) primers [45] in a 25-μL PCR reaction volume containing 25 ng of DNA template, 1x standard Taq reaction buffer (New England Biolabs, Ipswich, MA, USA), 0.2 mM of each dNTP, 0.5 μM of each primer, and 1.25 U Taq DNA polymerase (New England Biolabs). DNA amplification was performed in a MyCycler thermal cycler (Bio-Rad, Hercules, CA, USA) under the following conditions: 5 min of initial denaturation at 95 °C; 30 cycles of 30 s at 94 °C, 30 s at 55 °C, and 30 s at 72 °C; and 1 cycle of final extension of 7 min at 72 °C. The 5.8S-ITS region of the 26S rRNA gene was amplified using the forward ITS1 (5′-TCC GTA GGT GAA CCT GCG G-3′) and reverse ITS4 (5′-GCA TAT CAA TAA GCG GAG GA-3′) primers [46] in a 25-μL PCR reaction volume as described above. DNA amplification was performed under the following conditions: 5 min of initial denaturation at 95 °C; 35 cycles of 1 min s at 94 °C, 2 min at 55.5 °C, and 2 min at 72 °C; and 1 cycle of final extension of 10 min at 72 °C. All the amplicons were analyzed by electrophoresis in 1.2% (*w*/*v*) agarose gels stained with GelRed. After purification with the DNA Clean and Concentrator kit (Zymo Research), the D1/D2 and ITS-5.8S PCR amplification products were sent to Macrogen (Seoul, Republic of Korea) for sequencing. The D1/D2 sequences were edited with BioEdit 7.0.5 [47] and compared with the sequences in the GenBank database using the BLASTN online tool [48] for identification. For the ITS-5.8S phylogenetic analysis, the sequences were aligned using BioEdit, and the phylogenetic reconstruction was performed with the neighbor-joining (NJ) algorithm in MEGA7 [49] using 10,000 bootstrap replicates.

### 2.3. Microsatellites Analysis

Microsatellite PCR fingerprinting was carried out in a 25-μL reaction volume containing 25 ng of DNA template, 1× standard Taq reaction buffer (New England Biolabs), 0.2 mM of each dNTP, 0.5 μM of the $(GTG)_5$ microsatellite primer, 2.5 μg of bovine serum albumin (BSA), and 1.25 U Taq DNA polymerase (New England Biolabs) according to [50]. DNA amplification was performed in a MyCycler thermal cycler (Bio-Rad) under the following conditions: 5 min of initial denaturation at 94 °C; 40 cycles of 30 s at 94 °C, 45 s at 55 °C, and 1 min and 30 s at 72 °C; and 1 cycle of final extension of 6 min at 72 °C. The amplification products were separated on 2% (*w*/*v*) agarose gels containing GelRed in 0.5× TBE buffer at 7.5 V/cm for 150 min. A 100 pb DNA ladder (New England Biolabs) was used as the molecular weight marker in these gels.

### 2.4. Morphological and Phenotypical Characterization

The colony and ascospore morphologies, as well as carbon source utilization profiles, were described and tested according to already described methodologies [51,52]. The morphological tests and most of the biochemical tests were performed at the Micromycetes Laboratory (C006) of the Institute of Biology of the Universidad Nacional Autónoma de México. Some biochemical tests were realized at the BCCM/IHEM Fungi Collection (Mycology and Aerobiology Section, Sciensano, Brussels, Belgium).

### 2.5. Thermotolerance and Calcofluor White Tolerance Evaluation

Tolerance to temperature and calcofluor white (CFW) was tested by the spot inoculation of liquid cultures previously grown at 30 °C onto YPD agar, as already described [34]. For this, the isolates were grown for 16 h at 30 °C and 150 rpm in YPD broth and re-inoculated into fresh medium and grown to the exponential phase at 30 °C. The obtained cultures were adjusted to an optical density of 0.2 at 600 nm ($OD_{600}$) in YPD broth, serially diluted into saline solution (NaCl 9‰), and spot inoculated (2.5 µL) onto the different test media and incubated at the temperatures listed below. The plates were sealed using parafilm before incubation, and growth was observed after 24 and 48 h. For thermotolerance, the plates were incubated at 30, 37, 42, 45, and 48 °C. For CFW tolerance, YPD agar was supplemented with 0.02, 0.05, 0.1, 0.15, or 0.2 mM of CFW and incubated at 30, 37, 42, and 45 °C. The CFW (F3543, Sigma-Aldrich, St. Louis, MO, USA) stock solution was prepared by dissolving CFW in 0.5% (*w*/*v*) KOH and 83% (*v*/*v*) glycerol according to [53]. The YPD agar was supplemented with this stock solution to obtain the CFW concentrations listed above.

### 2.6. Growth Kinetics at Different Temperatures

The inoculum were prepared in YPD broth from one colony inoculated in 5 mL of medium and grown at 30 °C and 200 rpm for 16 h. The obtained cultures were used to inoculate 125 mL Erlenmeyer flasks containing 25 mL of YPD broth at an initial $OD_{600}$ of 0.25. The flasks were then incubated at different temperatures (30, 37, 42, 45, and 48 °C) for 24 h at 200 rpm. Cell growth was followed by measuring the $OD_{600}$ using a BioPhotometer Plus (Eppendorf, Hamburg, Germany).

The modified Gompertz model (Equation (1)) was used to evaluate the effect of temperature on the shape of each isolate growth curve and determine the maximum growth rate ($\mu_{max}$) and the lag phase duration [54]. The model parameters were determined by fitting the growth curves to the Gompertz equation using the Levenberg–Marquardt nonlinear least squares method programmed in the MATLAB software (MathWorks, Inc., Natick, MA, USA).

$$Y_t = Y_{max} \cdot exp\left[-exp\left(\frac{\mu_m \cdot \exp(1)}{Y_{max}} \cdot (\lambda - t) + 1\right)\right] \quad (1)$$

where $Y_t$ is the $OD_{600}$ value at a time t, $Y_{max}$ the maximum $OD_{600}$ value, $\mu_m$ the maximum growth ($OD_{600} \cdot h^{-1}$), and $\lambda$ is the lag phase time (h). In addition, $\mu_{max}$ ($h^{-1}$) was calculated with Equation (2). The accuracy of the model was assessed through the coefficient of determination ($R^2$), according to Equation (3).

$$\mu_{max} = \frac{\mu_m \cdot \exp(1)}{Y_{max}} \quad (2)$$

$$R^2 = \frac{\sum(Y_e - Y_m)^2}{\sum(Y_e - Y_{avg})^2} \quad (3)$$

*2.7. Tolerance to Stress Conditions Relevant for Lignocellulosic Hydrolysates Fermentation*

Tolerance to stressful conditions was assessed by spot inoculation (as described in Section 2.5) of liquid cultures inoculated onto YPD agar plates supplemented with the different test compounds and observing their growth after 24–48 h [55]. The plates were incubated at 30 and 42 °C. The isolates were evaluated for tolerance (i) to glucose (2–50%, $w/v$); (ii) to ethanol (2.5–10%, $v/v$); (iii) to NaCl and KCl (25–75 g/L and 25–100 g/L, respectively); (iv) to acetic acid (1.5–4.5 g/L), furfural (1–2.5 g/L), and coniferyl aldehyde (0.25–4 mM); and (v) to metals ($ZnCl_2$, 0.625–10 mM; $CuCl_2$, 0.1–4 mM; $CdCl_2$, 0.25–1 mM; and $MnCl_2$, 2–6 mM).

*2.8. SSF Tests*

The lignocellulosic substrate tested for bioethanol production was a pretreated corncob residue. The pretreatment consisted of a thermochemical treatment at a moderate temperature with diluted sulfuric acid to hydrolyze hemicellulose, and the obtained solid fraction, primarily constituted by cellulose, was used as the substrate, as described in [56]. The SSF tests were conducted in duplicate in 125 mL Erlenmeyer flasks sealed with rubber stoppers. Each flask contained 8% ($w/v$) of pretreated corncob solids in a total volume of 62.5 mL of fermentation medium (5 g/L of yeast extract, 2 g/L of $NH_4Cl$, 1 g/L of $KH_2PO_4$, and 0.3 g/L of $MgSO_4$ $7H_2O$) without glucose and supplemented with 0.1 M of sodium citrate buffer at pH 5.5. The pH of the medium was finally adjusted to 5.5 with NaOH 2N after adding all the components. The pretreated corncobs and the medium were sterilized in an autoclave prior to SSF. The *K. marxianus* isolates entering the early stationary phase in YPD broth were inoculated at an initial $OD_{600}$ of 0.5. 9% ($w/w$) of the commercial cellulase cocktail CelliCtec 2 (Novozymes Latin America, Araucária, Paraná, Brazil) with respect to the cellulose content were then added, and the flasks were incubated at 42 °C with an orbital agitation of 150 rpm for 72 h. Glucose, xylose, acetic acid, glycerol, and ethanol were determined by high-performance liquid chromatography (HPLC) as described in [57].

*2.9. Nucleotide Sequence Accession Numbers*

The sequences obtained in this study were deposited in the GenBank database under the accession numbers shown in Table 2.

**Table 2.** GenBank accession numbers and percent of nucleotide similarity in the D1/D2 domains and ITS regions between the isolates and type strain CBS 712.

| Isolate | D1/D2 Domain | % Identity with Type Strain (KY108075) | ITS-5.8S Region | % Identity with Type Strain (NR_111251) |
|---|---|---|---|---|
| •Kmx11 | OP010195 | 99.6 | OP021667 | 99.55 |
| •Kmx14 | OP010196 | 100 | OP021667 | 99.85 |
| •Kmx15 | OP010197 | 99.6 | OP021667 | 99.55 |
| •Kmx16 | OP010198 | 100 | OP021667 | 99.40 |
| •Kmx21 | OP010199 | 100 | OP021667 | 99.70 |
| •Kmx22 | OP010200 | 100 | OP021667 | 100 |
| •Kmx24 | OP010201 | 100 | MW193124 [&] | 99.85 |

[&] Castillo-Plata et al. (2022).

## 3. Results

*3.1. Molecular Identification and Phylogenetic Analysis*

The identity of the *K. marxianus* isolates was obtained by sequence analysis of the D1/D2 domain of the 26S rRNA gene. All sequences had a similarity value of 99.6 to 100% compared to the type strain sequence (*K. marxianus* CBS 712) (Table 2). Sequencing of the

ITS-5.8S region confirmed the *K. marxianus* species assignation of all the isolates with above 99% identity with *K. marxianus* CBS 712 (Table 2).

A phylogenetic tree was constructed with the ITS-5.8S sequences of the isolates showing that they grouped according to their origin (see Table 1 for the origin and color code of the different isolates), except Kmx14 from the base of a henequen leaf and Kmx24 from a cooked agave core, which grouped together despite their different origins (Figure 1).

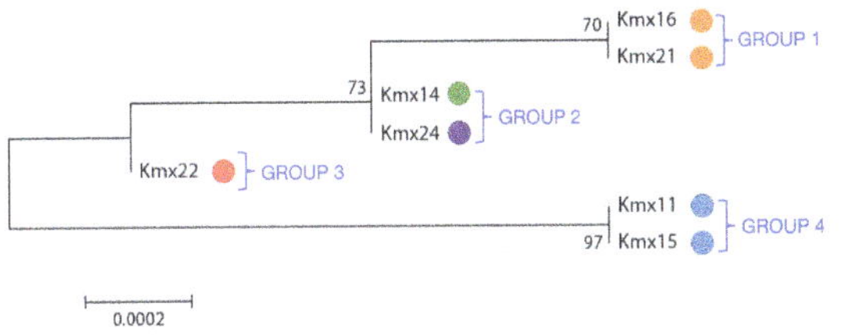


**Figure 1.** Phylogenetic tree of the *K. marxianus* isolates based on their ITS-5.8S sequences. The tree was constructed using 690 bp sequences. Colored circles indicate the origin of the isolates as described in Table 1. Numbers on nodes indicate bootstrap values. Branch lengths are proportional to the number of nucleotide substitutions and are measured using the bar scale (0.0002). Sequence accession numbers are shown in Table 2.

### 3.2. Microsatellites

To assess the genetic diversity among the *K. marxianus* isolates, microsatellite PCR fingerprinting with the $(GTG)_5$ primer was performed. The observed patterns were simple and consisted of three to five bands with sizes between 500 and 2000 bp (Figure 2). As in the ITS-5.8S phylogenetic tree, the isolates clustered according to their origin, except for Kmx14 and Kmx24. The pulque isolates Kmx11 and Kmx15 produced the same $(GTG)_5$ banding pattern (pattern I in Figure 2 and Group 4 in Figure 1), while the henequen mezcal isolates clustered into three different $(GTG)_5$ banding patterns: pattern II in Figure 2 (Group 2 in Figure 1) for Kmx14 and Kmx24 isolated from a henequen leaf base and from a cooked agave stem, respectively; pattern III in Figure 2 (Group 1 in Figure 1) for Kmx16 and Kmx21, both from non-fermented juice extracted from cooked henequen stems; and pattern IV (Group 3 in Figure 1) for Kmx22 from fermented (48 h) henequen juice.

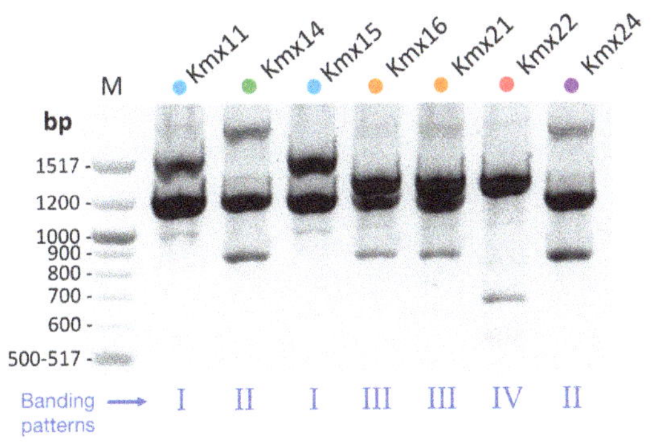


**Figure 2.** PCR-fingerprinting of the *K. marxianus* isolates obtained with the $(GTG)_5$ microsatellite primer.

### 3.3. Morphology and Carbon Sources Utilization

All strains formed cream-colored colonies in YPD agar, except Kmx16, which produced a pinkish coloration. Concerning an ascospore shape, all isolates formed reniform ascospores, except Kmx22, which produced round ones. Table 3 presents the results of the carbon sources fermentation and assimilation tests of compounds relevant for biotechnological applications. Table S1 shows additional carbon and nitrogen assimilation tests.

All the isolates were able to grow at 37 °C, ferment glucose, assimilate raffinose, and did not assimilate maltose, which are the key characteristics of *K. marxianus*, according to [58]. All the isolates were clearly positive for lactose, inulin, and xylitol assimilation and could ferment galactose, lactose, and inulin but not xylose and cellobiose. Positive, delayed, or weak assimilation profiles were recorded for xylose and cellobiose, while negative, delayed, or weak profiles were observed for sugar alcohols other than xylitol, gluconolactone, and citrate. Finally, a positive or weak succinate assimilation pattern was found instead of the positive or delayed profiles reported for this species. All the isolates could assimilate lactate, but Kmx24 gave a weak instead of a positive response. A positive ethanol assimilation was also observed for all the isolates, as expected for *K. marxianus*.

**Table 3.** Carbon sources' fermentation and assimilation patterns of the *K. marxianus* isolates. The two last columns show the results reported for *K. marxianus* in the literature for comparison.

| Carbon Source | Isolate | | | | | | | [58] | [51] |
|---|---|---|---|---|---|---|---|---|---|
| | Kmx11 | Kmx14 | Kmx15 | Kmx16 | Kmx21 | Kmx22 | Kmx24 | | |
| Fermentation | | | | | | | | | |
| Glucose | + | + | + | + | + | + | + | + | + |
| Fructose | + | + | + | + | + | + | + | n.d. | n.d. |
| Galactose | + | w | + | + | + | + | + | d | +, d |
| Xylose | - | - | - | - | - | - | - | n.d. | - |
| Sucrose | + | + | + | + | + | + | + | + | + |
| Maltose | - | - | - | - | - | - | - | - | +, - |
| Lactose | + | + * | + | + | + | + | + * | v | +, - |
| Cellobiose | - | - | - | - | - | - | - | n.d. | d, - |
| Raffinose | + | + | + | + | + | + | + | + | +, - |
| Inulin | + | + | + | + | + | + | + | d | +, - |
| Assimilation | | | | | | | | | |
| Glucose | + | + | + | + | + | + | + | + | + |
| Raffinose | + | + | + | + | + | + | + | + | + |
| Maltose | - | - | - | - | - | - | - | +,- | - |
| Rhamnose | - | - | - | - | - | - | - | - | - |
| Sucrose | + | + | + | + | + | + | + | + | + |
| Xylose | d | + | w | + | + | + | + | v | +, - |
| Ribose | - | w | - | - | - | w | d | v | +, - |
| Lactose | + | + | + | + | + | + | + | v | +, - |
| Cellobiose | + | + | + | + | + | d | + | v | +, - |
| Trehalose | - | - | - | - | - | - | - | - | +, - |
| Inulin | + | + | + | + | + | + | + | v | +, - |
| Glycerol | - | w | w | - | - | w | w | v | +, d |
| Ribitol | d | w | d | - | d | w | d | v | +, - |
| Xylitol | + | + | + | + | + | + | + | v | +, - |
| Mannitol | - | w | w | - | w | w | w | v | +, - |
| Sorbitol | d | + | + | w | d | w | + | v | +, - |
| Gluconolactone | - | - | - | - | - | - | - | v | +, - |
| Citrate | - | - | - | w | - | - | - | v | +, - |
| Succinate | w | w | w | + | + | + | w | + | +, d |
| DL-Lactate | + | + | + | + | + | + | w | + | + |
| Ethanol | + | + | + | + | + | + | + | + | +, d |

+, positive; -, negative; d, delayed: positive response > 14 days; w, weak positive response; v, variable response; n.d., not determined; and *, positive response after a week.

### 3.4. Thermotolerance and Growth Profiles at Different Temperatures

The growth of all isolates was tested at 30, 37, 42, 45, and 48 °C in YPD agar plates (Figure 3). All the isolates failed to grow at 48 °C under these conditions. Kmx14, Kmx16, Kmx21, Kmx22, and Kmx24 could grow up to 45 °C, but Kmx22 presented less growth at this temperature. Kmx11 and Kmx15, isolated from pulque, grew well up to 42 °C.

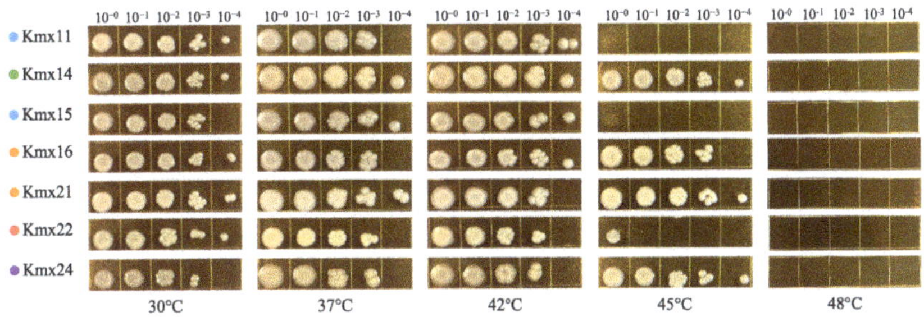

**Figure 3.** Growth of the *K. marxianus* isolates at different temperatures after 48 h of incubation in YPD agar. The dilutions are indicated at the top of the pictures.

The experimental growth curves of the isolates cultured in YPD broth at different temperatures were fitted to the modified Gompertz model using growth data up to 42 and 45 °C for the pulque and henequen isolates, respectively. The calibrated parameters for each fermentation are detailed in Table S2, where the relative standard deviation for both the $\mu_{max}$ and lag phase duration ($\lambda$) showed average values of 9.9% and 5%, respectively, demonstrating that the kinetic parameters were accurately determined. Figure 4 displays the $\mu_{max}$ of the isolates as a function of the temperature (i.e., the thermal growth curves). Lower $\mu_{max}$ values were observed for the pulque isolates (Figure 4A) compared to the henequen isolates (Figure 4B–D). The optimum growth temperatures of the pulque isolates (37–42 °C) were also lower than those of the henequen isolates (42 °C). Isolate Kmx22 from henequen showed a different behavior with an almost flat pattern over the range of tested temperatures and intermediate $\mu_{max}$ values with respect to the pulque and henequen isolates (Figure 4D). The shapes of the thermal curves were similar for isolates belonging to the same ITS-5.8S group/microsatellite pattern.

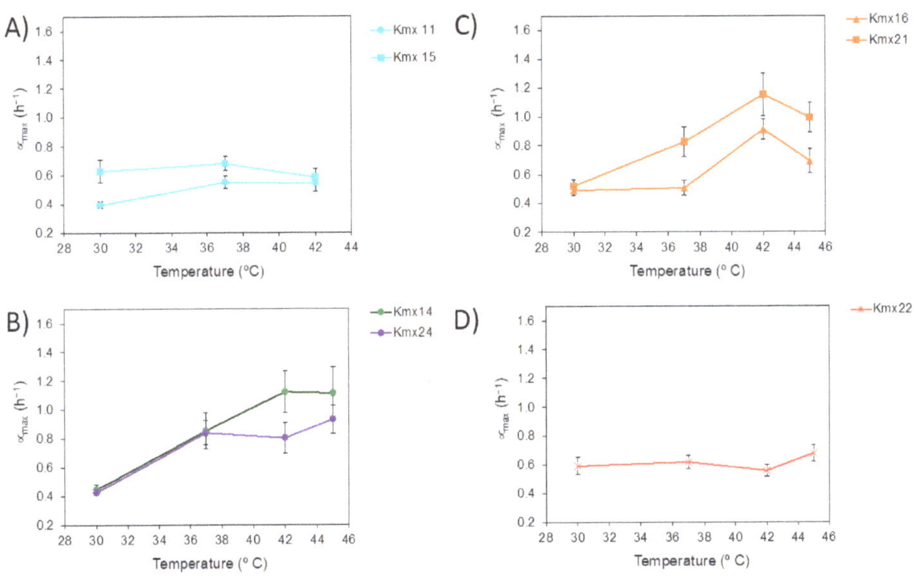

**Figure 4.** Plots of the maximum growth rate ($\mu_{max}$) of the isolates versus temperature. (**A**) Pulque isolates. (**B–D**) Henequen isolates grouped according to their ITS-5.8S group/microsatellite pattern.

### 3.5. Tolerance to CFW at Different Temperatures

Tolerance to the cell wall-perturbing agent CFW was tested to evaluate possible differences in the cell wall structure among the isolates and at different temperatures.

CFW interferes with the cell wall assembly by binding to chitin [53] and has been used to measure the chitin content of yeast cell walls [59]. Cells with a high chitin content bind more CFW and have a low tolerance to this dye and a high staining index. The growth results are shown in Figure 5. Most of the isolates were sensitive to CFW, and in general, good growth was observed with only 0.05 mM of this compound at 30 °C, except for Kmx11 and Kmx15, which were clearly more tolerant and could grow at up to 0.2 and 0.5 mM CFW at this temperature. Kmx22 was the least tolerant isolate and presented a limited growth at the lowest CFW concentration tested (0.02 mM) at 30 °C. In all cases, the tolerance to CFW decreased at higher incubation temperatures, and Kmx11 and Kmx15 were the most tolerant isolates at 42 °C.

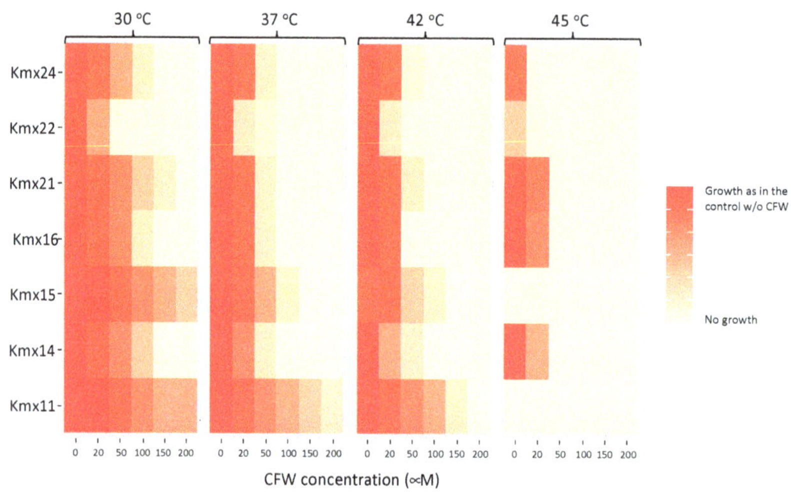


**Figure 5.** Tolerance to CFW at different temperatures.

### 3.6. Tolerance to Stress Conditions Relevant for Bioethanol Production

Table 4 shows the tolerance limits of the *K. marxianus* isolates for glucose, ethanol, salts, fermentation inhibitors, and metals, both at 30 and 42 °C. In general, the isolates were tolerant to high concentrations of glucose (30%) at both temperatures, except one isolate from pulque (Kmx15), which was less tolerant (15%) at 42 °C. All the isolates were less tolerant to ethanol (2.5–7.5%) at 42 than at 30 °C (7.5–10%), with Kmx11 and Kmx15 from pulque being the least tolerant. These two isolates did not grow in the presence of NaCl at 42 °C, while Kmx22 from fermented cooked henequen must was the most tolerant isolate (50 g/L) at 30 °C. Concerning KCl, again, the two isolates from pulque presented the lowest tolerance limit (50 g/L) to this salt at 42 °C.

All the isolates tolerated 3 g/L of acetic acid at 30 °C and were all less tolerant to this acid (1.5 g/L) at 42 °C. In this case, Kmx11 from pulque did not grow in the presence of acetic acid at 42 °C. Contrary to acetic acid, the isolates were slightly more tolerant to furfural at 42 (1.5–2.0 g/L) than at 30 °C, except the pulque isolates that were equally tolerant to furfural at 30 and 42 °C. In the case of CA, contrary to furfural, the isolates from henequen were equally tolerant (2 mM) at both temperatures, while the pulque isolates presented the lowest tolerance (1 mM) at 42 °C. Concerning Zn tolerance, Kmx22 from henequen and Kmx15 from pulque were the less tolerant at 30 (2.5 mM) and 45 °C (1.25 mM), respectively. For Cd, all henequen isolates were less tolerant at 42 °C (0.25 mM) than at 30 °C (0.5 mM), except Kmx22, which did not tolerate any of the Cd concentrations tested at both temperatures. Pulque isolates Kmx11 and Kmx15 were less tolerant at 30 (0.25 mM) than at 42 °C. Concerning Cu, all the isolates showed the same tolerance (4 mM) at both temperatures tested, except for the pulque isolates that presented a lower tolerance at 42 °C (2 mM).

**Table 4.** Tolerance limits of the *K. marxianus* isolates to different concentrations of GLU, glucose; EtOH, ethanol; AA, acetic acid; FUR, furfural; and CA, coniferyl aldehyde at 30 and 42 °C. The concentrations tested are described in Section 2.7.

| Isolate | GLU (% *w*/*v*) | | EtOH (% *v*/*v*) | | NaCl (g/L) | | KCl (g/L) | | AA (g/L) | | FUR (g/L) | | CA (mM) | | Zn (mM) | | Cd (mM) | | Cu (mM) | | Mn (mM) | |
|---|---|---|---|---|---|---|---|---|---|---|---|---|---|---|---|---|---|---|---|---|---|---|
| | 30 °C | 42 °C | 30 °C | 42 °C | 30 °C | 42 °C | 30 °C | 42 °C | 30 °C | 42 °C | 30 °C | 42 °C | 30 °C | 42 °C | 30 °C | 42 °C | 30 °C | 42 °C | 30 °C | 42 °C | 30 °C | 42 °C |
| •Kmx11 | 30 | 30 | 7.5 | 2.5 | 25 | 0 | 75 | 50 | 3 | 0 | 2 | 2 | 2 | 1 | 5 | 2.5 | 0.25 | 0 | 4 | 2 | 2 | 2 |
| •Kmx14 | 30 | 30 | 10 | 7.5 | 37.5 | 37.5 | 75 | 75 | 3 | 1.5 | 1.5 | 2 | 2 | 2 | 5 | 2.5 | 0.5 | 0.25 | 4 | 4 | 2 | 2 |
| •Kmx15 | 30 | 15 | 5 | 2.5 | 25 | 0 | 75 | 50 | 3 | 1.5 | 2 | 2 | 2 | 1 | 5 | 1.25 | 0.25 | 0 | 4 | 2 | 2 | 0 |
| •Kmx16 | 30 | 30 | 10 | 5 | 37.5 | 37.5 | 75 | 75 | 3 | 1.5 | 1.5 | 2 | 2 | 2 | 5 | 2.5 | 0.5 | 0.25 | 4 | 4 | 2 | 2 |
| •Kmx21 | 30 | 30 | 10 | 5 | 37.5 | 37.5 | 75 | 75 | 3 | 1.5 | 1.5 | 2 | 2 | 2 | 5 | 2.5 | 0.5 | 0.25 | 4 | 4 | 2 | 2 |
| •Kmx22 | 30 | 30 | 7.5 | 5 | 50 | 37.5 | 75 | 75 | 3 | 1.5 | 1.5 | 1.5 | 2 | 2 | 2.5 | 2.5 | 0 | 0 | 4 | 4 | 4 | 2 |
| •Kmx24 | 30 | 30 | 7.5 | 5 | 37.5 | 37.5 | 75 | 75 | 3 | 1.5 | 1.5 | 2 | 2 | 2 | 5 | 2.5 | 0.5 | 0.25 | 4 | 4 | 2 | 2 |

### 3.7. SSF of a Corncob Hydrolysate

The thermotolerance observed for the seven isolates allowed to perform SSF tests at 42 °C, the usual temperature of SSF with *K. marxianus* [12]. The unwashed solid from thermochemically pretreated corncob was used as the substrate. The solid contained 63% cellulose on a dry weight basis, corresponding to a potential concentration of glucose of 55.8 g/L. The initial concentration of acetic acid in the culture medium was 3.1 g/L. The ethanol, acetic acid, and glycerol production and glucose consumption were determined at the end of the SSF trials (72 h) (Table 5). All the *K. marxianus* isolates were able to convert glucose into ethanol, with isolate Kmx11 being the most productive one (15.8 g ethanol/L), followed by Kmx21, Kmx24, Kmx22, Kmx16, Kmx14, and Kmx15. The increase in ethanol production was accompanied by a decrease in glucose concentration, which was almost zero at the end of the trial, except for isolate Kmx15, in which the glucose concentration remained high (15 g/L). In the case of the most and less productive isolates (Kmx11 and Kmx15, both from pulque), the ethanol yields were 56.4 and 43.3%, respectively, while, for Kmx14, Kmx16, Kmx21, Kmx22, and Kmx24, the calculated ethanol yields were 45.9 to 49.4% (Table 5).

**Table 5.** Ethanol yields and glucose (GLU), acetic acid (AA), glycerol (GLY), and ethanol (ETOH) concentrations obtained in the SSF of a corncob hydrolysate with the *K. marxianus* isolates.

| Isolate | GLU at 72 h (g/L) | AA at 72 h (g/L) | GLY at 72 h (g/L) | EtOH at 72 h (g/L) | $Y_{P/S}$ [(a)] | EtOH Yield [(b)] |
|---|---|---|---|---|---|---|
| •Kmx11 | 0.8 | 3.5 | 0.8 | 15.8 | 0.3 | 56.4% |
| •Kmx14 | 3.7 | 3.1 | 0.6 | 12.5 | 0.2 | 47.0% |
| •Kmx15 | 15.0 | 3.1 | 0.2 | 9.0 | 0.2 | 43.3% |
| •Kmx16 | 0.3 | 3.4 | 0.7 | 13.0 | 0.2 | 45.9% |
| •Kmx21 | 0.4 | 3.3 | 0.5 | 13.9 | 0.3 | 49.3% |
| •Kmx22 | 0.3 | 3.4 | 0.9 | 13.8 | 0.2 | 48.8% |
| •Kmx24 | 0.6 | 3.6 | 0.5 | 13.9 | 0.3 | 49.4% |

[(a)] Yield product/substrate, g ethanol/g (potential GLU in pretreated corncob − GLU at 72 h). [(b)] As percentage of the maximum theoretical ethanol yield: 0.51 g ethanol/g glucose.

## 4. Discussion

Here, the genotypic and phenotypic characterizations of two *K. marxianus* isolates from the elaboration process of pulque (Kmx11 and Kmx15) and five *K. marxianus* isolates from the elaboration process of henequen mezcal (Kmx14, Kmx16, Kmx21, Kmx22, and Kmx24) are reported. It has been suggested that strains with more than 1% nucleotide substitutions in their D1/D2 domain sequence are most likely to belong to different yeast species, while strains with 1% or less substitutions are conspecific or sister species [60]. Therefore, according to the base pair sequence analysis of the D1/D2 domain of the 26S rRNA gene, all the isolates belonged to the *K. marxianus* species. The identity of all the isolates was further confirmed by sequencing the ITS-5.8S region.

The phylogenetic tree constructed with the ITS-5.8S sequences separated the isolates into four groups, most of them in accordance with their origin, except Kmx14 isolated from the base of a henequen leaf and Kmx24 from a cooked agave stem, which clustered in the same group in the phylogenetic tree. As agave stems are obtained after cutting the leaves and comprise both the stem of the plant and leaf bases, it is somehow not surprising that isolates Kmx14 and Kmx24 grouped together, although Kmx24 was isolated from a cooked agave stem and not from a fresh plant. The ITS-5.8S region was therefore a useful marker for the differentiation of agave isolates within the *K. marxianus* species and showed that a significant intraspecific genetic diversity was present among these isolates. By examining the ITS-5.8S sequences of eleven *K. marxianus* strains from different origins, [61] observed that four different ITS-5.8S sequence haplotypes were present among these strains, pointing out the high intraspecific diversity within this species. On the other hand, the polymorphism in the ITS-5.8S sequence was discriminative enough to distinguish terroir *Saccharomyces cerevisiae* wine yeasts with specific fermentative properties [62].

In addition to the ITS-5.8S analysis, the genetic diversity of the isolates was also assessed by microsatellite PCR fingerprinting. Microsatellites are short DNA motifs repeated in tandem present in eukaryotic genomes. Although originally designed to study genetic variations in humans due to their high degree of variability, microsatellites have also become a powerful tool to study intraspecific diversity in yeasts, enabling, for example, to discriminate between *S. cerevisiae* strains from wine and beer [63] or from artisanal versus industrial bread-making processes [64]. Interestingly, a microsatellites analysis of the *K. marxianus* isolates from henequen and pulque produced four different patterns that corresponded to the four groups already detected in the ITS-5.8S phylogenetic tree, suggesting a link between the genotype and the substrate of origin. The pulque isolates were discriminated from the henequen isolates, which were, in turn, separated into three distinct populations according to their substrate of origin.

The colony and spore morphologies were homogenous among the isolates, except for Kmx16 and Kmx22, respectively. Kmx16, from non-fermented cooked henequen juice, presented a reddish color on YPD agar. This color has been attributed to the production of the siderophore pulcherrimin [58]. Kmx22, from fermented cooked henequen must, formed round instead of reniform ascospores as the rest of the isolates. Both reniform and round ascospores have been described in *K. marxianus* [51,58]. An ascospore shape has been used as an important character in ascomycetous yeast taxonomy, but its functional role has not yet been studied in detail. One study with the budding yeast *Dipodascus albidus* suggests that the spore shape may aid the dispersal and survival of yeasts, reniform over round spores being favored for an efficient release [65]. Here, the Kmx22 isolate from fermented henequen juice was the only isolate with round ascospores, and interestingly, it was located on a separate branch in the ITS-5.8S phylogenetic tree and produced a unique microsatellite pattern. More studies are needed to understand the possible biological meaning of these results.

Concerning the utilization of carbon sources, the biochemical tests consisted mainly of testing the carbon sources relevant for possible industrial yeast processes. The fermentation tests results were almost homogenous among the isolates, except for Kmx14, which presented a weak response for galactose. All the isolates were able to ferment lactose and inulin under the conditions described in [52] for yeast systematic studies. Inulin and lactose fermentation are considered variable traits in *K. marxianus*. Inulin is a type of fructose polymer (fructan) that serves as a storage carbohydrate in agave plants [66]. It is therefore not surprising that isolates from agave can use inulin. The ability to use inulin is due to the presence of extracellular inulinase enzymes that break down inulin into fructose, an easily assimilable and fermentable sugar.

In the Kmx16 and Kmx24 isolates, a positive response for lactose fermentation was observed after a week, which could indicate that these two isolates were slower in fermenting lactose. Important differences in the kinetics of growth and ethanol production from lactose have been reported in *K. marxianus* strains from dairy or unknown environments, soil, fermented corn dough (pozol), and rooting agave leaves from South Africa. In particular, agave strain CBS 745 presented the lowest biomass and ethanol yields on lactose [34]. Recent studies based on genomic analyses have indicated that the lactose fermentative metabolism of *K. marxianus* is related to the presence and expression of functional alleles of the lactose permease gene (LAC12) in dairy strains [38,67]. Detailed physiological and genomic studies are needed to characterize the lactose fermentative metabolism of the agave-related isolates reported here.

Regarding carbon source assimilation, the isolates showed more phenotypic diversity. All isolates could assimilate lactose, inulin, and cellobiose, while xylose was only efficiently assimilated by henequen-derived isolates. This may not be surprising, as agave fresh sap used for pulque elaboration mainly contains fructose, glucose, fructo-oligosaccharides, and sucrose [68,69], while noticeable amounts of xylose are found in cooked agave juices used for mezcal elaboration [70]. The isolates related to henequen plants and mezcal elaboration may have adapted to a xylose-containing environment. As mentioned above, inulin assimilation by all the isolates is not surprising, as this sugar is present in agave plants. Cellobiose can be assimilated by some *K. marxianus* strains that possess a specific cellobiose transporter or a dual lactose transporter, depending on the genetic backgrounds of the strains, and a cellobiase enzyme that hydrolyzes cellobiose to glucose [71]. As xylose is a major component of hemicellulose, the ability to use this sugar as a carbon source is relevant for lignocellulosic biomass utilization. The assimilation of cellobiose, a disaccharide produced during the partial hydrolysis of cellulose, is also relevant for the integral use of a lignocellulosic biomass.

Concerning polyols assimilation, all the isolates could efficiently assimilate xylitol and some isolates, sorbitol, ribitol, mannitol, and glycerol, under the conditions described in [52]. Polyols transport and metabolism in yeasts has been poorly studied, although the ability of yeasts to assimilate polyols is part of the physiological tests for yeasts' phenotypic characterization. According to the analysis performed in [72], most yeasts can assimilate at least one polyol; 10% of the described species can assimilate four polyols (arabitol, ribitol, sorbitol, and xylitol); and ascomycetous yeasts preferably assimilate glycerol, followed by sorbitol and mannitol. As *K. marxianus*, most *Kluyveromyces* species give a variable response to xylitol [52,60]. This polyol is a low-calory sweetener with a growing demand in the food sector. Both wild type and engineered *K. marxianus* strains have been used for xylitol production from lignocellulosic biomass, and the main challenges to improve xylitol production in *K. marxianus* are related to xylose uptake and the NADP supply [11]. However, it has also been reported that xylitol consumption or reassimilation by xylitol-producing yeasts also tends to reduce xylitol yields [73], so another limiting factor in the final xylitol yield and productivity may also be the reassimilation of the produced xylitol by *K. marxianus*. Finally, regarding carboxylic acid assimilation, as expected, all the isolates were clearly positive for DL-lactate, except Kmx24, which showed a weak response. Six of the isolates were negative for citrate assimilation, with a weak response obtained for Kmx16. Three and four of the seven isolates gave a positive and weak responses for succinate, respectively. The results obtained for these two Krebs substrates were consistent with the positive and delayed responses described in [51,58], confirming the Krebs-positive status of these *K. marxianus* isolates [74].

The ability of *K. marxianus* to grow at high temperatures is one of the remarkable characteristics of this species that is not present in the other *Kluyveromyces* species described so far. All the *K. marxianus* isolates reported here were able to grow at up to 42 °C on YPD agar; however, only the isolates from henequen grew at up to 45 °C, and none of the isolates could grow at 48 °C. These temperature limits had already been reported for *K. marxianus* [34,37]. According to studies performed in *S. cerevisiae*, the physiological base of yeast thermotolerance is complex and influenced by multiple genes [75]. Differences in thermotolerance limits have been frequently observed in industrial strains of *S. cerevisiae* [76] and linked to the presence of superior alleles in specific genomic loci in more thermotolerant strains [77]. No conserved thermotolerance mechanism has been found in thermotolerant yeasts [78]. It has been reported that an evolutionary young gene with an unknown function was required for the competitive growth of *K. marxianus* at a high temperature [79]. However, this gene did not confer thermotolerance to *Kluyveromyces lactis*, indicating that thermotolerance might be linked to the de novo emergence of species-specific genes.

The isolates exhibited different growth patterns in terms of the optimum growth temperature, maximum growth rate, and thermal growth curve shape, and interestingly, these differences corresponded to the groups and patterns observed in the ITS-5.8S and microsatellite patterns, indicating that the thermal adaptation of these yeasts was different according to their origin. For example, the isolates from pulque had a lower thermotolerance limit compared to all the henequen isolates and were located on a separate branch in the ITS-5.8S phylogenetic tree, meaning that they were genetically more distant. These less thermotolerant isolates also had different microsatellites patterns. The observed differences in thermotolerance and thermal growth curve shape might be explained by the different environmental and elaboration conditions of pulque and henequen mezcal. The Central Mexican plateau where pulque is produced has a warm, temperate, subtropical climate with mild winters, while Yucatan has a tropical climate with high temperatures throughout the year. Moreover, contrary to pulque, henequen mezcal elaboration involves high temperatures. Although further studies are required to understanding the physiological basis of these differences in thermotolerance, it can be speculated that *K. marxianus* populations derived from henequen mezcal's elaboration process have evolved to adapt to higher temperature niches and other harsh conditions. More studies are needed to understand the basis of these differences.

Tolerance of the isolates to CFW at different temperatures was tested to infer the relationship between thermotolerance and the cell wall structure. Here, Kmx11 and Kmx15 from pulque were highly resistant to CFW at 30 and 42 °C, indicating that they had a lower chitin content that might explain their lowest thermotolerance. On the contrary, the isolates from henequen presented a lower CFW tolerance, indicating they had more chitin in their cell walls. The fact that the CFW tolerance of these isolates decreased at 42 °C may indicate that the chitin content in the cell wall was higher at 42 °C or that chitin was more accessible. Kmx22 from fermented cooked henequen juice presented the lowest tolerance to CFW, indicating it had the highest chitin content among all the isolates or that the chitin was more exposed to the dye. The results obtained somehow indicated that there was a relationship between the cell wall structure and thermotolerance in *K. marxianus*, although the involvement of cell wall in the adaptation and tolerance of yeasts to temperature have received a lot of attention [80]. Detailed cell wall structural studies are needed to confirm the role of chitin and its cross-linking with other polysaccharides in the thermotolerance of *K. marxianus*.

The isolates were also evaluated against stress conditions relevant for the use of lignocellulosic hydrolysates as carbon sources to produce bioethanol and other products (Table 4) [55,81,82]. The osmotolerance was evaluated using glucose as the solute. All the isolates, except Kmx15 from pulque, which tolerance limit was lower at 42 °C, tolerated a maximum of 30% glucose. This tolerance limit is low compared to *S. cerevisiae* and other yeast species that can at least tolerate 50% glucose at 30 °C [81]. Concerning ethanol tolerance, the isolates from pulque and henequen mezcal tolerated between 5 and 10% ethanol at 30 °C, which was similar to other non-conventional yeasts but low in comparison to *S. cerevisiae*, the most ethanol tolerant yeast species (14% ethanol) [81]. Contrary to osmotolerance, ethanol tolerance decreased at 42 °C in all the isolates, confirming that temperature and ethanol exert a synergistic toxic effect on yeast cells, mainly by disrupting the plasma membrane integrity [83]. As previously reported, the isolates were more tolerant to $K^+$ than to $Na^+$ [55,81]. The two isolates from pulque, Kmx11 and Kmx15, were completely intolerant to NaCl at 42 °C. It could be assumed that this effect was mainly due to the $Na^+$ ion toxicity rather than to general osmotic stress, as these isolates were tolerant to 30% glucose at 42 °C. As shown by CFW susceptibility assays, Kmx11 and Kmx15 had different cell wall structures, which might limit their mechanical resistance to ionic osmotic stress.

Regarding fermentation inhibitors, it has been reported that acetic acid tolerance is variable among *K. marxianus* and that this yeast, such as *S. cerevisiae*, is more sensitive to acetic acid as the temperature increases [57,80]. The acetic acid tolerance of the isolates reported here was low (1.5 g/L at 42 °C) compared to the most tolerant strain reported to date, *K. marxianus* CECT 10875, which can tolerate up to 10 g/L acetic acid at 42 °C. The acetic acid tolerance of *S. cerevisiae* is also highly variable, from 0.6 to 12 g/L, depending on the pH and composition of the culture medium [84]. All the isolates of this study displayed a good furfural tolerance, between 1.5 and 2 g/L, considering that the most furfural tolerant *S. cerevisiae* strains found in a collection of 71 environmental and industrial isolates were able to grow in the presence of up to 3 g/L furfural [85]. High temperatures did not affect the furfural tolerance in a negative way; on the contrary, the tolerance to furfural was maintained or even increased at 42 °C. This effect may not be surprising considering that furfural detoxification to the less toxic furfuryl alcohol is NADPH-dependent [86] and that *K. marxianus* produces more NADPH at high temperatures to fuel antioxidant systems that scavenge the reactive oxygen species formed under heat stress [87]. A similar effect was observed for the phenolic fermentation inhibitor CA, except for the two pulque isolates Kmx11 and Kmx15, which tolerated less CA at 42 °C than at 30 °C. CA has been identified as the most toxic phenolic compound derived from lignin in lignocellulosic hydrolysates [82]. Although they have significant toxicity at low concentrations, phenolic compounds are the less-studied group of inhibitors [88]. The mechanisms of yeast tolerance to phenolic compounds are an in situ reduction to less-toxic alcohols, cell wall remodeling, and protein homeostasis [88]. It has also been shown that CA can act as a cell wall active agent with antifungal activity [89]. Differences in the cell wall structures of pulque isolates may explain their lower tolerance to CA.

Some biomass feedstocks also contain heavy metals, such as zinc, cadmium, manganese, and copper [82], which are essential elements in small quantities but toxic for cells in excessive amounts. The tolerance of the isolates was comparable to that of *S. cerevisiae* in the case of zinc and cadmium [55] but higher in the case of copper. Considerable intra- and interspecific variations in Cd and Cu tolerance have been found among 15 yeast species (not *K. marxianus*) from water, soil, and plant environments [90]. Little is known so far about the different mechanisms underlying metal tolerance in yeasts. In the case of Cd, the pulque isolates and strain Kmx22 were the least tolerant to this heavy metal. These isolates also presented different behaviors in the presence of the cell wall-perturbing agent CFW, so differences in the cell wall structure may be related to their lowest Cd tolerance.

The *K. marxianus* isolates were finally evaluated for their ability to produce ethanol from the cellulosic fraction of a corncob hydrolysate by SSF. The obtained ethanol yields were similar to those already described for *K. marxianus* CECT 10875 in the SSF of different lignocellulosic biomass hydrolysates in shake flask fermentations [12]. The acetic acid concentration was almost constant during the SSF and corresponded to the concentration initially present in the hydrolysate (around 3 g/L). Although the isolates could not tolerate 3 g/L of acetic acid in the assays performed on YPD agar at 42 °C, they could efficiently ferment the corncob hydrolysate. It has been reported that the results obtained from assays in laboratory media and synthetic hydrolysates generally cannot be extrapolated to the fermentation of real lignocellulosic hydrolysates due to the complexity of their compositions and the presence of solids in high concentrations, minerals, and other nutrients or antinutrients [91]. It has been reported that, despite the inhibitory conditions present in wheat straw and sugarcane bagasse hydrolysates, a *K. marxianus* strain (SLP1) isolated from an *Agave salmiana* mezcal must showed a better performance than the *S. cerevisiae* Ethanol Red strain currently utilized in second-generation ethanol plants to produce ethanol from these hydrolysates [92]. Therefore, the isolates described here could be used as cell factories for the valorization of lignocellulosic biomass hydrolysates.

## 5. Conclusions

This study provides a general survey of the genetic and physiological diversity of *K. marxianus* isolates obtained from the elaboration processes of pulque and henequen mezcal, two agave-derived alcoholic beverages. Significant genotypic and phenotypic diversities were found between the pulque and henequen mezcal isolates and among the henequen mezcal isolates, suggesting that local selective pressure may originate different *K. marxianus* populations in these environments. The differences in thermotolerance, cell wall structure, sugar assimilation profiles, and stress tolerance between the pulque and henequen mezcal isolates could be related to differences in the climatic and process conditions, as well as available carbon sources in the sampled substrates. The ability to ferment lactose and inulin; assimilate xylose, lactose, inulin, and cellobiose; and produce ethanol at high temperatures represent interesting features for industrial applications. Further genomic and physiological studies are required to expand our knowledge on the intraspecific diversity and physiology of *K. marxianus* incorporating more non-dairy specimens and expanding the industrial use of this interesting species.

**Supplementary Materials:** The following supporting information can be downloaded at https://www.mdpi.com/article/10.3390/jof9080795/s1: Table S1: Carbon and nitrogen compounds assimilation patterns of the *K. marxianus* isolates obtained from the elaboration process of henequen mezcal and pulque; Table S2: Calibrated values of the Gompertz model parameters for each isolate grown at different temperatures.

**Author Contributions:** Conceptualization, P.L.-O., M.A., G.B. and S.L.B.; methodology, P.L.-O., M.A., S.D.S.-R., A.K.C.-P., L.P., G.B. and S.L.B.; software, M.A., S.D.S.-R. and G.B.; validation, M.A., G.B. and S.L.B.; formal analysis, P.L.-O., M.A., G.B. and S.L.B.; investigation, P.L.-O., M.A., S.D.S.-R., A.K.C.-P., L.P., G.B. and S.L.B.; resources, P.L.-O. and S.L.B.; data curation, P.L.-O., M.A., G.B. and S.L.B.; writing—original draft preparation, P.L.-O., M.A., G.B. and S.L.B.; writing—review and editing, P.L.-O. and S.L.B.; supervision, S.L.B.; project administration, S.L.B.; and funding acquisition, P.L.-O., L.P. and S.L.B. All authors have read and agreed to the published version of the manuscript.

**Funding:** This work was supported by the Consejo Nacional de Humanidades, Ciencias y Tecnologías (CONAHCyT): grant CB-2010-156451 to S.L.B (Principal Investigator) and P.L.-O. (Participant); master and PhD scholarships to S.D.S.-R. (CVU 1147245) and A.K.C.-P. (CVU 411865), respectively. This research was also supported by the Universidad Autónoma Metropolitana-Unidad Cuajimalpa (DCNI project 87 S210-21).

**Institutional Review Board Statement:** Not applicable.

**Informed Consent Statement:** Not applicable.

**Data Availability Statement:** Not applicable. All data is contained within the article.

**Acknowledgments:** The authors thank Karina Maldonado for the initial studies on inhibitors tolerance and SSF of corncobs.

**Conflicts of Interest:** The authors declare no conflict of interest.

## References

1. Fonseca, G.G.; Heinzle, E.; Wittmann, C.; Gombert, A.K. The Yeast *Kluyveromyces marxianus* and Its Biotechnological Potential. *Appl. Microbiol. Biotechnol.* **2008**, *79*, 339–354. [CrossRef] [PubMed]
2. Lane, M.M.; Morrissey, J.P. *Kluyveromyces marxianus*: A Yeast Emerging from Its Sister's Shadow. *Fungal Biol. Rev.* **2010**, *24*, 17–26. [CrossRef]
3. Bilal, M.; Ji, L.; Xu, Y.; Xu, S.; Lin, Y.; Iqbal, H.M.N.; Cheng, H. Bioprospecting *Kluyveromyces marxianus* as a Robust Host for Industrial Biotechnology. *Front. Bioeng. Biotechnol.* **2022**, *10*, 851768. [CrossRef]
4. Qiu, Y.; Lei, P.; Wang, R.; Sun, L.; Luo, Z.; Li, S.; Xu, H. *Kluyveromyces* as Promising Yeast Cell Factories for Industrial Bioproduction: From Bio-Functional Design to Applications. *Biotechnol. Adv.* **2023**, *64*, 108125. [CrossRef] [PubMed]
5. Morrissey, J.P.; Etschmann, M.M.W.; Schrader, J.; de Billerbeck, G.M. Cell Factory Applications of the Yeast *Kluyveromyces marxianus* for the Biotechnological Production of Natural Flavour and Fragrance Molecules. *Yeast* **2015**, *32*, 3–16. [CrossRef] [PubMed]

6. Jach, M.E.; Serefko, A.; Ziaja, M.; Kieliszek, M. Yeast Protein as an Easily Accessible Food Source. *Metabolites* **2022**, *12*, 63. [CrossRef] [PubMed]
7. Nag, D.; Goel, A.; Padwad, Y.; Singh, D. In vitro Characterisation Revealed Himalayan Dairy *Kluyveromyces marxianus* PCH397 as Potential Probiotic with Therapeutic Properties. *Probiotics Antimicrob. Proteins* **2022**, *15*, 761–773. [CrossRef]
8. Rollero, S.; Zietsman, A.J.J.; Buffetto, F.; Schückel, J.; Ortiz-Julien, A.; Divol, B. *Kluyveromyces marxianus* Secretes a Pectinase in Shiraz Grape Must That Impacts Technological Properties and Aroma Profile of Wine. *J. Agric. Food Chem.* **2018**, *66*, 11739–11747. [CrossRef]
9. Pendón, M.D.; Madeira, J.V., Jr.; Romanin, D.E.; Rumbo, M.; Gombert, A.K.; Garrote, G.L. A Biorefinery Concept for the Production of Fuel Ethanol, Probiotic Yeast, and Whey Protein from a by-Product of the Cheese Industry. *Appl. Microbiol. Biotechnol.* **2021**, *105*, 3859–3871. [CrossRef]
10. Leandro, M.J.; Marques, S.; Ribeiro, B.; Santos, H.; Fonseca, C. Integrated Process for Bioenergy Production and Water Recycling in the Dairy Industry: Selection of *Kluyveromyces* Strains for Direct Conversion of Concentrated Lactose-Rich Streams into Bioethanol. *Microorganisms* **2019**, *7*, 545. [CrossRef] [PubMed]
11. Baptista, M.; Domingues, L. *Kluyveromyces marxianus* as a Microbial Cell Factory for Lignocellulosic Biomass Valorisation. *Biotechnol. Adv.* **2022**, *60*, 108027. [CrossRef]
12. Ballesteros, M.; Oliva, J.M.; Negro, M.J.; Manzanares, P.; Ballesteros, I. Ethanol from Lignocellulosic Materials by a Simultaneous Saccharification and Fermentation Process (SFS) with *Kluyveromyces marxianus* CECT 10875. *Process Biochem.* **2004**, *39*, 1843–1848. [CrossRef]
13. Suryawati, L.; Wilkins, M.R.; Bellmer, D.D.; Huhnke, R.L.; Maness, N.O.; Banat, I.M. Simultaneous Saccharification and Fermentation of Kanlow Switchgrass Pretreated by Hydrothermolysis Using *Kluyveromyces marxianus* IMB4. *Biotechnol. Bioeng.* **2008**, *101*, 894–902. [CrossRef] [PubMed]
14. Charoensopharat, K.; Thanonkeo, P.; Thanonkeo, S.; Yamada, M. Ethanol Production from Jerusalem Artichoke Tubers at High Temperature by Newly Isolated Thermotolerant Inulin-Utilizing Yeast *Kluyveromyces marxianus* Using Consolidated Bioprocessing. *Antonie Leeuwenhoek* **2015**, *108*, 173–190. [CrossRef] [PubMed]
15. Panelli, S.; Brambati, E.; Bonacina, C.; Feligini, M. Diversity of Fungal Flora in Raw Milk from the Italian Alps in Relation to Pasture Altitude. *SpringerPlus* **2013**, *2*, 405. [CrossRef] [PubMed]
16. Tofalo, R.; Fasoli, G.; Schirone, M.; Perpetuini, G.; Pepe, A.; Corsetti, A.; Suzzi, G. The Predominance, Biodiversity and Biotechnological Properties of *Kluyveromyces marxianus* in the Production of Pecorino Di Farindola Cheese. *Int. J. Food Microbiol.* **2014**, *187*, 41–49. [CrossRef]
17. Han, X.; Zhang, L.J.; Wu, H.Y.; Wu, Y.F.; Zhao, S.N. Investigation of Microorganisms Involved in Kefir Biofilm Formation. *Antonie Leeuwenhoek* **2018**, *111*, 2361–2370. [CrossRef]
18. Lachance, M.A. Yeasts from Natural Tequila Formation. *Antonie Leeuwenhoek* **1995**, *68*, 151–160. [CrossRef]
19. Buenrostro-Figueroa, J.; Tafolla-Arellano, J.C.; Flores-Gallegos, A.C.; Rodríguez-Herrera, R.; De la Garza-Toledo, H.; Aguilar, C.N. Native Yeasts for Alternative Utilization of Overripe Mango Pulp for Ethanol Production. *Rev. Argent Microbiol.* **2018**, *50*, 173–177. [CrossRef]
20. Schroeder, B.K.; Rogers, J.D.; Johnson, D.A.; Pelter, G. Occurrence of *Kluyveromyces marxianus* var. *marxianus* Causing Onion Soft Rot in the Columbia Basin of Washington State. *Plant Dis.* **2007**, *91*, 1059. [CrossRef]
21. Limtong, S.; Sringiew, C.; Yongmanitchai, W. Production of Fuel Ethanol at High Temperature from Sugar Cane Juice by a Newly Isolated *Kluyveromyces marxianus*. *Bioresour. Technol.* **2007**, *98*, 3367–3374. [CrossRef] [PubMed]
22. Suzuki, T.; Hoshino, T.; Matsushika, A. Draft Genome Sequence of *Kluyveromyces marxianus* Strain DMB1, Isolated from Sugarcane Bagasse Hydrolysate. *Genome Announc.* **2014**, *2*, e00733-14. [CrossRef] [PubMed]
23. Trapala, J.; Bustos-Jaimes, I.; Manzanares, P.; Bárzana, E.; Montiel, C. Purification and Characterization of an Inulinase Produced by a *Kluyveromyces marxianus* Strain Isolated from Blue Agave Bagasse. *Protein Expr. Purif.* **2020**, *176*, 105718. [CrossRef] [PubMed]
24. Serrat, M.; Bermúdez, R.C.; Villa, T.G. Production, Purification, and Characterization of a Polygalacturonase from a New Strain of *Kluyveromyces marxianus* Isolated from Coffee Wet-Processing Wastewater. *Appl. Biochem. Biotechnol.* **2002**, *97*, 193–208. [CrossRef] [PubMed]
25. Avchar, R.; Lanjekar, V.; Baghela, A. Bioprospecting Thermotolerant Yeasts from Distillery Effluent and Molasses for High-Temperature Ethanol Production. *J. Appl. Microbiol.* **2022**, *132*, 1134–1151. [CrossRef]
26. Atter, A.; Diaz, M.; Tano-Debrah, K.; Kunadu, A.P.H.; Mayer, M.J.; Colquhoun, I.J.; Nielsen, D.S.; Baker, D.; Narbad, A.; Amoa-Awua, W. Microbial Diversity and Metabolite Profile of Fermenting Millet in the Production of Hausa Koko, a Ghanaian Fermented Cereal Porridge. *Front. Microbiol.* **2021**, *12*, 681983. [CrossRef]
27. Coton, E.; Coton, M.; Levert, D.; Casaregola, S.; Sohier, D. Yeast Ecology in French Cider and Black Olive Natural Fermentations. *Int. J. Food Microbiol.* **2006**, *108*, 130–135. [CrossRef]
28. Vigentini, I.; Maghradze, D.; Petrozziello, M.; Bonello, F.; Mezzapelle, V.; Valdetara, F.; Failla, O.; Foschino, R. Indigenous Georgian Wine-Associated Yeasts and Grape Cultivars to Edit the Wine Quality in a Precision Oenology Perspective. *Front. Microbiol.* **2016**, *7*, 352. [CrossRef]
29. Freitas Schwan, R.; Mendonça, A.T.; Da Silva, J.J.; Rodrigues, V.; Wheals, A.E. Microbiology and Physiology of Cachaça (Aguardente) Fermentations. *Antonie Leeuwenhoek* **2001**, *79*, 89–96. [CrossRef]

30. Escalante-Minakata, P.; Blaschek, H.P.; Barba De La Rosa, A.P.; Santos, L.; De León-Rodríguez, A. Identification of Yeast and Bacteria Involved in the Mezcal Fermentation of *Agave salmiana*. *Lett. Appl. Microbiol.* **2008**, *46*, 626–630. [CrossRef]
31. Verdugo Valdez, A.; Segura Garcia, L.; Kirchmayr, M.; Ramírez Rodríguez, P.; González Esquinca, A.; Coria, R.; Gschaedler Mathis, A. Yeast Communities Associated with Artisanal Mezcal Fermentations from *Agave salmiana*. *Antonie Leeuwenhoek* **2011**, *100*, 497–506. [CrossRef] [PubMed]
32. Nolasco-Cancino, H.; Santiago-Urbina, J.A.; Wacher, C.; Ruíz-Terán, F. Predominant Yeasts during Artisanal Mezcal Fermentation and Their Capacity to Ferment Maguey Juice. *Front. Microbiol.* **2018**, *9*, 2900. [CrossRef] [PubMed]
33. Belloch, C.; Barrio, E.; García, M.D.; Querol, A. Inter- and Intraspecific Chromosome Pattern Variation in the Yeast Genus *Kluyveromyces*. *Yeast* **1998**, *14*, 1341–1354. [CrossRef]
34. Lane, M.M.; Burke, N.; Karreman, R.; Wolfe, K.H.; O'Byrne, C.P.; Morrissey, J.P. Physiological and Metabolic Diversity in the Yeast *Kluyveromyces marxianus*. *Antonie Leeuwenhoek* **2011**, *100*, 507–519. [CrossRef] [PubMed]
35. Fasoli, G.; Barrio, E.; Tofalo, R.; Suzzi, G.; Belloch, C. Multilocus Analysis Reveals Large Genetic Diversity in *Kluyveromyces marxianus* Strains Isolated from Parmigiano Reggiano and Pecorino Di Farindola Cheeses. *Int. J. Food Microbiol.* **2016**, *233*, 1–10. [CrossRef]
36. Tittarelli, F.; Varela, J.A.; Gethins, L.; Stanton, C.; Ross, R.P.; Suzzi, G.; Grazia, L.; Tofalo, R.; Morrissey, J.P. Development and Implementation of Multilocus Sequence Typing to Study the Diversity of the Yeast *Kluyveromyces marxianus* in Italian Cheeses. *Microb. Genom.* **2018**, *4*, e000153. [CrossRef]
37. Martini, S.; Bonazzi, M.; Malorgio, I.; Pizzamiglio, V.; Tagliazucchi, D.; Solieri, L. Characterization of Yeasts Isolated from Parmigiano Reggiano Cheese Natural Whey Starter: From Spoilage Agents to Potential Cell Factories for Whey Valorization. *Microorganisms* **2021**, *9*, 2288. [CrossRef] [PubMed]
38. Ortiz-Merino, R.A.; Varela, J.A.; Coughlan, A.Y.; Hoshida, H.; da Silveira, W.B.; Wilde, C.; Kuijpers, N.G.A.; Geertman, J.M.; Wolfe, K.H.; Morrissey, J.P. Ploidy Variation in *Kluyveromyces marxianus* Separates Dairy and Non-Dairy Isolates. *Front. Genet.* **2018**, *9*, 94. [CrossRef]
39. Escalante, A.; López Soto, D.R.; Velázquez Gutiérrez, J.E.; Giles-Gómez, M.; Bolívar, F.; López-Munguía, A. Pulque, a Traditional Mexican Alcoholic Fermented Beverage: Historical, Microbiological, and Technical Aspects. *Front. Microbiol.* **2016**, *7*, 1026. [CrossRef] [PubMed]
40. Villarreal-Morales, S.L.; Montañez-Saenz, J.C.; Aguilar-González, C.N.; Rodriguez-Herrera, R. Chapter 11—Metagenomics of Traditional Beverages. In *Advances in Biotechnology for Food Industry*; Holban, A.M., Grumezescu, A.M., Eds.; Academic Press: Cambridge, MA, USA, 2018; pp. 301–326. ISBN 978-0-12-811443-8.
41. Arellano-Plaza, M.; Paez-Lerma, J.B.; Soto-Cruz, N.O.; Kirchmayr, M.R.; Gschaedler Mathis, A. Mezcal Production in Mexico: Between Tradition and Commercial Exploitation. *Front. Sustain. Food Syst.* **2022**, *6*, 832532. [CrossRef]
42. Colunga-Garcíamarín, P.; May-Pat, F. Agave Studies in Yucatan, Mexico. I. Past and Present Germplasm Diversity and Uses. *Econ. Bot.* **1993**, *47*, 312–327. [CrossRef]
43. Herrera Solórzano, M.C. *Identificación Polifásica de Levaduras y Bacterias Ácido Aisladas de Aguamiel, Pulque y Semilla*; Centro de Investigación Científica y de Educación Superior de Ensenada: Ensenada, Baja California, México, 2008.
44. Pérez-Brito, D. Molecular Characterization of *Kluyveromyces marxianus* Strains Isolated from *Agave fourcroydes* (Lem.) in Yucatan, Mexico. *Mol. Biotechnol.* **2007**, *37*, 181–186. [CrossRef] [PubMed]
45. O'Donnell, K. Fusarium and Its near Relatives. In *The fungal Holomorph: Mitotic, Meiotic and Pleomorphic Speciation in Fungal Systematics*; Reynolds, D.R., Taylor, J.W., Eds.; CABI Publishing: Wallingford, UK, 1993; pp. 225–232.
46. White, T.J.; Bruns, T.; Lee, S.; Taylor, J. Amplification and Direct Sequencing of Fungal Ribosomal RNA Genes for Phylogenetics. In *PCR Protocols*; Innis, M.A., Gelfand, D.H., Sninsky, J.J., White, T.J., Eds.; Academic Press: Cambridge, MA, USA, 1990; pp. 315–322.
47. Hall, T.A. BIOEDIT: A User-Friendly Biological Sequence Alignment Editor and Analysis Program for Windows 95/98/NT. *Nucleic Acids Symp. Ser.* **1999**, *41*, 95–98.
48. Altschul, S.F.; Gish, W.; Miller, W.; Myers, E.W.; Lipman, D.J. Basic Local Alignment Search Tool. *J. Mol. Biol.* **1990**, *215*, 403–410. [CrossRef] [PubMed]
49. Kumar, S.; Stecher, G.; Tamura, K. MEGA7: Molecular Evolutionary Genetics Analysis Version 7.0 for Bigger Datasets. *Mol. Biol. Evol.* **2016**, *33*, 1870–1874. [CrossRef] [PubMed]
50. da Silva-Filho, E.A.; dos Santos, S.K.B.; do Monte Resende, A.; de Morais, J.O.F.; de Morais, M.A.; Simões, D.A. Yeast Population Dynamics of Industrial Fuel-Ethanol Fermentation Process Assessed by PCR-Fingerprinting. *Antonie Leeuwenhoek* **2005**, *88*, 13–23. [CrossRef]
51. Barnett, J.A.; Payne, R.W.; Yarrow, D.; Barnett, L. *Yeasts: Characteristics and Identification*; Cambridge University Press: Cambridge, UK, 2000; ISBN 9780521573962.
52. Kurtzman, C.; Fell, J.W.; Boekhout, T. *The Yeasts: A Taxonomic Study*, 5th ed.; Elsevier Science: Amsterdam, The Netherlands, 2010; ISBN 9780444521491.
53. Ram, A.F.J.; Klis, F.M. Identification of Fungal Cell Wall Mutants Using Susceptibility Assays Based on Calcofluor White and Congo Red. *Nat. Protoc.* **2006**, *1*, 2253–2256. [CrossRef]
54. Soro, A.B.; Oliveira, M.; O'Donnell, C.P.; Tiwari, B.K. Ultrasound Assisted Modulation of Yeast Growth and Inactivation Kinetics. *Ultrason. Sonochem.* **2021**, *80*, 105819. [CrossRef]

55. Mukherjee, V.; Steensels, J.; Lievens, B.; Van de Voorde, I.; Verplaetse, A.; Aerts, G.; Willems, K.A.; Thevelein, J.M.; Verstrepen, K.J.; Ruyters, S. Phenotypic Evaluation of Natural and Industrial *Saccharomyces* Yeasts for Different Traits Desirable in Industrial Bioethanol Production. *Appl. Microbiol. Biotechnol.* **2014**, *98*, 9483–9498. [CrossRef]
56. Pedraza, L.; Flores, A.; Toribio, H.; Quintero, R.; Le Borgne, S.; Moss-Acosta, C.; Martinez, A. Sequential Thermochemical Hydrolysis of Corncobs and Enzymatic Saccharification of the Whole Slurry Followed by Fermentation of Solubilized Sugars to Ethanol with the Ethanologenic Strain *Escherichia coli* MS04. *Bioenergy Res.* **2016**, *9*, 1046–1052. [CrossRef]
57. Castillo-Plata, A.K.; Sigala, J.C.; Lappe-Oliveras, P.; Le Borgne, S. KCL/KOH Supplementation Improves Acetic Acid Tolerance and Ethanol Production in a Thermotolerant Strain of *Kluyveromyces marxianus* Isolated from Henequen (*Agave fourcroydes*). *Rev. Mex. Ing. Quim.* **2022**, *21*, Bio2567. [CrossRef]
58. Lachance, M.A. Kluyveromyces van Der Walt (1971). In *The Yeasts*; Kurtzman, C.P., Fell, J.W., Boekhout, T., Eds.; Elsevier: Amsterdam, The Netherlands, 2011; Volume 2, pp. 471–481. ISBN 9780444521491.
59. Costa-de-Oliveira, S.; Silva, A.P.; Miranda, I.M.; Salvador, A.; Azevedo, M.M.; Munro, C.A.; Rodrigues, A.G.; Pina-Vaz, C. Determination of Chitin Content in Fungal Cell Wall: An Alternative Flow Cytometric Method. *Cytom. A* **2013**, *83 A*, 324–328. [CrossRef]
60. Kurtzman, C.P.; Robnett, C.J. Identification and Phylogeny of Ascomycetous Yeasts from Analysis of Nuclear Large Subunit (26S) Ribosomal DNA Partial Sequences. *Antonie Leeuwenhoek* **1998**, *73*, 331–371. [CrossRef] [PubMed]
61. Belloch, C.; Fernández-Espinar, T.; Querol, A.; Dolores García, M.; Barrio, E. An Analysis of Inter- and Intraspecific Genetic Variabilities in the *Kluyveromyces marxianus* Group of Yeast Species for the Reconsideration of the *K. lactis* Taxon. *Yeast* **2002**, *19*, 257–268. [CrossRef] [PubMed]
62. Tofalo, R.; Perpetuini, G.; Fasoli, G.; Schirone, M.; Corsetti, A.; Suzzi, G. Biodiversity Study of Wine Yeasts Belonging to the "Terroir" of Montepulciano d'Abruzzo "Colline Teramane" Revealed *Saccharomyces cerevisiae* Strains Exhibiting Atypical and Unique 5.8S-ITS Restriction Patterns. *Food Microbiol.* **2014**, *39*, 7–12. [CrossRef]
63. Baleiras Couto, M.M.; Hartog, B.J.; Huis In't Veld, J.H.J.; Hofstra, H.; Van Der Vossen, J.M.B.M. Identification of Spoilage Yeasts in a Food-Production Chain by Microsatellite Polymerase Chain Reaction Fingerprinting. *Food Microbiol.* **1996**, *13*, 59–67. [CrossRef]
64. Bigey, F.; Segond, D.; Friedrich, A.; Guezenec, S.; Bourgais, A.; Huyghe, L.; Agier, N.; Nidelet, T.; Sicard, D. Evidence for Two Main Domestication Trajectories in *Saccharomyces cerevisiae* Linked to Distinct Bread-Making Processes. *Curr. Biol.* **2021**, *31*, 722–732.e5. [CrossRef]
65. Van Heerden, A.; Kock, J.L.F.; Botes, P.J.; Pohl, C.H.; Strauss, C.J.; Van Wyk, P.W.J.; Nigam, S. Ascospore Release from Bottle-Shaped Asci in *Dipodascus albidus*. *FEMS Yeast Res.* **2005**, *5*, 1185–1190. [CrossRef]
66. Hughes, S.R.; Qureshi, N.; López-Núñez, J.C.; Jones, M.A.; Jarodsky, J.M.; Galindo-Leva, L.Á.; Lindquist, M.R. Utilization of Inulin-Containing Waste in Industrial Fermentations to Produce Biofuels and Bio-Based Chemicals. *World J. Microbiol. Biotechnol.* **2017**, *33*, 78. [CrossRef]
67. de Paiva, L.C.; Diniz, R.H.S.; Vidigal, P.M.P.; de Oliveira Mendes, T.A.; Santana, M.F.; Cerdán, M.E.; González-Siso, M.I.; da Silveira, W.B. Genomic Analysis and Lactose Transporter Expression in *Kluyveromyces marxianus* CCT 7735. *Fungal Biol.* **2019**, *123*, 687–697. [CrossRef]
68. Ortiz-Basurto, R.I.; Pourcelly, G.; Doco, T.; Williams, P.; Dormer, M.; Belleville, M.P. Analysis of the Main Components of the Aguamiel Produced by the Maguey-Pulquero (*Agave mapisaga*) throughout the Harvest Period. *J. Agric. Food Chem.* **2008**, *56*, 3682–3687. [CrossRef] [PubMed]
69. Romero-López, M.R.; Osorio-Díaz, P.; Flores-Morales, A.; Robledo, N.; Mora-Escobedo, R. Chemical Composition, Antioxidant Capacity and Prebiotic Effect of Aguamiel (*Agave atrovirens*) during in vitro Fermentation. *Rev. Mex. Ing. Quim.* **2015**, *14*, 281–292.
70. Michel-Cuello, C.; Juárez-Flores, B.I.; Aguirre-Rivera, J.R.; Pinos-Rodríguez, J.M. Quantitative Characterization of Nonstructural Carbohydrates of Mezcal Agave (*Agave salmiana* Otto Ex Salm-Dick). *J. Agric. Food Chem.* **2008**, *56*, 5753–5757. [CrossRef]
71. Varela, J.A.; Puricelli, M.; Ortiz-Merino, R.A.; Giacomobono, R.; Braun-Galleani, S.; Wolfe, K.H.; Morrissey, J.P. Origin of Lactose Fermentation in *Kluyveromyces lactis* by Interspecies Transfer of a Neo-Functionalized Gene Cluster during Domestication. *Curr. Biol.* **2019**, *29*, 4284–4290.e2. [CrossRef] [PubMed]
72. Pereira, I.; Madeira, A.; Prista, C.; Loureiro-Dias, M.C.; Leandro, M.J. Characterization of New Polyol/H+ Symporters in *Debaryomyces hansenii*. *PLoS ONE* **2014**, *9*, e88180. [CrossRef]
73. Hor, S.; Kongkeitkajorn, M.B.; Reungsang, A. Sugarcane Bagasse-Based Ethanol Production and Utilization of Its Vinasse for Xylitol Production as an Approach in Integrated Biorefinery. *Fermentation* **2022**, *8*, 340. [CrossRef]
74. Barnett, J.A.; Kornberg, H.L. The Utilization by Yeasts of Acids of the Tricarboxylic Acid Cycle. *J. Gen. Microbiol.* **1960**, *23*, 65–82. [CrossRef]
75. Gao, L.; Liu, Y.; Sun, H.; Li, C.; Zhao, Z.; Liu, G. Advances in Mechanisms and Modifications for Rendering Yeast Thermotolerance. *J. Biosci. Bioeng.* **2016**, *121*, 599–606. [CrossRef]
76. Lip, K.Y.F.; García-Ríos, E.; Costa, C.E.; Guillamón, J.M.; Domingues, L.; Teixeira, J.; van Gulik, W.M. Selection and Subsequent Physiological Characterization of Industrial *Saccharomyces cerevisiae* Strains during Continuous Growth at Sub- and- Supra Optimal Temperatures. *Biotechnol. Rep.* **2020**, *26*, e00462. [CrossRef]
77. Wang, Z.; Qi, Q.; Lin, Y.; Guo, Y.; Liu, Y.; Wang, Q. QTL Analysis Reveals Genomic Variants Linked to High-Temperature Fermentation Performance in the Industrial Yeast. *Biotechnol. Biofuels* **2019**, *12*, 59. [CrossRef]

78. Lehnen, M.; Ebert, B.E.; Blank, L.M. Elevated Temperatures Do Not Trigger a Conserved Metabolic Network Response among Thermotolerant Yeasts. *BMC Microbiol.* **2019**, *19*, 100. [CrossRef]
79. Montini, N.; Doughty, T.W.; Domenzain, I.; Fenton, D.A.; Baranov, P.V.; Harrington, R.; Nielsen, J.; Siewers, V.; Morrissey, J.P. Identification of a Novel Gene Required for Competitive Growth at High Temperature in the Thermotolerant Yeast *Kluyveromyces marxianus*. *Microbiology (Reading)* **2022**, *168*, 001148. [CrossRef] [PubMed]
80. Ribeiro, R.A.; Bourbon-Melo, N.; Sá-Correia, I. The Cell Wall and the Response and Tolerance to Stresses of Biotechnological Relevance in Yeasts. *Front. Microbiol.* **2022**, *13*, 953479. [CrossRef]
81. Mukherjee, V.; Radecka, D.; Aerts, G.; Verstrepen, K.J.; Lievens, B.; Thevelein, J.M. Phenotypic Landscape of Non-Conventional Yeast Species for Different Stress Tolerance Traits Desirable in Bioethanol Fermentation. *Biotechnol. Biofuels* **2017**, *10*, 216. [CrossRef]
82. Deparis, Q.; Claes, A.; Foulquié-Moreno, M.R.; Thevelein, J.M. Engineering Tolerance to Industrially Relevant Stress Factors in Yeast Cell Factories. *FEMS Yeast Res.* **2017**, *17*, fox036. [CrossRef]
83. Piper, P.W. The Heat Shock and Ethanol Stress Responses of Yeast Exhibit Extensive Similarity and Functional Overlap. *FEMS Microbiol. Lett.* **1995**, *134*, 12–27. [CrossRef]
84. Guaragnella, N.; Bettiga, M. Acetic Acid Stress in Budding Yeast: From Molecular Mechanisms to Applications. *Yeast* **2021**, *38*, 391–400. [CrossRef] [PubMed]
85. Field, S.J.; Ryden, P.; Wilson, D.; James, S.A.; Roberts, I.N.; Richardson, D.J.; Waldron, K.W.; Clarke, T.A. Identification of Furfural Resistant Strains of *Saccharomyces cerevisiae* and *Saccharomyces paradoxus* from a Collection of Environmental and Industrial Isolates. *Biotechnol. Biofuels* **2015**, *8*, 33. [CrossRef] [PubMed]
86. Ujor, V.C.; Okonkwo, C.C. Microbial Detoxification of Lignocellulosic Biomass Hydrolysates: Biochemical and Molecular Aspects, Challenges, Exploits and Future Perspectives. *Front. Bioeng. Biotechnol.* **2022**, *10*, 1061667. [CrossRef] [PubMed]
87. Kosaka, T.; Tsuzuno, T.; Nishida, S.; Pattanakittivorakul, S.; Murata, M.; Miyakawa, I.; Lertwattanasakul, N.; Limtong, S.; Yamada, M. Distinct Metabolic Flow in Response to Temperature in Thermotolerant *Kluyveromyces marxianus*. *Appl. Environ. Microbiol.* **2022**, *88*, e020062. [CrossRef]
88. Haclsalihoglu, B.; Holyavkin, C.; Topaloglu, A.; Klsakesen, H.I.; Cakar, Z.P. Genomic and Transcriptomic Analysis of a Coniferyl Aldehyde-Resistant *Saccharomyces cerevisiae* Strain Obtained by Evolutionary Engineering. *FEMS Yeast Res.* **2019**, *19*, foz021. [CrossRef] [PubMed]
89. Bang, K.H.; Lee, D.W.; Park, H.M.; Rhee, Y.H. Inhibition of Fungal Cell Wall Synthesizing Enzymes by Trans-Cinnamaldehyde. *Biosci. Biotechnol. Biochem.* **2000**, *64*, 1061–1063. [CrossRef] [PubMed]
90. Vadkertiová, R.; Sláviková, E. Metal Tolerance of Yeasts Isolated from Water, Soil and Plant Environments. *J. Basic Microbiol.* **2006**, *46*, 145–152. [CrossRef] [PubMed]
91. Tang, X.; Sousa, L.D.C.; Jin, M.; Chundawat, S.P.S.; Chambliss, C.K.; Lau, M.W.; Xiao, Z.; Dale, B.E.; Balan, V. Designer Synthetic Media for Studying Microbial-catalyzed Biofuel Production. *Biotechnol. Biofuels* **2015**, *8*, 1. [CrossRef] [PubMed]
92. Sandoval-Nuñez, D.; Arellano-Plaza, M.; Gschaedler, A.; Arrizon, J.; Amaya-Delgado, L. A Comparative Study of Lignocellulosic Ethanol Productivities by *Kluyveromyces marxianus* and *Saccharomyces cerevisiae*. *Clean Technol. Environ. Policy* **2018**, *20*, 1491–1499. [CrossRef]

MDPI AG
Grosspeteranlage 5
4052 Basel
Switzerland
Tel.: +41 61 683 77 34

*Journal of Fungi* Editorial Office
E-mail: jof@mdpi.com
www.mdpi.com/journal/jof

www.ingramcontent.com/pod-product-compliance
Lightning Source LLC
Chambersburg PA
CBHW040828250926
51850CB00007BA/1289
* 9 7 8 3 7 2 5 8 4 6 2 9 0 *